Lecture Notes in Mathematics 2129

More information about this series at
http://www.springer.com/series/304

Sigrun Bodine • Donald A. Lutz

Asymptotic Integration of Differential and Difference Equations

Sigrun Bodine
Dept. of Mathematics and Computer Science
University of Puget Sound
Tacoma, WA, USA

Donald A. Lutz
Department of Mathematics and Statistics
San Diego State University
San Diego, CA, USA

ISSN 0075-8434 ISSN 1617-9692 (electronic)
Lecture Notes in Mathematics
ISBN 978-3-319-18247-6 ISBN 978-3-319-18248-3 (eBook)
DOI 10.1007/978-3-319-18248-3

Library of Congress Control Number: 2015940541

Mathematics Subject Classification (2010): 34E10, 34A30, 39A06, 34N05, 34C41, 39A22

Springer Cham Heidelberg New York Dordrecht London

Printed on acid-free paper

Springer International Publishing AG Switzerland is part of Springer Science+Business Media (www.springer.com)

Dedicated to

Roy

and

Margaret

Preface

This book is concerned with the asymptotic behavior for solutions of certain classes of linear differential equations as well as corresponding results for solutions of linear difference equations. The type of asymptotic analysis we employ is based on some fundamental principles attributed to Levinson [100]. While he applied them to a special class of differential equations, subsequent work has shown that the same principles lead to asymptotic results for much wider classes of differential and also difference equations. For differential equations, his approach, which is now referred to as "asymptotic integration," was utilized in an excellent monograph by Eastham [53] and covers many results up through the mid-1980s. We extend the usage of "asymptotic integration" to describe similar results for difference equations.

Our purpose is twofold: First, in Chaps. 2–7, broad types of theoretical results will be discussed which apply to rather general classes of systems. In the case of differential equations, these extend and supplement Eastham's work by including many newer contributions and some earlier results not found there. For difference equations, a parallel approach yields a similar bank of general results and extends a much earlier treatment by Rapoport [131]. Second, we consider in Chaps. 8 and 9 applications to some special situations, usually scalar equations with coefficients having particular properties, and compare what can be achieved by applying results from the general approach with what was previously known using ad hoc methods.

The idea for developing asymptotic representation results for both differential and difference equations in parallel came from a series of lectures in 2000 by the second author at Kumamoto University, Japan. This was supported by a grant from the Japanese government and hosted by Prof. M. Kohno. In 2002, we began collaborating on several research projects which followed this pattern of parallel development. Then while on sabbatical in Provence, we decided to begin assembling our results and others from many authors which either directly or indirectly apply Levinson's principles and which can be treated from the same perspective. A final push to complete work on this book came in 2013 while the first author was on sabbatical in Olomouc, Czech Republic, and we were able to work together again for several months, especially concentrating on applying the general results to special classes of scalar equations. We thank Profs. J. Andres and D. Livingstone at Palackého University for their support during this time.

We acknowledge the contributions of many authors whose work is cited in this book as well as other colleagues or collaborators who have influenced our mathematical development. In particular, we thank our Doktorvaeter, R.J. Sacker (University of Southern California), and W.B. Jurkat (Syracuse and Ulm) for introducing us to differential equations and giving us an appreciation of and respect for research. Also, we remember W.A. Harris, Jr., who participated in some of the earlier developments.

Finally, we thank our spouses, Roy and Margaret, for their encouragement and support during the long process of writing this book as well as their understanding for the time taken away from family.

Tacoma, WA, USA Sigrun Bodine
San Diego, CA, USA Donald A. Lutz
February 2015

Contents

Chapter 1
Introduction, Notation, and Background

1.1 Introduction

This book is concerned with the problem of determining the asymptotic behavior of solutions of non-autonomous systems of linear differential and linear difference equations. It has been observed from the work by Poincaré and Perron that there is a very close and symbiotic relationship between many results for differential and difference equations, and we wish to further demonstrate this by treating the asymptotic theories here in parallel. In Chaps. 2, 4, 6, and 8 we will discuss topics related to asymptotic behavior of solutions of differential equations, and in Chaps. 3, 5, 7, and 9 some corresponding results for difference equations. In Chap. 10 we will show how some of these results can be simultaneously treated within the framework of so-called dynamic equations on time scales.

The general procedure we follow is based on some pioneering results of Levinson [100] on what is now called *asymptotic integration* of linear systems of differential equations. Levinson's approach was later shown to provide a framework for a unified treatment of many other related results on asymptotic integration of differential equations and especially in the last 40 years has resulted in many new ones. The excellent monograph by Eastham [53] documents many of these basic results for differential equations as well as applications to self-adjoint systems and oscillatory behavior. The exposition on differential equations in this book complements and augments that found in Eastham, naturally including some overlap, but focusing especially on results from the past 25 years.

It had been known since the early days of the last century that solutions of perturbed systems $y' = [A + R(t)]y$ for all $t \geq t_0$, behave asymptotically like solutions of $x' = Ax$, if A is a constant matrix with *distinct* eigenvalues and $R \in L^1$. Levinson generalized this to the case when R tends to 0 as t tends to infinity, $R' \in L^1$, and the eigenvalues of $A + R(t)$ satisfy certain separation or *dichotomy conditions* (see [100, Theorem 1] or [42, Theorem 8.1]). This result involved two main ideas;

S. Bodine, D.A. Lutz, *Asymptotic Integration of Differential and Difference Equations*, Lecture Notes in Mathematics 2129,
DOI 10.1007/978-3-319-18248-3_1

one concerned the construction of a certain kind of linear transformation to reduce the system to an L^1 perturbation of a non-constant diagonal system whose entries satisfy certain dichotomy conditions; the second consisted in showing that solutions of such perturbed non-constant diagonal systems are asymptotically equivalent to solutions of the diagonal system.

While Levinson combined these two aspects into one statement, to extend his approach to systems satisfying quite different types of assumptions it has been found to be advantageous to treat them as independent contributions to the final asymptotic integration of the system. That is, depending upon certain assumptions, many different types of linear transformations could be constructed, whose role is to condition the system into one whose perturbation is smaller in an appropriate sense. Also, depending upon particular assumptions on the diagonal terms and the perturbation, various results can be proven showing the so-called asymptotic equivalence of solutions. During the past 40 years, this has led to a wide variety of new asymptotic integration results which were inspired by Levinson's approach.

In Chaps. 2 and 3, we first consider the second aspect above, which has been called *Levinson's fundamental theorem* and involves constructing solutions of certain perturbed diagonal systems that behave asymptotically as the solutions of the unperturbed system. This solution arises as a fixed point of an operator equation which can also be obtained via successive approximations (and traditionally was). Another way to see this uses a general result attributed to W.A. Coppel [43], using dichotomies to compare the asymptotic behavior of certain bounded solutions of perturbed and unperturbed linear systems. In this process, one can clearly observe the distinct and critical roles played by the dichotomy conditions and growth (or decay) conditions on the perturbation terms. Thus several types of corresponding results can be easily obtained by suitably modifying the procedure. These involve strengthening the dichotomy conditions while weakening the growth conditions or visa versa. By estimating the operator, one can also obtain various kinds of estimates on the error or remainder terms in the asymptotic representation.

In Chaps. 4 and 5, we discuss the role of conditioning transformations on a system, corresponding to several different types of assumptions related to the first aspect of Levinson's approach. These will have the form $y = [I + Q(t)]z$, where entries of Q are solutions of certain types of equations depending upon the particular properties of the perturbations and the diagonal terms. In some cases one such conditioning transformation is sufficient to bring the system into a form where a result from Chaps. 2 or 3 can be applied; in other cases, especially ones encountered in Chaps. 8 and 9, several transformations might be required. Results obtained in this manner include asymptotic integration theorems due to Hartman–Wintner, Levinson–Eastham, and Trench, to name a few. While most asymptotic integration results concern separate dichotomy assumptions on the diagonal terms and other, growth or decay conditions on the perturbations, some coupled conditions have been introduced by Elias and Gingold (see Sect. 2.9), which generalize results with separated conditions.

In Chaps. 6 and 7 we discuss perturbations of Jordan systems, in which the principal diagonal part of the linear system is replaced by one in Jordan form.

Here one can also utilize the basic theory developed in the prior chapters to obtain asymptotic behavior of solutions of such systems.

Several different kinds of applications can be made using the general asymptotic integration results for matrix systems in Chaps. 2–7. Among these, the ones that concern us in this book involve applications to some special classes of scalar equations, usually of second-order, which have been taken from the literature. In Chaps. 8 and 9 we discuss several such situations involving special assumptions on the coefficients, corresponding to certain types of applications. Depending upon those assumptions in a given situation, certain special concrete transformations are constructed to bring about the asymptotic behavior of a fundamental solution. This information can then be compared with what was achieved in the original treatment, which often involved ad hoc methods especially suited to particular scalar cases. In most cases what can be achieved using the general approach compares favorably with what was known using more special techniques, and it often leads to some improvements with respect to weakening assumptions, strengthening conclusions, or extending to more general situations.

Some other types of applications of asymptotic integration not discussed in this book concern:

(a) Determining the stability of solutions linear systems.
    We refer to Coppel [43, Chap. 2] for a discussion of various types of stability and how these are characterized as properties of a fundamental solution matrix.
(b) Using linearization methods to determine the asymptotic behavior of certain solutions of nonlinear systems.
    We refer the reader to [42, Chap. 6 and 7].
(c) Developing formulas for the analysis and computation of special functions, often arising from mathematical physics.
    See [113] for a treatment of such applications. (The asymptotic results presented here such as the Liouville–Green formulas are usually a good, first approximation on which to base further, more exact, and effective formulas which would also take into account validity near turning points and investigating ranges for parameter dependence.)

The treatment here will emphasize a "systems approach" to the study of the asymptotic behavior, using matrix methods whenever possible. Even for higher order scalar differential or difference equations, this approach is often both notationally as well as conceptually simpler and sometimes even leads to better results.

With respect to historical remarks and the development of the subject up to 1989, we refer to the excellent "Notes and References" at the end of each chapter in Eastham [53]. We will only make such remarks when they supplement what Eastham has given or when they concern more recent papers.

For the basic theory of differential equations, we refer the reader to the classical books by Coddington and Levinson [42] and Hartman [72]. Coppel's book [43] on *Stability and Asymptotic Behavior of Differential Equations* has been especially useful for us and treats many other topics, also related to nonlinear equations. The more recent book by Hsieh and Sibuya [77] also contains results on asymptotic behavior and especially a discussion of blocked-matrix methods.

For the basic theory of difference equations, we recommend Elaydi [55], Lakshmikantham and Trigiante [99], Kelley and Peterson [88], and Agarwal [1]. Earlier classical treatises on difference equations and the calculus of finite differences (see [86, 109, 114]) contain many results applicable to difference methods for numerical calculations and approximations of derivatives.

1.2 Relationship to Other Asymptotic Theories

The classical case of differential equations with a coefficient matrix that is *meromorphic at infinity* is a special case of the theory we consider, where the asymptotic integration takes place along a single ray to infinity in the complex plane. We note that the meromorphic case yields what could be called a "complete" theory of asymptotic integration. By this we mean that so-called "formal fundamental solutions" can always be algorithmically constructed and these can be shown to be asymptotic representations for actual solutions. Moreover, these actual "asymptotic" solutions exist not only on individual rays, but in certain open, unbounded sectors of the complex plane. Using them, the behavior of solutions of meromorphic differential equations in a full neighborhood at infinity can be analyzed. See Balser [3] and Jurkat [87] for comprehensive treatments of the modern asymptotic theory of solutions for meromorphic differential equations.

For the wider classes of equations studied in this book, there is not such a complete theory as for the meromorphic case. However, this meromorphic theory is a good model for the more general theory of asymptotic integration, and some of the techniques we apply are motivated by analogies. In spite of the fact that there appears no algorithm for treating more general situations, the methods of Chap. 4 apply to ever-widening classes of equations.

A particularly interesting discussion of difference equations, especially ones of second order analogues to the classical differential equations of mathematical physics, can be found in Batchelder [5]. This treatment also involves the extension to difference equations in sectors of the complex plane and expands on ideas of G.D. Birkhoff which have not found their way into the modern theory. For analytic difference equations, we refer the reader to G. Immink's book [82] involving representations of formal solutions using power series and their interpretation as asymptotic expansions in sectors of the complex plane. Also see the very interesting treatise by Kohno [95] which contains some asymptotic results for solutions of certain analytic, linear difference equations and representations for solutions using generalized Gamma functions.

Another very powerful and useful approach for obtaining asymptotic representations for solutions of differential equations involves representations not with respect to the large independent variable t and containing error terms $o(1)$, but instead with respect to a large parameter in the equation. There is an extensive literature connected with this approach and many interesting applications, especially to numerical calculations, but we will not discuss such results here as they appear to have little in common with the kinds of asymptotic representations we consider. On the other hand, certain asymptotic integration results can be applied to yield results for equations with large parameters, notably the classical Liouville–Green formulas. See Olver [116, Chap. 6] for an interesting account of such an application.

1.3 Common Asymptotic Terminology

In the literature concerned with the quantitative analysis of solutions of differential and difference equations, many terms have been introduced containing the adjective "asymptotic" for the objects and processes involved. We use the term *asymptotic integration* as it has become to be known in a broad sense (for both differential as well as difference equations) to describe the general process of starting with a given equation and finding an asymptotic representation for either a fundamental solution matrix or a single solution vector, often shortened to *asymptotic solution*.

This process is embodied in what we call an *asymptotic factorization* for a solution (see Sect. 4.1 for a more detailed discussion). This involves deriving a representation for a fundamental solution matrix $X(t)$ as a product of matrices that displays the asymptotic nature of solutions and consists of certain explicit or computable quantities together with a factor $I + E(t)$ which is *asymptotically constant* such as

$$X(t) = P(t)\,[I + Q(t)]\,[I + E(t)]\exp\left[\int^t \Lambda(\tau)\,d\tau\right].$$

Here $E(t)$, called the *remainder* or *error*, is a matrix-valued function that tends to zero as t tends to infinity. In case the reduced equation has a diagonal matrix $\Lambda(t)$ as coefficient, this last process is often called *asymptotic diagonalization* (see, e.g., [20] or [80]) and is usually the result of applying Levinson's fundamental theorem (see Sects. 2.2 and 3.2). The matrix $E(t)$ is typically constructed implicitly as the solution of a certain operator equation, which also can be used for estimates (see Sects. 2.7 and 3.7). Viewed as a linear transformation, $I + E(t)$ is also sometimes called an *asymptotic transformation* of a perturbed system since it performs the essential step of asymptotically eliminating the perturbation.

The other matrices in the asymptotic factorization consist of certain *preliminary transformations* $P(t)$ (see Sects. 4.7 and 5.7) and *conditioning transformations* $I + Q(t)$ (see Chaps. 4 and 5). Their role is to bring the equation into a suitable form so that the final step, an asymptotic transformation, can be applied.

In Sects. 2.5 and 2.6, 3.5 and 3.6, we also discuss briefly another type of asymptotic integration which has been called *asymptotic integration on a logarithmic scale*. This is especially appropriate for scalar equations which are treated in Sects. 8.2 and 9.2, where an asymptotic factorization is replaced by a significantly weaker statement involving the limit of the logarithm of a solution. This weaker conclusion comes about due to corresponding weaker assumptions on the perturbations and contains results originating with Poincaré and Perron.

Finally, in Sects. 2.10 and 3.8, the notion of *asymptotic equivalence* between two linear systems is discussed. There are several natural interpretations, some of which correspond to asymptotic integration.

1.4 Notation and Background

We consider systems of differential equations, resp. difference equations which are expressed in the form

$$x' = A(t)x, \qquad t \geq t_0, \tag{1.1}$$

resp.

$$x(n+1) = A(n)x(n), \qquad n \in \mathbb{N}, \quad n \geq n_0. \tag{1.2}$$

Here A denotes a $d \times d$ matrix and x a d-dimensional vector of complex-valued functions defined on the appropriate domain $[t_0, \infty)$ or $n \in \mathbb{N}$, $n \geq n_0$. In (1.1), the continuity of $A(t)$ on $[t_0, \infty)$ implies the existence of a unique solution satisfying $x(t_0) = x_0$ for any prescribed initial value x_0 and also the existence of a *fundamental matrix* $X(t)$ whose columns are linearly independent solution vectors so that $x(t) = X(t)X^{-1}(t_0)x_0$. We note that if A is just locally Lebesgue integrable, then solutions exist almost everywhere (a.e.) and fundamental solution matrices are also invertible a.e. [42, Chap. 2]. This assumption could be weakened to A being just gauge-integrable (see, e.g., Bartle [4]). We will usually assume continuity, but we remark that under the weaker assumptions we would have all the same asymptotic representations that will be developed with the only proviso that solutions exist a.e.

We will frequently use the Lebesgue spaces

$$L^p[t_0, \infty) = \left\{ f : \int_{t_0}^{\infty} |f(t)|^p \, dt < \infty \right\}$$

in order to specify certain growth/decay assumptions. In case we would not also require f to be continuous on $[t_0, \infty)$, the assumption $f \in L^p[t_0, \infty)$ will always include the implicit assumption that f is at least locally integrable (in the Lebesgue sense) on $[t_0, \infty)$. In this regard, we sometimes use the notation $f(t) = g(t) + (L^p)$ to denote $f - g \in L^p[t_0, \infty)$ in the above sense.

For systems of difference equations, we prefer to work with (1.2) rather the equivalent

$$\Delta x(n) := x(n+1) - x(n) = [A(n) - I]\, x(n).$$

A solution matrix of (1.2) can always be represented as the ordered product

$$X(n) = \prod_{n_0}^{n-1} A(k) = A(n-1)A(n-2)\cdots A(n_0) \qquad \text{for } n > n_0, \tag{1.3}$$

and $X(n_0) = I$. However, except for simple cases (such as autonomous systems, i.e., $A(n) \equiv A$ for all $n \geq n_0$ and hence $X(n) = A^{n-1}$), such a product representation is not useful for determining the asymptotic behavior of solutions.

Recall that a solution matrix $X(n)$ of (1.2) is fundamental if and only if $A(n)$ is invertible for all $n \geq n_0$. Without the assumption of invertibility, solutions may be all eventually trivial as shown by the example

$$x(n+1) = \begin{bmatrix} 0 & 1 \\ 0 & 0 \end{bmatrix} x(n).$$

To avoid such situations, we normally will assume invertibility except in Sects. 3.5 and 3.10. If invertibility is not assumed, certain additional assumptions are required to discuss subspaces of non-trivial solutions and their asymptotic behavior.

One may discuss difference equations (1.2) with n_0 as an integer and $n \geq n_0$, or the equation may equally be considered for any real sequence $\{x_0, x_0 + 1, \ldots\}$ or even when x_0 is complex. In the product representation (1.3) we have made implicitly the assumption that n is an integer, but one could also write $X(x_0 + n) = \prod_{k=0}^{n-1} A(k + x_0)$ for any integer $n > 0$. In questions concerning the asymptotic behavior of solutions of (3.1) as $n \to \infty$, formulas apply equally well for $n \in \mathbb{N}$ or $n \in x_0 + \mathbb{N}$, but observe that this does not mean that the asymptotic behavior is necessarily independent of x_0. An example for the dependence on x_0 is the scalar equation $x(t+1) = (t + 1 - [t])x(t)$ for $t \in \mathbb{R}$.

For linear differential equations, all solutions can be generated from a fundamental solution matrix by multiplying on the right by constant vectors. For linear difference equations on $\mathbb{R}$, such constants must be replaced by periodic functions of period 1. Working on $\mathbb{N}$ or $\{x_0 + \mathbb{N}\}$ one can of course just deal with constant right hand factors.

In order to analyze the asymptotic behavior of solutions, we employ three basic tools. The first one involves linear matrix transformations of the equations. For differential equations a linear, invertible transformation $x = T(t)y$ takes (1.1) into

$$y' = [T^{-1}(t)A(t)T(t) - T^{-1}(t)T'(t)]y,$$

while for difference equations a linear and invertible transformation $x(n) = T(n)y(n)$ takes (1.2) into

$$y(n+1) = T^{-1}(n+1)A(n)T(n)y(n).$$

In both cases we see that the transformation is close to, but not the same, a similarity transformation of A, except of course when T is constant. This suggests that the eigenvalues of A are somewhat responsible for the asymptotic behavior of solutions, but may need to be "corrected" by terms arising from T' or $T(n+1) - T(n)$.

A second frequently used tool is a scalar normalizing transformation on the system. While the transformation $y = f(t)\hat{y}$ takes (1.1) into $\hat{y}' = [A(t) - f(t)I]\hat{y}$, the corresponding transformation $y(n) = f(n)\hat{y}(n)$ takes (1.2) into $\hat{y}(n+1) = \frac{f(n)}{f(n+1)}A(n)\hat{y}(n)$. This multiplicative factor instead of the additive one will manifest itself frequently (see e.g., Theorem 3.4, a discrete version of Levinson's fundamental theorem).

The third tool we use is the so-called variations-of-constants formula. In its simplest form, one sees that if $X(t)$ is a fundamental solution of (1.1), then the solution of the nonhomogeneous equations $x' = A(t)x + b(t)$ satisfying $x(t_0) = x_0$ is given by

$$x(t) = X(t)X^{-1}(t_0)x_0 + \int_{t_0}^{t} X(t)X^{-1}(s)b(s)\,ds.$$

This also applies to the homogeneous system $x' = [A(t) + B(t)]\,x$, where one just replaces $b(t)$ by $B(t)x$ to obtain

$$x(t) = X(t)X^{-1}(t_0)x_0 + \int_{t_0}^{t} X(t)X^{-1}(s)B(s)x(s)\,ds, \tag{1.4}$$

which is now an integral equation for x. Changing the initial value, one can go further and choose different lower limits of integration for the various components, especially letting the lower limit be ∞ (i.e., forming $-\int_t^\infty$) whenever it can be shown that the resulting integrals converge. Thus, the integral in (1.4) can be decomposed into the sum of two integrals, one along a path from t_0 to t and the other from ∞ to t. We will see this used for the first time in Theorem 2.2.

In order to establish the existence of solutions of such integral equations as (1.4), we will use the contraction mapping principle (Banach's Fixed Point Theorem) on appropriately defined Banach spaces and then obtain the asymptotic formulas by analyzing the integrals. We mention that in lieu of the contraction mapping principle, we could instead choose to construct successive approximations and estimate them and their differences to obtain the same result (see [42, 53]). But we find that what we call the "T-operator" approach (see Sect. 2.2) is more economical and yields the desired formulas more immediately.

For difference equations, an analogous formula for a solution of $x(n+1) = A(n)x(n) + b(n)$ is

$$x(n) = X(n)X^{-1}(n_0)x_0 + \sum_{k=n_0}^{n-1} X(n)X^{-1}(k+1)b(k)$$

and for the system $x(n+1) = A(n)x(n) + B(n)x(n)$ we have

$$x(n) = X(n)X^{-1}(n_0)x_0 + \sum_{k=n_0}^{n-1} X(n)X^{-1}(k+1)B(k)x(k). \tag{1.5}$$

Again, we can use the standard operating procedure of choosing different lower limits of summation in (1.5) (especially including infinity) whenever appropriate. This decomposes the sum into two parts which can be estimated independently. To show the existence of solutions, we also use the contraction mapping principle on appropriate Banach spaces.

For the estimates we shall make on vectors $x(t)$ or $x(n)$ in $\mathbb{R}^n$ or $\mathbb{C}^n$, we use the notation $|x(\cdot)|$ for any of the equivalent finite-dimensional norms and $|A(\cdot)|$ for their induced matrix norms. Other norms on function spaces will be defined as required and any other special notation will be deferred until we use them.

For future use, we remark here that systems (1.1) and (1.2) also include scalar d^{th} order equations

$$y^{(d)} + a_1(t)y^{(d-1)} + \cdots + a_d(t)y = 0$$

and

$$y(n+d) + a_1(n)y(n+d-1) + \cdots + a_d(n)y(n) = 0,$$

as special cases, namely when A in (1.1) and (1.2) are companion matrices of the form

$$A = \begin{pmatrix} 0 & 1 & 0 & \cdots & 0 \\ 0 & 0 & 1 & & \vdots \\ \vdots & & & \ddots & 0 \\ 0 & & & & 1 \\ -a_d & & \cdots & & -a_1 \end{pmatrix}$$

and $x = [y, y', \ldots, y^{(d-1)}]^T$ in (1.1), resp. $x(n) = [y(n), y(n+1), \ldots, y(n+d-1)]^T$ in (1.2).

Thus the systems results also apply to higher order scalar differential or difference equations. In Chaps. 8 and 9 we will exploit this approach and compare the results with those usually obtained using ad hoc procedures, especially in the second-order case.

Chapter 2
Asymptotic Integration for Differential Systems

2.1 Chapter Overview

In this chapter we are concerned with the following general problem: If we are given a linear system

$$y' = [A(t) + R(t)]\, y, \qquad t \geq t_0, \tag{2.1}$$

and "know" a fundamental solution matrix $X(t)$ for the (unperturbed) system $x' = A(t)x$, how "small" should the perturbation $R(t)$ be so that we can determine an asymptotic behavior for solutions of (2.1)? This question is intentionally vague because depending upon the particular circumstances, there are many possible answers.

More specifically, in the case

$$A(t) = \Lambda(t) = \operatorname{diag}\{\lambda_1(t), \ldots, \lambda_d(t)\}$$

and $X(t) = \exp\left[\int^t \Lambda(\tau)\, d\tau\right]$, one would like to know under what conditions there exists a fundamental solution matrix of (2.1) of the form

$$Y(t) = [I + E(t)] \exp\left[\int^t \Lambda(\tau)\, d\tau\right], \qquad E(t) \to 0, \tag{2.2}$$

as $t \to \infty$. If this occurs, we say that

$$y' = [\Lambda(t) + R(t)]y \tag{2.3}$$

is in *L-diagonal form.* An interpretation of (2.2) that we will focus on is that the linear transformation $y = [I + E(t)]x$ takes (2.3) into $x' = \Lambda(t)x$, hence it

S. Bodine, D.A. Lutz, *Asymptotic Integration of Differential and Difference Equations*, Lecture Notes in Mathematics 2129,
DOI 10.1007/978-3-319-18248-3_2

diagonalizes the perturbed system, and this process is often referred to as *asymptotic diagonalization.*

Observe that $E(t)$ in (2.2) satisfies a so-called *transformation equation*

$$E' = \Lambda E - E\Lambda + R + RE, \tag{2.4}$$

and the goal is to find sufficient conditions involving $\Lambda(t)$ and $R(t)$ such that it has a solution $E(t) = o(1)$ as $t \to \infty$.

In this chapter we first present one of the most important contributions which give very sharp sufficient conditions for a system to be in L-diagonal form. These conditions (see Theorem 2.7) were identified by N. Levinson in the course of dealing with a more special case (namely when $\Lambda(t) \to \Lambda_0$ as $t \to \infty$). Since these conditions and his proof apply to the more general situation, he is given credit for it, and we frequently refer to Theorem 2.7 as *Levinson's fundamental theorem* on asymptotic integration.

As we shall see in Sect. 2.2, the two main assumptions for an asymptotic integration result of (2.3) concern a *dichotomy condition* involving $\Lambda(t)$ and a *growth condition* on $R(t)$. Levinson's conditions are, in some sense, the most "balanced" in a wide range of weaker dichotomy conditions with stronger growth conditions, or stronger dichotomy conditions with weaker growth conditions. We will also give examples of such results. Another reason why Levinson's result has been called "fundamental" is because many other asymptotic integration results can be obtained by transforming a given system so that Theorem 2.7 does apply. Results of this nature will be discussed in Chap. 4.

Another question that will be addressed in Sect. 2.4 concerns what happens if $\Lambda(t)$ does not satisfy Levinson's dichotomy conditions, but some weaker version. We show there how the growth condition on $R(t)$ should be suitably strengthened to allow for an asymptotic integration result.

Then in Sects. 2.5 and 2.6 we consider much stronger dichotomy conditions than Levinson's and investigate how weaker growth conditions lead to some corresponding statements such as (2.2), but with modified $\Lambda(t)$, or to a weaker type of asymptotic statement which Hartman refers to as "asymptotic integration on a logarithmic scale." A particularly interesting special case concerns classical results of H. Poincaré and O. Perron who treated asymptotically constant higher-order scalar equations (see Chap. 8). We give both an extension and generalization of these results for systems, also modifying similar results of Hartman, and show how they can be obtained using a similar T-operator approach as in Sect. 2.2.

Returning to (2.2), one could also ask what can be said about the "error terms" $E(t)$, in particular how they depend on the data $\Lambda(t)$ and $R(t)$. In Sect. 2.7 we discuss some results along these lines which follow from a closer examination of the key ingredient to the proof of Theorem 2.2, namely what we call the T-operator that is responsible for the terms in $E(t)$.

In Sect. 2.9 we consider results of Elias/Gingold where dichotomy and growth conditions were not considered separately, but studied as a "coupled" conditions.

Such results may be interpreted as generalization of some results in previous sections.

Finally in Sect. 2.10 we return to the general question we began with for general, not necessarily diagonal matrices $A(t)$ and present a perturbation result in the context of asymptotic equivalence between systems.

2.2 Ordinary Dichotomies, L^1-Perturbations, and Levinson's Fundamental Theorem

In this section we will discuss ideas which result in a proof of what we call *Levinson's fundamental theorem* on asymptotic integration of perturbed linear systems. There are several approaches that can be used, all of which involve the construction of an integral equation, which in effect solves the transformation (2.4). The approaches differ mainly on how to use pre-normalizations to simplify the calculations, whether to use successive approximations directly or to think of the integral equation as defining an operator on a certain Banach space.

We prefer an approach attributed to Coppel [43], which uses a normalization so that the basic asymptotic correspondence is reduced to one between bounded solutions of a perturbed and unperturbed system. Also his approach emphasizes the role played by a so-called T-operator, which avoids working with successive approximations. It instead focuses on direct estimations of the operator itself.

We begin with a rather general result by Coppel concerning L^1-perturbations of a system satisfying a so-called ordinary dichotomy.

Definition 2.1 (Ordinary Dichotomy) A linear differential system

$$x' = A(t)x\,, \qquad t \geq t_0, \tag{2.5}$$

is said to have an *ordinary dichotomy* if there exist a fundamental solution matrix $X(t)$, a projection P ($P^2 = P$), and a positive constant K such that

$$\left.\begin{array}{ll} |X(t)PX^{-1}(s)| \leq K & \forall \;\; t_0 \leq s \leq t, \\ |X(t)[I-P]X^{-1}(s)| \leq K & \forall \;\; t_0 \leq t \leq s. \end{array}\right\} \tag{2.6}$$

It follows that (2.5) has an ordinary dichotomy with $P = I$ if and only if it is uniformly stable (see, e.g., [43, p. 54]). To help understand the meaning of this definition in the general case, we also state an equivalent formulation which was given in Coppel [44, p. 11]: there exist positive constants K_1, K_2, and M such that for all $\xi \in \mathbb{R}^d$

$$\begin{array}{lll} \text{(i)} & |X(t)P\xi| \leq K_1\,|X(s)P\xi| & \forall\, t \geq s \geq t_0, \\ \text{(ii)} & |X(t)[I-P]\xi| \leq K_2\,|X(s)[I-P]\xi| & \forall\, s \geq t \geq t_0, \\ \text{(iii)} & |X(t)PX^{-1}(t)| \leq M & \forall\, t \geq t_0. \end{array}$$

Put $k = \operatorname{rank} P$. Then an interpretation is that

1. $\exists$ a k-dimensional subspace of solutions that are uniformly stable (forward in time)
2. $\exists$ a supplemental $(n-k)$-dimensional subspace of solutions that are uniformly stable "backwards in time"
3. The "angle" between the two subspaces remains bounded away from zero.

Theorem 2.2 (Coppel [43, Chap. 3, Thm. 11]) *For continuous $d \times d$ matrices $A(t)$ and $R(t)$, consider the unperturbed system* (2.5) *and the perturbed system*

$$y'(t) = [A(t) + R(t)]\, y, \qquad t \geq t_0, \tag{2.7}$$

where $R \in L^1[t_0, \infty)$, i.e.,

$$\int_{t_0}^{\infty} |R(t)|\, dt < \infty. \tag{2.8}$$

Assume that (2.5) *satisfies an ordinary dichotomy* (2.6) *with projection matrix P. Then there exists a one-to-one and bicontinuous correspondence between the bounded solutions of* (2.5) *and* (2.7) *on $[t_0, \infty)$. Moreover, the difference between corresponding solutions of* (2.5) *and* (2.7) *tends to zero as $t \to \infty$ if $X(t)P \to 0$ as $t \to \infty$.*

Proof Fix $t_1 \geq t_0$ such that

$$\theta := K \int_{t_1}^{\infty} |R(t)|\, dt < 1. \tag{2.9}$$

Let $\mathscr{B}$ be the Banach space of bounded (for $t \geq t_1$) d-dimensional vector-valued functions with the supremum norm

$$\|y\| = \sup_{t \geq t_1} |y(t)|. \tag{2.10}$$

For $t \geq t_1$, define an operator T on $\mathscr{B}$ by

$$\begin{aligned}(Ty)(t) = & \int_{t_1}^{t} X(t)PX^{-1}(s)R(s)y(s)\, ds \\ & - \int_{t}^{\infty} X(t)[I-P]X^{-1}(s)R(s)y(s)\, ds. \end{aligned} \tag{2.11}$$

By (2.6), (2.9) and (2.10),

$$|(Ty)(t)| \leq \|y\| K \int_{t_1}^{\infty} |R(s)|\, ds \leq \theta \|y\|,$$

which shows that T maps $\mathscr{B}$ into $\mathscr{B}$. A similar computation shows that for $y_i \in \mathscr{B}$ $(i = 1, 2)$,

$$\|T(y_2 - y_1)\| \le \theta \|y_2 - y_1\|,$$

i.e., $T : \mathscr{B} \to \mathscr{B}$ is a contraction. Hence for every function $x \in \mathscr{B}$, the operator equation

$$x = y - Ty \tag{2.12}$$

has a unique solution $y \in \mathscr{B}$. It can easily be checked that if x is a solution of (2.5), then $y = x + Ty$ is a solution of (2.7). Therefore, if $x(t)$ is a bounded solution of (2.5), then the solution $y(t)$ of (2.12) is a bounded solution of (2.7). Conversely, if $y(t)$ is a bounded solution of (2.7), the function $x(t)$ defined by (2.12) is a bounded solution of (2.5).

The equation $y = x + Ty$ therefore establishes for $t \ge t_1$ a one-to-one correspondence between the bounded solutions of (2.5) and (2.7). The bicontinuity of this correspondence on $[t_1, \infty)$ follows from the inequalities

$$\frac{1}{1+\theta}\|x_1 - x_2\| \le \|y_1 - y_2\| \le \frac{1}{1-\theta}\|x_1 - x_2\|.$$

Using continuous dependence of solutions on initial conditions, the bicontinuity can be extended to $[t_0, \infty)$. Finally, given a solution y of (2.7) and $\varepsilon > 0$, fix $t_2 = t_2(\varepsilon) > t_1$ such that

$$K \int_{t_2}^{\infty} |R(t)|\,|y(t)|\,dt \le K\|y\| \int_{t_2}^{\infty} |R(t)|\,dt < \frac{\varepsilon}{2}.$$

Then for $t \ge t_2$

$$\begin{aligned}
|y(t) - x(t)| &= |(Ty)(t)| \\
&\le |X(t)P| \left| \int_{t_1}^{t_2} X^{-1}(s)R(s)y(s)\,ds \right| + K\|y\| \int_{t_2}^{\infty} |R(s)|\,ds \\
&\le |X(t)P| \int_{t_1}^{t_2} |X^{-1}(s)R(s)y(s)|\,ds + \frac{\varepsilon}{2} < \varepsilon,
\end{aligned}$$

for all t sufficiently large since $X(t)P \to 0$ as $t \to \infty$. □

Remark 2.3 We chose to write (2.5) and (2.7) as the linear homogeneous systems given above since this is precisely the form we will be interested in. However, Theorem 2.2 can be generalized by adding a continuous inhomogeneous term $b(t)$ to both (2.5) and (2.7) as well as by replacing $R(t)y(t)$ in (2.7) by "mildly nonlinear perturbations" $f(t, y)$ such that $\int_{t_0}^{\infty} |f(t, 0)|\,dt < \infty$ and $|f(t, y_1) - f(t, y_2)| \le$

$\gamma(t)\|y_1 - y_2\|$ for all $y_i \in \mathscr{B}$ $(i = 1, 2)$ where $\int_{t_0}^{\infty} \gamma(t)\,dt < \infty$ (see Coppel [43, Chap. 3, Thm. 11]).

Remark 2.4 It is interesting to mention that the proof of Theorem 2.2 did not use that the matrices P and $I - P$ are projections. Only the property $P + [I - P] = I$ was used. The reason is that projections correspond to a decomposition of the solutions into complementary subspaces and this is a useful way to interpret such a dichotomy condition. In many applications, $A(t)$ in (2.5) will be a diagonal matrix and the matrices $P, I - P$ will be supplementary diagonal matrices with zeroes and ones on the diagonal and therefore projection matrices.

Remark 2.5 In what follows, we will not need the full strength of Theorem 2.2. What will be important for our purposes is that in the presence of an ordinary dichotomy (2.6), for every bounded solution $x(t)$ $(t \geq t_0)$ of the unperturbed system (2.5) there exists a bounded solution $y(t)$ $(t \geq t_0)$ of the perturbed system (2.7) and, moreover, the difference between these two bounded solutions will vanish at infinity provided that $X(t)P \to 0$ as $t \to \infty$.

Continuing toward a proof of Levinson's fundamental theorem, we now state *Levinson's dichotomy conditions* for a diagonal matrix $\Lambda(t) = \{\lambda_1(t), \ldots, \lambda_d(t)\}$. We will later see that these are equivalent to certain shifted systems [cf. (2.20)] satisfying the ordinary dichotomy (2.6) in Coppel's sense.

Definition 2.6 A diagonal matrix $\Lambda(t) = \operatorname{diag}\{\lambda_1(t), \ldots, \lambda_d(t)\}$ defined for $t \geq t_0$ is said to satisfy *Levinson's dichotomy conditions* if there exists $K > 0$ such that for each pair of integers (i, j), $1 \leq i \neq j \leq d$
either

$$\left.\begin{array}{rl} & \int_{t_0}^{t} \operatorname{Re}\{\lambda_j(\tau) - \lambda_i(\tau)\}\,d\tau \to -\infty \text{ as } t \to \infty, \\ \text{and} & \int_{s}^{t} \operatorname{Re}\{\lambda_j(\tau) - \lambda_i(\tau)\}\,d\tau \leq K \;\forall\; t_0 \leq s \leq t, \end{array}\right\} \tag{2.13}$$

or

$$-\int_{t}^{s} \operatorname{Re}\{\lambda_j(\tau) - \lambda_i(\tau)\}\,d\tau \leq K \qquad \forall \;\; t_0 \leq t \leq s. \tag{2.14}$$

We now apply Theorem 2.2 to prove

Theorem 2.7 (Levinson's Fundamental Theorem) *Suppose that the diagonal and continuous $d \times d$ matrix $\Lambda(t) = \operatorname{diag}\{\lambda_1(t), \ldots, \lambda_d(t)\}$ satisfies Levinson's dichotomy conditions* (2.13), (2.14) *for $t \geq t_0$. Furthermore, assume that the $d \times d$ matrix $R(t)$ is continuous for $t \geq t_0$ and that*

$$R(t) \in L^1[t_0, \infty). \tag{2.15}$$

Then the linear differential system

$$y' = [\Lambda(t) + R(t)]y, \qquad t \geq t_0, \tag{2.16}$$

has for $t \geq t_0$ a fundamental matrix satisfying

$$Y(t) = [I + E(t)] \exp\left[\int^t \Lambda(s)\, ds\right], \qquad \textit{where } E(t) = \mathrm{o}(1) \textit{ as } t \to \infty. \tag{2.17}$$

Proof Fix $i \in \{1, \ldots, d\}$ and put

$$y(t) = z(t) \exp\left[\int^t \lambda_i(\tau)\, d\tau\right]. \tag{2.18}$$

Then (2.16) implies that

$$z'(t) = [\Lambda(t) - \lambda_i(t)I + R(t)]\, z. \tag{2.19}$$

We also consider the unperturbed shifted system

$$w' = [\Lambda(t) - \lambda_i(t)I]\, w, \tag{2.20}$$

and we will apply Theorem 2.2 to these two shifted systems. For that purpose, let $P = \mathrm{diag}\,\{p_1, \ldots, p_d\}$, where

$$p_j = \begin{cases} 1 & \text{if } (i,j) \text{ satisfies (2.13)} \\ 0 & \text{otherwise.} \end{cases}$$

Then (2.20) satisfies the ordinary dichotomy condition 2.1 and $W(t)P \to 0$ as $t \to \infty$. Now, by Theorem 2.2 and since $w_i(t) = e_i$ (where $e_i \in \mathbb{R}^d$ is the ith Euclidean vector) is trivially a bounded solution of (2.20), there exists a bounded solution $z_i(t)$ on $[t_0, \infty)$ of (2.19) with

$$z_i(t) = e_i + \mathrm{o}(1) \qquad \text{as } t \to \infty.$$

By (2.18), (2.16) has therefore for $t \geq t_0$ a solution

$$y_i(t) = [e_i + \mathrm{o}(1)] \exp\left[\int^t \lambda_i(\tau)\, d\tau\right] \qquad \text{as} \quad t \to \infty. \tag{2.21}$$

Repeating this process for all $1 \leq i \leq d$, this set of solutions is clearly linearly independent for t sufficiently large by the structure of the solutions given in (2.21). Being solutions of the linear differential equation (2.16), this linear independence is given on all of $[t_0, \infty)$, which establishes (2.17). □

While one could argue that all the ingredients for the above proof of Levinson's fundamental theorem are more or less contained within Levinson's original ideas, the organization and explicit use of the shifted system help to significantly reduce the number and difficulty of the ensuing calculations and estimations and does so in a manner which makes it also extendable to other situations. Eastham [53, Sect. 1.4] has also given a simplified proof, but still involving explicit estimates for successive approximations instead of applying the contraction mapping and estimating the integrals directly. While again one could argue that these two ways to produce an asymptotic solution are equivalent, we prefer the simplicity and flexibility of what we like to call the "T-operator approach" [see (2.11)].

As noted in the proof of Theorem 2.2, the operator T [which is used to establish (2.17)] was only shown to be a contraction on $[t_1, \infty)$ with t_1 chosen to be sufficiently large ($t_1 \geq t_0$). But because solutions of (2.7) can be continued back to t_0, the function $E(t)$ in (2.17) exists on the full interval $[t_0, \infty)$. It would have been possible instead of (2.10) to use a weighted (or Bielecki) norm (see [43, Chap. 1]) to make T a contraction on $[t_0, \infty)$. But for simplicity in the proofs involving T-operators and because the only essential information concerning $E(t)$ at present deals with its asymptotic behavior as $t \to \infty$, we will not normally employ such weighted norms. However, as an exception in Chap. 8 when dealing with error bounds for $E(t)$ on the entire interval, we will use such a weighted norm (see Sect. 8.3.3).

Remark 2.8 Written in the form (2.13) and (2.14), Levinson's dichotomy conditions are mutually exclusive. As shown in Eastham [53, p. 9], they can be re-written in a more symmetric, yet not mutually exclusive form. Namely, (2.13) and (2.14) can be stated in the following equivalent way: for each pair of integers (i, j), $1 \leq i \neq j \leq d$ either

$$\int_s^t \operatorname{Re}\{\lambda_j(\tau) - \lambda_i(\tau)\}\, d\tau \leq K_1 \qquad \forall \quad t_0 \leq s \leq t, \tag{2.22}$$

or

$$\int_s^t \operatorname{Re}\{\lambda_j(\tau) - \lambda_i(\tau)\}\, d\tau \geq K_2 \qquad \forall \quad t_0 \leq s \leq t, \tag{2.23}$$

for some constants K_1 and K_2. To establish equivalence, first note that (2.13) and (2.14) clearly imply (2.22) and (2.23). On the other hand, setting

$$I_{ij}(t) = \int_{t_0}^t \operatorname{Re}\{\lambda_j(\tau) - \lambda_i(\tau)\}\, d\tau,$$

(2.22) becomes

$$I_{ij}(t) - I_{ij}(s) \leq K_1 \qquad \text{for all } t_0 \leq s \leq t.$$

It follows that either $I_{ij}(t) \to -\infty$ as $t \to \infty$ or not, in which case this "bounded variation" condition (2.22) of the function $I_{ij}(t)$ implies that $I_{ij}(t)$ is bounded from below, i.e., there is a constant κ such that $I_{ij}(t) \geq \kappa$ for all $t \geq t_0$. Putting this latter alternative together with (2.23) and noting that then

$$I_{ij}(t) - I_{ij}(s) \geq \kappa - I_{ij}(s) \geq \kappa - K_1 - I_{ij}(t_0) \qquad \forall\ t \geq s \geq t_0$$

proves the equivalence.

Remark 2.9 Every constant diagonal matrix Λ satisfies these dichotomy conditions without any conditions on its entries: they might be distinct, have the same real part, or even be identical. For non-autonomous $\Lambda(t)$, its entries should be "separated" in a certain sense in order to satisfy (2.13) and (2.14). For example, if $\mathrm{Re}\left[\lambda_i(t) - \lambda_j(t)\right]$ does not change sign for all t sufficiently large, then (2.13) and (2.14) hold (this result can also be found in Coppel [43, p. 88]). However, this rather strong restriction is not necessary as can be seen in the example $\lambda_i = \sin t$ and $\lambda_j = 0$. On the other hand, the example $\lambda_i = \sin t$ and $\lambda_j = -t\cos t$ does not satisfy (2.13) and (2.14) as $\mathrm{Re} \int_0^t \{\lambda_i(\tau) - \lambda_j(\tau)\}\, d\tau = t \sin t$. More generally, the situation where

$$\liminf_{t\to\infty} \int_{t_0}^t \mathrm{Re}\{\lambda_j(\tau) - \lambda_i(\tau)\}\, d\tau = -\infty,$$

and

$$\limsup_{t\to\infty} \int_{t_0}^t \mathrm{Re}\{\lambda_j(\tau) - \lambda_i(\tau)\}\, d\tau = +\infty,$$

for the same pair (i, j) violates Levinson's dichotomy conditions.

If Levinson's dichotomy conditions (2.13) and (2.14) do not hold for *all* i and j, but only for *one fixed* value of i and all $1 \leq j \leq d$, then it is apparent from the proof that there exists *one* solution of (2.16) of the form (2.21) for this particular value of i.

It is natural to investigate quantitative estimates of the "error term" o(1) in (2.17). We refer to [53, p. 14] for a general discussion. In Sect. 2.7, we will study the error term in more detail using stronger dichotomy conditions and a slow decay condition on the perturbation R.

In Sect. 2.3, we discuss examples which will show why a dichotomy condition on the entries of Λ is necessary in general if $R \in L^1$ and also why a growth condition on R is essential.

For completeness sake, we want to mention that Kimura [91] gave another proof of a special case of Levinson's fundamental theorem where $\Lambda(t)$ is diagonal and tends to a constant matrix with distinct entries. To do this, he applied an asymptotic existence theorem for systems of nonlinear differential equations due to Hukuhara [81]. At first glance, Hukuhara's theorem appears to be a statement concerning a solution of an initial value problem. To relate this to an asymptotic representation one needs to think of an o(1) asymptotic solution as satisfying a kind of initial value problem at infinity. In order to verify the main hypothesis of Hukahara's theorem, Levinson's dichotomy conditions were needed. Ensuing computations are in some ways analogous to arguments in Coppel's theorem (Theorem 2.2). The main difference, however, is that instead of using a scalar shift, Kimura used an exponential matrix shift which leads to more delicate calculations.

We conclude this section by noting that Bodine and Sacker [20] derived sufficient conditions for the diagonalizability and the asymptotic diagonalizability of differential systems of the form $x' = [\Lambda(t) + R(t)]x$, with $\Lambda(t)$ being diagonal. Their approach was based on the theory of linear skew-product flows. They assumed that the omega-limit set of Λ in an appropriate function space is compact and that it satisfies the so-called full spectrum property. Growth condition were phrased in terms of L^p-conditions. Moreover, Bodine [11] derived an asymptotic integration result where both dichotomy and growth conditions were phrased in terms of omega-limit sets.

2.3 Sharpness of Dichotomy and Growth Conditions

In a generic sense, one can show with examples that Levinson's dichotomy conditions on Λ and the L^1 growth condition on R are "necessary" for Theorem 2.7 to hold. By this we mean the following:

(i) There exists examples where $\Lambda(t)$ satisfies Levinson's dichotomy conditions (2.13), (2.14) and $R(t) \notin L^1[t_0, \infty)$, where the result (2.17) fails.
(ii) There exist examples where $\Lambda(t)$ does not satisfy Levinson's dichotomy conditions and $R(t) \in L^1[t_0, \infty)$, where the result (2.17) fails.

Examples of type (i) are easy to construct. The two-dimensional system

$$y' = \begin{bmatrix} a & t^{-1} \\ 0 & a \end{bmatrix} y = A(t)y,$$

shows this with a fundamental matrix given by

$$Y(t) = \begin{bmatrix} 1 & \ln t \\ 0 & 1 \end{bmatrix} e^{at} \neq [I + o(1)]e^{at}C \qquad (t \to \infty),$$

where C is any constant, invertible matrix. But Λ might be considered too simple in this example by being diagonal with equal eigenvalues. Also note that in this trivial example, it holds that $Y(t) = \exp\left[\int^t A(\tau)\, d\tau\right]$, which is not true in general.

An example where Λ has unequal eigenvalues requires more thought. It can be shown, however, that for the two-dimensional system

$$y' = \begin{bmatrix} i & \frac{e^{2it}}{t} \\ \frac{e^{-2it}}{t} & -i \end{bmatrix} y,$$

a fundamental matrix is given by

$$Y(t) = \begin{bmatrix} e^{it} & 0 \\ 0 & e^{-it} \end{bmatrix} \begin{bmatrix} 1 & 1 \\ -1 & 1 \end{bmatrix} \begin{bmatrix} t^{-1} & 0 \\ 0 & t \end{bmatrix} \neq [I + \mathrm{o}(1)] \begin{bmatrix} e^{it} & 0 \\ 0 & e^{-it} \end{bmatrix} C \qquad (t \to \infty),$$

for any constant invertible matrix C. In this example the eigenvalues of Λ are distinct, but with equal real parts. Such an example with $\{i, -i\}$ replaced by $\{1, -1\}$ would not work since Theorem 2.19 would apply. In Chap. 4 we will present many other results and examples showing to what extent the condition $R \in L^1$ can be weakened (while the dichotomy condition is strengthened), and either (2.17) or a modified statement with $\Lambda(t)$ replaced by an explicitly computable $\Lambda_1(t)$ still holds.

Examples of systems that do not satisfy Levinson's dichotomy conditions [i.e., case (ii)] are more challenging to construct. Eastham [53, p. 10] provides the following two-dimensional system where

$$\Lambda(t) = \operatorname{diag}\left\{0, \frac{\rho'(t)}{\rho(t)}\right\}, \qquad \text{and } \rho(t) = t^2(1 - \sin t) + 1.$$

If

$$R(t) = \begin{bmatrix} 0 & t^{-2} \\ 0 & 0 \end{bmatrix},$$

it can be shown that a fundamental matrix is given by

$$Y(t) = \begin{bmatrix} 1 & t + \cos t - \frac{1}{t} \\ 0 & \rho(t) \end{bmatrix} \neq [I + \mathrm{o}(1)]\operatorname{diag}\{1, \rho(t)\}C, \tag{2.24}$$

for any constant, invertible matrix C.

In the above example, however, Λ is unbounded as $t \to \infty$. An example with bounded $\Lambda(t)$ was first introduced by Perron [122] and later generalized by Coppel [43, p. 71] to

$$y' = \begin{bmatrix} -a & 0 \\ e^{-bt} & \sin \ln t + \cos \ln t - 2a \end{bmatrix} y = [\Lambda(t) + R(t)]y, \tag{2.25}$$

where a and b are positive. Perron and Coppel were concerned with perturbations of asymptotically, but not uniformly stable systems. There, even exponentially small perturbations of asymptotically stable systems may be unstable. Using that

$$\lambda_2(t) - \lambda_1(t) = \sqrt{2}\sin(\ln t + \pi/4) - a,$$

one can show that if $a \geq \sqrt{2}$, $\Lambda(t) = \operatorname{diag}\{-a, \sin\ln t + \cos\ln t - 2a\}$ satisfies Levinson's dichotomy conditions, but not if $0 < a < \sqrt{2}$. It was shown in [18, Ex. 8] that for each $a \in (0, \sqrt{2})$, there is a unique $\gamma(a) > 0$ such that for $0 < b < \gamma(a)$ the result (2.17) of Theorem 2.7 fails to hold, but for $b > \gamma(a)$ (2.17) holds. Here $\gamma(a)$ is uniquely determined by a, but not as an elementary function. So this is an example where even exponentially small perturbations are not sufficient for (2.17) to hold.

This example can be further modified by setting in (2.25) $t = \ln\tau$ and $\tilde{y}(\tau) = y(t)$, which yields

$$\frac{d\tilde{y}}{d\tau} = \frac{1}{\tau}\begin{bmatrix} -a & 0 \\ \tau^{-b} & \sin\ln\ln\tau + \cos\ln\ln\tau - 2a \end{bmatrix}\tilde{y} = \left[\tilde{\Lambda}(\tau) + \tilde{R}(\tau)\right]\tilde{y}. \tag{2.26}$$

Now for $0 < a < \sqrt{2}$, Levinson's dichotomy conditions are also not satisfied and $\tilde{R} \in L^1[1, \infty)$ for all $b > 0$. Because the two systems are related just by the change of variable $t = \ln\tau$, (2.26) also violates the conclusions of Theorem 2.7 in the case $0 < a < \sqrt{2}$ and $0 < b < \gamma(a)$. One can modify this even further by applying the transformation

$$\tilde{y} = \operatorname{diag}\{e^{i\tau}, e^{-i\tau}\}\hat{y}$$

to obtain

$$\frac{d\hat{y}}{dt} = \left[\hat{\Lambda}(t) + \hat{R}(t)\right]\hat{y}, \tag{2.27}$$

where

$$\begin{aligned} \hat{\Lambda}(t) &= \operatorname{diag}\{-i - \frac{a}{\tau}, i + \frac{\sin\ln\ln\tau + \cos\ln\ln\tau - 2a}{\tau}\} \\ &= \operatorname{diag}\{-i, i\} + \mathrm{o}(1) \qquad \text{as } \tau \to +\infty, \end{aligned}$$

and

$$\hat{R}(t) = \begin{bmatrix} 0 & 0 \\ e^{2i\tau}\tau^{-1-b} & 0 \end{bmatrix} \in L^1[1, \infty).$$

Note that $\hat{\Lambda}$ does not satisfy Levinson's dichotomy conditions [(2.13) and (2.14)] if $0 < a < \sqrt{2}$. It follows that since Theorem 2.7 does not apply to (2.26) for $0 < b < \gamma(a)$, it can be shown that it does not apply to (2.27). From these examples, one sees that the validity of the conclusion (2.17) depends upon whether the parameter $b > \gamma(a)$ or $b < \gamma(a)$ and this indicates the possibility that if one further restricts the growth of $R(t)$ in cases where $\Lambda(t)$ does not satisfy Levinson's dichotomy conditions, a positive result could be obtained. This is the subject of the next section.

2.4 Weak Dichotomies

In this section, we will show that a Levinson-like result holds provided that the perturbation is sufficiently small even when the diagonal part does not satisfy Levinson's dichotomy conditions. These recent results can be found in Bodine and Lutz [18] and were motivated by earlier work of Chiba and Kimura [38].

We begin with a reformulation of Coppel's Theorem 2.2, allowing weaker dichotomy conditions on the unperturbed system, but imposing a more restrictive requirement on the perturbation $R(t)$. The proof is a modification of Coppel's original proof, and we will just point out the necessary modifications here. We emphasize that Theorem 2.2 is a special case of the following theorem corresponding to bounded $\beta(t)$.

Theorem 2.10 *For continuous $d \times d$ matrices $A(t)$ and $R(t)$, consider the unperturbed system* (2.5) *and the perturbed system* (2.7)*. Let $X(t)$ be a fundamental matrix of* (2.5)*. Assume that there exist a projection matrix P and a continuous function $\beta(t) \geq 1$ such that*

$$|X(t)PX^{-1}(s)| \leq \beta(s) \qquad \text{for all} \quad t_0 \leq s \leq t, \tag{2.28}$$

$$\text{and} \quad |X(t)[I-P]X^{-1}(s)| \leq \beta(s) \qquad \text{for all} \quad t_0 \leq t \leq s. \tag{2.29}$$

Suppose that

$$\beta R \in L^1[t_0, \infty).$$

Then there exists a one-to-one and bicontinuous correspondence between the bounded solutions of (2.5) *and* (2.7) *on* $[t_0, \infty)$*. Moreover, the difference between corresponding bounded solutions of* (2.5) *and* (2.7) *tends to zero as $t \to \infty$ if $X(t)P \to 0$ as $t \to \infty$.*

Proof Fix $t_1 \geq t_0$ such that

$$\theta = \int_{t_1}^{\infty} \beta(t)|R(t)|\,dt < 1. \tag{2.30}$$

Let $\mathscr{B}$ be the Banach space of bounded d-dimensional vector-valued functions defined on $[t_1, \infty)$ with the supremum norm $\|y\| = \sup_{t \geq t_1} |y(t)|$. For $t \geq t_1$, let

T be the operator defined in (2.11). Then (2.28)–(2.30) imply for $t \geq t_1$ that

$$|(Ty)(t)| \leq \|y\| \int_{t_1}^{\infty} \beta(s)|R(s)|\, ds \leq \theta \|y\|,$$

and, similarly,

$$\|Ty_1 - Ty_2\| \leq \theta \|y_1 - y_2\|.$$

Hence $T: \mathscr{B} \to \mathscr{B}$ is a contraction and for any $x \in \mathscr{B}$, the operator equation

$$y = x + Ty \tag{2.31}$$

has a unique solution $y \in \mathscr{B}$. It can easily be checked that if x is a solution of (2.5), then $y = x + Ty$ is a solution of (2.7). Therefore, if $x(t)$ is a bounded solution of (2.5), then the solution $y(t)$ of (2.31) is a bounded solution of (2.7). Conversely, if $y(t)$ is a bounded solution of (2.7), the function $x(t)$ defined by (2.31) is a bounded solution of (2.5). Therefore (2.31) establishes a one-to-one correspondence between bounded (for $t \geq t_1$) solutions of (2.5) and (2.7) [using (2.11)]. Using continuous dependence of solutions on initial conditions, the bicontinuity can be extended to $[t_0, \infty)$. Finally, given corresponding bounded solutions x and y of (2.5) and (2.7), respectively, and $\varepsilon > 0$, fix $t_2 = t_2(\varepsilon) > t_1$ such that

$$\int_{t_2}^{\infty} \beta(t)|R(t)|\; |y(t)|\, dt \leq \|y\| \int_{t_2}^{\infty} \beta(t)|R(t)|\, dt < \frac{\varepsilon}{2}.$$

Then for $t \geq t_2$

$$\begin{aligned} |y(t) - x(t)| &= |(Ty)(t)| \\ &\leq |X(t)P| \int_{t_1}^{t_2} \left|X^{-1}(s)R(s)y(s)\right|\, ds + \|y\| \int_{t_2}^{\infty} \beta(s)|R(s)|\, ds \\ &\leq |X(t)P| \int_{t_1}^{t_2} |X^{-1}(s)R(s)y(s)|\, ds + \frac{\varepsilon}{2} < \varepsilon, \end{aligned}$$

for all t sufficiently large if $X(t)P \to 0$ as $t \to \infty$. □

Remark 2.11 In Theorem 2.10, the existence of a function β satisfying (2.28) and (2.29) was assumed. However, it can easily be shown that such a β always exists. For example, one may put for $s \geq t_0$

$$\beta(s) = \sup_{t_0 \leq t \leq s} \left|X(t)X^{-1}(s)\right|,$$

where $X(t)$ is the fundamental matrix of the differential system (2.5) satisfying $X(t_0) = I$. Then $\beta(s)$ is well-defined and continuous for $s \geq t_0$, $\beta(s) \geq 1$ for all $s \geq s_0$, and (2.29) holds with $P = 0$ while (2.28) is trivial. It is clear, however, that this choice for $\beta(s)$ will not yield, in general, an optimal bound for the perturbation. Instead, one should take into account the special structure of $X(t)$ and choose P and β using that information.

Theorem 2.10 will be used to prove the following theorem on the asymptotic integration of diagonal systems not necessarily satisfying Levinson's dichotomy conditions.

Theorem 2.12 *Let $\Lambda(t) = \operatorname{diag}\{\lambda_1(t), \ldots, \lambda_d(t)\}$ be a diagonal and continuous $d \times d$ matrix for $t \geq t_0$. Fix $h \in \{1, 2, \ldots, d\}$. Assume that there exists a continuous function $\beta_h(t) \geq 1$ for $t \geq t_0$ such that for all $1 \leq i \leq d$*
either

$$\left.\begin{aligned} &\exp\left[\int_{t_0}^t \operatorname{Re}\{\lambda_i(\tau) - \lambda_h(\tau)\}\, d\tau\right] \to 0 \quad \text{as } t \to \infty, \\ \text{and} \quad &\exp\left[\int_s^t \operatorname{Re}\{\lambda_i(\tau) - \lambda_h(\tau)\}\, d\tau\right] \leq \beta_h(s) \qquad \forall\ t_0 \leq s \leq t, \end{aligned}\right\} \tag{2.32}$$

or

$$\exp\left[\int_s^t \operatorname{Re}\{\lambda_i(\tau) - \lambda_h(\tau)\}\, d\tau\right] \leq \beta_h(s) \qquad \forall\ t_0 \leq t \leq s. \tag{2.33}$$

Furthermore, assume that the $d \times d$ matrix $R(t)$ is continuous for $t \geq t_0$ and that $\beta_h(t)R(t) \in L^1[t_0, \infty)$. Then the linear differential system (2.16) has for $t \geq t_0$ a solution satisfying

$$y_h(t) = [e_h + o(1)] \exp\left[\int^t \lambda_h(s)\, ds\right] \qquad \text{as } t \to \infty, \tag{2.34}$$

where e_h is the hth column of the identity matrix.

Proof For fixed $h \in \{1, \ldots, d\}$, put

$$y(t) = z(t) \exp\left[\int^t \lambda_h(\tau)\, d\tau\right]. \tag{2.35}$$

Then (2.16) implies that

$$z'(t) = [\Lambda(t) - \lambda_h(t)I + R(t)]\, z. \tag{2.36}$$

We also consider the unperturbed shifted system

$$w' = [\Lambda(t) - \lambda_h(t)I]\, w, \tag{2.37}$$

and we will apply Theorem 2.10 to these shifted systems. For that purpose, let $P = \text{diag}\{p_1, \ldots, p_d\}$, where

$$p_i = \begin{cases} 1 & \text{if } (i,h) \text{ satisfies (2.32)} \\ 0 & \text{if } (i,h) \text{ satisfies (2.33).} \end{cases}$$

By (2.32) and (2.33), the unperturbed shifted system (2.37) satisfies the dichotomy conditions (2.28), (2.29) with $\beta(s)$ replaced by $\beta_h(s)$ and, moreover, $W(t)P \to 0$ as $t \to \infty$. Now, by Theorem 2.10 and since $w_h(t) = e_h$ is a bounded solution of (2.37), there exists for $t \geq t_0$ a bounded solution $z_h(t)$ of (2.36) with

$$z_h(t) = e_h + o(1) \qquad \text{as } t \to \infty.$$

By (2.35), (2.16) has a solution of the form (2.34) as $t \to \infty$. □

Remark 2.13 If the positive function $\beta_h(s)$ satisfying (2.32) and (2.33) is bounded above, then $\Lambda(t)$ satisfies Levinson's dichotomy conditions (2.13), (2.14), and the statement of Theorem 2.12 reduces to the result of Theorem 2.7 for one vector-valued solution $y_h(t)$.

Although in Theorem 2.12 the existence of β_h was assumed, it can again be shown that such a function always exists. For example, for fixed h and for each $i \in \{1, \ldots, d\}$, one may put

$$\gamma_i(s) = \sup_{t_0 \leq t \leq s} \left(\exp\left[\int_s^t \text{Re}\{\lambda_i(\tau) - \lambda_h(\tau)\}\, d\tau \right] \right),$$

and

$$\beta_h(s) = \max_{1 \leq i \leq d} \gamma_i(s),$$

i.e., the point-wise maximum over all the γ_i. It follows that (2.33) holds for all $1 \leq i \leq d$. However, this construction of β_h should be avoided in applications, since $\beta_h(t)$ derived this way is likely neither the smallest possible choice nor easy to compute.

As an example, we consider a modification of Eastham's example illustrating the necessity of Levinson's dichotomy condition for absolutely integrable perturbations which was discussed in Sect. 2.3.

Example 2.14 (Modified Eastham Example) For $t \geq 0$, consider

$$y' = \left[\begin{pmatrix} 0 & \\ & \frac{\rho'(t)}{\rho(t)} \end{pmatrix} + \begin{pmatrix} 0 & r(t) \\ 0 & 0 \end{pmatrix} \right] y, \qquad \rho(t) = t^2(1 - \sin t) + 1. \tag{2.38}$$

Recall that it was shown in Sect. 2.3 that the unperturbed diagonal system does not satisfy Levinson's dichotomy conditions (2.13), (2.14) and that for the absolutely integrable perturbation $r(t) = 1/t^2$, the conclusion of Theorem 2.7 does not follow [see (2.24)].

To find a measure on the magnitude of the perturbation $r(t)$ to ensure that (2.38) has a fundamental matrix of the form (2.17), it suffices to find in Theorem 2.12 a suitable function $\beta_2(t)$ (due to the triangular structure of the perturbation in (2.38), $\beta_1(t)$ is not needed).

Since $\exp\left[\int_0^t \operatorname{Re}\{\lambda_1(\tau) - \lambda_2(\tau)\}\,d\tau\right]$ does not go to zero as $t \to \infty$, one needs to find $\beta_2(t)$ such that (2.33) holds. Observe that for $0 \le t \le s$,

$$\exp\left[\operatorname{Re}\int_s^t (\lambda_1 - \lambda_2)\,d\tau\right] = \frac{\rho(s)}{\rho(t)} = \frac{s^2(1-\sin s)+1}{t^2(1-\sin t)+1}$$
$$\le s^2(1-\sin s)+1 \le 2s^2+1,$$

which one can choose for $\beta_2(s)$. Assuming then, for example, that

$$r(t) = t^{-p}, \quad p = 3 + \varepsilon \text{ for some } \varepsilon > 0,$$

(thus $\beta r \in L^1[1,\infty)$), Theorem 2.12 implies that (2.38) has a fundamental matrix of the form

$$Y(t) = [I + o(1)]\begin{bmatrix} 1 & 0 \\ 0 & \rho(t) \end{bmatrix} \qquad \text{as } t \to \infty. \tag{2.39}$$

In fact, integration of (2.38) by quadrature (integrating from t to infinity) shows that for this choice of $r(t)$, there exists a fundamental matrix, satisfying, as $t \to \infty$, $Y(t) = \begin{bmatrix} 1 & O(t^{-\varepsilon}) \\ 0 & \rho(t) \end{bmatrix}$ which is indeed of the form (2.39).

See [18, Ex. 8] for another example (a generalization of Example 2.25) which shows the sharpness of the condition $\beta_h R \in L^1[t_0,\infty)$.

Behncke [6] considered L^1-perturbations of asymptotically constant systems not necessarily satisfying Levinson's dichotomy conditions. He does not assume a stronger growth condition such as for example in Theorem 2.12. In light of Eastham's example in Sect. 2.3, it is not possible, in general, to obtain an asymptotic integration result. Instead Behncke shows the existence of certain solutions and provides upper and lower estimates of their norms. To obtain these results, he uses estimates of successive approximations of solutions of the T-operator equation. We refer to [6] for details.

2.5 Perturbations of Systems with an Exponential Dichotomy

In this section, we want to discuss systems of the form

$$y' = [A(t) + R(t)]\, y, \qquad t \geq t_0,$$

where $A(t)$ is a $d \times d$ matrix having the block structure

$$A(t) = \begin{pmatrix} A_{-1}(t) & & \\ & \lambda_j(t) & \\ & & A_1(t) \end{pmatrix}. \tag{2.40}$$

Here $A_{-1}(t)$ and $A_1(t)$ are square blocks and $\lambda_j(t)$ is a complex-valued scalar function in the (j,j) position of $A(t)$. We first observe that if $R \in L^1[t_0, \infty)$ and if the shifted system $w' = [A(t) - \lambda_j(t)I]w$ satisfies an ordinary dichotomy condition (2.6), then an application of Theorem 2.2 shows there exists a solution vector of the perturbed system satisfying

$$y_j(t) = \left[e_j + o(1)\right] \exp\left[\int_{t_0}^{t} \lambda_j(s)\, ds\right] \qquad \text{as } t \to \infty. \tag{2.41}$$

In what follows we want to significantly weaken the assumption on the perturbation $R(t)$ to include functions which are either in some L^p class with $1 < p \leq 2$ or perhaps just tend to zero in some averaged sense [see (2.56)]. To obtain asymptotic integration results in such cases we will impose a much stronger "exponential dichotomy" condition. These results also involve corresponding modifications to $\lambda_j(s)$ in (2.41).

We begin with a modification of Coppel's Theorem 2.2, replacing the ordinary dichotomy condition in that theorem by an exponential dichotomy assumption. We also phrase it in terms of a "mildly nonlinear system" because this is the form needed in a subsequent application.

Definition 2.15 (Exponential Dichotomy) A linear differential system

$$x' = A(t)x\,, \qquad t \geq t_0, \tag{2.42}$$

is said to have an *exponential dichotomy*, if there exist a fundamental matrix $X(t)$, a projection P ($P^2 = P$), and positive constants K and δ such that

$$\left.\begin{aligned} |X(t)PX^{-1}(s)| &\leq Ke^{-\delta(t-s)} \quad \forall \;\; t_0 \leq s \leq t, \\ |X(t)[I-P]X^{-1}(s)| &\leq Ke^{\delta(t-s)} \;\; \forall \;\; t_0 \leq t \leq s. \end{aligned}\right\} \tag{2.43}$$

It follows that (2.5) has an exponential dichotomy with $P = I$ if and only if it is uniformly asymptotically stable (see, e.g., [43, p. 54]).

Theorem 2.16 *For a continuous $d \times d$ matrix-valued function $A(t)$ and continuous vector-valued functions $b : [t_0, \infty) \to \mathbb{C}^d$ and $f : [t_0, \infty) \times \mathbb{C}^d \to \mathbb{C}^d$, consider the unperturbed system* (2.42) *and the perturbed system*

$$y' = A(t)y + b(t) + f(t, y(t)), \qquad t \geq t_0. \tag{2.44}$$

Suppose that the unperturbed system (2.42) *has an exponential dichotomy of the form* (2.43)*. Assume that $|f(t, y_1) - f(t, y_2)| \leq \gamma(t)|y_1 - y_2|$ for all $y_1, y_2 \in \mathbb{C}^d$ and all $t \geq t_0$. Suppose that*

$$r(t) := |b(t)| + \gamma(t) + |f(t, 0)|, \tag{2.45}$$

satisfies

$$g(t) := \sup_{s \geq t} \frac{1}{s - t + 1} \int_t^s r(\tau)\, d\tau \to 0 \qquad \text{as } t \to \infty. \tag{2.46}$$

Then, for t sufficiently large, there exists a one-to-one correspondence between the bounded solutions of (2.42) *and* (2.44)*, and the difference between corresponding solutions of* (2.42) *and* (2.44) *tends to zero as $t \to \infty$.*

Proof Recalling (2.46), fix $t_1 \geq t_0$, $t_1 \in \mathbb{N}$, such that

$$g(t) \leq \frac{1}{2K} \frac{\delta}{(2e^{\delta} + 3\delta + 1)} =: M_{t_1} \qquad \text{for all } t \geq t_1, \tag{2.47}$$

where δ and K are given in (2.43). Let $\mathscr{B}$ be the Banach space of bounded d-dimensional vector-valued functions defined on $[t_1, \infty)$ with the supremum norm $\|y\| = \sup_{t \geq t_1} |y(t)|$, and let $\mathscr{W} = \{y \in \mathscr{B} : \|y\| \leq 1\}$. Hence $\mathscr{W}$ is a closed subset of $\mathscr{B}$. For $t \geq t_1$, define an operator T on $\mathscr{W}$ by

$$\begin{aligned} (Ty)(t) = &\int_{t_1}^t X(t)PX^{-1}(s)[b(s) + f(s, y(s))]\, ds \\ &- \int_t^{\infty} X(t)[I - P]X^{-1}(s)[b(s) + f(s, y(s))]\, ds. \end{aligned} \tag{2.48}$$

We first claim that T maps $\mathscr{W}$ into $\mathscr{W}$. To substantiate this claim, note that for $y \in \mathscr{W}$, i.e., $\|y\| \leq 1$,

$$\begin{aligned} |f(s, y(s))| &\leq |f(s, y(s)) - f(s, 0)| + |f(s, 0)| \\ &\leq \gamma(s)\|y\| + |f(s, 0)| \leq \gamma(s) + |f(s, 0)|. \end{aligned}$$

Thus it follows from (2.43), (2.45), and (2.48) that

$$|(Ty)(t)| \leq K \int_{t_1}^{t} e^{-\delta(t-s)} r(s)\, ds + K \int_{t}^{\infty} e^{\delta(t-s)} r(s)\, ds. \tag{2.49}$$

Using first integration by parts and then (2.46), one can see that

$$\begin{aligned}
&\lim_{T\to\infty} \int_t^T e^{\delta(t-s)} r(s)\, ds \\
&= \lim_{T\to\infty} \left[e^{\delta(t-T)} \int_t^T r(\tau)\, d\tau + \delta \int_t^T e^{\delta(t-s)} \int_t^s r(\tau)\, d\tau\, ds \right] \\
&\leq \lim_{T\to\infty} g(t) \left[e^{\delta(t-T)} [T-t+1] + \delta \int_t^T e^{\delta(t-s)} [s-t+1]\, ds \right] \\
&= g(t) \frac{\delta+1}{\delta}.
\end{aligned} \tag{2.50}$$

For fixed $t \geq t_1$, fix $n \in \mathbb{N}$ such that $n \leq t < n+1$. Then it follows that

$$\begin{aligned}
\int_{t_1}^{t} e^{-\delta(t-s)} r(s)\, ds &= \sum_{k=t_1}^{n-1} \int_k^{k+1} e^{-\delta(t-s)} r(s)\, ds + \int_n^t e^{-\delta(t-s)} r(s)\, ds \\
&\leq \sum_{k=t_1}^{n-1} e^{\delta(k+1-t)} \int_k^{k+1} r(s)\, ds + \int_n^t r(s)\, ds \\
&\leq \sum_{k=t_1}^{n-1} e^{\delta(k+1-t)} 2g(k) + 2g(n) \\
&\leq 2M_{t_1} \left[e^{\delta(1-t)} \sum_{k=t_1}^{n-1} e^{\delta k} + 1 \right] \leq 2M_{t_1} \left[\frac{e^\delta}{e^\delta - 1} + 1 \right] \\
&\leq 2M_{t_1} \left[\frac{e^\delta}{\delta} + 1 \right] = 2M_{t_1} \frac{e^\delta + \delta}{\delta},
\end{aligned} \tag{2.51}$$

where we used (2.47) and a standard geometric series argument. Adding up the two estimates in (2.50) and (2.51), it follows from (2.47) for $t \geq t_1$ that

$$|(Ty)(t)| \leq KM_{t_1} \left[\frac{\delta+1}{\delta} + 2\frac{e^\delta + \delta}{\delta} \right] = \frac{1}{2}, \tag{2.52}$$

which establishes the first claim.

To show that T is a contraction on $\mathscr{W}$, a similar argument shows that

$$\begin{aligned}|(Ty_1)(t)-(Ty_2)(t)| &\le \|y_1-y_2\|K\left[\int_{t_1}^{t} e^{-\delta(t-s)}\gamma(s)\,ds+\int_t^{\infty} e^{\delta(t-s)}\gamma(s)\,ds\right]\\ &\le \|y_1-y_2\|K\left[\int_{t_1}^{t} e^{-\delta(t-s)}r(s)\,ds+\int_t^{\infty} e^{\delta(t-s)}r(s)\,ds\right]\\ &\le \frac{\|y_1-y_2\|}{2}.\end{aligned}$$

Therefore

$$\|T(y_1-y_2)\|\le\frac{1}{2}\|y_1-y_2\|.$$

Hence for any $x\in\mathscr{W}$, the operator equation

$$x=y-Ty \tag{2.53}$$

has a unique solution $y\in\mathscr{W}$. It can easily be checked that if x is a solution of (2.42), then $y=x+Ty$ is a solution of (2.44). Therefore, if $x(t)$ is a bounded solution of (2.42), then the solution $y(t)$ of (2.53) is a bounded solution of (2.44). Conversely, if $y(t)$ is a bounded solution of (2.44), the function $x(t)$ defined by (2.53) is a bounded solution of (2.42). The equation $y=x+Ty$ therefore establishes for $t\ge t_1$ a one-to-one correspondence on $[t_1,\infty)$ between the bounded in norm by 1 solutions of (2.42) and (2.44). Finally we claim that $(Ty)(t)\to 0$ as $t\to\infty$, which implies that $|x(t)-y(t)|\le|(Ty)(t)|\to 0$ as $t\to\infty$.

To show that $(Ty)(t)$ vanishes at infinity, note that (2.49) implies for $t_1<t_2<t$ that

$$|(Ty)(t)|\le K\left[\int_{t_1}^{t_2} e^{-\delta(t-s)}r(s)\,ds+\int_{t_2}^{t} e^{-\delta(t-s)}r(s)\,ds+\int_t^{\infty} e^{\delta(t-s)}r(s)\,ds\right].$$

Given $\varepsilon>0$, fix $t_2\in\mathbb{N}$, sufficiently large such that $g(s)\le\frac{\varepsilon}{6K}\frac{\delta}{e^\delta+\delta}=:M_{t_2}$ for all $s\ge t_2$. Then similar calculations to the ones above leading to (2.51) show that

$$\int_{t_2}^{t} e^{-\delta(t-s)}r(s)\,ds\le 2M_{t_2}\frac{e^\delta+\delta}{\delta},$$

and therefore

$$|(Ty)(t)|\le K\left[\int_{t_1}^{t_2} e^{-\delta(t-s)}r(s)\,ds+\frac{\varepsilon}{3}+g(t)\frac{\delta+1}{\delta}\right].$$

By making t sufficiently large ($t>t_2$), it follows $|(Ty)(t)|<\varepsilon$. □

Note that the value "1/2" of the contraction constant was chosen for convenience and definiteness. Introducing a suitable multiplicative factor in (2.47), we could have used any other value $\theta \in (0, 1)$ instead of 1/2.

We now apply the previous result to systems $y' = [A(t) + R(t)]y$, where $A(t)$ as in (2.40) satisfies a kind of "exponential dichotomy with respect to $\lambda_j(t)$," and $R(t)$ on the average tends to zero. This very weak condition on the perturbation is reflected in corresponding rather weak asymptotic statements (2.57), (2.58) about solutions. It was motivated by a result appearing in Hartman's classical book (see [72, Chap. X, Thm. 17.4]), but first published by Hartman and Wintner [73], concerning the case when A is a constant matrix. The proof of the following extension appears to be more direct, and the conclusion contains a somewhat more explicit asymptotic representation. Both Theorem 2.17 as well as Hartman's Theorem 17.4 can be thought of as generalizations of classical results of Poincaré and Perron (see Sects. 2.6 and 8.2).

Theorem 2.17 *Consider*

$$y' = [A(t) + R(t)]y, \tag{2.54}$$

Here $A(t)$ is supposed to be continuous for $t \geq t_0$ and in block-diagonal form (2.40)*, where $A_{-1}(t)$ is a $(j-1) \times (j-1)$ matrix and $A_1(t)$ is a $(d-j) \times (d-j)$ matrix. Assume that the $(d-1) \times (d-1)$ system*

$$\hat{x}' = \hat{A}(t)\hat{x}, \qquad \hat{A}(t) = \begin{pmatrix} A_{-1}(t) - \lambda_j(t) & \\ & A_1(t) - \lambda_j(t) \end{pmatrix} \tag{2.55}$$

satisfies an exponential dichotomy (2.43) *with projection $P = \mathrm{diag}\{I_{j-1}, 0_{d-j}\}$ (and positive constants δ and K). Suppose that $R(t)$ is a $d \times d$ continuous matrix satisfying for $t_0 \leq t \leq s$*

$$g_R(t) := \sup_{s \geq t} \frac{1}{s-t+1} \int_t^s |R(\tau)|\, d\tau \to 0 \qquad \text{as } t \to \infty. \tag{2.56}$$

Then there exists a solution $y_j(t)$ of (2.54) *satisfying*

$$y_j(t) = [e_j + s_j(t)] \exp\left[\int^t \{\lambda_j(s) + l_j(s)\}\, ds\right], \tag{2.57}$$

where

$$l_j(t) = r_{jj}(t) + \mathrm{o}\left(\sum_{\nu \neq j} |r_{j\nu}(t)|\right), \tag{2.58}$$

and $s_j(t) = \{s_{1j}(t), \ldots, s_{dj}(t)\}^T = \mathrm{o}(1)$ as $t \to \infty$.

Proof By introducing $\hat{y} = y \exp[-\int^t \lambda_j(s)\, ds]$ as the new dependent variable, we can assume without loss of generality that $\lambda_j(t) \equiv 0$ in (2.55). We first remark that Theorem 2.16 does not immediately apply to (2.54), because the unperturbed system $x' = A(t)x$ (with $\lambda_j(t) \equiv 0$) does not have an exponential dichotomy (see Definition 2.15). The reason is that $x_j(t) \equiv e_j$ is a solution that is neither exponentially increasing nor decreasing.

We note that (2.54) has a solution of the form (2.57) if and only if

$$s_j'(t) - A(t)s_j(t) = -l_j(t)[e_j + s_j(t)] + r_j(t) + R(t)s_j(t), \tag{2.59}$$

where $r_j(t)$ is the jth column of $R(t)$. In this system of d equations for $d + 1$ unknowns, there is a lack of uniqueness due to the fact that in (2.57), the diagonal term $s_{jj}(t)$ can be modified by adjusting $l_j(t)$ by a suitable o(1) perturbation. To remove this ambiguity in the representation, we choose to normalize $s_j(t)$ so that

$$s_{jj}(t) \equiv 0. \tag{2.60}$$

Then we obtain for the elements in the jth position of (2.59),

$$l_j(t) = r_{jj}(t) + \sum_{\nu=1,\nu\neq j}^{d} r_{j\nu}(t)s_{\nu j}(t). \tag{2.61}$$

Substituting (2.61) back into (2.59), we obtain a (nonlinear) system of d–1 equations for $d-1$ unknowns, the remaining components of the vector $s_j(t)$.

In order to express this system in vector form, we use the following notation:

$$\sigma(t) = \left[s_{1j}(t), \dots, s_{j-1,j}(t),\, s_{j+1,j}(t), \dots, s_{dj}(t)\right]^T, \tag{2.62}$$

$$\hat{r}_j(t) = \left[r_{1j}(t), \dots, r_{j-1,j}(t),\, r_{j+1,j}(t), \dots, r_{dj}(t)\right]^T,$$

$$\hat{\rho}_j(t) = \left[r_{j1}(t), \dots, r_{j,j-1}(t),\, r_{j,j+1}(t), \dots, r_{jd}(t)\right].$$

Then (2.61) can be expressed as

$$l_j(t) = r_{jj}(t) + \hat{\rho}_j(t)\sigma(t),$$

and substituting this into the remaining $(d - 1)$ equations (2.59), we obtain the system

$$\sigma'(t) = \hat{A}(t)\sigma(t) + V(t, \sigma(t)), \tag{2.63}$$

where $\hat{A}(t)$ was defined on (2.55) and

$$V(t, \sigma(t)) = \hat{r}_j(t) + \hat{R}(t)\sigma(t) - \left\{r_{jj}(t) + \hat{\rho}_j(t)\sigma(t)\right\} \sigma(t). \tag{2.64}$$

Here $\hat{R}(t)$ is the $(d-1)\times(d-1)$-dimensional matrix obtained by deleting the jth row and jth column from $R(t)$, Thus one can write

$$V(t,\sigma(t)) = b(t) + f(t,\sigma(t)),$$

where $b(t) = \hat{r}_j(t), f(t,0) \equiv 0$, and $|f(t,\sigma_1(t)) - f(t,\sigma_2(t))| = O(R(t))|\sigma_1(t) - \sigma_2(t)|$. Since the unperturbed system (2.55) has an exponential dichotomy and $R(t)$ satisfies (2.56), Theorem 2.16 now implies that for the trivial solution $\hat{x} \equiv 0$ of (2.55) (in particular, $\hat{x} \in \mathscr{W}$, see Theorem 2.16), there exists a unique solution $\sigma(t)$ of (2.63) such that $\sigma(t) \to 0$ as $t \to \infty$. Recalling (2.62) and the normalization $s_{jj}(t) \equiv 0$, this implies that (2.59) has a solution $s_j(t) \to 0$. Now (2.61) immediately implies (2.58), which completes the proof. □

While it is clear $R(t) = o(1)$ implies that (2.56) holds, the example

$$|R(t)| = \begin{cases} 1 & n \leq t \leq n + 1/n, \qquad n \in N, \\ 0 & \text{else} \end{cases}$$

shows that the converse does not hold. It is also clear that $R \in L^1$ implies (2.56). Moreover, Hölder's inequality shows that $R \in L^p$ for $p > 1$ also implies (2.56).

In case that the block-matrix $A(t)$ in (2.54) is replaced by a diagonal matrix $\Lambda(t)$, Theorem 2.17 implies the following result.

Corollary 2.18 *Suppose that the matrix $A(t)$ in (2.54) is replaced by a matrix $\Lambda(t) = \operatorname{diag}\{\lambda_1(t), \ldots \lambda_d(t)\}$, and assume that there exists $K > 0$ and $\delta > 0$ such that for all $1 \leq i \neq j \leq d$*

$$\begin{array}{lll} \text{either} & e^{\mathrm{Re}\int_s^t \{\lambda_i(\tau)-\lambda_j(\tau)\}\,d\tau} \leq Ke^{-\delta(t-s)}, & t_0 \leq s \leq t \\ \text{or} & e^{\mathrm{Re}\int_s^t \{\lambda_i(\tau)-\lambda_j(\tau)\}\,d\tau} \leq Ke^{-\delta(s-t)}, & t_0 \leq t \leq s. \end{array} \tag{2.65}$$

Suppose that $R(t)$ satisfies (2.56). Then

$$y' = [\Lambda(t) + R(t)]\,y, \qquad t \geq t_0,$$

has a fundamental matrix satisfying as $t \to \infty$

$$Y(t) = [I + o(1)] \exp\left[\int^t \{\Lambda(s) + L(s)\}\,ds\right],$$

$$\text{where } L(s) = \operatorname{diag} R(s) + o\left(|R(s)|\right).$$

We now turn to perturbations $R(t)$ which do not only satisfy (2.56), but the stronger condition that $R(t) \in L^p$ for $1 < p \leq 2$. Hartman and Wintner [73, pp. 71–72] studied such perturbations of non-constant Jordan systems $x' = J(t)x$

and assumed that there exists a 1×1 Jordan block $\lambda_j(t)$ and $\delta > 0$ such that

$$|\operatorname{Re}(\lambda_j(t) - \lambda_i(t))| \geq \delta \text{ for all } i \neq j \quad \text{and} \quad t \geq t_1. \tag{2.66}$$

They derived the asymptotic behavior of solutions of the perturbed system corresponding to $\lambda_j(t)$ for this fixed value of j. Assuming that $J(t)$ is actually diagonal, say $J(t) = \Lambda(t)$, and that (2.66) holds for every pair (i,j), $1 \leq i \neq j \leq d$, and not only for a fixed value of j, Harris and Lutz [71] simplified the proof to derive an equivalent statement about a fundamental solution matrix (see Sect. 4.4), and their approach allows for extensions to L^p-perturbations with $p > 2$ (see Corollary 4.12). Their proof-technique was later adjusted to replace the pointwise dichotomy assumption (2.66) by an "averaged" exponential dichotomy condition (2.65) (see [80]).

While these results on fundamental matrices will be discussed in detail in Sect. 4.4, we now return to the original Hartman–Wintner setting of perturbations of Jordan systems containing a 1×1-Jordan block $\lambda_j(t)$.

Theorem 2.19 (Hartman–Wintner) *For fixed $j \in \{1, \ldots, d\}$, let $A(t)$ and $\hat{A}(t)$ satisfy the assumptions of Theorem 2.17. Suppose that $R(t)$ is a continuous $d \times d$ matrix and satisfies*

$$R(t) \in L^p[t_0, \infty), \qquad 1 < p \leq 2. \tag{2.67}$$

Then there exists a solution $y_j(t)$ of (2.54) satisfying

$$y_j(t) = [e_j + o(1)] \exp\left[\int^t \{\lambda_j(s) + r_{jj}(s)\} \, ds\right]. \tag{2.68}$$

Proof Note that (2.67) and Hölder's inequality show that (2.56) holds and therefore Theorem 2.17 applies. That is, (2.54) has a solution $y_j(t)$ of the form (2.57), with $l_j(t)$ given in (2.58). We also recall from the proof of Theorem 2.17 that (2.58) was derived by considering $l_j(t)$ in (2.61) and showing that $s_j(t) \to 0$ as $t \to \infty$. We will now show, that under the stronger condition (2.67) on the perturbation $R(t)$, it follows that $s_j(t)$ not only vanishes at infinity, but additionally satisfies that $s_j(t) \in L^p[t_0, \infty)$, which will be the main argument to establish (2.68). To show that $s_j(t) \in L^p$ is by (2.60) and (2.62) equivalent to showing that $\sigma(t) \in L^p$, which we will do by revisiting the contraction mapping (2.48) in Theorem 2.16.

Let $1/p + 1/p' = 1$. Returning to (2.49) (with $r(s)$ replaced by $|R(s)|$), we first observe that from Hölder's inequality follows that

$$\begin{aligned}
&\int_{t_1}^{\infty} \left| \int_{t_1}^{t} e^{-\delta(t-s)} |R(s)| \, ds \right|^p dt \\
&= \int_{t_1}^{\infty} \left[\int_{t_1}^{t} |R(s)| e^{-\frac{\delta(t-s)}{p}} e^{-\frac{\delta(t-s)}{p'}} \, ds \right]^p dt
\end{aligned}$$

$$\leq \int_{t_1}^{\infty} \left\{ \int_{t_1}^{t} |R(s)|^p e^{-\delta(t-s)}\, ds \right\} \left\{ \int_{t_1}^{t} e^{-\delta(t-s)}\, ds \right\}^{p/p'} dt$$
$$\leq \left(\frac{1}{\delta}\right)^{p-1} \int_{t_1}^{\infty} \int_{t_1}^{t} e^{-\delta(t-s)} |R(s)|^p\, ds\, dt$$
$$= \left(\frac{1}{\delta}\right)^{p} \int_{t_1}^{\infty} |R(s)|^p\, ds < \infty, \tag{2.69}$$

where the theorem of Fubini was used in the last inequality. It can be shown similarly that

$$\int_{t_1}^{\infty} \left| \int_{t}^{\infty} e^{\delta(t-s)} |R(s)|\, ds \right|^p dt \leq \left(\frac{1}{\delta}\right)^{p} \int_{t_1}^{\infty} |R(s)|^p\, ds < \infty. \tag{2.70}$$

Thus (2.69) and (2.70) together with Minkowski's inequality show that the right-hand side of (2.49) is in L^p. Therefore, as in Theorem 2.17, corresponding to the trivial solution $\hat{x} \equiv 0$ of the unperturbed system (2.55), the corresponding fixed point $y((t) = (Ty)(t) \in L^p$. Recalling that the corresponding fixed point in Theorem 2.17 was denoted $\sigma(t)$, this shows that $\sigma(t)$ and therefore $s_j(t)$ are in L^p.

In (2.61), we see that for all $1 \leq \nu \neq j \leq d$, $r_{j\nu}(t)s_{\nu j}(t)$ is then the product of two L^p functions. Moreover, $s_{\nu j}(t)$ is also bounded since $s_j(t) \to 0$ as $t \to \infty$. Noting that $p' \geq p$ for $1 < p \leq 2$, it follows that $s_{\nu j}(t) \in L^{p'}$ and hence the product $r_{j\nu}(t)s_{\nu j} \in L^1$. Thus $l_j(t)$ in (2.58) is of the form $l_j(t) = r_{jj}(t) + \delta_j(t)$, with $\delta_j(t) \in L^1$. Equation (2.57) can therefore be re-written

$$y_j(t) = [e_j + s_j(t)] \exp\left[\int^t \{\lambda_j(s) + r_{jj}(s)\} \right] \exp\left[\int^t \delta_j(s)\, ds\}\, ds \right]$$
$$= [e_j + o(1)]\, [c + o(1)] \exp\left[\int^t \{\lambda_j(s) + r_{jj}(s)\}\, ds \right],$$

for some $c \neq 0$, which concludes the proof. □

We note that Castillo and Pinto [34, Thm. 1] proved a result very similar to Theorem 2.19 directly, without building upon Theorem 2.17.

In the special case that the block-matrix $A(t)$ in Theorem 2.19 is replaced by a diagonal matrix, an application of Theorem 2.19 yields the following asymptotic diagonalization result:

Theorem 2.20 *Suppose that* $A(t)$ *in* (2.54) *has the special form* $A(t) = \Lambda(t) = \operatorname{diag}\{\lambda_1(t), \ldots, \lambda_d(t)\}$, *and assume that there exist* $K > 0$ *and* $\delta > 0$ *such that* (2.65) *holds for* $1 \leq i \neq j \leq d$. *Suppose that* $R(t)$ *satisfies* (2.67). *Then*

$$y'(t) = [\Lambda(t) + R(t)]y, \qquad t \geq t_0 \tag{2.71}$$

has a fundamental matrix satisfying as $t \to \infty$

$$Y(t) = [I + o(1)] \exp\left[\int^t \{\Lambda(s) + \operatorname{diag} R(s)\}\, ds\right].$$

The last theorem in this section is due to Bodine and Lutz [15, Thm. 12] and considers what could be called an "interpolation" between the results of Levinson and Hartman–Wintner.

Theorem 2.21 *Let $\Lambda(t)$ and $R(t)$ be continuous $d \times d$-matrices for all $t \geq t_0$. Assume that $\Lambda(t) = \operatorname{diag}\{\lambda_1(t), \ldots, \lambda_d(t)\}$ satisfies the following dichotomy condition: There exist constants $\delta > 0$, $K \geq 1$ and $\alpha < 1$ such that for each pair of indices (i, j), $1 \leq i \neq j \leq d$, the entries of $\Lambda(t)$ satisfy*

$$\text{either } e^{\operatorname{Re} \int_s^t \{\lambda_i(\tau) - \lambda_j(\tau)\}\, d\tau} \leq K e^{-\delta\left(t^{1-\alpha} - s^{1-\alpha}\right)}, \quad t_0 \leq s \leq t, \tag{2.72}$$

$$\text{or } \quad e^{\operatorname{Re} \int_s^t \{\lambda_i(\tau) - \lambda_j(\tau)\}\, d\tau} \leq K e^{-\delta\left(s^{1-\alpha} - t^{1-\alpha}\right)}, \quad t_0 \leq t \leq s. \tag{2.73}$$

Assume $R(t)$ satisfies that

$$\int^{\infty} t^{\alpha(p-1)}\, |R(t)|^p\, dt < \infty \qquad \text{for some } 1 < p \leq 2. \tag{2.74}$$

Then (2.71) *has a fundamental matrix satisfying as $t \to \infty$*

$$Y(t) = [I + o(1)] \exp\left[\int^t \{\Lambda(s) + \operatorname{diag} R(s)\}\, ds\right].$$

Proof The change of the independent variable $u = t^{1-\alpha}$ and setting $\hat{y}(u) = y(t)$ transforms (2.71) into

$$\begin{aligned}
\frac{d\hat{y}(u)}{du} &= \left[\Lambda\left(u^{\frac{1}{1-\alpha}}\right) + R\left(u^{\frac{1}{1-\alpha}}\right)\right] \frac{u^{\frac{\alpha}{1-\alpha}}}{1-\alpha}\, \hat{y}(u) \\
&=: \left[\quad \hat{\Lambda}(u) \quad + \quad \hat{R}(u) \quad\right] \hat{y}(u).
\end{aligned} \tag{2.75}$$

An application of the chain rule shows that the entries of $\hat{\Lambda}(u)$ satisfy the exponential dichotomy condition (2.65) and that $\int^{\infty} |\hat{R}(u)|^p\, du < \infty$. Now Theorem 2.20 implies that there exists a fundamental matrix of (2.75) satisfying, as $u \to \infty$,

$$\hat{Y}(u) = [I + o(1)] \exp\left[\int^u \{\hat{\Lambda}(\sigma) + \operatorname{diag} \hat{R}(\sigma)\}\, d\sigma\right],$$

which in turn implies the result. □

A sufficient condition on $R(t)$ to satisfy (2.74) is that $\exists\ \varepsilon > 0$ such that $\|R(t)\| = O\left(t^{-\alpha-\frac{1}{p}(1-\alpha)-\varepsilon}\right)$ as $t \to \infty$.

As an example, consider

$$y' = \left[\begin{pmatrix} \frac{1}{\sqrt{t}} & 0 \\ 0 & 0 \end{pmatrix} + \frac{C}{t^{4/5}}\right] y = [\Lambda(t) + R(t)]y,$$

where C is a constant matrix with entries c_{ij}. Note that $x' = \Lambda(t)x$ satisfies an ordinary, but not an exponential dichotomy, and that R in L^2, but not in L^1. Since $R \notin L^1$, Theorem 2.7 cannot be applied, but Theorem 2.21 with $\alpha = 1/2$ and $p = 2$ implies the existence of a fundamental matrix satisfying as $t \to \infty$

$$Y(t) = [I + o(1)] \begin{pmatrix} e^{\int^t \left(\frac{1}{\sqrt{\tau}} + \frac{c_{11}}{\tau^{4/5}}\right) d\tau} & \\ & e^{\int^t \frac{c_{22}}{\tau^{4/5}} d\tau} \end{pmatrix}.$$

2.6 Perturbations of Constant Systems

While we treated a more general case in the last section, we summarize in this section various asymptotic integration results for the important class of perturbations of constant systems

$$y' = [A + R(t)]y, \qquad t \geq t_0, \tag{2.76}$$

where A is a constant matrix and the perturbation $R(t)$ is small in some sense.

- In case A is diagonalizable, say $P^{-1}AP = \Lambda = \operatorname{diag}\{\lambda_1, \ldots, \lambda_d\}$, and $R(t) \in L^1[t_0, \infty)$, then it follows from Theorem 2.7 that (2.76) has a fundamental matrix satisfying

$$Y(t) = P[I + o(1)] \exp[\Lambda t].$$

- If A is not diagonalizable, then we will discuss in Chap. 6 some stronger conditions on $R(t)$ required to obtain an asymptotic integration result for a fundamental matrix of such a perturbed Jordan system (2.76).
- The classical case of asymptotically constant linear scalar differential systems was treated by O. Perron [119, 120] following a related result by Poincaré [130]. Perron's result for systems of the form (2.76) assumes that A has all eigenvalues with distinct real parts (and hence $x' = Ax$ satisfying an exponential dichotomy) and the perturbation

$$R(t) = o(1) \qquad \text{as } n \to \infty. \tag{2.77}$$

Then it follows that (2.76) has a fundamental matrix satisfying

$$Y(t) = [P + o(1)] \exp\left[\int^t \{\Lambda + D(t)\}\, d\tau\right], \qquad D(t) = o(1) \text{ as } t \to \infty.$$

Here $P^{-1}AP = \Lambda$, and $D(t)$ is a diagonal matrix.

Note that all four of the above mentioned results provide an asymptotic representation of a fundamental matrix of (2.76) and not just of some vector-valued solution.

Hartman and Wintner found a still weaker growth condition on $R(t)$ by replacing the assumption (2.77) by (2.56). In the proof [72, Chap. X, Thm. 17.4], Hartman treated a general differential system (2.76) with A being a constant $d \times d$ matrix of the form

$$A = \begin{pmatrix} A_{-1} & & \\ & \lambda_j & \\ & & A_1. \end{pmatrix}, \tag{2.78}$$

and obtained a representation for a single solution vector corresponding to λ_j. The next theorem, which is a special case of Theorem 2.17, gives a slightly improved version of their result.

Theorem 2.22 *Assume A is of the form* (2.78), *where A_{-1} is a $(j-1) \times (j-1)$ matrix whose eigenvalues λ satisfy $Re(\lambda) < Re(\lambda_j)$ and A_1 is a $(d-j) \times (d-j)$ matrix whose eigenvalues λ satisfy $Re(\lambda) > Re(\lambda_j)$. Suppose $R(t)$ satisfies* (2.56). *Then there exists a solution $y_j(t)$ of* (2.76) *satisfying*

$$y_j(t) = [e_j + s_j(t)] \exp\left[\int^t \{\lambda_j + l_j(s)\}\, ds\right], \tag{2.79}$$

where

$$l_j(t) = r_{jj}(t) + o\left(\sum_{\nu \neq j} |r_{j\nu}(t)|\right), \tag{2.80}$$

and $s_j(t) = \{s_{1j}(t), \ldots, s_{dj}(t)\}^T = o(1)$ as $t \to \infty$.

Remark 2.23 While the approach in Sect. 2.5 has a few elements in common with the proof given by Hartman in [72, Chap. 10, Thm. 17.4], his proof is based on what he calls a "topological principle" related to a theorem of Wazewski.

In the final theorem of this section, we consider perturbation of diagonal matrices. We are interested in a kind of "hybrid" result which in a sense interpolates between the very weak Hartman–Wintner assumption (2.56) for some rows of a system and the much stronger Levinson L^1 assumption on other rows. The rows in question correspond to eigenvalues satisfying a certain strong dichotomy condition or ones that satisfy the weaker Levinson type. We decompose the index set

$$S = \{1, \ldots, d\} = \mathcal{U}_1 \cup \mathcal{U}_2$$

as follows:

$$i \in \mathcal{U}_1 \iff \operatorname{Re}\lambda_i \neq \operatorname{Re}\lambda_j \text{ for all } j \in S,\ j \neq i, \tag{2.81}$$

$$i \in \mathcal{U}_2 \iff \operatorname{Re}\lambda_i = \operatorname{Re}\lambda_j \text{ for some } j \neq i. \tag{2.82}$$

Theorem 2.24 *Let* $\Lambda = \operatorname{diag}\{\lambda_1, \ldots, \lambda_d\}$. *Consider* $\mathcal{U}_1$, $\mathcal{U}_2$ *be defined as above and assume that the rows of* $R(t)$, $r_{ik}(t)$, $1 \leq k \leq d$, *satisfy conditions* (2.56) *for* $i \in \mathcal{U}_1$, *while for* $i \in \mathcal{U}_2$, *the rows* $r_{ik}(t)$, $1 \leq k \leq d$, *are in* $L^1[t_0, \infty)$. *Then for each* $i \in \mathcal{U}_1$,

$$y'(t) = [\Lambda + R(t)]y(t), \tag{2.83}$$

has a solution $y_i(t)$ *satisfying* (2.79), *where* $l_j(t)$ *is given in* (2.80), *while for each* $i \in \mathcal{U}_2$, (2.83) *has a solution* $y_i(t)$ *satisfying*

$$y_j(t) = [e_j + o(1)]e^{\lambda_j t}.$$

Proof For $j \in \mathcal{U}_1$, the assertion follows from Theorem 2.22 since the matrix $R(t)$ satisfies (2.56). Hence it suffices to consider $j \in \mathcal{U}_2$. In the following, fix $j \in \mathcal{U}_2$. We define diagonal matrices $P_k = \operatorname{diag}\{p_{k1}, \ldots p_{kd}\}$, $k \in \{1, 2, 3, \}$ as follows: For $i \in \{1, \ldots, d\}$, let

$$p_{1i} = 1 \text{ if } \operatorname{Re}[\lambda_i - \lambda_j] < 0; \qquad p_{1i} = 0 \text{ else,}$$

$$p_{2i} = 1 \text{ if } \operatorname{Re}[\lambda_i - \lambda_j] = 0; \qquad p_{2i} = 0 \text{ else,}$$

$$p_{3i} = 1 \text{ if } \operatorname{Re}[\lambda_i - \lambda_j] > 0; \qquad p_{3i} = 0 \text{ else.}$$

Clearly, $P_1 + P_2 + P_3 = I$. Setting $y = e^{\lambda_j t}z$, (2.83) implies that $z' = [\Lambda - \lambda_j I + R(t)]z$. We claim that this has a solution $z_j(t)$ given as a fixed point of the following contraction mapping. We consider the integral equation $z(t) = e_j + (Tz)(t)$, where for $t \geq t_1$ sufficiently large

$$(Tz)(t) = \int_{t_1}^{t} P_1 e^{[\Lambda - \lambda_j I](t-s)} R(s)z(s)\, ds$$
$$- \int_{t_1}^{\infty} [P_2 + P_3] e^{[\Lambda - \lambda_j I](t-s)} R(s)z(s)\, ds.$$

Looking at the entries, one finds for $1 \leq \nu, \mu \leq d$ and $1 \leq k \leq 3$ that

$$\left(P_k e^{[\Lambda - \lambda_j I](t-s)} R(s)\right)_{\nu\mu} = p_{k\nu} e^{[\lambda_\nu - \lambda_j](t-s)} r_{\nu\mu}(s),$$

i.e., $P_k e^{[\Lambda-\lambda_j I](t-s)}$ acts on the *rows* of the perturbation matrix $R(t)$. Therefore, for z being in the Banach space $\mathscr{B}$ of functions bounded on $[t_1, \infty)$, one finds that there exists constants $\delta > 0$ and $K > 0$ such that

$$\left|(Tz)_{\nu\mu}(t)\right| \leq K\|z\| \left[\int_{t_1}^{t} e^{-\delta(t-s)} \underbrace{|p_{1\nu} r_{\nu\mu}(s)|}_{(2.56)} ds + \int_{t}^{\infty} \underbrace{|p_{2\nu} r_{\nu\mu}(s)|}_{\text{in } L^1[t_0,\infty)} ds \right.$$
$$\left. + \int_{t}^{\infty} e^{\delta(t-s)} \underbrace{|p_{3\nu} r_{\nu\mu}(s)|}_{(2.56)} ds \right].$$

Using estimates as the ones following (2.49), it can be shown that T is a contraction mapping on $\mathscr{B}$ and, moreover, $(Tz)(t) \to 0$ as $t \to \infty$. Therefore $z' = [\Lambda - \lambda_j I + R(t)]z$ has a solution of the form $z(t) = e_j + o(1)$ and the assertion follows. □

As an example for Theorem 2.24, consider

$$y' = \left[\begin{pmatrix} 1 & & \\ & 1 & \\ & & 2 \end{pmatrix} + \begin{pmatrix} 1/t^2 & 1/t^3 & 1/t^4 \\ 1/t^2 & 1/t^3 & 1/t^4 \\ 1/\ln t & 1/\ln(\ln t) & 1/t^2 \end{pmatrix} \right] y,$$

which has solutions

$$y_i(t) = [e_i + o(1)]e^t \qquad \text{for } i = 1, 2;$$
$$y_3(t) = [e_3 + o(1)] \exp\left[\int^t \left\{ 2 + \frac{1}{s^2} + o\left(\frac{1}{\ln(\ln s)} \right) \right\} ds \right].$$

2.7 Estimate of the Error Term o(1)

This section is concerned with a more quantitative estimate of the error term $E(t)$ in (2.17), specifically its dependence on $R(t)$. Eastham [53, pp. 14–15] recognized that any such quantitative knowledge can only stem from a more careful analysis of the T-operator since it is the term $(Ty)(t)$ given in (2.11) which represents the error term o(1). He investigated this term under additional assumptions on the real parts of the differences of entries of $\Lambda(t)$. More specifically, he considered the cases $\text{Re}\{\lambda_j(t) - \lambda_i(t)\} = c$ and $\text{Re}\{\lambda_j(t) - \lambda_i(t)\} = c\rho'(t)/\rho(t)$, where c is a constant and $\rho(t) \to \infty$ as $t \to \infty$.

We wish to discuss more general results for estimates of the error term o(1). The first result is based on previous work by Bodine and Lutz [13]. We regret that at the time that paper was written, we were unaware of and neglected to mention Eastham's contributions.

Just assuming Levinson's dichotomy conditions appears to be too weak to provide an estimate of the error term o(1) in terms of the perturbation R. In comparison with Theorem 2.7, we will strengthen the dichotomy condition (2.13) and, moreover, require that the perturbation R does not decrease too quickly.

Theorem 2.25 *Let $\Lambda(t) = \operatorname{diag}\{\lambda_1(t), \dots, \lambda_d(t)\}$ be a diagonal and continuous $d \times d$ matrix for $t \geq t_0$. Fix $i \in \{1, \dots, d\}$ and assume that there exist positive constants K_1, K_2, and α such that for each $1 \leq j \leq d$ either*

$$\exp\left\{\operatorname{Re}\int_s^t [\lambda_j(\tau) - \lambda_i(\tau)]\,d\tau\right\} \leq K_1 e^{-\alpha(t-s)} \text{ for } \quad t \geq s \geq t_0, \tag{2.84}$$

or that (2.14) *holds. Assume that the $d \times d$ matrix $R(t)$ is continuous for $t \geq t_0$, and suppose that there exists a scalar-valued majorant $\phi(t) \geq |R(t)|$ for $t \geq t_0$ satisfying*

$$\int_{t_0}^{\infty} \phi(t)\,dt < \infty.$$

Moreover, if (2.84) *is satisfied for at least one j, assume that there exists a positive constant $\beta \in (0, \alpha)$ such that*

$$\phi(t_1)e^{\beta t_1} \leq \phi(t_2)\,e^{\beta t_2} \qquad \text{for all } t_0 \leq t_1 \leq t_2. \tag{2.85}$$

Then (2.16) *has a solution vector satisfying*

$$y_i(t) = \left[e_i + O\left(\int_t^{\infty} \phi(\tau)\,d\tau\right)\right] e^{\int^t \lambda_i(s)\,ds} \qquad \text{as } t \to \infty. \tag{2.86}$$

Proof As in the proof of Theorem 2.7, we make, for a fixed value of i, the preliminary transformation (2.18), leading to the perturbed and unperturbed shifted systems (2.19) and (2.20), respectively. We want apply Theorem 2.2 to these two shifted systems. For that purpose, we define a projection matrix $P = \operatorname{diag}\{p_1, \dots, p_n\}$, where

$$p_j = \begin{cases} 1 & \text{if } (i,j) \text{ satisfies (2.84)}, \\ 0 & \text{if } (i,j) \text{ satisfies (2.14)}. \end{cases}$$

With t_1 defined in Theorem 2.2, we consider

$$(Tz)(t) = \int_{t_1}^{t} W(t)PW^{-1}(\tau)R(\tau)z(\tau)d\tau \\ - \int_t^{\infty} W(t)[I-P]W^{-1}(\tau)R(\tau)z(\tau)\,d\tau. \tag{2.87}$$

Since $w_i(t) = e_i$ is a bounded solution of (2.20), it follows from applying the Banach contraction principle (see Theorem 2.2) that there exist a bounded solution $z_i(t)$ of (2.19) and nonnegative constants M_1 and M_2 (at least one of them being positive) such that for $t \geq t_1$,

$$|z_i(t) - e_i| = |(Tz_i)(t)| \leq M_1 \int_{t_1}^{t} e^{-\alpha(t-\tau)}\phi(\tau)\,d\tau + M_2 \int_{t}^{\infty} \phi(\tau)\,d\tau. \qquad (2.88)$$

If all ordered pairs (i,j) satisfy (2.14), then $P = 0$, $M_1 = 0$, and

$$|z_i(t) - e_i| = O\left(\int_t^{\infty} \phi(\tau)\,d\tau\right),$$

and, by (2.18), the system (2.16) has a solution of the form (2.86).

If there is at least one value of j such that (i,j) satisfies (2.84), then by (2.85)

$$\int_{t_1}^{t} e^{-\alpha(t-\tau)}\phi(\tau)\,d\tau \leq \phi(t) \int_{t_1}^{t} e^{(\beta-\alpha)(t-\tau)}\,d\tau \leq \frac{\phi(t)}{\alpha - \beta}.$$

On the other hand, (2.85) also implies that $\phi(t) = O\left(\int_t^{\infty} \phi(\tau)\,d\tau\right)$ since

$$\int_t^{\infty} \phi(\tau)\,d\tau \geq \phi(t) \int_t^{\infty} e^{-\beta(\tau-t)}\,d\tau = \frac{\phi(t)}{\beta}.$$

This leads to that (2.88) can be re-written as

$$|(Tz_i)(t)| = O\left(\int_t^{\infty} \phi(\tau)\,d\tau\right) \qquad \text{as } t \to \infty,$$

and, arguing as before, (2.16) has again a solution of the form (2.86) for $t \geq t_1$. Using continuous dependence on initial conditions, this solution can be continued to $[t_0, \infty)$. □

The final theorem of this section is motivated by Hartman and Wintner [73, Thm. (**)]. This result, a refinement of work by Dunkel [50], investigated the asymptotic behavior of solutions of perturbations of constant matrices in Jordan form $y' = [J + R(t)]y$. Dunkel had established an asymptotic integration result containing an error term o(1) provided that $t^m|R(t)| \in L^1$, where m is a non-negative integer determined by the size of certain Jordan blocks (cf. Corollary 6.3). Hartman and Wintner showed that the error term o(1) can be improved to $o(t^{-\gamma})$ ($\gamma > 0$) by strengthening the hypothesis $t^m|R(t)| \in L^1$ to $t^{(m+\gamma)}|R(t)| \in L^1$. Common with Dunkel, they were considering perturbations of constant systems in Jordan form.

We are interested next in similar results for perturbations of diagonal systems, where the diagonal matrix is not necessarily constant, but satisfies the dichotomy conditions of Theorem 2.25. In what follows, we will generalize these powers of

t to suitable "weight functions" $w(t)$. To differentiate from Theorem 2.25, we note that we will neither require the existence of the majorant ϕ nor the "slow-decay condition" (2.85) of this majorant. We do, however, sharpen the condition on the perturbation $R(t)$. To the best of our knowledge, this result seems to be new here.

Theorem 2.26 *For fixed $i \in \{1, \dots, d\}$, suppose that $\Lambda(t) = \operatorname{diag}\{\lambda_1(t), \dots, \lambda_d(t)\}$ satisfies the hypotheses of Theorem 2.25. Assume that the $d \times d$ matrix $R(t)$ is continuous for $t \geq t_0$, and suppose that there exists a function $w(t) \geq 1$ for all $t \geq t_0$ such that*

$$w(\tau) \leq w(t) \qquad \text{for all} \quad t_0 \leq \tau \leq t, \tag{2.89}$$

and

$$\int_{t_0}^{\infty} w(t)|R(t)|\, dt < \infty. \tag{2.90}$$

Moreover, with $\alpha > 0$ given in (2.84), suppose that there exists $\rho \in (0, \alpha)$ such that

$$\frac{e^{-(\alpha-\rho)(t-\tau)}}{w(\tau)} \leq \frac{1}{w(t)} \qquad \text{for all} \quad t \geq \tau \geq t_0. \tag{2.91}$$

Then (2.16) has a solution vector satisfying

$$y_i(t) = \left[e_i + o\left(\frac{1}{w(t)}\right)\right] \exp\left\{\int^t \lambda_i(s)\, ds\right\} \qquad \text{as } t \to \infty. \tag{2.92}$$

Proof Fix $i \in \{1, \dots, d\}$. $\Lambda(t)$ and $R(t)$ satisfy the condition of Theorem 2.7 for this fixed value of i, hence Theorem 2.7 implies the existence of a solution $y_i(t) = [e_i + o(1)] \exp\left\{\int_{t_0}^t \lambda_i(s)\, ds\right\}$ as $t \to \infty$. To establish the sharper error estimate in (2.92), we recall that $\Lambda(t)$ satisfies the hypotheses of Theorem 2.25, which allows us to follow its proof up to (2.87). We will show that, under the given hypotheses, the T-operator given in (2.87) satisfies

$$|(Tz_i)(t)| = o\left(\frac{1}{w(t)}\right) \qquad \text{as } t \to \infty, \tag{2.93}$$

which will suffice to establish this result. To this end, we note that there exist nonnegative constants L_1 and L_2 (at least one of them positive) such that for $t \geq t_1 \geq t_0$ (t_1 sufficiently large) and $(Tz_i)(t)$ given in (2.87)

$$|(Tz_i)(t)| \leq L_1 \int_{t_1}^{t} e^{-\alpha(t-\tau)}|R(\tau)|\, d\tau + L_2 \int_t^{\infty} |R(\tau)|\, d\tau. \tag{2.94}$$

Let $\tilde{\varepsilon} > 0$ be given. Put $L = \max\{L_1, L_2\} > 0$ and, recalling (2.90), fix $t_2 \geq t_1$ such that

$$L \int_{t_2}^{\infty} w(\tau)|R(\tau)|\, d\tau < \frac{\tilde{\varepsilon}}{2}.$$

Then for the fixed value $\rho \in (0, \alpha)$ and $t \geq t_2$ it follows from (2.91) and (2.89) that

$$\begin{aligned}
|(Tz_i)(t)| &\leq L \left[\int_{t_1}^{t} e^{-\rho(t-\tau)} \frac{e^{-(\alpha-\rho)(t-\tau)}}{w(\tau)} (w(\tau)|R(\tau)|)\, d\tau + \int_{t}^{\infty} \frac{w(\tau)|R(\tau)|}{w(\tau)}\, d\tau \right] \\
&\leq \frac{L}{w(t)} \left[\int_{t_1}^{t} e^{-\rho(t-\tau)} w(\tau)|R(\tau)|\, d\tau + \int_{t}^{\infty} w(\tau)|R(\tau)|\, d\tau \right] \\
&\leq \frac{L}{w(t)} \left[e^{-\rho t} \int_{t_1}^{t_2} e^{\rho\tau} w(\tau)|R(\tau)|\, d\tau + \int_{t_2}^{\infty} w(\tau)|R(\tau)|\, d\tau \right] \\
&\leq \frac{\tilde{\varepsilon}}{w(t)},
\end{aligned}$$

for all $t \geq t_2$ sufficiently large. Hence $(Tz_i)(t) = \mathrm{o}\left(\frac{1}{w(t)}\right)$ as $t \to \infty$, which establishes (2.92). □

Important in applications is the special case

Corollary 2.27 *For fixed $i \in \{1, \ldots, d\}$, assume that $\Lambda(t) = \mathrm{diag}\,\{\lambda_1(t), \ldots, \lambda_d(t)\}$ satisfies the hypotheses of Theorem 2.26. Suppose that the $d \times d$ matrix $R(t)$ is continuous for $t \geq t_0$, and suppose that there exists a constant $\beta \in (0, \alpha)$ such that*

$$\int_{t_0}^{\infty} t^{\beta} |R(t)|\, dt < \infty.$$

Then (2.16) *has a solution vector satisfying*

$$y_i(t) = \left[e_i + \mathrm{o}\left(\frac{1}{t^{\beta}}\right) \right] \exp\left\{ \int^{t} \lambda_i(s)\, ds \right\} \qquad \text{as } t \to \infty. \tag{2.95}$$

A possible interpretation of Corollary 2.27 is the fact that if the perturbation, even multiplied by a positive power still remains in L^1, that information leads to a corresponding improvement of the error estimate. As mentioned above, this idea was motivated by work of Hartman–Wintner [73, Thm. (**)] in the context of perturbed constant Jordan systems, where powers of t show up naturally.

Example 2.28 As an example, consider

$$y' = \left[\begin{pmatrix} \frac{1}{t} & \\ & 1 \end{pmatrix} + R(t) \right] y, \qquad t \geq 1, \tag{2.96}$$

where $R(t) = O\left(\frac{1}{t^{1+\varepsilon}}\right)$ for some $\varepsilon > 0$. One can show that the unperturbed system

$$x' = \begin{pmatrix} \frac{1}{t} & \\ & 1 \end{pmatrix} x,$$

satisfies (2.84), (2.14) for every $\alpha \in (0,1)$ and for $i = 1,2$ (with appropriately chosen constants K_1, K_2). An application of Corollary 2.27 shows that (2.96) has a fundamental matrix satisfying

$$Y(t) = \left[I + o\left(t^{-\beta}\right)\right]\begin{pmatrix} t & \\ & e^t \end{pmatrix} \tag{2.97}$$

for every $0 < \beta < \min\{\alpha, \varepsilon\}$. In the case that $R(t) = O\left(\frac{1}{t^{3/2}}\right)$, one can fix $\alpha \in [1/2, 1)$ to find that (2.96) has a fundamental matrix satisfying (2.97) for any $\beta \in (0, 1/2)$.

Examples for other weight functions $w(t)$ satisfying (2.89) and (2.91) for $t \geq \tau$ sufficiently large include function with "sub-exponential" or at most "small exponential growth" such as $w(t) = e^{t^p}$ for $0 < p < 1$, or $w(t) = e^{\delta t}$ for some $0 \leq \delta < \alpha$.

2.8 (h, k)-Dichotomies

M. Pinto (see [124] or [111]) generalized the concept of a dichotomy for a linear system $x' = A(t)x$ in the following way.

Definition 2.29 Let $h(t)$ and $k(t)$ be two continuous and positive functions defined on $[t_0, \infty)$. Let $X(t)$ be a fundamental matrix of (2.5). Assume that there exist a projection P and a positive constant c such that

$$\left.\begin{array}{ll} |X(t)PX^{-1}(s)| \leq ch(t)h^{-1}(s) & \forall \;\; t_0 \leq s \leq t, \\ |X(t)[I-P]X^{-1}(s)| \leq ck^{-1}(t)k(s) & \forall \;\; t_0 \leq t \leq s. \end{array}\right\} \tag{2.98}$$

Then (2.5) is said to have an *(h, k)-dichotomy*. Furthermore, the case $h = k^{-1}$ is called an *h-dichotomy*, if also $X(t)P/h(t) \to 0$ as $t \to \infty$.

Examples of (h, k)-dichotomies include the ordinary dichotomy (see Definition 2.1) with $h = k = 1$, the exponential dichotomy (see Definition 2.15) with $h(t) = k(t) = e^{-\delta t}$ for a positive constant δ, and so-called exponential-ordinary dichotomies with $h(t) = e^{-\alpha t}$ and $k(t) = 1$.

In [126, Thm. 1], Pinto established a result on the asymptotic behavior of solutions of "mildly nonlinear perturbations" of linear systems, and we give now a linear version of his work which fits the setting of this book.

We will use the notation that $y_+ \in C_+$ (resp., $y_- \in C_-$) means

$$\|y\|_+ := \sup_{t\geq t_0}\{|y_+(t)|/h(t)\} < \infty \qquad (\text{resp. } \sup_{t\geq t_0}\{|y_-(t)|\, k(t)\} < \infty.)$$

Theorem 2.30 (Pinto) *Assume that the $d \times d$ matrix $A(t)$ is continuous and that* (2.5) *has an (h,k)-dichotomy satisfying the (so-called) compensation law*

$$\frac{h(t)\,k(t)}{h(s)\,k(s)} \leq c_1 \qquad \forall \;\; t_0 \leq s \leq t, \tag{2.99}$$

where c_1 is a positive constant. Furthermore, assume that $R(t)$ is a continuous $d \times d$ matrix for $t \geq t_0$ satisfying $R(t) \in L^1[t_0, \infty)$. Then there exists a one-to-one, bicontinuous correspondence between the solutions $x_+ \in C_+$ (respectively, $x_- \in C_-$) of the linear system (2.5) *and the solutions $y_+ \in C_+$ (respectively, $y_- \in C_-$) of the perturbed system* (2.7)*. Moreover, if*

$$\frac{X(t)P}{h(t)} \to 0 \qquad as \;\; t \to \infty$$

then

$$y_\pm = x_\pm + o\,(h_\pm) \qquad as\; t \to \infty, \tag{2.100}$$

where $h_+ = h$ and $h_- = k^{-1}$.

The proof, based on the Contraction Mapping Principle in the Banach space C_+, resp. C_-, establishes the bi-continuous correspondence between solutions x_+ and y_+ (resp. x_- and y_-). Using (2.98), (2.99), and the definitions of C_+ and C_-, it is a modification of the proof of Theorem 2.2 and we just indicate the main steps here. To show (2.100), for example for $y \in C_+$, one sees for fixed t_1 sufficiently large and $t \geq t_1$ that

$$y = x + Ty,$$

where the operator T was defined in (2.11). It follows that

$$|(Ty)(t)| \leq |X(t)P| \int_{t_0}^{t_1} |X^{-1}(s)R(s)y(s)|\, ds$$
$$+ \int_{t_1}^{t} c\frac{h(t)}{h(s)}|R(s)|\,|y(s)|\, ds + \int_{t}^{\infty} c\frac{k(s)}{k(t)}|R(s)|\,|y(s)|\, ds.$$

Using the compensation law (2.99), this can be reduced to

$$|(Ty)(t)| \leq |X(t)P| \int_{t_0}^{t_1} |X^{-1}(s)R(s)y(s)|\, ds + c(1+c_1)\|y\|_+ h(t) \int_{t_1}^{\infty} |R(s)|\, ds.$$

Given a solution y and $\varepsilon > 0$, choose t_1 sufficiently large such that

$$c(1+c_1)\|y\|_+ \int_{t_1}^{\infty} |R(s)|\, ds < \frac{\varepsilon}{2}$$

for all $t \geq t_1$. Then, using $|X(t)P|/h(t) \to 0$ as $t \to \infty$, choose t sufficiently large such that

$$|X(t)P| \int_{t_0}^{t_1} |X^{-1}(s)R(s)y(s)|\, ds < \frac{\varepsilon}{2} h(t).$$

This establishes (2.100) in the case $y \in C_+$. The proof for $y \in C_-$ is similar, requiring use of the compensation law (2.99) one more time.

We note that this result is potentially more applicable than Theorem 2.2 since the latter result could be vacuous in case the unperturbed system has no bounded solutions. In such a case, (2.100) still could give information about the growth or decay of solutions, depending upon choices of h and k.

For diagonal unperturbed systems $A(t) = \Lambda(t) = \text{diag}\,\{\lambda_1(t), \ldots, \lambda_d(t)\}$ satisfying Levinson's dichotomy conditions (2.13), (2.14) for a fixed value of i, one can select

$$h(t) = \exp\left(\int^t \text{Re}\,[\lambda_i(\tau)]\, d\tau\right) \quad \text{and} \quad k(t) = \exp\left(\int^t -\text{Re}\,[\lambda_i(\tau)]\, d\tau\right).$$

Note that this choice implies that $C_+ = C_-$. Then, choosing $P = \text{diag}\,\{p_1, \ldots, p_d\}$ with $p_j = 1$ if (2.13) holds and $p_j = 0$ in the case of (2.14), it is straightforward to show that $x' = \Lambda(t)x$ satisfies an (h,k)-dichotomy. In particular, one can take $x_+(t) = e_i \exp\left(\int^t \lambda_i(\tau)\, d\tau\right)$ and it follows from (2.100) that (2.7) has a solution satisfying

$$y = [e_i + o(1)] \exp\left[\int^t \lambda_i(\tau)\, d\tau\right] \qquad \text{as } t \to \infty,$$

which is equivalent to the conclusion of Theorem 2.7 for the ith column of (2.17).

Remark 2.31 On the other hand, as Pinto remarks in [126], it is not necessary to optimally select (h,k) in order to obtain useful bounds on solutions from Theorem 2.30. See [126] for some examples along these lines.

For completeness' sake, we want to point out that sufficient conditions for (h,k) dichotomies were studied in [102].

2.9 Coupled Dichotomy-Growth Conditions

This section is concerned with another result concerning the asymptotic diagonalization of a perturbed system

$$y' = [\Lambda(t) + R(t)]\, y, \qquad t \geq t_0. \tag{2.101}$$

In previous such asymptotic diagonalization results, e.g., Theorems 2.7 and 2.20, the hypotheses involved two separate kinds of assumptions, i.e., dichotomy conditions for $\Lambda(t)$ and growth conditions for $R(t)$. In this section we discuss results of Elias and Gingold [58], where these separate assumptions on Λ and R are replaced by combined or *coupled* integral conditions which are weaker than the separated conditions, but still imply both Theorems 2.7 and 2.20 as well as other variations of dichotomy/growth conditions. In what follows, we give a somewhat modified treatment of their result in [58], more in line with the techniques used above.

Consider (2.101), where

$$\Lambda(t) = \text{diag}\,\{\lambda_1(t), \ldots, \lambda_d(t)\},$$

and

$$\text{diag}\, R(t) \equiv 0.$$

The goal is to establish sufficient conditions for (2.101) to have a fundamental solution matrix of the form

$$Y(t) = [I + E(t)] \exp\left[\int^t \Lambda(s)\, ds\right], \quad \text{where} \lim_{t\to\infty} E(t) = 0. \tag{2.102}$$

That is, we want to find, for large t, sufficient conditions for the existence of $E(t) \to 0$ as $t \to \infty$ such that

$$y = [I + E(t)]z$$

reduces (2.101) to

$$z' = \Lambda(t)z, \qquad t \geq t_0. \tag{2.103}$$

This is equivalent to E satisfying the linear, two-sided, matrix differential equation

$$E' = \Lambda E - E\Lambda + R[I + E]. \tag{2.104}$$

It suffices to establish for sufficiently large t the existence of a matrix-valued function $E(t) = o(1)$ as $t \to \infty$ satisfying the associated integral equation

$$E(t) = \Phi(t) \int_{(L)}^{t} \Phi^{-1}(s)R(s)[I + E(s)]\Phi(s)\, ds\, \Phi^{-1}(t), \tag{2.105}$$

where

$$\Phi(t) = \exp\left[\int_{t_0}^{t} \Lambda(\tau)\, d\tau\right],$$

and $\int_{(L)}^{t} F(s)\, ds = \left[\int_{l_{ij}}^{t} f_{ij}(s)\, ds\right]_{i,j=1}^{d}$, and l_{ij} is either finite or ∞, with $\int_{\infty}^{t} = -\int_{t}^{\infty}$. Using the same notation as in [58], put

$$\hat{M}_1(t) = \Phi(t) \int_{(L)}^{t} \Phi^{-1}(s)R(s)\Phi(s)\, ds\, \Phi^{-1}(t). \tag{2.106}$$

Then (2.105) can be re-written as

$$\begin{aligned} E(t) &= \hat{M}_1(t) + \Phi(t) \int_{(L)}^{t} \Phi^{-1}(s)R(s)E(s)\Phi(s)\, ds\, \Phi^{-1}(t) \\ &=: \hat{M}_1(t) + A[E](t). \end{aligned} \tag{2.107}$$

First, one should assume that $\hat{M}_1(t)$ vanishes at infinity. Next, consider

$$(A[E])_{ik}(t) = \sum_{l} \int_{l_{ik}}^{t} e^{\int_s^t [\lambda_i(u) - \lambda_k(u)]\, du} r_{il}(s) e_{lk}(s)\, ds,$$

and, in particular,

$$(A[E])_{ii}(t) = \sum_{l} \int_{l_{ik}}^{t} r_{il}(s) e_{li}(s)\, ds.$$

A problem with the diagonal terms is that they are independent of Λ. To determine what extra conditions are sufficient to show that $E(t) = o(1)$ as $t \to \infty$, it is useful to first re-write (2.107) as

$$E(t) = \hat{M}_1(t) + \Phi(t) \int_{(L)}^{t} \underbrace{\Phi^{-1}(s)R(s)\Phi(s)}_{\tilde{R}(s)}\ \underbrace{\Phi^{-1}(s)E(s)\Phi(s)}_{\tilde{E}(s)}\, ds\, \Phi^{-1}(t).$$

Using (2.104) we see

$$\tilde{E}' = \Phi^{-1}R[I + E]\Phi,$$

and then integration by parts implies that

$$\begin{aligned}
E(t) &= \hat{M}_1(t) \\
&\quad + \Phi(t)\left[\int_{(L)}^{t} \tilde{R}(s)\, ds\, \tilde{E}(t) - \int_{(L)}^{t}\left(\int_{(L)}^{s} \tilde{R}(\tau)\, d\tau\right)\tilde{E}'(s)\, ds\right]\Phi^{-1}(t) \\
&= \hat{M}_1(t)[I + E(t)] \\
&\quad - \Phi(t)\int_{(L)}^{t}\left(\int_{(L)}^{s} \Phi^{-1}(\tau)R(\tau)\Phi(\tau)\, d\tau\right)\Phi^{-1}(s)R(s)[I+E(s)]\Phi(s)\, ds\, \Phi^{-1}(t) \\
&= \hat{M}_1(t)[I + E(t)] - \hat{M}_2(t) \\
&\quad - \Phi(t)\int_{(L)}^{t} \Phi^{-1}(s)\hat{M}_1(s)R(s)E(s)\Phi(s)\, ds\, \Phi^{-1}(t),
\end{aligned} \tag{2.108}$$

where

$$\hat{M}_2(t) = \Phi(t)\int_{(L)}^{t}\left[\int_{(L)}^{s} \Phi^{-1}(\tau)R(\tau)\Phi(\tau)\, d\tau\right]\Phi^{-1}(s)R(s)\Phi(s)\, ds\, \Phi^{-1}(t).$$

We now rewrite (2.108) as

$$E(t) = U(t) + (\mathscr{T}E)(t), \tag{2.109}$$

where we define

$$U(t) = \hat{M}_1(t) - \hat{M}_2(t), \tag{2.110}$$

and

$$(\mathscr{T}E)(t) = \hat{M}_1(t)E(t) - \Phi(t)\int_{(L)}^{t} \Phi^{-1}(s)\hat{M}_1(s)R(s)E(s)\Phi(s)\, ds\, \Phi^{-1}(t). \tag{2.111}$$

Looking at (2.109), we want that $U(t) \to 0$ as $t \to \infty$ and $(\mathscr{T}E)$ to be a contraction on a suitable Banach space and, moreover, $(\mathscr{T}E)(t) \to 0$ as $t \to \infty$. Elias and Gingold [58] gave sufficient conditions for this, and we now state their result.

Theorem 2.32 *Let $\Lambda(t) = \operatorname{diag}\{\lambda_1(t), \cdots, \lambda_d(t)\}$ be continuous for $t \geq t_0$. Let $R(t) = \{r_{ij}(t)\}$ be a continuous $d \times d$ matrix such that $r_{ii}(t) = 0$ for all $1 \leq i \leq d$ and all $t \geq t_0$. If there exist constants $l_{ik} \leq \infty$ such that for all $1 \leq i, j, k, m \leq d$, $i \neq k$, $k \neq j$*

$$(\hat{M}_1)_{ik}(t) = \int_{l_{ik}}^{t} r_{ik}(s)e^{\int_s^t [\lambda_i - \lambda_k]\, d\tau}\, ds \to 0 \quad \text{as } t \to \infty, \tag{2.112}$$

and

$$\int_{l_{im}}^{t} \left| \left[\int_{l_{ik}}^{s} r_{ik}(\tau) e^{\int_\tau^s [\lambda_i - \lambda_k]\, du}\, d\tau \right] r_{kj}(s) e^{\int_s^t [\lambda_i - \lambda_m]\, du} \right| ds \to 0 \ \text{as}\ t \to \infty, \tag{2.113}$$

then (2.101) *has a fundamental solution matrix satisfying* (2.102).

Proof Note that the restrictions $i \neq k$ and $k \neq j$ would not need to be stated explicitly because of the hypothesis that $\operatorname{diag} R(t) = 0$ for all $t \geq t_0$. It can be seen that (2.112) ensures that $\hat{M}_1$ vanishes at infinity and that (2.113) guarantees that both $\hat{M}_2$ and the integral in (2.111) go to zero for bounded $E(t)$ as $t \to \infty$.

Using (2.112) and (2.113), fix $t_1 \geq t_0$ sufficiently large such that

$$|\hat{M}_1(t)| + d^2 N(t) \leq \delta < 1 \qquad \text{for all } t \geq t_1, \tag{2.114}$$

where

$$N(t) = \max_{i,j,k,m} \left| \int_{l_{im}}^{t} \left| \left[\int_{l_{ik}}^{s} r_{ik}(\tau) e^{\int_\tau^s [\lambda_i - \lambda_k]\, du}\, d\tau \right] r_{kj}(s) e^{\int_s^t [\lambda_i - \lambda_m]\, du} \right| ds \right|.$$

Let $\mathscr{B}$ be the Banach space of continuous and bounded $d \times d$ matrix-valued functions equipped with the supremum norm $\|W\| = \sup_{t \geq t_1} |W(t)|$. First, we claim that $U \in \mathscr{B}$, where U was defined in (2.110). It is immediate that $\hat{M}_1 \in \mathscr{B}$ by (2.112). Also, since

$$\left(\hat{M}_2(t)\right)_{im} = \sum_{k \neq i,m}^{d} \int_{l_{im}}^{t} \left[\int_{l_{ik}}^{s} e^{\int_\tau^s [\lambda_i - \lambda_k]\, du} r_{ik}(\tau)\, d\tau \right] e^{\int_s^t [\lambda_i - \lambda_m]\, d\tau} r_{km}(s)\, ds,$$

(2.113) (with $j = m$) implies that $\hat{M}_2 \in \mathscr{B}$. Hence $\mathscr{U} \in \mathscr{B}$.

Next, we will show that $\mathscr{T} : \mathscr{B} \to \mathscr{B}$ is a contraction for t sufficiently large, where $\mathscr{T}$ was defined in (2.111). For that purpose, let $E(t) \in \mathscr{B}$. Then

$$\begin{aligned} |(\mathscr{T}E)_{im}| &\leq \sum_k \left| (\hat{M}_1)_{ik}(t) E_{km}(t) \right| \\ &\quad + \left| \sum_{k,j} \int_{l_{im}}^{t} e^{\int_s^t [\lambda_i - \lambda_m]} (\hat{M}_1)_{ik}(s)\, r_{kj}(s) E_{jm}(s)\, ds \right| \\ &\leq \|E\| \left(\sum_k \left| (\hat{M}_1)_{ik} \right| + \sum_{k,j} \left| \int_{l_{im}}^{t} \left| e^{\int_s^t [\lambda_i - \lambda_m]} (\hat{M}_1)_{ik}(s)\, r_{kj}(s) \right| ds \right| \right), \end{aligned}$$

which is bounded by (2.112) and (2.113). Hence $\mathscr{T}$ maps $\mathscr{B}$ into $\mathscr{B}$. To show that $(\mathscr{T}E)$ is a contraction for t sufficiently large, observe that for $E,\ \hat{E} \in \mathscr{B}$

$$[\mathscr{T}(E-\hat{E})]_{im}(t) = [\hat{M}_1(t)(E(t)-\hat{E}(t))]_{im} - \sum_{k,j}\int_{l_{im}}^{t} e^{\int_s^t[\lambda_i-\lambda_m]}\left[\int_{l_{ik}}^{s} r_{ik}(\tau)e^{\int_\tau^s[\lambda_i-\lambda_k]}d\tau\right] r_{kj}(s)[E_{jm}(s)-\hat{E}_{jm}(s)]\,ds$$

hence

$$\left|\mathscr{T}(E-\hat{E})]_{im}(t)\right| \le \|E-\hat{E}\|\left(\|\hat{M}_1\| + \sum_{k,j}\left|\int_{l_{im}}^{t}\left|e^{\int_s^t[\lambda_i-\lambda_m]}\left[\int_{l_{ik}}^{s} r_{ik}(\tau)e^{\int_\tau^s[\lambda_i-\lambda_k]}d\tau\right] r_{kj}(s)\right| ds\right|\right)$$
$$\le \|E-\hat{E}\|\left(\|\hat{M}_1\| + d^2|N(t)|\right).$$

It follows from (2.114) that

$$\|\mathscr{T}(E-\hat{E})\| \le \delta\|E-\hat{E}\| < \|E-\hat{E}\|,$$

hence $\mathscr{T}$ is a contraction and for given $U \in \mathscr{B}$, there exists a unique solution $E \in \mathscr{B}$ of (2.109). By (2.112) and (2.113), it is straightforward to show that $U(t) = \mathrm{o}(1)$ and $(\mathscr{T}E)(t) = \mathrm{o}(1)$ as $t \to \infty$, and therefore $E(t)$ also vanishes as $t \to \infty$, and the proof is complete. □

It is shown in [58] that Theorem 2.32 implies the results of Levinson (Theorem 2.7), Hartman–Wintner (Theorem 2.20), and Behncke–Remling (Theorem 2.39). See [58, pp. 289–293] for the details involved to show that the assumptions in those theorems satisfy the conditions (2.112) and (2.113) of Theorem 2.32.

Remark 2.33 Even though Theorem 2.32 is very general and encompasses many special cases, it is not capable of handling all situations of "asymptotically diagonal" equations. An example of this is the triangular system

$$y' = \left[\frac{1}{t}\begin{pmatrix}1 & & \\ & 0 & \\ & & 2\end{pmatrix} + \frac{1}{t\sqrt{\ln t}}\begin{pmatrix}0 & 1 & 0\\ & 0 & 1\\ & & 0\end{pmatrix}\right] y = [\Lambda(t)+V(t)]y,$$

or an arbitrary L^1-perturbation of it. Note that it satisfies the assumptions of Corollary 4.17 (with $\beta = 1$) and hence is asymptotically diagonal in the sense of

Elias/Gingold. However, the condition (2.113) is not satisfied (put, e.g., $i = m = 1$, $k = 2$ and $j = 3$).

2.10 Asymptotic Equivalences of Linear Systems

In the literature, there are several quite different concepts of *asymptotic equivalence* between solutions of systems of linear differential equations. Each type of asymptotic equivalence has its own motivation in the way of applications, and for each type it is natural to ask under what condition on the coefficient matrices are two systems asymptotically equivalent. Moreover, it is also natural to ask how, if at all, asymptotic equivalence results are related to asymptotic integration, and we will take a very brief look at these questions here.

We define two systems

$$x' = A(t)x, \qquad t \geq t_0, \tag{2.115}$$

and

$$y' = B(t)y, \qquad t \geq t_0, \tag{2.116}$$

to be *asymptotically left-equivalent* (as $t \to \infty$) if there exist fundamental solution matrices $X(t)$, $Y(t)$ satisfying

$$Y(t) = [I + \mathrm{o}(1)]X(t) \qquad \text{as } t \to +\infty. \tag{2.117}$$

It is immediate that (2.117) satisfies the requirements of being an equivalence relation on the class of linear differential systems and therefore partitions it into disjoint equivalence classes. Theorem 2.7 gives sufficient conditions for a perturbed diagonal system to be asymptotically left-equivalent to the corresponding unperturbed system and thus for both to be in the same equivalence class.

Another type of equivalence can be defined by the property that (2.115) and (2.116) have fundamental matrices satisfying

$$Y(t) = X(t) + \mathrm{o}(1) \qquad \text{as } t \to +\infty, \tag{2.118}$$

(see, e.g., [39]). This has been called simply "asymptotic equivalence" in the literature, but a more suitable name in the context of this discussion might be *asymptotic central equivalence*. To compare this type of equivalence with that implied by in the special cases of perturbations of *diagonal* systems, $x' = \Lambda(t)x$ and $y' = [\Lambda(t) + R(t)]y$, note that Theorem 2.7 yields

$$y_i(t) = x_i(t) + \mathrm{o}\left(|x_i(t)|\right) \qquad \text{as } t \to +\infty, \tag{2.119}$$

where $x_i(t) = e_i \exp\left(\int^t \lambda_i(\tau)\, d\tau\right)$. Compare this with the result

$$y_i(t) = x_i(t) + o(1) \tag{2.120}$$

following from (2.118). The relative error in (2.119) may coincide with (2.120) in case that $\int^t \lambda_i(\tau)\, d\tau$ is bounded from above and below or maybe stronger and weaker depending upon whether $\int^t \lambda_i(\tau)\, d\tau$ tends to minus or plus infinity as $t \to +\infty$. Hence this type of equivalence does not seem to be directly related to asymptotic integration.

A further type of asymptotic equivalence can be defined by (2.115) and (2.116) having fundamental matrices satisfying

$$Y(t) = X(t)\,[I + o(1)] \qquad \text{as } t \to +\infty, \tag{2.121}$$

which could be called *asymptotic right-equivalence*. Locke [103] gave the following characterization of asymptotically right-equivalent systems:

Theorem 2.34 (Locke) *The two systems* (2.115) *and* (2.116) *have fundamental matrices satisfying*

$$Y(t) = X(t)[I + E(t)] \qquad \textit{as } t \to \infty, \tag{2.122}$$

with

$$\lim_{t\to\infty} E(t) = 0 \quad \textit{and} \quad E' \in L^1[t_0, \infty), \tag{2.123}$$

if and only if

$$X^{-1}\,[B - A]\,X \in L^1[t_0, \infty). \tag{2.124}$$

Proof Assume first that (2.124) holds. Putting $y = X\tilde{y}$ in (2.116) implies that $\tilde{y}' = X^{-1}[B{-}A]X\tilde{y}$, which can be considered as a perturbation of the trivial system $z' = 0$. It follows from Theorem 2.7 that this perturbed system has a fundamental matrix $\tilde{Y}(t) = I + E(t)$, where $E(t) \to 0$ as $t \to \infty$. Moreover, $E' \in L^1[t_0, \infty)$ since $E' = \tilde{Y}' = X^{-1}[B - A]X\tilde{Y}$ with $X^{-1}[B - A]X \in L^1[t_0, \infty)$ and $\tilde{Y}$ bounded.

Now suppose (2.115) and (2.116) have fundamental matrices X and Y, respectively, such that (2.122) and (2.123) hold. Since $Y' = BX(I{+}E) = AX(I{+}E){+}XE'$, it follows that

$$E' = X^{-1}[B - A]X(I + E),$$

hence (2.124) holds recalling that $I + E \to I$ as $t \to \infty$. □

The condition (2.124) seems to be too strong of an assumption for applications, except if $X(t)$ and $X^{-1}(t)$ are bounded for all $t \geq t_0$. Note, however, that in this special case, all of the three concepts above coincide.

Finally, we discuss a result on asymptotic left-equivalence of perturbations of not necessarily diagonal systems. To that end, we consider the linear system

$$y' = [A(t) + R(t)]y, \qquad t \geq t_0, \tag{2.125}$$

as a perturbation of (2.115). Here A and R are continuous $d \times d$ matrices. We are interested in finding conditions sufficient for (2.125) having a fundamental matrix of the form

$$Y(t) = [I + E(t)]X(t), \text{ where } \lim_{t\to\infty} E(t) = 0. \tag{2.126}$$

This problem was first considered in the context of linear difference systems by Naulin and Pinto [112] and we give a similar, although not identical approach (see also Chap. 3.8).

The statement (2.126) is equivalent to requiring that there is a solution $E(t)$ of

$$E' = AE - EA + R(I + E), \text{ such that } \lim_{t\to\infty} E(t) = 0. \tag{2.127}$$

We note that (2.127) might be difficult to analyze because of its two-sided nature. To avoid this difficulty, we follow an alternative approach suggested by Naulin and Pinto and work with a "vectorized equation." To that end, we let $\mathbf{q}(t) = \text{vec}(E(t))$, i.e., $\mathbf{q}$ is a column vector of length d^2 consisting of columns of E. Similarly, $\mathbf{r}(t) = \text{vec}(R(t))$. Then (2.127) is equivalent to

$$\mathbf{q}' = \left(I \otimes A - A^T \otimes I\right)\mathbf{q} + (I \otimes R)\,\mathbf{q} + \mathbf{r}, \text{ such that } \lim_{t\to\infty} \mathbf{q}(t) = 0. \tag{2.128}$$

Here $\otimes$ denotes the Kronecker product, i.e.,

$$A \otimes B = \begin{pmatrix} a_{11}B & \cdots & a_{1d}B \\ \vdots & \ddots & \vdots \\ a_{d1}B & \ldots & a_{dd}B \end{pmatrix}. \tag{2.129}$$

We will make use of the following identities for $d \times d$ matrices (see, e.g., [76, Chap. 4])

$$(A \otimes B)(C \otimes D) = AC \otimes BD, \tag{2.130}$$

$$(A \otimes B)' = A' \otimes B + A \otimes B',$$

and

$$\det(A \otimes B) = [(\det A)(\det B)]^d. \tag{2.131}$$

Theorem 2.35 *Let $X(t)$ be a fundamental solution matrix of* (2.115) *and define*

$$Z(t) = (X^{-1}(t))^T \otimes X(t). \tag{2.132}$$

Let a constant matrix P and a function $\beta(t) \geq 1$ for $t \geq t_0$ be given such that

$$Z(t)P \to 0 \qquad \text{as } t \to \infty,$$

$$\left|Z(t)PZ^{-1}(s)\right| \leq \beta(s) \qquad \text{for } t_0 \leq s \leq t$$

and

$$\left|Z(t)[I - P]Z^{-1}(s)\right| \leq \beta(s) \qquad \text{for } t_0 \leq t \leq s.$$

Suppose that

$$\beta\,|R| \in L^1[t_0, \infty). \tag{2.133}$$

Then (2.125) *has a fundamental matrix satisfying* (2.126).

Remark 2.36 It can be shown that such a matrix P and function β always exist. For example, one can choose $P = 0$ and

$$\beta(s) = \sup_{t_0 \leq t \leq s} \left|Z(t)Z^{-1}(s)\right|.$$

This choice, however, will frequently lead to a non-optimal β, making the growth condition (2.133) unnecessarily restrictive.

Proof We first note that $Z(t)$ defined in (2.132) is a fundamental matrix for the homogeneous unperturbed differential equation associated with (2.128),

$$Z' = \left(I \otimes A - A^T \otimes I\right) Z. \tag{2.134}$$

Fix $t_1 \geq t_0$ such that

$$\theta = \int_{t_1}^{\infty} \beta(s)|R(s)|\,ds < 1.$$

Let $\mathscr{B}$ be the Banach space of bounded d^2-dimensional vector-valued functions with the supremum norm $\|\cdot\|$ and consider, for $t \geq t_1$, the operator

$$
\begin{aligned}
(T\mathbf{q})(t) = & \int_{t_1}^{t} Z(t)PZ^{-1}(s)\left[(I \otimes R(s))\,\mathbf{q}(s) + \mathbf{r}(s)\right] ds \\
& - \int_{t}^{\infty} Z(t)[I - P]Z^{-1}(s)\left[I \otimes R(s)\mathbf{q}(s) + \mathbf{r}(s)\right] ds.
\end{aligned}
$$

It follows for $t \geq t_1$ that

$$
\begin{aligned}
|(T\mathbf{q})(t)| & \leq \int_{t_1}^{\infty} \beta(s)\left\{|R(s)|\|q\| + |R(s)|\right\} ds \\
& \leq (\|q\| + 1) \int_{t_1}^{\infty} \beta(s)|R(s)|\, ds = (\|q\| + 1)\,\theta.
\end{aligned}
$$

Similarly,

$$
\|(T\mathbf{q}_1)(t) - (T\mathbf{q}_2)(t)\| \leq \theta\,\|\mathbf{q}_1 - \mathbf{q}_2\|.
$$

Hence $T : \mathscr{B} \to \mathscr{B}$ is a contraction, and arguments analog to those used in Theorem 2.10 show that there exists a one-to-one and bicontinuous correspondence between the bounded solutions of (2.134) and (2.128) and, moreover, the difference between corresponding bounded solutions of (2.134) and (2.128) tends to zero as $t \to \infty$. In particular, for every $\mathbf{z} \in \mathscr{B}$, there is a unique solution $\mathbf{q}$ of (2.128) in $\mathscr{B}$ given by the fixed point of $\mathbf{q} = \mathbf{z} + T\mathbf{q}$ and $(T\mathbf{q}) = o(1)$ as $t \to \infty$. Putting $\mathbf{z} = 0$, this shows that there exists a solution $\mathbf{q}$ of (2.128) that vanishes at infinity, which establishes (2.127). □

We conclude this section with an example. Consider (2.125) with

$$
A(t) = \begin{pmatrix} 0 & 1 \\ 0 & 0 \end{pmatrix} \quad \text{and} \quad R(t) = \begin{pmatrix} 0 & 0 \\ r(t) & 0 \end{pmatrix}.
$$

Putting $X(t) = \begin{pmatrix} 1 & t \\ 0 & 1 \end{pmatrix}$, $Z(t)$ in (2.132) can be written as the blocked 4×4 matrix

$$
Z(t) = \begin{bmatrix} X(t) & 0 \\ -tX(t) & X(t) \end{bmatrix}.
$$

Using straightforward matrix multiplication, it can be seen that

$$
|Z(t)Z^{-1}(s)| \leq Ms^2 = \beta(s), \qquad 1 \leq t \leq s,
$$

where the value of the positive constant M depends on particular choice of a matrix norm. Put $P = 0$ and $\beta(s) = s^2$. Then Theorem 2.35 implies that for $t^2 r(t) \in L^1$, (2.125) has a solution of the form

$$Y(t) = [I + o(1)]X(t). \tag{2.135}$$

2.11 Other Asymptotic Integration Results

2.11.1 A Row-Wise Generalization of Levinson's Fundamental Theorem

Chiba [37] published in 1965 a result concerning perturbations of diagonal system,

$$y' = [\Lambda(t) + R(t)]\,y, \qquad t \geq t_0, \tag{2.136}$$

which was a generalization of Theorem 2.7. While he used an existence theorem of solutions of initial value problems due to Hukuhara [81] as the major tool in his proof, we will give a proof here using the usual T-operator method, which allows for minor simplifications.

Theorem 2.37 (Chiba) *Let $\Lambda(t) = \{\lambda_1(t), \ldots, \lambda_d(t)\}$ and $R(t)$ be continuous $d \times d$ matrices for $t \geq t_0$. Let h be a fixed index in $\{1, \ldots, d\}$. Define $\rho_i(t)$ $(1 \leq i \leq d)$ by*

$$\rho_i(t) = \sum_{j=1}^{d} |r_{ij}(t)|, \qquad (1 \leq i \leq d).$$

Suppose there exists a partition of the index set

$$\{1, \ldots, d\} = \{1, \ldots, m\} \cup \{m+1, \ldots, d\}$$

such that

$$\rho_i(t) e^{\int_{t_0}^{t} \mathrm{Re}\,[\lambda_h(\tau) - \lambda_i(\tau)]\,d\tau} \in L^1[t_0, \infty), \qquad m+1 \leq i \leq d.$$

Assume the functions $\sigma_i(t)$ defined for $t \geq \hat{t}$ by

$$\sigma_i(t) = \begin{cases} \int_{\hat{t}}^{t} e^{\mathrm{Re}\int_s^t [\lambda_i(\tau) - \lambda_h(\tau)]\,d\tau} \rho_i(s)\,ds & 1 \leq i \leq m \\ \int_t^{\infty} e^{\mathrm{Re}\int_t^s [\lambda_h(\tau) - \lambda_i(\tau)]\,d\tau} \rho_i(s)\,ds & m+1 \leq i \leq d \end{cases} \tag{2.137}$$

satisfy

$$\lim_{t\to+\infty} \sigma_i(t) = 0 \qquad (1 \le i \le d), \tag{2.138}$$

where $\hat{t} \ge t_0$ *is a suitably chosen number. Then* (2.136) *has a solution satisfying*

$$y_h = [e_h + o(1)]\, e^{\int^t \lambda_h(\tau)\, d\tau}, \qquad as \quad t \to \infty. \tag{2.139}$$

Proof For the fixed value of $h \in \{1, \dots, d\}$, make in (2.136) the transformation

$$y(t) = z(t) \exp\left[\int^t \lambda_h(\tau)\, d\tau\right], \tag{2.140}$$

leading to

$$z' = [\Lambda(t) - \lambda_h(t)I + R(t)]\, z, \qquad t \ge t_0. \tag{2.141}$$

Let $W(t)$ be a fundamental matrix of the unperturbed system

$$w' = [\Lambda(t) - \lambda_h(t)I]\, w, \qquad t \ge t_0. \tag{2.142}$$

Put $P = \operatorname{diag}\{I_{m\times m}, 0_{d-m\times d-m}\}$, and fix t_1 $(t_1 \ge t_0)$ such that

$$\max_{1\le i\le d} \sigma_i(t) \le \theta < 1 \qquad \text{for all } t \ge t_1. \tag{2.143}$$

By (2.138), t_1 exists and is well defined. Let $\mathscr{B}$ be the Banach space of bounded d-dimensional vector-valued functions with the supremum norm $\|z\| = \sup_{t\ge t_1} |z(t)|$. Here $|\cdot|$ denote the infinity vector norm in $\mathbb{C}^2$, that is $|z| = \max_{1\le i\le d} |z_i|$. Note that the induced matrix norm is the maximum of the "absolute row sum."

We claim that (2.141) has for $t \ge t_1$ a solution provided

$$\begin{aligned} z(t) &= e_h + \int_{t_1}^{t} W(t)PW^{-1}(s)R(s)z(s)\, ds \\ &\quad - \int_t^\infty W(t)[I - P]W^{-1}(s)R(s)z(s)\, ds \\ &=: e_h + (Tz)(t). \end{aligned} \tag{2.144}$$

Observe that for $z \in \mathscr{B}$,

$$|(Tz)(t)| \le \|z\| \left(\int_{t_1}^{t} \left|W(t)PW^{-1}(s)R(s)\right|\, ds \right.$$

$$+ \int_t^\infty \left|W(t)[I-P]W^{-1}(s)R(s)\right| ds \Big)$$

$$\le \|z\| \left[\max_{1\le i\le m} \int_{t_1}^t e^{\mathrm{Re}\int_s^t\{\lambda_i(\tau)-\lambda_h(\tau)\}\,d\tau} \sum_{j=1}^d |r_{ij}(s)|\, ds \right.$$

$$\left. + \max_{m+1\le i\le d} \int_t^\infty e^{\mathrm{Re}\int_s^t\{\lambda_i(\tau)-\lambda_h(\tau)\}\,d\tau} \sum_{j=1}^d |r_{ij}(s)|\, ds \right]$$

$$\le \|z\| \left[\max_{1\le i\le m} \int_{t_1}^t e^{\mathrm{Re}\int_s^t\{\lambda_i(\tau)-\lambda_h(\tau)\}\,d\tau} \rho_i(s)\, ds \right.$$

$$\left. + \max_{m+1\le i\le d} \int_t^\infty e^{\mathrm{Re}\int_t^s\{\lambda_h(\tau)-\lambda_i(\tau)\}\,d\tau} \rho_i(s)\, ds \right]$$

$$\le \|z\| \left[\max_{1\le i\le m} \sigma_i(t) + \max_{m+1\le i\le d} \sigma_i(t) \right] = \|z\| \max_{1\le i\le d} \sigma_i(t)$$

$$\le \theta \|z\| < \|z\|,$$

where the functions σ_i were defined in (2.137) and we chose $\hat{t} = t_1$. Thus T maps $\mathscr{B}$ into $\mathscr{B}$ and (2.138) implies that $(Tz)(t) \to 0$ as $t \to \infty$. One can show in the same way that T is a contraction. Hence (2.141) has a solution of the form (2.144), and (2.139) now follows from (2.140). □

Remark 2.38 To relate Theorem 2.7 to Theorem 2.37, re-arrange, if necessary, the entries in the diagonal matrix $\Lambda(t)$ such that for $1 \le i \le m$

$$\left.\begin{array}{ll} \int_{t_0}^t \mathrm{Re}\{\lambda_i(\tau)-\lambda_h(\tau)\}\,d\tau \to -\infty & \text{as } t\to\infty \\ \text{and} \quad \int_s^t \mathrm{Re}\{\lambda_i(\tau)-\lambda_h(\tau)\}\,d\tau < K & \forall\ t_0 \le s \le t \end{array}\right\} \tag{2.145}$$

and for $m+1 \le i \le d$

$$\int_t^s \mathrm{Re}\{\lambda_i(\tau)-\lambda_h(\tau)\}\,d\tau > -K \qquad \forall\ t_0 \le t \le s, \tag{2.146}$$

for some constant K, and that

$$R \in L^1[t_0, \infty).$$

It is now fairly straightforward to show that the assumptions of Theorem 2.37 are satisfied.

As an example, consider

$$y' = \left[\begin{pmatrix} 0 & \\ & -1 \end{pmatrix} + \begin{pmatrix} & \frac{1}{\ln t} \\ \frac{1}{t^2} & \end{pmatrix}\right] y = [\Lambda(t) + R(t)]y, \qquad t \geq 1.$$

Since $R(t) \notin L^1[1, \infty)$, Theorem 2.7 does not apply. The assumptions of Theorem 2.37 are satisfied for $h = 2$ (and $m = 0$), and Theorem 2.37 implies the existence of a solution of the form

$$y_2(t) = [e_2 + o(1)]e^{-t}, \qquad \text{as } t \to \infty.$$

2.11.2 A Row-Wise L^p Result

In [7, Thm. 2.1], Behncke and Remling established the following interesting asymptotic integration result. The leading matrix is diagonal, which allows for a "row-wise link" between the dichotomy and the growth condition. As the authors remark, this result can be considered an "interpolation" between the theorems of Levinson and Hartman–Wintner (cf. Thm. 2.21 above in case $p = 2$).

Theorem 2.39 (Behncke/Remling) *Consider*

$$y' = [\Lambda(t) + R(t) + U(t)]\, y, \qquad t \geq t_0, \tag{2.147}$$

where all matrices are $d \times d$ and continuous, $\Lambda(t) = \operatorname{diag}\{\lambda_1(t), \cdots, \lambda_d(t)\}$ and $U(t) \in L^1$. Suppose that there exist constants $c > 0$ and $\alpha_{ij} < 1$ for all $1 \leq i \neq j \leq d$ such that

$$\left|\operatorname{Re}\left[\lambda_i(t) - \lambda_j(t)\right]\right| \geq \frac{c}{t^{\alpha_{ij}}}. \tag{2.148}$$

Assume without loss of generality that $\alpha_{ij} = \alpha_{ji}$, and define

$$\alpha_i = \max_{1 \leq i \neq j \leq d} \alpha_{ij}.$$

Suppose that

$$r_{ij}(t) \cdot t^{\beta_{ij}} \in L^p[t_0, \infty), \qquad p > 1, \tag{2.149}$$

for all $1 \leq i \neq j \leq d$. *Here* $\beta_{ij} \in \mathbb{R}$ *and satisfy*

$$\beta_{ij} \geq \frac{\alpha_i}{p'}, \qquad \frac{1}{p} + \frac{1}{p'} = 1. \tag{2.150}$$

Then (2.147) *has d linearly independent solutions of the form*

$$y_j(t) = [e_j + o(1)] \exp\left[\int^t \{\lambda_j(s) + r_{jj}(s) + o(s^{-\alpha_j})\}\, ds\right], \qquad (t \to \infty).$$

We remark that (2.150) couples the dichotomy and growth conditions. While the original proof is based on so-called conditioning transformations (see Chap. 4), we will give here the outline of a proof in the spirit of the techniques used in Sect. 2.5. In particular, the beginning of the proof follows closely the techniques used in the proof of Theorem 2.17.

Proof Fix $j \in \{1, \ldots, d\}$. We begin by observing that (2.147) has a solution of the form (2.57) if and only if

$$\begin{aligned} s_j'(t) + [\lambda_j(t)I - \Lambda(t)]s_j(t) &= -l_j(t)[e_j + s_j(t)] \\ &\quad + [R(t) + U(t)]\{e_j + s_j(t)\}. \end{aligned} \tag{2.151}$$

We normalize $s_j(t)$ so that $s_{jj}(t) \equiv 0$. Then from the jth position follows that

$$l_j(t) = r_{jj}(t) + u_{jj}(t) + \sum_{\nu=1,\nu\neq j}^{d} \left[r_{j\nu}(t) + u_{j\nu}(t)\right] s_{\nu j}(t).$$

Substituting this back into (2.151), we obtain a (nonlinear) system of $d-1$ equations for $d-1$ unknowns, the remaining components of the vector $s_j(t)$.

We use the following notation in vector form

$$\begin{aligned} \sigma(t) &= \left[s_{1j}(t), \ldots, s_{j-1,j}(t), s_{j+1,j}(t), \ldots, s_{dj}(t)\right]^T, \\ \hat{r}_j(t) &= \left[r_{1j}(t), \ldots, r_{j-1,j}(t), r_{j+1,j}(t), \ldots, r_{dj}(t)\right]^T, \\ \hat{\rho}_j(t) &= \left[r_{j1}(t), \ldots, r_{j,j-1}(t), r_{j,j+1}(t), \ldots, r_{jd}(t)\right]. \end{aligned}$$

Similarly, $\hat{u}_j(t)$ and $\tilde{u}_j(t)$ are the jth column and row of $U(t)$ without the element $u_{jj}(t)$, respectively.

Then $l_j(t)$ can be re-written as

$$l_j(t) = r_{jj}(t) + u_{jj}(t) + [\hat{\rho}_j(t) + \tilde{u}_j(t)]\sigma(t), \tag{2.152}$$

and substituting this into the remaining $(d-1)$ equations of (2.151), we obtain the system

$$\sigma'(t) = \left[\hat{\Lambda}(t) - \lambda_j(t)I_{d-1}\right]\sigma(t) + V(t,\sigma(t)). \tag{2.153}$$

Here and in what follows $\hat{A}$ denotes a $(d-1)\times(d-1)$-dimensional matrix obtained by deleting the jth row and jth column from a $d\times d$ matrix A. With this notation

$$V(t,\sigma(t)) = \hat{r}_j(t) + \hat{u}_j(t) + [\hat{R}(t) + \hat{U}(t)]\sigma(t) - l_j(t)\sigma(t),$$

with $l_j(t)$ defined in (2.152). We note that the unperturbed system

$$x' = \left[\hat{\Lambda}(t) - \lambda_j(t)I_{d-1}\right]x$$

does not satisfy an exponential dichotomy, which is the reason we cannot simply apply results from Sect. 2.5, but need to analyze a corresponding integral equation. Since we used similar arguments before, we will indicate the main steps without showing every detail.

To that end, let $\mathscr{B}$ be the Banach space of bounded $d-1$-dimensional vector-valued functions with the supremum norm $\|\sigma\| = \sup_{t\geq t_1}|\sigma(t)|$, and let $\mathscr{W} = \{\sigma \in \mathscr{B} : \|y\| \leq 1\}$. Hence $\mathscr{W}$ is a closed subset of $\mathscr{B}$. Here t_1 is chosen sufficiently large (see below). For $t \geq t_1$, define an operator T on $\mathscr{W}$ by

$$(T\sigma)(t) = \int_{t_1}^{t} X(t)PX^{-1}(s)V(s,\sigma(s))\,ds$$
$$- \int_{t}^{\infty} X(t)[I-P]X^{-1}(s)V(s,\sigma(s))\,ds.$$

Here $P = \operatorname{diag}\{p_1,\ \dots\ ,p_{j-1},p_{j+1},\ \dots\ ,p_d\}$, with

$$p_i = \begin{cases} 0 \text{ if } \operatorname{Re}[\lambda_i(t)-\lambda_j(t)] \geq ct^{-\alpha_{ij}} \\ 1 \text{ if } \operatorname{Re}[\lambda_i(t)-\lambda_j(t)] < -ct^{-\alpha_{ij}}. \end{cases}$$

If $p_i = 1$, then it follows for the ith component of $(T\sigma)(t)$ that

$$|(T\sigma)_i(t)| = \int_{t_1}^{t} e^{\int_s^t \operatorname{Re}[\lambda_i(\tau)-\lambda_j(\tau)]\,d\tau}\left[\sum_{\nu=1}^{d}\{|r_{i\nu}(s)| + |r_{j\nu}(s)|\} + |u(s)|\right] ds,$$

where $u(s)$ is some L^1-function. Now applying Hölders inequality yields that

$$\int_{t_1}^{t} e^{\int_s^t \mathrm{Re}\,[\lambda_i(\tau)-\lambda_j(\tau)]\,d\tau} |r_{iv}(s)|$$

$$\leq \int_{t_1}^{t} \frac{e^{-\frac{c}{1-\alpha_i}\left(t^{1-\alpha_i}-s^{1-\alpha_i}\right)}}{s^{\beta_{iv}}} \left(|r_{iv}(s)|s^{\beta_{iv}}\right) ds$$

$$\leq \left(\int_{t_1}^{t} \frac{e^{-\frac{cp'}{1-\alpha_i}\left(t^{1-\alpha_i}-s^{1-\alpha_i}\right)}}{s^{p'\beta_{iv}}} ds\right)^{1/p'} \left(\int_{t_1}^{t} \left(|r_{iv}(s)|s^{\beta_{iv}}\right)^p ds\right)^{1/p}$$

$$\leq \left(\frac{1}{cp'}\right)^{1/p'} \left(\int_{t_1}^{t} \left(|r_{iv}(s)|s^{\beta_{iv}}\right)^p ds\right)^{1/p},$$

where the substitution $u = s^{1-\alpha_i}$ and (2.150) are used to show the last inequality. Similarly,

$$\int_{t_1}^{t} e^{\int_s^t \mathrm{Re}\,[\lambda_i(\tau)-\lambda_j(\tau)]\,d\tau} |r_{jv}(s)| \leq \left(\frac{1}{cp'}\right)^{1/p'} \left(\int_{t_1}^{t} \left(|r_{jv}(s)|s^{\beta_{jv}}\right)^p ds\right)^{1/p}.$$

Finally,

$$\int_{t_1}^{t} e^{\int_s^t \mathrm{Re}\,[\lambda_i(\tau)-\lambda_j(\tau)]\,d\tau} |u(s)|\, ds \leq \int_{t_1}^{t} |u(s)|\, ds.$$

Similar estimates follow if $p_1 = 0$. Now (2.149) and the assumption $U(t) \in L^1$ can be used to show that there exists $t_1 \geq t_0$ such that T is a contraction on $\mathscr{W}$ and, moreover, $(T\sigma)(t) \to 0$ as $t \to \infty$. It follows that (2.153) has a solution $\sigma(t) = \mathrm{o}(1)$ as $t \to \infty$. Thus (2.152) takes on the form

$$l_j(t) = r_{jj}(t) + \mathrm{o}\left(\sum_{v\neq j} |r_{jv}(t)|\right) + l^1,$$

and hence (2.147) has a solution for t sufficiently large satisfying

$$y_j(t) = [e_j + s_j(t)] \exp\left[\int^t \left\{\lambda_j(s) + r_{jj}(s) + \mathrm{o}\left(\sum_{v\neq j} |r_{jv}(s)|\right)\right\} ds\right].$$

It remains to show that $|r_{jv}(s)| = \mathrm{O}\left(s^{-\alpha_j}\right)$, and we refer to [7, Lemmas 4.2, 4.3] for a proof of this fact. □

As remarked on [7, p. 596], the pointwise dichotomy condition (2.148) could be weakened to the averaged condition that

$$\left|\mathrm{Re} \int_s^t [\lambda_i(\tau) - \lambda_j(\tau)]\, d\tau\right| \geq C[t^{1-\alpha} - s^{1-\alpha}] - K, \; t_0 \leq s \leq t < \infty.$$

2.11.3 Uniform Asymptotic Integration

One natural extension of Theorem 2.7 concerns systems of the form

$$\frac{dy(t,z)}{dt} = [\Lambda(t,z) + R(t,z)]\, y(t,z), \qquad t \geq t_0, \; z \in S \subset \mathbb{C}, \tag{2.154}$$

where $\Lambda(t,z) = \mathrm{diag}\,\{\lambda_1(t,z), \ldots, \lambda_d(t,z)\}$ satisfies certain dichotomy conditions and $R(t,z) \in L^1[t_0), \infty$, and where z is a complex parameter. If the dichotomy and growth conditions of Theorem 2.7 hold uniformly for $z \in S \subset \mathbb{C}$, then it follows immediately from the T-operator proof of Theorem 2.7 that there exist solutions

$$y_j(t) = [e_j + s_j(t,z)] \exp\left[\int^t \lambda_j(\tau,z)\, d\tau\right],$$

where also $s_j(t,z) \to 0$ as $t \to \infty$ uniformly in S.

This result, however, is not applicable in cases where there is a change of dominance in the $\lambda_j(t,z)$ as z approaches a critical interval. For example, if $\Lambda(t,z) = iz^{1/2}\mathrm{diag}\,\{1,-1\}$ and $S = \{z \in \mathbb{C} : 0 < a \leq \mathrm{Re}\, z \leq b, \; 0 \leq \mathrm{Im} z \leq 1\}$, then for $\mathrm{Im}\, z > 0$ there is an exponential dichotomy while for $\mathrm{Im} z = 0$ this becomes just an ordinary dichotomy. This example is discussed by Behncke and Remling [8] in a very interesting and quite deep extension of Levinson's fundamental theorem. In an example concerning the one-dimensional Schrödinger equation $y'' + [z + V(t)]y = 0$, where they want to apply such a uniform asymptotic integration to analyze the spectrum of the operator, a situation as with the above $\Lambda(t,z)$ occurs and they point out the inadequacy of the obvious uniform results mentioned above. Their main result, simply stated, is that a uniform asymptotic result is possible, if the elements of $\Lambda(t,z)$ satisfy certain mutually exclusive dichotomy conditions on S, if $|\Lambda(t,z)| \leq a(t)$ and if $|R(t,z)| \leq \rho(t)$ hold.

Their proof uses a very clever modification of the argument used for Theorem 2.7 in which a certain linear combination of solutions (with coefficients $c_j(z)$) is used to modify the usual asymptotic ones in order that when a critical interval is approached, the remainder estimate stays uniform. The authors point out that the argument is related to the so-called Jost solutions in scattering theory. It is also somewhat related to arguments in the meromorphic theory of linear equations, when certain linear combinations of subdominant solutions are used to allow asymptotic estimates to remain valid as one wants to analytically continue solutions across Stokes rays (where a change of dominance occurs) while maintaining the same asymptotics.

For the precise statement of their main result, the proof, and interesting applications to the Schrödinger equation, see [8].

Chapter 3
Asymptotic Representation for Solutions of Difference Systems

3.1 Chapter Overview

In this chapter we first consider systems of linear difference equations

$$y(n+1) = A(n)y(n), \qquad n \geq n_0, \tag{3.1}$$

which are in what we call l^1-diagonal form

$$y(n+1) = [\Lambda(n) + R(n)]y(n), \qquad n \geq n_0 \tag{3.2}$$

and establish results similar to those in Sects. 2.2–2.4. This includes discrete versions of Coppel's theorem and Levinson's fundamental theorem from Sect. 2.2 for l^1-diagonal systems, and some results for weak dichotomies.

As it was pointed out in Sect. 1.4, there are many similarities between the methods used for differential equations and difference equations. This permits some statements and proofs to be more or less just be "translated" from one case to the other. But there can also arise some marked differences. One such difference concerns the effect of a normalization of a system using a scalar transformation, which manifests in a multiplicative rather than an additive factor. Such a factor appears, for example, in the growth condition in a discrete version of Levinson's fundamental theorem (Theorem 3.4). More specifically, the growth condition [see (3.10)] carries the factor $1/\lambda_i(n)$. If $\lambda_i(n)$ has a constant, nonzero limit, such a factor has no effect on the growth condition, whereas when it would tend to zero, infinity, or not have a limit, it may have a significant consequence.

Theorem 3.4 is a natural analogue of Theorem 2.7 and pre-dates an improved version presented here in Theorem 3.8 with a relaxed growth condition. In the case when Theorem 3.4 could be applied to obtain a fundamental solution, the growth condition (3.10) would require that every element of R divided by every element

S. Bodine, D.A. Lutz, *Asymptotic Integration of Differential and Difference Equations*, Lecture Notes in Mathematics 2129,
DOI 10.1007/978-3-319-18248-3_3

in Λ has to be in l^1. The improvement by R.J. Kooman in Theorem 3.8 replaces those conditions by significantly weaker ones [either (3.20) or (3.26)] involving just row or column conditions on the elements of R and certain of the elements of Λ. This improvement, which can be important for applications, comes about due to a clever and more detailed study of the T-operator which performs the asymptotic diagonalization. We include both types of results since Theorem 3.4 might involve weaker conditions for a single solution vector than those for the complete diagonalization.

Another difference between the discrete and continuous cases, already mentioned in Chap. 1 concerns the fact that for the existence of a fundamental solution matrix in (3.1), it is necessary and sufficient that $A(n)$ be invertible for all $n \geq n_0$. In the case of the perturbed system (3.2), we usually assume the invertibility of $\Lambda(n)$ to insure the existence of fundamental solutions for both the perturbed and unperturbed equations (the latter for n sufficiently large), but in some cases we will allow non-invertibility to occur. These are notable in Sects. 3.5 and 3.10.

The results in Sect. 3.5 concern perturbations of systems having an exponential dichotomy. Various classes of perturbations are studied and this results in some analogues of well known results for differential equations, e.g., of the Hartman–Wintner type. Theorem 3.12 can be interpreted as an extension of a classical result of O. Perron [117, 118] and it will be applied in Sect. 9.2 in connection with the classical Poincaré/Perron results for asymptotically constant scalar equations. For such applications it is important to allow for certain kinds of non-invertibility to occur. In this case the formulation of the T-operator is accomplished through the introduction of a state transition matrix for a special class of not necessarily invertible systems.

Other results in this chapter concern systems having an (h, k) dichotomy, which have been studied extensively by M. Pinto et al., some results on asymptotic equivalence of not necessarily diagonal systems, and error estimates for the o(1) terms in the asymptotic representations.

Some of the results in this chapter, including an early version of Theorem 3.4, appeared in Rapoport [131], but were unfortunately neglected and he did not receive proper credit for them. The proofs he gave were inspired by Levinson's treatment for differential equations.

3.2 Ordinary Dichotomies and l^1-Perturbations

In this section, we will discuss an analogue of Levinson's fundamental theorem for perturbed systems of difference equations. This approach parallels the treatment in Sect. 2.2 and begins with a modification of Theorem 2.2 for difference systems, which was first given in [9, Thm. 1.1].

Definition 3.1 A linear difference system (3.1′) where $A(n)$ is invertible for all $n \geq n_0$ is said to have an *discrete ordinary dichotomy* if there exist a fundamental matrix

$X(n)$, a projection matrix P, and a positive constant K such that

$$\left.\begin{array}{l} |X(n)PX^{-1}(k)| \leq K \qquad \forall \quad n_0 \leq k \leq n \\ |X(n)[I-P]X^{-1}(k)| \leq K \;\forall \quad n_0 \leq n \leq k \end{array}\right\} . \tag{3.3}$$

For noninvertible matrices $A(n)$, see Sect. 3.10 for a corresponding definition.

Theorem 3.2 (Discrete Version of Coppel's Theorem) *Consider* (3.1)*, where $A(n)$ are nonsingular for all $n \geq n_0$ and a perturbed system*

$$y(n+1) = [A(n) + R(n)]\, y(n)\,, \qquad n \geq n_0, \tag{3.4}$$

where

$$\sum_{n=n_0}^{\infty} |R(n)| < \infty.$$

Assume that (3.1) *satisfies an discrete ordinary dichotomy* (3.3) *with projection P and constant $K > 0$. Fix n_1 sufficiently large ($n_1 \geq n_0$) such that*

$$\theta := K \sum_{n=n_1}^{\infty} |R(n)| < 1. \tag{3.5}$$

Then, for $n \geq n_1$, there exists a one-to-one and bicontinuous correspondence between the bounded solutions of (3.1) *and* (3.4)*. Moreover, the difference between corresponding solutions of* (3.1) *and* (3.4) *tends to zero as $n \to \infty$ if $X(n)P \to 0$ as $n \to \infty$.*

Proof With n_1 given in (3.5), let l_∞ be the Banach space of bounded d-dimensional vector-valued sequences defined on $[n_1, \infty)$ with the supremum norm

$$\|y\| = \sup_{n \geq n_1} |y(n)|.$$

For $n \geq n_1$, define an operator T on l_∞ by

$$(Ty)(n) = \sum_{k=n_1}^{n-1} X(n)PX^{-1}(k+1)R(k)y(k)$$
$$- \sum_{k=n}^{\infty} X(n)[I-P]X^{-1}(k+1)R(k)y(k). \tag{3.6}$$

Given $y \in l_\infty$, it follows from (3.3) and (3.6) for $n \geq n_1$ that

$$|(Ty)(n)| \leq K\|y\| \sum_{k=n_1}^{\infty} |R(k)| = \theta \|y\|,$$

and therefore also

$$\|Ty\| \leq \theta \|y\|.$$

Similarly, for y and $\tilde{y} \in l_\infty$, one can show that

$$\|(T(y-\tilde{y})\| \leq K\|y-\tilde{y}\| \sum_{n_1}^{\infty} |R(k)| = \theta \|y-\tilde{y}\|,$$

i.e., T maps l_∞ into l_∞ and is a contraction.

Hence, for every function $x \in l_\infty$, the operator equation

$$x = y - Ty \tag{3.7}$$

has a unique solution $y \in l_\infty$. One can check that if x is a solution of (3.1) for $n \geq n_1$, then $y = x + Ty$ is a solution of (3.4) for $n \geq n_1$. Therefore, if x is a bounded solution of (3.1), then the solution y of (3.7) is a bounded solution of (3.4). Conversely, if y is a bounded solution of (3.4), the function x defined by (3.7) is a bounded solution of (3.1).

Equation (3.7) establishes therefore a one-to-one correspondence between the bounded solutions of (3.1) and (3.4) on $[n_1, \infty)$. The bicontinuity of this correspondence as well as that the difference between corresponding solutions tends to zero if $X(n)P \to 0$ as $n \to \infty$ can be shown as in Theorem 2.2. □

Remark 3.3 In what follows, we will not need the full strength of Theorem 3.2. What will be important for our purposes is that in the presence of an ordinary dichotomy (3.3), for every bounded solution of the unperturbed system (2.5) there exists a bounded solution of the perturbed system (2.7) for $n \geq n_1$ and, moreover, the difference between these two bounded solutions will vanish at infinity provided that $X(n)P \to 0$ as $n \to \infty$. We emphasize that without assuming the invertibility of $\Lambda(n) + R(n)$, this bounded solution of (2.7) can in general not be continued to $[n_0, \infty)$.

We again stated the theorem for linear homogeneous systems instead of the more general case of an inhomogeneous, "mildly nonlinear" system analogous to the continuous case as outlined in Remark 2.3.

Applying Theorem 3.2, we now consider a discrete analogue of Theorem 2.7.

Theorem 3.4 (Discrete Version of Levinson's Fundamental Theorem) *Let the* $d \times d$ *matrix* $\Lambda(n) = \operatorname{diag}\{\lambda_1(n), \ldots, \lambda_d(n)\}$ *be diagonal and invertible for* $n \geq$

n_0. *Suppose that Λ satisfies the following dichotomy condition: There exist positive constants K_1 and K_2 such that for each pair of integers (i,j), $1 \leq i \neq j \leq d$ either*

$$\left.\begin{array}{lll} \prod_{k=n_0}^{n} \left|\frac{\lambda_j(k)}{\lambda_i(k)}\right| \to 0 & & \text{as } n \to \infty \\ \text{and} \quad \prod_{k=n_1}^{n_2} \left|\frac{\lambda_j(k)}{\lambda_i(k)}\right| \leq K_1 & & \forall \; n_0 \leq n_1 \leq n_2 \end{array}\right\}, \tag{3.8}$$

or

$$\prod_{k=n_1}^{n_2} \left|\frac{\lambda_j(k)}{\lambda_i(k)}\right| \geq K_2 \qquad \forall \;\; n_0 \leq n_1 \leq n_2. \tag{3.9}$$

Let the $d \times d$ matrix $R(n)$ be defined for $n \geq n_0$ such that

$$\sum_{n=n_0}^{\infty} \frac{|R(n)|}{|\lambda_i(n)|} < \infty \qquad \text{for all } 1 \leq i \leq d. \tag{3.10}$$

Then the linear difference system

$$y(n+1) = [\Lambda(n) + R(n)]\, y(n) \tag{3.11}$$

has for n sufficiently large a fundamental matrix satisfying

$$Y(n) = [I + \mathrm{o}(1)] \prod_{k=n_0}^{n-1} \Lambda(k) \qquad \text{as} \;\; n \to \infty. \tag{3.12}$$

Proof Fix $i \in \{1, \ldots, d\}$ and make the change of variables

$$y(n) = z(n) \prod_{k=n_0}^{n-1} \lambda_i(k). \tag{3.13}$$

Then (3.11) implies that

$$z(n+1) = \frac{1}{\lambda_i(n)} [\Lambda(n) + R(n)]\, z(n). \tag{3.14}$$

We also consider the unperturbed normalized system

$$w(n+1) = \frac{\Lambda(n)}{\lambda_i(n)} w(n), \tag{3.15}$$

and we will apply Theorem 3.2 to these normalized diagonal systems. For that purpose, let P be a diagonal projection matrix given by $P = \operatorname{diag}\{p_1, \ldots, p_d\}$, where

$$p_j = \begin{cases} 1 & \text{if } (i,j) \text{ satisfies (3.8)} \\ 0 & \text{otherwise} \end{cases} \tag{3.16}$$

Then (3.15) satisfies the discrete ordinary dichotomy condition (3.3) and $W(n)P \to 0$ as $n \to \infty$. Now, by Theorem 3.2 and since $w_i(n) = e_i$ is a bounded solution of (3.15), there exists for n sufficiently large a bounded solution $z_i(n)$ of (3.14) with

$$z_i(n) = e_i + o(1) \qquad \text{as } n \to \infty.$$

By (3.13), (3.11) has for n sufficiently large a solution

$$y_i(n) = [e_i + o(1)] \prod_{k=n_0}^{n-1} \lambda_i(k) \qquad \text{as } n \to \infty. \tag{3.17}$$

Repeating this process for all $1 \leq i \leq d$ and noting that this set of solutions is linearly independent for n sufficiently large gives the desired result (3.12). □

It is clear from the proof that if the conditions of Theorem 3.4 are satisfied for one fixed value of $i \in \{1, \ldots, d\}$, then (3.12) has to be replaced by (3.17) in the conclusion of the theorem. This becomes important when the $\lambda_i(n)$ are of different orders of magnitude when $n \to \infty$.

Remark 3.5 Theorem 3.4 can be found in a slightly weaker form in the book of Rapoport [131] and in the current form in Benzaid and Lutz [9].

Since $\prod_{n_0}^{n-1} \lambda(k)$ is a nontrivial solution of

$$y(n+1) = \lambda(n)y(n), \qquad \lambda(n) \neq 0 \quad \text{for all } n \geq n_0,$$

Theorem 3.4 may be used even in the scalar case to provide a more explicit asymptotic structure for the product in case $\lambda_i(n)$ also has a corresponding asymptotic behavior as $n \to \infty$. For example, to find the asymptotic behavior of a solution of

$$y(n+1) = \frac{2n^2 + 4n - 3}{n+7} y(n) = 2n \left[1 - \frac{5}{n} + O\left(\frac{1}{n^2} \right) \right] y(n) \qquad (n \geq 1),$$

one may put

$$y(n) = 2^n \Gamma(n) n^{-5} z(n)$$

to find that

$$z(n+1) = \left[1 + O\left(\frac{1}{n^2}\right)\right] z(n) \qquad \text{as } n \to \infty.$$

Viewing this as a perturbation of the trivial system $w(n+1) = w(n)$, Theorem 3.4 yields the asymptotic solution $z(n) = 1 + o(1)$ and therefore there exists a solution satisfying

$$y(n) = [1 + o(1)]n^{-5}2^n\Gamma(n) \qquad \text{as } n \to \infty.$$

In more complicated cases, it might be possible to use, for example, the Euler–Maclaurin formula (see, e.g., [45, p. 40]) for an explicit asymptotic representation.

Remark 3.6 Written in the form (3.8) and (3.9), Levinson's discrete dichotomy conditions are mutually exclusive. Analogous to Remark 2.8, they can be re-written in a symmetric, yet not mutually exclusive form. Namely, (3.8) and (3.9) can be stated in the following equivalent way: for each pair of integers (i,j), $1 \le i \ne j \le d$ either

$$\prod_{k=n_1}^{n_2} \left|\frac{\lambda_j(k)}{\lambda_i(k)}\right| \le K_1 \qquad \forall\ n_0 \le n_1 \le n_2, \tag{3.18}$$

or

$$\prod_{k=n_1}^{n_2} \left|\frac{\lambda_j(k)}{\lambda_i(k)}\right| \ge K_2 > 0 \qquad \forall\ n_0 \le n_1 \le n_2, \tag{3.19}$$

for some positive constants K_1 and K_2.

To establish equivalence, first note that (3.8) and (3.9) clearly imply (3.18) and (3.19). On the other hand, assume that (3.18) or (3.19) holds. Now (3.19) implies trivially (3.9), so it suffices to assume that (3.18) holds. Setting $I_{ij}(n_0) = 1$ and

$$I_{ij}(n) = \prod_{k=n_0}^{n-1} \left|\frac{\lambda_j(k)}{\lambda_i(k)}\right| \qquad \text{for } n > n_0,$$

(3.18) can be re-written as

$$\frac{I_{ij}(n_2+1)}{I_{ij}(n_1)} \le K_1 \qquad \text{for all } n_0 \le n_1 \le n_2.$$

It follows that either $I_{ij}(n) \to 0$ as $n \to \infty$ or not. In the first case, (3.8) holds and we are done. In the latter case, the "bounded variation" condition (3.18) implies that $I_{ij}(n)$ is bounded away from zero, i.e., there is a constant $\delta > 0$ such that $I_{ij}(n) \ge \delta$

for all $n \geq n_0$. This together with using (3.18) once more shows that

$$\frac{I_{ij}(n_2+1)}{I_{ij}(n_1)} \geq \frac{\delta}{I_{ij}(n_1)} \geq \frac{\delta}{K_1 I_{ij}(n_0)} =: K_2 \qquad \forall\ n_2 \geq n_1 \geq n_0.$$

Thus (3.9) holds, which proves the equivalence.

Remark 3.7 Every constant invertible diagonal matrix Λ satisfies dichotomy conditions (3.8) and (3.9). For nonautonomous $\Lambda(n)$, its entries should be "separated" in a certain sense in order to satisfy (3.8) and (3.9). A sufficient condition is, for example, that either $|\lambda_i(n)| \geq |\lambda_j(n)|$ for all n sufficiently large or $|\lambda_i(n)| \leq |\lambda_j(n)|$ for all n sufficiently large. For a more detailed discussion of conditions on $\Lambda(n)$ to satisfy the dichotomy conditions (3.8), (3.9), we refer the reader to [9, Rem. 2.2]. A situation where

$$\liminf_{n\to\infty} \prod_{k=n_0}^{n} \left|\frac{\lambda_j(k)}{\lambda_i(k)}\right| = 0,$$

and

$$\limsup_{n\to\infty} \prod_{k=n_0}^{n} \left|\frac{\lambda_j(k)}{\lambda_i(k)}\right| = \infty,$$

for the same pair (i,j), violates Levinson's dichotomy conditions.

While Theorem 3.4 is widely referred to in the literature and used in applications, a recent and important improvement is due to R.J. Kooman [98, Thm. 2.1]. The idea is to examine more carefully the effect the matrix multiplication of the *diagonal* matrices $X(n)PX^{-1}(k+1)$ and $X(n)[I-P]X^{-1}(k+1)$ has on the perturbation matrix in the operator given in (3.6). This allows one to weaken the growth condition on the perturbation matrix (3.10) to a row- or column-wise condition [see (3.20) and (3.26)].

Theorem 3.8 (Kooman) *Let the $d \times d$ matrix $\Lambda(n) = \operatorname{diag}\{\lambda_1(n), \ldots, \lambda_d(n)\}$ be a diagonal and invertible for $n \geq n_0$. Suppose that $\Lambda(n)$ satisfies the dichotomy conditions* (3.8), (3.9). *Let the $d \times d$ matrix $R(n)$ be defined for $n \geq n_0$ such that*

$$\sum_{n=n_0}^{\infty} \left|\Lambda^{-1}(n)R(n)\right| < \infty. \tag{3.20}$$

Then the linear difference system (3.11) *has for n sufficiently large a fundamental matrix satisfying* (3.12).

Proof The proof begins as the proof of Theorem 3.4. In particular, for fixed $i \in \{1, \ldots, d\}$, we make the change of variables (3.13), leading to (3.14) and (3.15).

We also define a projection matrix P by (3.16). Instead of just applying Theorem 3.2, one takes a more careful look at the contraction mapping defined in this theorem.

With l_∞ being the Banach space of bounded d-dimensional vector-valued sequences for $n \geq n_2$ with the supremum norm

$$\|y\| = \sup_{n \geq n_2} |y(n)|, \qquad |y(n)| = \max_{1 \leq i \leq d} |y_i(n)|,$$

where n_2 is fixed sufficiently large such that

$$\hat{\theta} = \hat{K} \sum_{n=n_2}^{\infty} |\Lambda^{-1}(n) R(n)| < 1, \tag{3.21}$$

where $\hat{K} = \max\left\{K_1, \frac{1}{K_2}\right\}$, and K_i were defined in (3.8) and (3.9). Note that the induced matrix norm is $|B| = \max_j \sum_{\nu=1}^{d} \left|b_{j\nu}\right|$. To establish the existence of a solution of (3.11) corresponding to $\lambda_i(n)$, define an operator $\hat{T}$ on l_∞ for $n \geq n_2$, by

$$\begin{aligned}(\hat{T}z)(n) = &\sum_{k=n_2}^{n-1} W(n) P W^{-1}(k+1) \frac{R(k)}{\lambda_i(k)} z(k) \\ &- \sum_{k=n}^{\infty} W(n)[I-P] W^{-1}(k+1) \frac{R(k)}{\lambda_i(k)} z(k),\end{aligned} \tag{3.22}$$

where $W(n)$ is a fundamental matrix for (3.15). Hence

$$\begin{aligned}\left|(\hat{T}z)(n)\right| \leq &\|z\| \sum_{k=n_2}^{n-1} \left|W(n) P W^{-1}(k+1) \frac{R(k)}{\lambda_i(k)}\right| \\ &+ \|z\| \sum_{k=n}^{\infty} \left|W(n)[I-P] W^{-1}(k+1) \frac{R(k)}{\lambda_i(k)}\right|,\end{aligned}$$

and we will now more carefully study the entries in these matrices. To that end, let $1 \leq j, \nu \leq d$. If p_j defined in (3.16) satisfies $p_j = 1$, then it follows that $\left(W(n)[I-P] W^{-1}(k+1) \frac{R(k)}{\lambda_i(k)}\right)_{j\nu} = 0$ and from (3.8) that

$$\begin{aligned}&\sum_{k=n_2}^{n-1} \left|\left(W(n) P W^{-1}(k+1) \frac{R(k)}{\lambda_i(k)}\right)_{j\nu}\right| \\ &= \sum_{k=n_2}^{n-1} \left|\frac{r_{j\nu}(k)}{\lambda_i(k)}\right| \prod_{l=k+1}^{n-1} \left|\frac{\lambda_j(l)}{\lambda_i(l)}\right|\end{aligned}$$

$$= \sum_{k=n_2}^{n-1} \left| \frac{r_{j\nu}(k)}{\lambda_j(k)} \right| \prod_{l=k}^{n-1} \left| \frac{\lambda_j(l)}{\lambda_i(l)} \right|$$

$$\leq K_1 \sum_{k=n_2}^{n-1} \left| \frac{r_{j\nu}(k)}{\lambda_j(k)} \right| \leq \hat{K} \sum_{k=n_2}^{n-1} \left| \frac{r_{j\nu}(k)}{\lambda_j(k)} \right|. \tag{3.23}$$

Note that the change in the lower bound of the product from $k+1$ to k caused the change from $\lambda_i(k)$ to $\lambda_j(k)$ in the denominator of the perturbation terms.

Similarly, if $p_j = 0$, then $\left(W(n)PW^{-1}(k+1)\frac{R(k)}{\lambda_i(k)} \right)_{j\nu} = 0$ and

$$\sum_{k=n}^{\infty} \left| \left(W(n)[I-P]W^{-1}(k+1)\frac{R(k)}{\lambda_i(k)} \right)_{j\nu} \right|$$

$$= \sum_{k=n}^{\infty} \left| \frac{r_{j\nu}(k)}{\lambda_i(k)} \right| \prod_{l=n}^{k} \left| \frac{\lambda_i(l)}{\lambda_j(l)} \right|$$

$$= \sum_{k=n}^{\infty} \left| \frac{r_{j\nu}(k)}{\lambda_j(k)} \right| \prod_{l=n}^{k-1} \left| \frac{\lambda_i(l)}{\lambda_j(l)} \right|$$

$$\leq \frac{1}{K_2} \sum_{k=n}^{\infty} \left| \frac{r_{j\nu}(k)}{\lambda_j(k)} \right|$$

$$\leq \hat{K} \sum_{k=n}^{\infty} \left| \frac{r_{j\nu}(k)}{\lambda_j(k)} \right|. \tag{3.24}$$

With $\hat{\theta}$ defined in (3.21), it follows for $n \geq n_2$ that (recalling the choice of the matrix norm made above)

$$\left| (\hat{T}z)(n) \right| \leq \max_j \sum_{\nu=1}^{d} \|z\| \hat{K} \left[\sum_{k=n_2}^{n-1} \left| \frac{r_{j\nu}(k)}{\lambda_j(k)} \right| + \sum_{k=n}^{\infty} \left| \frac{r_{j\nu}(k)}{\lambda_j(k)} \right| \right]$$

$$\leq \|z\| \hat{K} \sum_{k=n_2}^{\infty} \left| \Lambda^{-1}(k)R(k) \right| = \hat{\theta} \|z\| < \|z\|.$$

Similar calculations show that for $z, \tilde{z} \in l_\infty$ it follows that

$$\left| (\hat{T}z)(n) - (\hat{T}\tilde{z})(n) \right| \leq \hat{\theta} \|z - \tilde{z}\| < \|z - \tilde{z}\|.$$

Thus $\hat{T}$ maps l_∞ into l_∞ and is a contraction. Hence

$$z(n) = e_i + (Tz)(n) \tag{3.25}$$

has a unique solution $z_i \in l_\infty$ (i.e., $n \geq n_2$) which is also a solution of (3.11). To show that $\lim_{n\to\infty}(Tz_i)(n) = 0$, let $\varepsilon > 0$ be given. Fix $n_3 > n_2$ such that

$$\hat{K}\|z_i\| \sum_{k=n_3}^{\infty} \left|\Lambda^{-1}(k)R(k)\right| < \frac{\varepsilon}{2}.$$

Then it follows for $n > n_3$ that

$$\begin{aligned}\left|(\hat{T}z_i)(n)\right| &\leq \|z_i\|\hat{K}\left[|W(n)P| \sum_{k=n_2}^{n_3-1} \left|W^{-1}(k+1)\frac{R(k)}{\lambda_i(k)}\right| + \sum_{k=n_3}^{\infty} \left|\Lambda^{-1}(n)R(n)\right|\right] \\ &\leq |W(n)P|\, \|z_i\|\hat{K} \sum_{k=n_2}^{n_3-1} \left|W^{-1}(k+1)\frac{R(k)}{\lambda_i(k)}\right| + \frac{\varepsilon}{2}.\end{aligned}$$

Since $W(n)P \to 0$ as $n \to \infty$ by (3.8), $\left|(\hat{T}z_i)(n)\right| < \varepsilon$ for all n sufficiently large ($n > n_3$) and therefore $(\hat{T}z_i)(n)$ vanishes at infinity. Repeating this process for all $1 \leq i \leq d$ and noting that this set of solutions is linearly independent for n sufficiently large by (3.25), gives the desired result (3.12). □

To illustrate how the weaker conditions of Theorem 3.8 may be applied, consider

$$\begin{aligned} y(n+1) &= \begin{pmatrix} \frac{1}{\sqrt{n}} & \frac{1}{n^2} \\ \frac{1}{n} & \sqrt{n} \end{pmatrix} y(n) = \left[\begin{pmatrix} \frac{1}{\sqrt{n}} & \\ & \sqrt{n} \end{pmatrix} + \begin{pmatrix} & \frac{1}{n^2} \\ \frac{1}{n} & \end{pmatrix}\right] y(n) \\ &=: [\Lambda(n) + R(n)]y(n). \end{aligned}$$

Since $R(n)/\lambda_1(n) \notin l^1$, Theorem 3.4 does not apply. However, $\Lambda^{-1}(n)R(n) \in l^1$ and hence Theorem 3.8 implies or n sufficiently large the existence of a fundamental matrix of the form

$$Y(n) = [I + o(1)] \begin{pmatrix} \prod^{n-1} \sqrt{k} & \\ & \prod^{n-1} \frac{1}{\sqrt{k}} \end{pmatrix} \qquad \text{as } n \to \infty.$$

In Theorem 3.8, condition (3.20), which is a row-wise condition on the perturbation $R(n)$, can be replaced by an equivalent condition on the columns of $R(n)$. This is formalized in

Corollary 3.9 *Let $\Lambda(n) = \operatorname{diag}\{\lambda_1(n), \ldots, \lambda_d(n)\}$ satisfy the hypotheses of Theorem 3.8. Let the $d \times d$ matrix $\tilde{R}(n)$ be defined for $n \geq n_0$ such that*

$$\sum_{n=n_0}^{\infty} \left|\tilde{R}(n)\Lambda^{-1}(n)\right| < \infty. \tag{3.26}$$

Then the linear difference system

$$\tilde{y}(n+1) = [\Lambda(n) + \tilde{R}(n)]\tilde{y}(n), \qquad n \geq n_0, \tag{3.27}$$

has for n sufficiently large a fundamental matrix satisfying

$$\tilde{Y}(n) = [I + o(1)] \prod^{n-1} \Lambda(k). \tag{3.28}$$

Proof For $n \geq n_0 + 1$, define

$$R(n) := \Lambda(n)\tilde{R}(n-1)\Lambda^{-1}(n-1). \tag{3.29}$$

Observe that (3.26) implies that (3.20) holds and Theorem 3.8 applies to (3.11) with $R(n)$ defined in (3.29). Hence $y(n+1) = [\Lambda(n) + R(n)]y(n)$ has or n sufficiently large a fundamental matrix of the form

$$Y(n) = F(n) \prod^{n-1} \Lambda(k), \qquad F(n) = I + o(1) \qquad \text{as } n \to \infty.$$

In (3.27), put

$$\tilde{y}(n) = G(n)u(n),$$

where

$$\begin{aligned} G(n) &:= [I + \Lambda^{-1}(n)R(n)]F(n) \\ &= [I + \tilde{R}(n-1)\Lambda^{-1}(n-1)]F(n), \end{aligned}$$

and note that $G(n)$ is nonsingular for large n by (3.26) and the invertibility of $I + F(n)$, and that $G(n) = I + o(1)$. It follows from (3.27) for n sufficiently large that

$$\begin{aligned} u(n+1) &= G^{-1}(n+1)[\Lambda(n) + \tilde{R}(n)]G(n)u(n) \\ &= F^{-1}(n+1)[I + \Lambda^{-1}(n+1)R(n+1)]^{-1} \\ &\quad \times \left[\Lambda(n) + \Lambda^{-1}(n+1)R(n+1)\Lambda(n)\right]\left[I + \Lambda^{-1}(n)R(n)\right]F(n)u(n) \\ &= F^{-1}(n+1)\left[\Lambda(n) + R(n)\right]F(n)u(n) = \Lambda(n)u(n). \end{aligned}$$

It is clear that this system has for large n a fundamental matrix $U(n) = \prod^{n-1} \Lambda(k)$ and hence (3.27) has a fundamental matrix $\tilde{Y}(n) = G(n)\prod^{n-1} \Lambda(k)$, which establishes (3.28). □

In his original proof, Kooman proved his "column-wise" improvement of Theorem 3.4 using an induction based on successive partial diagonalization of the system. For his argument, the column-wise assumption $\tilde{R}(n)\Lambda^{-1}(n) \in l^1$ is more appropriate whereas with the T-operator proof given here, the row-wise assumption $\Lambda^{-1}(n)R(n) \in l^1$ is natural. To show that each of these results implies the other, Kooman used a clever algebraic argument that we repeated in the proof of Corollary 3.9.

We end by noting that Kooman also gave error estimates [98, Thm. 2.1]. In Sect. 3.7, we will derive some related results by estimating the T-operator used in the proof of Theorem 3.8.

As a final remark, we mention that Bodine and Sacker [21] derived sufficient conditions for the diagonalizability and asymptotic diagonalizability of difference systems of the form $x(n+1) = [A(n)+R(n)]x(n)$, where the $|\det A(n)| \geq \delta > 0$ and where $R(n) = o(1)$ as $n \to \infty$. Their approach was based on the theory of linear skew-product flows. They assumed that the omega-limit set of A in an appropriate function space is compact and that it satisfies the so-called full spectrum property.

3.3 Sharpness of Dichotomy and Growth Conditions

To illustrate the generic *necessity* of the *dichotomy condition* (with respect to l_1-perturbations), we now give a discrete version of Eastham's example for differential equations from Sect. 2.3. For that purpose, consider the following example of a triangular difference system of the form (3.11):

$$y(n+1) = \left[\begin{pmatrix} 1 & \\ & \frac{\rho(n+1)}{\rho(n)} \end{pmatrix} + \begin{pmatrix} 0 & r(n) \\ 0 & 0 \end{pmatrix}\right] y(n), \qquad n \geq 0, \tag{3.30}$$

where

$$\rho(n) = n^2\left[1 + \sin\left(\frac{n\pi}{2}\right)\right] + 1,$$

and

$$r(n) = \frac{\tilde{r}(n)}{\rho(n)}, \qquad \text{where } \tilde{r}(n) = \begin{cases} 0 & n = 2 \bmod 4 \text{ and } n = 3 \bmod 4 \\ 1 & n = 0 \bmod 4 \text{ and } n = 1 \bmod 4. \end{cases}$$

Then a short computation shows that the growth condition (3.10) is satisfied, but the dichotomy conditions (3.8), (3.9) do not hold since the partial products

$$\prod_{k=n_1}^{n_2} \left| \frac{\lambda_1(k)}{\lambda_2(k)} \right| = \frac{\rho(n_1)}{\rho(n_2+1)}$$

are neither bounded from above nor bounded away from zero for all $1 \leq n_1 \leq n_2$.

A fundamental matrix of (3.30) is given by

$$\hat{Y}(n) = \begin{bmatrix} 1 & \sum_{k=0}^{n-1} \rho(k) r(k) \\ 0 & \rho(n) \end{bmatrix}$$

and hence any fundamental matrix of (3.11) is of the form

$$Y(n) = \hat{Y}(n) \begin{bmatrix} \alpha & \beta \\ \gamma & \delta \end{bmatrix}, \tag{3.31}$$

where $\alpha, \beta, \gamma, \delta$ are constants with $\alpha\delta - \beta\gamma \neq 0$. Note that

$$\limsup_{n\to\infty} \rho(n) = +\infty,$$

and

$$\sum_{k=0}^{n-1} \rho(k) r(k) = \sum_{k=1}^{n-1} \tilde{r}(k) \to \infty \qquad \text{as } n \to \infty.$$

It is easy to see that for $Y(n)$ given in (3.31) to be of the form

$$[I + \mathrm{o}(1)] \operatorname{diag}\{1, \rho(n)\},$$

it is necessary that $\alpha = \delta = 1$, $\gamma = 0$, and that

$$\beta + \sum_{k=0}^{n-1} \tilde{r}(k) = \mathrm{o}(\rho(n)) \qquad \text{as } n \to \infty.$$

However,

$$\limsup_{n\to\infty} \frac{\beta + \sum_{k=0}^{n-1} \tilde{r}(k)}{\rho(n)} = +\infty,$$

and therefore (3.30) does not possess a fundamental matrix of the form (3.12).

Another somewhat intricate and less explicit example demonstrating the necessity of a dichotomy condition can be found in [9, Section 4].

That a *growth condition* related to l^1 perturbations is necessary for a Levinson-type result can be easily seen by considering the perturbed system

$$y(n+1) = \begin{pmatrix} 1 & n^{-1} \\ 0 & 1 \end{pmatrix} y(n) \qquad n \geq 1,$$

whose diagonal part satisfies Levinson's discrete dichotomy conditions (3.8), (3.9), but whose fundamental matrices

$$\begin{pmatrix} 1 & \sum_{k=1}^{n-1} \frac{1}{k} \\ 0 & 1 \end{pmatrix} C \qquad \det C \neq 0,$$

are not of the form (3.12). Details are left to the reader.

3.4 Weak Dichotomies

In this section, we are concerned about linear systems where the diagonal part does not satisfy Levinson's discrete dichotomy conditions (3.8) and (3.9). As with the case of differential equations (see Sect. 2.4), it is possible to obtain a Levinson-type result provided that the growth condition of the perturbation is strengthened accordingly. These results were published in [18]. We begin with a modification of Theorem 3.2.

Theorem 3.10 *For $d \times d$ matrices $A(n)$ and $R(n)$, consider the unperturbed system* (3.1) *and the perturbed system* (3.4)*, where $A(n)$ are nonsingular for all $n \geq n_0$. Let $X(n)$ be the fundamental matrix of* (3.1) *such that $X(n_0) = I$. Suppose that there exists a projection matrix P and a scalar-valued sequence $\beta(n) \geq 1$ such that*

$$\left.\begin{array}{lll} & |X(n)PX^{-1}(k+1)| \leq \beta(k) & \text{for all} \quad n_0 \leq k < n \\ \text{and} & |X(n)[I-P]X^{-1}(k+1)| \leq \beta(k) & \text{for all} \quad n_0 \leq n \leq k \end{array}\right\}. \tag{3.32}$$

Suppose that

$$\sum_{n=n_0}^{\infty} \beta(n)|R(n)| < \infty. \tag{3.33}$$

Then there exists for n sufficiently large a one-to-one and bicontinuous correspondence between the bounded solutions of (3.1) *and* (3.4)*. Moreover, the difference between corresponding bounded solutions of* (3.1) *and* (3.4) *tends to zero as $n \to \infty$ if $X(n)P \to 0$ as $n \to \infty$.*

Proof The proof is analogous to the proof of Theorem 2.10, and we will just outline the necessary modifications. Let l_∞ denote again the Banach space of bounded d-dimensional vector-valued sequences with the supremum norm

$$\|y\| = \sup_{n \geq n_1} |y(n)|,$$

where n_1 is chosen sufficiently large such that

$$\theta = \sum_{n=n_1}^{\infty} \beta(n)|R(n)| < 1.$$

For $n \geq n_1$, consider the operator T on l_∞ defined in (3.6). As in the proof of Theorem 2.10, one can show that T maps l_∞ into l_∞ and is a contraction.

Hence, for every function $x \in l_\infty$, the operator equation (3.7) has a unique solution $y \in l_\infty$. One can check that if x is a solution of (3.1), then $y = x + Ty$ is a solution of (3.4). Therefore, if x is a bounded solution of (3.1), then the solution y of (3.7) is a bounded solution of (3.4). Conversely, if y is a bounded solution of (3.4), the function x defined by (3.7) is a bounded solution of (3.1).

Equation (3.7) establishes therefore for n sufficiently large a one-to-one correspondence between the bounded solutions of (3.1) and (3.4) on $[n_1, \infty)$. The bicontinuity of this correspondence as well as that the difference between corresponding solutions tends to zero if $X(n)P \to 0$ as $n \to \infty$ can be shown as in Theorem 2.2. □

As in the case of differential equations, the existence of such a sequence $\beta(k)$ is easily shown. For example, corresponding to the projection matrix $P = 0$, one could choose

$$\beta(k) := \sup_{n_0 \leq n \leq k} |X(n)X^{-1}(k+1)|.$$

But this is generally far from optimal since it does not take into account any fine structure of $X(n)$.

We continue with a discrete version of Theorem 2.12.

Theorem 3.11 *Let $\Lambda(n) = \operatorname{diag}\{\lambda_1(n), \ldots, \lambda_d(n)\}$ be a diagonal and invertible $d \times d$ matrix for $n \geq n_0$. Fix $h \in \{1, \ldots, d\}$. Assume that there exists a scalar sequence $\beta_h(n) \geq 1$ for $n \geq n_0$ such that for each $1 \leq i \leq d$*

either

$$\left.\begin{array}{ll} \displaystyle\prod_{k=n_0}^{n} \left|\frac{\lambda_i(k)}{\lambda_h(k)}\right| \to 0 & \text{as } n \to \infty \\ \text{and} \quad \displaystyle\prod_{k=n_1+1}^{n_2-1} \left|\frac{\lambda_i(k)}{\lambda_h(k)}\right| \le \beta_h(n_1) & \forall \; n_0 \le n_1 < n_2 \end{array}\right\}, \tag{3.34}$$

or

$$\prod_{k=n_1}^{n_2} \left|\frac{\lambda_i(k)}{\lambda_h(k)}\right| \ge \frac{1}{\beta_h(n_2)} > 0 \qquad \forall \;\; n_0 \le n_1 \le n_2. \tag{3.35}$$

Furthermore, assume that $R(n)$ is a $d \times d$ matrix for $n \ge n_0$ such that

$$\sum_{n=n_0}^{\infty} \frac{\beta_h(n)|R(n)|}{|\lambda_h(n)|} < \infty \quad \text{for this fixed value of } h. \tag{3.36}$$

Then the linear difference system (3.11) has for n sufficiently large a solution satisfying

$$y_h(n) = [e_h + o(1)] \prod_{k=n_0}^{n-1} \lambda_h(k) \qquad \text{as} \quad n \to \infty, \tag{3.37}$$

where e_h is the hth column of the identity matrix.

Proof For fixed $h \in \{1, \dots, d\}$, put

$$y(n) = z(n) \prod_{k=n_0}^{n-1} \lambda_h(k). \tag{3.38}$$

Then (3.11) implies that

$$z(n+1) = \frac{1}{\lambda_h(n)} [\Lambda(n) + R(n)]\, z(n). \tag{3.39}$$

We also consider the unperturbed shifted system

$$w(n+1) = \frac{\Lambda(n)}{\lambda_h(n)} w(n), \tag{3.40}$$

and we will apply Theorem 3.10 to these shifted systems. Let $P = \text{diag}\{p_1, \dots, p_d\}$,

where

$$p_j = \begin{cases} 1 & \text{if } (i,h) \text{ satisfies (3.34)} \\ 0 & \text{if } (i,h) \text{ satisfies (3.35).} \end{cases}$$

Then (3.40) satisfies the dichotomy condition (3.32) with $\beta(k)$ replaced by $\beta_h(k)$, and $W(n)P \to 0$ as $n \to \infty$. Now, by Theorem 3.10 and since $w_h(n) = e_h$ is a bounded solution of (3.40), there exists for n sufficiently large a bounded solution $z_h(n)$ of (3.39) with

$$z_h(n) = w_h + o(1) \qquad \text{as } n \to \infty.$$

By (3.38), (3.11) has a solution of the form (3.37). □

In Theorem 3.11, such a sequence $\beta_h(n)$ always exists. For example, for this fixed value of h, one may put

$$\beta_h(n) = \sup_{1\le i\le d} \sup_{n_0\le k\le n} \prod_{l=k}^{n} \left| \frac{\lambda_h(l)}{\lambda_i(l)} \right|,$$

then (3.35) holds for all $1 \le i \le d$. As for differential equations, this choice usually leads to an unnecessarily large and non-optimal $\beta(n)$. However, it shows that any perturbed diagonal system has Levinson-type asymptotic behavior if the perturbation is "sufficiently small," i.e., it satisfies (3.36).

Returning to example (3.30) (which demonstrated in Sect. 3.3 the necessity of a dichotomy condition), we want to find a bound for the perturbation

$$r(n) = \hat{r}(n)/\rho(n),$$

with $\hat{r}(n)$ to be determined such that (3.30) has two linearly independent solutions of the form (3.37) for $h = 1, 2$. Note that, because of the triangular structure of the perturbation $R(n)$, we only have to be concerned with the case $h = 2$. Since

$$\prod_{k=0}^{n} \left| \frac{\lambda_1(k)}{\lambda_2(k)} \right| \ne o(1) \qquad \text{as } n \to \infty,$$

we need to find a sequence $\beta_2(n)$ such that (3.35) holds for $i = 1$ and $h = 2$. Note that for $0 \le n_1 \le n_2$

$$\prod_{k=n_1}^{n_2} \frac{\lambda_1(k)}{\lambda_2(k)} = \frac{\rho(n_1)}{\rho(n_2+1)} \ge \frac{1}{\rho(n_2+1)}.$$

Hence, for (3.35) to hold, it suffices to choose

$$\beta_2(n) = \rho(n+1) \qquad \forall\, n \geq 0. \tag{3.41}$$

For the growth condition (3.36) to be satisfied with $h = 2$, it must happen that

$$\sum_{n=0}^{\infty} \frac{\beta_2(n)|R(n)|}{|\lambda_2(n)|} = \sum_{n=0}^{\infty} \rho(n+1) \left|\frac{\hat{r}(n)}{\rho(n)}\right| \left|\frac{\rho(n)}{\rho(n+1)}\right| = \sum_{n=0}^{\infty} |\hat{r}(n)| < \infty,$$

i.e., $\hat{r} \in l^1[0,\infty)$. A straightforward computation then shows that (3.30) has a fundamental matrix

$$Y(n) = \begin{bmatrix} 1 & -\sum_{k=n}^{\infty} \hat{r}(k) \\ 0 & \rho(n) \end{bmatrix} = [I + o(1)] \begin{bmatrix} 1 & \\ & \rho(n) \end{bmatrix} \qquad \text{as } n \to \infty.$$

3.5 Perturbations of Systems with a "Column-Wise Exponential Dichotomy"

In this section, we want to discuss systems of the form

$$y(n+1) = [A(n) + R(n)]\, y, \qquad n \geq n_0,$$

where $A(n)$ is a $d \times d$ matrix having the block structure

$$A(n) = \begin{pmatrix} A_{-1}(n) & & \\ & \lambda_j(n) & \\ & & A_1(n) \end{pmatrix}, \tag{3.42}$$

satisfying some "exponential dichotomy" condition, and $R(n)$ is small in a sense to be made precise. We will always assume that $\lambda_j(n) \neq 0$ and $A_1(n)$ is invertible for all $n \geq n_0$.

In the case of differential systems, we were in Sect. 2.5 concerned with perturbations $R(t)$ that were weaker than $R(t)$ vanishing at infinity, but satisfied (2.56). We note that such a condition for perturbation sequences

$$\sup_{n_2 \geq n_1+1} \frac{1}{n_2 - n_1 + 1} \sum_{k=n_1}^{n_2} |R(k)| \to 0 \qquad \text{as } n_1 \to \infty,$$

with $n_0 \leq n_1 \leq n_2$ is equivalent to $R(n) \to 0$ as $n \to \infty$. In what follows, we will first be concerned with such perturbations $R(n)$ vanishing at infinity and then consider the case that $R(n)$ is in a certain l^p-class.

We deviate from the approach taken in Sect. 2.5. We will not give here a discrete version of an extended Coppel result for systems with a so-called exponential dichotomy and o(1)-perturbations in the spirit of Theorem 2.16. The main reason is that the usual definition of an exponential dichotomy involves a fundamental matrix of the unperturbed system. This is not given in this section since we make no assumptions on $A(n)$ being invertible. However, we refer the interested reader to [56, Thm. 4.1] for a version of such a result in the context of o(1) perturbations of nonsingular constant matrices possessing an exponential dichotomy.

Before we state the main result, we introduce the concept of a so-called state transition matrix as a convenient tool to address the fact that $A_{-1}(n)$ given in (3.42) might be singular. With $A_{\pm 1}(n)$ and $\lambda_j(n)$ defined in (3.42), we put

$$\hat{A}(n) = \begin{pmatrix} \frac{A_{-1}(n)}{\lambda_j(n)} & \\ & \frac{A_1(n)}{\lambda_j(n)} \end{pmatrix}. \tag{3.43}$$

We define the *state transition matrix* $\hat{\Phi}(n,m)$ of $x(n+1) = \hat{A}(n)x(n)$ by

$$\hat{\Phi}(n,k) = \begin{cases} \hat{A}(n-1)\hat{A}(n-2) \cdots \hat{A}(k) & \text{if } n > k \\ I & \text{if } n = k. \end{cases} \tag{3.44}$$

Thus $\hat{\Phi}(n,k)$ is in general not invertible and exists only for $n \geq k$. Let P be the $d-1 \times d-1$ matrix defined by

$$P = \begin{pmatrix} I_{j-1} & \\ & 0_{d-j} \end{pmatrix}. \tag{3.45}$$

Then since $A_1(n)$ is invertible, $\hat{\Phi}(n,k)\Big|_{\mathscr{R}(I-P)}$, the restriction of $\hat{\Phi}(n,k)$ to the range of $(I-P)$, is invertible for each $n \geq k$, and we define $\hat{\Phi}(k,n)\Big|_{\mathscr{R}(I-P)}$ as the inverse map of $\hat{\Phi}(n,k)\Big|_{\mathscr{R}(I-P)}$ from $\mathscr{R}(I-P)$ to $\mathscr{R}(I-P)$.

Theorem 3.12 *Consider*

$$y(n+1) = [A(n) + R(n)]y(n), \qquad n \geq n_0. \tag{3.46}$$

We assume that $A(n)$ is of the form (3.42), *where $A_{-1}(n)$ is a $(j-1)\times(j-1)$ matrix, $A_1(n)$ is a $(d-j)\times(d-j)$ invertible matrix, and $|\lambda_j(n)| \neq 0$ for all $n \geq n_0$. Let $\hat{A}(n)$, $\hat{\Phi}(n,r)$, and P be defined in* (3.43), (3.44), *and* (3.45), *respectively. Assume* *that there exist constants $K > 0$ and $q \in (0, 1)$ such that*

$$\left| \hat{\Phi}(n, k+1)\Big|_{\mathscr{R}(P)} \right| = \left| \prod_{l=k+1}^{n-1} \frac{A_{-1}(l)}{\lambda_j(l)} \right| \leq Kq^{n-k}, \qquad n_0 \leq k < n \tag{3.47}$$

$$\left| \hat{\Phi}(n, k+1)\Big|_{\mathscr{R}(I-P)} \right| = \left| \left(\prod_{l=n}^{k} \frac{A_1(l)}{\lambda_j(l)} \right)^{-1} \right| \leq Kq^{k-n}, \qquad n_0 \leq n \leq k. \tag{3.48}$$

Suppose that $R(n)$ satisfies

$$\lim_{n\to\infty} \frac{R(n)}{\lambda_j(n)} = 0. \tag{3.49}$$

Then there exists for n sufficiently large a solution $y_j(n)$ of (3.46) *satisfying*

$$y_j(n) = [e_j + s_j(n)] \prod^{n-1} [\lambda_j(k) + l_j(k)], \tag{3.50}$$

where

$$l_j(k) = r_{jj}(k) + \mathrm{o}\left(\sum_{\nu \neq j} |r_{j\nu}(k)| \right), \tag{3.51}$$

and

$$s_j(n) = \{s_{1j}(n), \ldots, s_{dj}(n)\}^T = \mathrm{o}(1) \text{ as } n \to \infty, \quad s_{jj}(n) \equiv 0. \tag{3.52}$$

Proof We note that (3.46) has a solution of the form (3.50) if and only if

$$\lambda_j(n)s_j(n+1) - A(n)s_j(n) = -l_j(n)[e_j + s_j(n+1)] + r_j(n) + R(n)s_j(n), \tag{3.53}$$

where $r_j(n)$ is the jth column of R. In this system of d equations for $d+1$ unknowns, there is a lack of uniqueness due to the fact that in (3.50), the diagonal term $s_{jj}(n)$ can be modified by adjusting $l_j(k)$ by a suitable $\mathrm{o}(1)$ perturbation. To remove this ambiguity in the representation, we choose to normalize $s_j(n)$ so that $s_{jj}(n) \equiv 0$. Then we obtain for the elements in the jth position of (3.53),

$$l_j(n) = r_{jj}(n) + \sum_{\nu=1, \nu\neq j}^{d} r_{j\nu}(n)s_{\nu j}(n). \tag{3.54}$$

Substituting this into (3.53), we obtain a (nonlinear) system of $d-1$ equations for $d-1$ unknowns, the remaining components of the vector $s_j(n)$.

In order to express this system in vector form, we use the following notation:

$$\sigma(n) = \left[s_{1j}(n), \ldots, s_{j-1,j}(n), s_{j+1,j}(n), \ldots, s_{dj}(n)\right]^T, \tag{3.55}$$
$$\hat{r}_j(n) = \left[r_{1j}(n), \ldots, r_{j-1,j}(n), r_{j+1,j}(n), \ldots, r_{dj}(n)\right]^T,$$
$$\hat{\rho}_j(n) = \left[r_{j1}(n), \ldots, r_{j,j-1}(n), r_{j,j+1}(n), \ldots, r_{jd}(n)\right].$$

Then (3.54) can be expressed as

$$l_j(n) = r_{jj}(n) + \hat{\rho}_j(n)\sigma(n)$$

and substituting this into the remaining $(d-1)$ equations (3.53) and dividing by $\lambda_j(n)$, we obtain the system

$$\sigma(n+1) - \hat{A}(n)\sigma(n) = V(n, \sigma(n), \sigma(n+1)), \tag{3.56}$$

where $\hat{A}(n)$ was defined in (3.43) and

$$V(n, \sigma(n), \sigma(n+1))$$
$$= \frac{1}{\lambda_j(n)}\left[\hat{r}_j(n) + \hat{R}(n)\sigma(n) - \left\{r_{jj}(n) + \hat{\rho}_j(n)\sigma(n)\right\}\sigma(n+1)\right]. \tag{3.57}$$

Here $\hat{R}(n)$ is the $(d-1) \times (d-1)$-dimensional matrix obtained by deleting the jth row and jth column from $R(n)$.

Recalling (3.49), we now fix n_1 sufficiently large ($n_1 \geq n_0$) such that

$$K\frac{q+1}{1-q}\left|\frac{R(n)}{\lambda_j(n)}\right| \leq \frac{1}{8}, \qquad \text{for all } n \geq n_1. \tag{3.58}$$

To show that (3.56) has a solution $\sigma(n) = \mathrm{o}(1)$ as $n \to \infty$, we define for $n \geq n_1$ the operator

$$(T\sigma)(n) = \sum_{k=n_1}^{n-1} \hat{\Phi}(n, k+1)\Big|_{\mathscr{R}(P)} V(k, \sigma(k), \sigma(k+1))$$
$$- \sum_{k=n}^{\infty} \hat{\Phi}(n, k+1)\Big|_{\mathscr{R}(I-P)} V(k, \sigma(k), \sigma(k+1)). \tag{3.59}$$

Let $\mathscr{B}$ be the Banach space of $d-1$-dimensional vector sequences $\sigma(n)$, $n \geq n_1$, which vanish at infinity, equipped with the norm $\|\sigma\| = \sup_{n \geq n_1} |\sigma(n)|$. Here $|\cdot|$ denotes the maximum vector norm, i.e., $|\sigma(n)| = \max_{1 \leq i \leq d-1} |\sigma_i(n)|$. The induced matrix norm for $m \times n$ matrices A is therefore $|A| = \max_{1 \leq i \leq m} \sum_{j=1}^{n} |a_{ij}|$. Since $\hat{R}(k)$ is a principal submatrix of $R(k)$, it follows that $|\hat{R}(k)| \leq |R(k)|$. Let $\mathscr{W} = \{w \in$

$\mathscr{B} : \|w\| \leq 1\}$. We will show that T maps $\mathscr{W}$ into itself and is a contraction, and it will then be easy to see that the fixed point of $(T\sigma)(n) = \sigma(n)$ in $\mathscr{W}$ is a solution of (3.56). Next, since $\sigma(n)$ vanishes at infinity, $l_j(n)$ defined by (3.54) satisfies (3.51), establishing for large n the existence of a solution $y_j(n)$ of (3.46) satisfying (3.50).

For $\sigma \in \mathscr{W}$, (3.57) leads to the estimate

$$
\begin{aligned}
|V(n,\sigma(n),\sigma(n+1))| &\leq \frac{1}{|\lambda_j(n)|}\left[|\hat{r}_j(n) + \hat{R}(n)\sigma(n)| + |r_{jj}(n) + \hat{\rho}_j(n)\sigma(n)|\right] \\
&\leq \max_{\substack{1 \leq i \leq d \\ i \neq j}} \left\{\frac{|r_{ij}(n)|}{|\lambda_j(n)|} + \sum_{\nu \neq j} \frac{|r_{i\nu}(n)|}{|\lambda_j(n)|}\right\} + \sum_{l=1}^{d} \frac{|r_{jl}(n)|}{|\lambda_j(n)|} \\
&\leq 2\frac{|R(n)|}{|\lambda_j(n)|}.
\end{aligned} \tag{3.60}
$$

Then it follows from (3.59), (3.47), (3.48), and (3.60), that for $n \geq n_1$

$$
\begin{aligned}
|(T\sigma)(n)| &\leq K\sum_{k=n_1}^{n-1} q^{n-k}|V(k,\sigma(k),\sigma(k+1))| \\
&\quad + K\sum_{k=n}^{\infty} q^{k-n}|V(k,\sigma(k),\sigma(k+1))| \\
&\leq 2K\left(\sum_{k=n_1}^{n-1} q^{n-k}\left|\frac{R(k)}{\lambda_j(k)}\right| + \sum_{k=n}^{\infty} q^{k-n}\left|\frac{R(k)}{\lambda_j(k)}\right|\right) \qquad (3.61) \\
&\leq \frac{1}{4}\left(\frac{1-q}{1+q}\right)\left(\sum_{k=n_1}^{n-1} q^{n-k} + \sum_{k=n}^{\infty} q^{k-n}\right) \leq \frac{1}{4}. \qquad (3.62)
\end{aligned}
$$

Also, for $n_1 < n_2 < n$,

$$
\begin{aligned}
|(T\sigma)(n)| &\leq 2K\left(\sum_{k=n_1}^{n_2-1} q^{n-k}\left|\frac{R(k)}{\lambda_j(k)}\right| + \sum_{k=n_2}^{n-1} q^{n-k}\left|\frac{R(k)}{\lambda_j(k)}\right| + \sum_{k=n}^{\infty} q^{k-n}\left|\frac{R(k)}{\lambda_j(k)}\right|\right) \\
&\leq 2K\left(\left\{\sup_{k\geq n_1}\left|\frac{R(k)}{\lambda_j(k)}\right|\right\}\sum_{k=n_1}^{n_2-1} q^{n-k} + \frac{2}{1-q}\sup_{k\geq n_2}\left|\frac{R(k)}{\lambda_j(k)}\right|\right).
\end{aligned}
$$

Given $\varepsilon > 0$, fix n_2 sufficiently large such that $4K\sup_{k\geq n_2}\left|\frac{R(k)}{\lambda_j(k)}\right|/(1-q) < \varepsilon/2$. By making n sufficiently large ($n > n_2$), $2K\left\{\sup_{k\geq n_1}\left|\frac{R(k)}{\lambda_j(k)}\right|\right\}\sum_{k=n_1}^{n_2-1} q^{n-k} < \varepsilon/2$. Hence $(T\sigma)(n)$ vanishes at infinity and we have shown that T maps $\mathscr{W}$ into $\mathscr{W}$.

To show that T is a contraction mapping, we first observe that from (3.57) follows that

$$\begin{aligned}
&V(k,\sigma(k),\sigma(k+1)) - V(k,\tilde{\sigma}(k),\tilde{\sigma}(k+1)) \\
&\quad = \frac{\hat{R}(k)[\sigma(k)-\tilde{\sigma}(k)] - r_{jj}(k)[\sigma(k+1)-\hat{\sigma}(k+1)]}{\lambda_j(k)} \\
&\quad\quad - \frac{\hat{\rho}_j(k)\,[\sigma(k)\sigma(k+1)-\tilde{\sigma}(k)\tilde{\sigma}(k+1)]}{\lambda_j(k)},
\end{aligned}$$

hence

$$\begin{aligned}
&|V(k,\sigma(k),\sigma(k+1)) - V(k,\tilde{\sigma}(k),\tilde{\sigma}(k+1))| \\
&\quad \le \frac{\left(|\hat{R}(k)| + |r_{jj}(k)|\right)\,\|\sigma-\tilde{\sigma}\|}{|\lambda_j(k)|} \\
&\quad\quad + \frac{\left|\hat{\rho}_j(k)\,\{\sigma(k)\,[\sigma(k+1)-\tilde{\sigma}(k+1)] - [\tilde{\sigma}(k)-\sigma(k)]\tilde{\sigma}(k+1)\}\right|}{|\lambda_j(k)|} \\
&\quad \le \frac{4|R(k)|\,\|\sigma-\tilde{\sigma}\|}{|\lambda_j(k)|}.
\end{aligned} \tag{3.63}$$

Therefore it follows from (3.59), (3.62), and (3.63) for $n \ge n_1$ that

$$\begin{aligned}
&|(T\sigma)(n) - (T\tilde{\sigma})(n)| \\
&\quad \le K\sum_{k=n_1}^{n-1} q^{n-k}|V(k,\sigma(k),\sigma(k+1)) - V(k,\tilde{\sigma}(k),\tilde{\sigma}(k+1))| \\
&\quad\quad + K\sum_{k=n}^{\infty} q^{k-n}|V(k,\sigma(k),\sigma(k+1)) - V(k,\tilde{\sigma}(k),\tilde{\sigma}(k+1))| \\
&\quad \le 4K\,\|\sigma-\tilde{\sigma}\|\left(\sum_{k=n_1}^{n-1} q^{n-k}\left|\frac{R(k)}{\lambda_j(k)}\right| + \sum_{k=n}^{\infty} q^{k-n}\left|\frac{R(k)}{\lambda_j(k)}\right|\right) \\
&\quad \le \frac{1}{2}\|\sigma-\tilde{\sigma}\|.
\end{aligned}$$

Hence,

$$\|T(\sigma-\tilde{\sigma})\| \le \frac{1}{2}\|\sigma-\tilde{\sigma}\|.$$

By the contraction mapping principle, there exists a unique fixed point $\sigma = T\sigma$ in $\mathscr{W}$. It is straightforward to show that σ satisfies (3.56). By the definition of $\mathscr{W}$,

$\sigma(n)$ vanishes at infinity. Recalling that $s_{jj}(n) \equiv 0$, this implies the existence of a unique d-dimensional sequence $s_j(n)$, $s_j(n) = o(1)$ as $n \to \infty$, satisfying (3.50). Hence, (3.51) follows directly from (3.54), which concludes the proof. □

Asymptotic representations of the form (3.50), while they can provide some useful growth information about solutions, are insufficient for other applications requiring finer estimates. For example, when $\lambda_j = 1$, the products $\prod^{n-1}\left[1 + \frac{a}{k^p}\right]$ for $0 < p \leq 1$ are not even bounded for $a > 0$. In order to derive more precise representations, it is generally necessary to have more detailed knowledge of $R(n)$ other than just (3.49). In what follows, we are concerned with perturbations $R(n) \in l^2$, which leads to a discrete version of a classical result (for differential equations) of Hartmann–Wintner. A first version of such a result was established by Benzaid/Lutz [9]. In 1997, M. Pituk [129] gave an interesting extension of this result for the case of a constant block-diagonal system A (see Theorem 3.20). Here we generalize this further to time-dependent block-diagonal matrices $A(n)$.

Theorem 3.13 *Assume that* $A(n)$ *and* $\hat{A}(n)$ *satisfy the assumptions of Theorem 3.12. Assume that* $R(n)$ *satisfies*

$$\frac{R(n)}{\lambda_j(n)} \in l^2[n_0, \infty). \tag{3.64}$$

Then there exists for n sufficiently large a solution $y_j(n)$ *of* (3.46) *satisfying*

$$y_j(n) = [e_j + o(1)] \prod^{n-1} [\lambda_j(k) + r_{jj}(k)], \qquad \text{as } n \to \infty. \tag{3.65}$$

Proof Since (3.64) implies that (3.49) holds, Theorem 3.12 implies the existence of a solution $y_j(n)$ satisfying (3.50) with $s_j(n) = o(1)$ as $n \to \infty$ and $l_j(n)$ given in (3.51). We will now use the l^2-nature of the perturbation to make the asymptotic behavior more precise by deriving sharper estimates of the T-operator defined in (3.59). Recalling that $s_{jj}(n) \equiv 0$ and the definition of $\sigma(n)$ in (3.55), we will derive more precise information about $s_j(n)$ by investigating $\sigma(n)$. Note that $\sigma(n)$ was the unique fixed point $\sigma(n) = (T\sigma)(n) = o(1)$, with $(T\sigma)(n)$ given in (3.59). We will use the estimate of the T-operator given in (3.61) as a starting point to establish that $\sigma(n) \in l^2$, and then apply this result to derive more precise information about $l_j(n)$.

For this purpose, we give the following argument (which was already given in [9]). Using the Cauchy-Schwarz inequality one finds that with n_1 defined in (3.58)

$$\begin{aligned}
\sum_{n=n_1}^{\infty} \left| \sum_{k=n_1}^{n-1} q^{n-k} \left| \frac{R(k)}{\lambda_j(k)} \right| \right|^2 &= \sum_{n=n_1}^{\infty} \left(\sum_{k=n_1}^{n-1} \left| \frac{R(k)}{\lambda_j(k)} \right| q^{\frac{n-k}{2}} q^{\frac{n-k}{2}} \right)^2 \\
&\leq \sum_{n=n_1}^{\infty} \left(\sum_{k=n_1}^{n-1} \left| \frac{R(k)}{\lambda_j(k)} \right|^2 q^{n-k} \right) \left(\sum_{k=n_1}^{n-1} q^{n-k} \right)
\end{aligned}$$

$$\leq \frac{1}{1-q} \sum_{n=n_1}^{\infty} \sum_{k=n_1}^{n-1} \left| \frac{R(k)}{\lambda_j(k)} \right|^2 q^{n-k}$$

$$= \frac{1}{1-q} \sum_{k=n_1}^{\infty} \left| \frac{R(k)}{\lambda_j(k)} \right|^2 \sum_{n=k+1}^{\infty} q^{n-k}$$

$$\leq \frac{1}{(1-q)^2} \sum_{k=n_1}^{\infty} \left| \frac{R(k)}{\lambda_j(k)} \right|^2 < \infty, \tag{3.66}$$

where Fubini's theorem was used to change the order of the sums. A similar argument shows that

$$\sum_{n=n_1}^{\infty} \left| \sum_{k=n}^{\infty} q^{k+1-n} \left| \frac{R(k)}{\lambda_j(k)} \right| \right|^2 \leq \frac{1}{(1-q)^2} \sum_{k=n_1}^{\infty} \left| \frac{R(k)}{\lambda_j(k)} \right|^2 < \infty. \tag{3.67}$$

Thus (3.61), (3.66), and (3.67) together with Minkowski's inequality show that

$$\sqrt{\sum_{n=n_1}^{\infty} |\sigma(n)|^2} = \sqrt{\sum_{n=n_1}^{\infty} |(T\sigma)(n)|^2} \leq 8K \frac{1}{1-q} \sqrt{\sum_{n=n_1}^{\infty} \left| \frac{R(n)}{\lambda_j(n)} \right|^2},$$

i.e., $\sigma(n) \in l^2$. Recalling that $s_{jj}(n) \equiv 0$, this implies the existence of a unique d-dimensional sequence $\{s_j(n)\} \in l^2$ satisfying (3.50). Observe that

$$\prod^{n-1} [\lambda_j(k) + l_j(k)] = \prod^{n-1} \left[\lambda_j(k) + r_{jj}(k) + \sum_{\nu \neq j} r_{j\nu}(n) s_{\nu j}(n) \right]$$

$$= \prod^{n-1} [\lambda_j(k) + r_{jj}(k)] \prod^{n-1} \left[1 + \sum_{\nu \neq j} \frac{r_{j\nu}(n) s_{\nu j}(n)}{\lambda_j(k) + r_{jj}(k)} \right].$$

Since $R(n)/\lambda_j(n) \to 0$ as $n \to \infty$ by (3.64), it follows that $|r_{jj}(n)| \leq |\lambda_j(n)|/2$ for all n sufficiently large and therefore

$$\left| \frac{r_{j\nu}(n) s_{\nu j}(n)}{\lambda_j(n) + r_{jj}(n)} \right| \leq 2 \left| \frac{r_{j\nu}(n)}{\lambda_j(n)} \right| \left| s_{\nu j}(n) \right|,$$

which is in l^1. Hence there exists a constant $c \neq 0$ such that

$$\prod^{n-1} [\lambda_j(k) + l_j(k)] = [c + o(1)] \prod^{n-1} [\lambda_j(k) + r_{jj}(k)] \qquad \text{as } n \to \infty,$$

which establishes (3.65), and concludes the proof. □

In the special case that the block-matrix $A(n)$ in (3.42) is replaced by a diagonal matrix $\Lambda(n)$, Theorem 3.13 implies the following.

Corollary 3.14 *Suppose that* $A(n)$ *in* (3.42) *is a diagonal matrix* $A(n) = \Lambda(n) = \operatorname{diag}\{\lambda_1(n), \dots, \lambda_d(n)\}$. *For a fixed value of* $j \in \{1, \dots, d\}$, *assume that* $\lambda_j(n) \neq 0$ *for all* n *and that there exist constants* $K > 0$ *and* $q \in (0, 1)$ *such that for all* $1 \le i \neq j \le d$

$$\text{either} \quad \prod_{k=n_1}^{n_2} \left| \frac{\lambda_i(k)}{\lambda_j(k)} \right| \le K q^{n_2 - n_1}, \qquad n_0 \le n_1 \le n_2, \tag{3.68}$$

$$\text{or} \quad \prod_{k=n_1}^{n_2} \left| \frac{\lambda_j(k)}{\lambda_i(k)} \right| \le K q^{n_2 - n_1}, \qquad n_0 \le n_1 \le n_2. \tag{3.69}$$

Suppose that $R(n)$ *satisfies* (3.64). *Then*

$$y(n+1) = [\Lambda(n) + R(n)]\, y, \qquad n \ge n_0, \tag{3.70}$$

has for n *sufficiently large a solution* $y_j(n)$ *satisfying* (3.65) *as* $n \to \infty$.

Proof This follows from Theorems 3.12 and 3.13, when replacing the projection matrix P in the proof of Theorem 3.12 by the $d-1 \times d-1$ diagonal projection matrix $\hat{P} = \operatorname{diag}\{p_1, \dots, p_{j-1}, p_{j+1}, \dots, p_d\}$. With j being fixed, put $p_i = 1$ if (i, j) satisfies (3.68), and $p_i = 0$ if (i, j) satisfies (3.69) for each index $1 \le i \neq j \le d$. □

As an easy example, consider

$$y(n+1) = \left[\begin{pmatrix} \frac{1}{n^2} & 0 \\ 0 & \sqrt{n} \end{pmatrix} + \begin{pmatrix} 0 & \frac{1}{\sqrt{n}} \\ \frac{1}{\sqrt{n}} & 0 \end{pmatrix} \right] y(n).$$

Then Corollary 3.14 implies the existence of a solution of the form $y_2(n) = [e_2 + o(1)] \prod^{n-1} \sqrt{k}$.

Corollary 3.15 *In Corollary 3.14, suppose that* $R(n)$ *satisfies* (3.64) *for all* $1 \le j \le d$. *Assume that* (3.68) *or* (3.69) *holds for all* $1 \le i \neq j \le d$. *Then* (3.70) *has for* n *sufficiently large a fundamental matrix satisfying as* $n \to \infty$

$$Y(n) = [I + o(1)] \prod^{n-1} \{\Lambda(k) + \operatorname{diag} R(k)\}.$$

Remark 3.16 We emphasize that a different approach to Corollary 3.15 will be discussed in Sect. 5.4 (see, in particular, Theorem 5.3). This approach, while restricted to assertions about fundamental matrices and not vector-valued solutions, allows for extensions to l^p-perturbations with $p > 2$.

3.6 Perturbations of Constant Systems

In this section we consider systems of the form

$$y(n+1) = [A + R(n)]y(n), \qquad n \geq n_0, \tag{3.71}$$

where $R(n) = o(1)$ as $n \to \infty$, and apply results of previous sections to investigate the asymptotic behavior of solutions.

In the case where $A + R(n)$ is a companion matrix

$$\begin{pmatrix} 0 & 1 & & \\ & 0 & \ddots & \\ & & \ddots & 1 \\ a_1 + r_1(n) & & \dots & a_d + r_d(n) \end{pmatrix}$$

(3.71) is equivalent to a dth order scalar difference equation and in a seminal work by H. Poincaré [130] one of the first general results concerning asymptotic behavior of solutions was established. See Sect. 9.2 for a statement of Poincaré's theorem, some historical remarks, and a discussion of the relations between solutions of general systems (3.71) and those for the special companion matrix situation.

Inspired by Poincaré's results, E.B. Van Vleck [159] treated systems (3.71) and proved two theorems which imply the following: Let A have all distinct eigenvalues and assume their moduli are also unequal. Then for any solution $y(n)$ of (3.71) which is nontrivial for all $n \geq n_0$, there exists an eigenvector $\mathbf{p}$ of A with corresponding eigenvalue λ such that

$$y(n) = [\mathbf{p} + o(1)] \prod^{n-1} [\lambda + \varepsilon(k)] \qquad \text{as } n \to \infty, \tag{3.72}$$

with $\varepsilon(k) = o(1)$ as $k \to \infty$. A more recent result of A. Máté and P. Nevai [105] contains another formulation about the asymptotic behavior of solutions which also can be shown to imply (3.72).

A different way to look at (3.72) is to consider a kind of converse statement: If λ is an eigenvalue of A with associated eigenvector $\mathbf{p}$, under what conditions does there exist a solution $y(n)$ satisfying (3.72)? This type of question was considered by O. Perron [117], who refined Poincaré's result by establishing a basis of asymptotic solutions. The following results provide rather weak conditions for statements of this kind as well as a somewhat improved conclusion which provides some quantitative information about $\varepsilon(k)$.

We begin with a special case of Theorem 3.12 concerning perturbations of constant matrices.

Theorem 3.17 *Consider* (3.71), *where A is a matrix of the form*

$$A = \begin{pmatrix} A_{-1} & & \\ & \lambda_j & \\ & & A_1 \end{pmatrix}. \tag{3.73}$$

Here $\lambda_j \neq 0$, A_{-1} is a $(j-1) \times (j-1)$ matrix with eigenvalues satisfying $|\lambda| < |\lambda_j|$ and A_1 is a $(d-j) \times (d-j)$ matrix with eigenvalues satisfying $|\lambda| > |\lambda_j|$. Suppose that $R(n)$ satisfies

$$\lim_{n\to\infty} R(n) = 0. \tag{3.74}$$

Then (3.71) *has for n sufficiently large a solution $y_j(n)$ satisfying*

$$y_j(n) = [e_j + s_j(n)] \prod^{n-1} [\lambda_j + l_j(k)],$$

where $l_j(k)$ and $s_j(n)$ satisfy (3.51) *and* (3.52), *respectively.*

For a similar result, we also refer to [96, Thm. 3.1 and Cor. 3.3].

In [128, Thms. 1–4], M. Pinto considered asymptotically constant systems (3.71) where A is an invertible matrix whose eigenvalues $\{\lambda_1, \ldots, \lambda_d\}$ are all different and also have distinct moduli, and $R(n) = o(1)$ as $n \to \infty$. Assuming w.l.o.g. that $A = \Lambda$ is already in diagonal form, he showed in Theorems 3 and 4 that (3.71) has a fundamental solution satisfying

$$Y(n) = [I + \hat{S}(n)] \prod^{n-1} \left[\Lambda + \text{diag}\, R(k) + \hat{\Lambda}(k)\right], \tag{3.75}$$

where $\hat{S}(n) = o(1)$ as $n \to \infty$, $\hat{\Lambda}(k)$ is diagonal and satisfies $\hat{\Lambda}(k) = o(|R(k))$ as $k \to +\infty$. We note that Pinto was the first to establish such an interesting result and remark that Theorem 3.17 is a generalization of his work. However, he continues to state without proof in [128, Thm. 1], that $\hat{S}(n) = O(|R|)$ and $\hat{\Lambda}(n) = O\left(|R|^2\right)$. This notation is ambiguous since $|R|$ could either be interpreted as $|R(n)|$ or $\|R(n)\| = \sup_{k\geq n} |R(k)|$. In either case such a statement is stronger than what follows from estimates given here and is, in general, incorrect, as the following example shows.

Consider the triangular system

$$y(n+1) = \begin{bmatrix} 1 & r(n) \\ 0 & 2 \end{bmatrix} y(n), \qquad n \geq 1,$$

which has a fundamental matrix

$$Y(n) = \begin{bmatrix} 1 & \sum_{k=1}^{n-1} r(k)2^{k-1} \\ 0 & 2^{n-1} \end{bmatrix}, \qquad Y(1) = I.$$

Theorem 3.17 implies that there exists an invertible constant matrix C such that

$$\begin{bmatrix} 1 & \sum_{k=1}^{n-1} r(k)2^{k-1} \\ 0 & 2^{n-1} \end{bmatrix} C = \begin{bmatrix} 1 & s_{12}(n) \\ s_{21}(n) & 1 \end{bmatrix} \begin{bmatrix} \prod_{k=1}^{n-1}[1 + l_1(k)] & 0 \\ 0 & \prod_{k=1}^{n-1}[2 + l_2(k)] \end{bmatrix},$$

where $l_1(k) = o(|r(k)|)$ and $l_2(k) \equiv 0$. Denote the elements of C by c_{ij}. We now choose the perturbation $r(k) = 2^{1-k}$, hence $|R(n)| = \sup_{k \geq n} |R(k)| = 2^{1-n}$. This last identity can be re-written as

$$\begin{bmatrix} c_{11} + c_{21}(n-1) & c_{12} + c_{22}(n-1) \\ c_{21}2^{n-1} & c_{22}2^{n-1} \end{bmatrix} = \begin{bmatrix} \prod^{n-1}[1 + l_1(k)] & s_{12}(n)2^{n-1} \\ s_{21}(n)\prod^{n-1}[1 + l_1(k)] & 2^{n-1} \end{bmatrix},$$

where $l_1(k) = o\left(2^{1-k}\right)$. It is not hard to show that it follows that

$$c_{11} = c_{22} = 1; \;\; l_1(n) = c_{21} = s_{21}(n) = 0 \quad \text{and} \quad c_{12} + (n-1) = s_{12}(n)\, 2^{n-1}.$$

That is,

$$s_{12}(n) = \frac{c_{12} + (n-1)}{2^{n-1}},$$

for some constant c_{12}. Clearly, $s_{12}(n) = o(1)$ as $n \to \infty$, but $s_{12}(n) \neq O(|R(n)|)$, which contradicts Pinto's statement.

The next result concerns o(1)-perturbations of diagonal constant matrices with one zero eigenvalue.

Theorem 3.18 *Consider*

$$y(n+1) = [\Lambda + R(n)]\, y(n), \qquad n \geq n_0 \tag{3.76}$$

where $\Lambda = \operatorname{diag}\{\lambda_1, \ldots, \lambda_d\}$. *Assume that*

$$0 = |\lambda_1| < |\lambda_2| \leq |\lambda_3| \leq \ldots \leq |\lambda_d|.$$

Suppose that $R(n)$ *satisfies* (3.74), $r_{11}(n) \neq 0$ *for all* $n \geq n_0$, *and for each* $\nu \in \{2, \ldots, d\}$

$$r_{1\nu}(n) = O(|r_{11}(n)|) \qquad \text{as } n \to \infty. \tag{3.77}$$

Then (3.76) *has for* n *sufficiently large a nontrivial solution of the form*

$$y_1(n) = [e_1 + s_1(n)] \prod_{n_1}^{n-1} [r_{11}(k) + l_1(k)], \tag{3.78}$$

where $l_1(n) = o\left(\sum_{\nu \neq 1} |r_{1\nu}(k)|\right)$, *and* $s_1(n) = \{s_{11}(n), \ldots, s_{d1}(n)\}^T = o(1)$ *as* $n \to \infty$.

Proof The proof is a modification of the proof of Theorem 3.12, with $\lambda_j(n)$ replaced by $r_{11}(n)$. We note that (3.54) is now re-written as

$$l_1(n) = 0 + \sum_{j=2}^{d} r_{1j}(n) s_{j1}(n) = \hat{\rho}_1(n)\sigma(n), \tag{3.79}$$

with $\hat{\rho}_1(n) = \{r_{12}(n), \ldots, r_{1d}(n)\}$. Also, (3.56) now takes on the form

$$\sigma(n+1) - \frac{\tilde{\Lambda}}{r_{11}(n)}\sigma(n) = V(n, \sigma(n), \sigma(n+1)), \tag{3.80}$$

with

$$V(n, \sigma(n), \sigma(n+1)) = \frac{1}{r_{11}(n)} \left[\hat{r}_1(n) + \hat{R}(n)\sigma(n) - \{\hat{\rho}_1(n)\sigma(n)\}\sigma(n+1)\right] \tag{3.81}$$

and $\tilde{\Lambda} = \{\lambda_2, \ldots, \lambda_d\}$. In lieu of (3.59), we define for $n \geq n_1$ (to be determined below) the following $\tilde{T}$-operator

$$(\tilde{T}\sigma)(n) = -\sum_{k=n}^{\infty} \left(\prod_{l=n}^{k} r_{11}(l) \tilde{\Lambda}^{-1} \right) V(k, \sigma(k), \sigma(k+1)),$$

and note that a fixed point $(T\sigma)(n) = \sigma(n)$, if it exists, satisfies (3.80).

Fix n_1 sufficiently large such that

$$|r_{11}(n)| \leq |R(n)| \leq \frac{|\lambda_2|}{5}, \qquad \text{for all } n \geq n_1, \tag{3.82}$$

where we choose the maximum norm. For this choice of n_1, let $\mathscr{B}$ be the Banach space given in Theorem 3.12 and $\mathscr{W}$ be again the closed subset $\mathscr{W} = \{w \in \mathscr{B} : \|w\| \leq 1\}$. Then for $\sigma \in \mathscr{W}$ follows that

$$|V(k, \sigma(k), \sigma(k+1))| \leq \frac{3}{|r_{11}(k)|}|R(k)|.$$

Noting that in the maximum norm $|\tilde{\Lambda}^{-1}| = \frac{1}{|\lambda_2|}$, one obtains

$$\left|(\tilde{T}\sigma)(n)\right| \leq \frac{3}{|\lambda_2|}\left\{|R(n)| + \sum_{k=n+1}^{\infty} |R(k)| \prod_{l=n}^{k-1} \frac{|r_{11}(l)|}{|\lambda_2|}\right\}.$$

It follows from (3.82) for $n \geq n_1$ that

$$\left|(\tilde{T}\sigma)(n)\right| \leq \frac{3}{|\lambda_2|} \sup_{k\geq n} |R(k)| \left\{1 + \sum_{k=n+1}^{\infty} \left(\frac{1}{5}\right)^{k-n}\right\} = \frac{15}{4\,|\lambda_2|} \sup_{k\geq n} |R(k)| \leq 3/4.$$

In particular, $(\tilde{T}\sigma)(n) = o(1)$ as $n \to \infty$ by (3.74) and T maps $\mathscr{W}$ into $\mathscr{W}$.

Similar estimates show that for $n \geq n_1$

$$\begin{aligned}
&|(\tilde{T}\sigma)(n) - (\tilde{T}\tilde{\sigma})(n)| \\
&\leq \sum_{k=n}^{\infty}\left(\prod_{l=n}^{k} \frac{|r_{11}(l)|}{|\lambda_2|}\right) |V(k, \sigma(k), \sigma(k+1)) - V(k, \tilde{\sigma}(k), \tilde{\sigma}(k+1))| \\
&\leq \sum_{k=n}^{\infty}\left(\prod_{l=n}^{k} \frac{|r_{11}(l)|}{|\lambda_2|}\right) \frac{3|\hat{R}(k)|\|\sigma - \tilde{\sigma}\|}{|r_{11}(k)|} \\
&\leq \frac{3\|\sigma - \tilde{\sigma}\|}{|\lambda_2|}\left\{|\hat{R}(n)| + \sum_{k=n+1}^{\infty} |\hat{R}(k)| \left(\frac{1}{5}\right)^{k-n}\right\} \\
&\leq \frac{3\|\sigma - \tilde{\sigma}\|}{|\lambda_2|} \sup_{k\geq n} |R(k)| \left\{1 + \sum_{k=n+1}^{\infty} \left(\frac{1}{5}\right)^{k-n}\right\} \leq \frac{3\|\sigma - \tilde{\sigma}\|}{4},
\end{aligned}$$

i.e., $\tilde{T}$ is a contraction. Note that by (3.77) and (3.79), $l_1(n) = o(\hat{\rho}_1(n))$. In particular, (3.78) is nontrivial for $n \geq n_1$, since $r_{11}(k) + l_1(k) \neq 0$ for all k sufficiently large by (3.77). The remainder of the proof is analogous to the proof of Theorem 3.12. □

Remark 3.19 We note that simply applying Theorem 3.12 with $\lambda_1(n) := r_{11}(n)$ and $r_{11}(n) := 0$ would have required the stronger hypothesis that $R(n)/\lambda_1(n) = R(n)/r_{11}(n) \to 0$ instead of $R(n)$ vanishing at infinity. The reason that the weaker hypothesis (3.74) on $R(n)$ suffices is that we have the advantage of dealing with a diagonal system where the projection matrix $P = 0$, i.e., where $\hat{\Phi}(n,k+1)\Big|_{\mathscr{R}(P)} \equiv 0$ (i.e., (3.47) is trivially satisfied) and $\left|\hat{\Phi}(n,k+1)\Big|_{\mathscr{R}(I-P)}\right| \leq \prod_{\nu=n}^{k} \frac{|\lambda_1(\nu)|}{|\lambda_2|}$. This not only implies that (3.48) holds, the term $|\lambda_1(k)|$ in the numerator $\hat{\Phi}(n,k+1)\Big|_{\mathscr{R}(I-P)}$ allows a cancellation with the term $1/r_{11}(n)$ in $V(k,\sigma(k),\sigma(k+1)$ [see (3.81)].

The last result in this section is due to Pituk [129].

Theorem 3.20 (Pituk) *Suppose that λ_0 is a simple nonzero eigenvalue of a $d \times d$ matrix A, and that if λ is any other eigenvalue of A, then $|\lambda| \neq |\lambda_0|$. Let ξ and η be nonzero vectors such that $A\xi = \lambda_0\xi$ and $A^*\eta = \overline{\lambda_0}\eta$. (Here A^* and $\overline{\lambda_0}$ denote the conjugate transpose of A and the conjugate of λ_0, respectively.) If*

$$\sum_{k=n_0}^{\infty} |R(n)|^2 < \infty, \tag{3.83}$$

then (3.71) *has for n sufficiently large a solution $y(n)$ satisfying*

$$y(n) = [\xi + o(1)] \prod^{n-1} [\lambda_0 + \beta(k)],$$

where

$$\beta(n) = \left(\eta^*\xi\right)^{-1} \eta^* R(n)\xi, \qquad n \geq n_0. \tag{3.84}$$

Proof We will not repeat Pituk's original proof here, which was also based on the Banach Fixed Point principle. Instead we will show that Theorem 3.20 follows from the more general Theorem 3.13. We refer to [129, Remark 1] for an algebraic argument showing that $\eta^*\xi \neq 0$.

With A satisfying the assumptions of Theorem 3.20, let P be any invertible matrix such that

$$P^{-1}AP = J = \operatorname{diag}\{J_{-1}, \lambda_0, J_1\},$$

blocked as in (3.73), with λ_0 in the jth diagonal position for some fixed $j \in \{1, \ldots, d\}$. Here the eigenvalues of J_{-1} and J_1 have modulus less and greater than $|\lambda_0|$, respectively. Putting $y(n) = Pz(n)$, we obtain

$$z(n+1) = [J + P^{-1}R(n)P]z(n) = [J + \tilde{R}(n)]z(n).$$

Since λ_0 is a constant, (3.83) implies that (3.64) holds for $\tilde{R}(n)$ and λ_0, that is, $\tilde{R}(n)/\lambda_0 \in l^2$. Theorem 3.13 now implies for large n the existence of a solution

$$z_j(n) = [e_j + o(1)] \prod^{n-1} [\lambda_0 + \tilde{r}_{jj}(k)], \qquad \text{as } n \to \infty.$$

Let $(P)_j$ denote the jth column of P. From $AP = PJ$ follows that

$$A(P)_j = (AP)_j = \lambda_0(P)_j,$$

so the jth column of P is a non-zero multiple of $\boldsymbol{\xi}$, say $(P)_j = c_1\boldsymbol{\xi}$ with $c_1 \neq 0$. Similarly, from $P^{-1}A = JP^{-1}$ one sees that the jth row of P^{-1} must be a nonzero scalar multiple of $\boldsymbol{\eta}$, say $_j(P^{-1}) = c_2\boldsymbol{\eta}^*$, with $c_2 \neq 0$. Since $P^{-1}P = I$, $c_1c_2(\boldsymbol{\eta}^*\boldsymbol{\xi}) = 1$, so $c_1c_2 = 1/(\boldsymbol{\eta}^*\boldsymbol{\xi})$. Therefore

$$\tilde{r}_{jj}(n) = \left(P^{-1}R(n)P\right)_{jj} = {}_j(P^{-1})R(n)P_j = \beta(n), \tag{3.85}$$

with $\beta(n)$ defined in (3.84). Recalling the transformation $y(n) = Pz(n)$ concludes the proof. □

We also refer to [17] for a study of exponentially small perturbations of constant linear systems of difference equations with an invertible coefficient matrix.

3.7 Estimate of the Error Term o(1)

The goal of this section is to give a quantitative estimate of the error term "o(1)" appearing in the asymptotic integration result (3.12) in both Theorems 3.4 and 3.8. As in the continuous case, it appears that the dichotomy condition needs to be somewhat strengthened in general to allow such a quantitative result. Results of this nature where first studied by Gel'fand and Kubenskaya [60] and later on by Bodine and Lutz [13], and we follow this later approach. However, taking into account Kooman's recent result of relaxing the growth condition on the perturbation matrix (see Theorem 3.8) allows for a corresponding relaxation of the assumptions. We first give these new results for fundamental solution matrices, followed by briefly referring to the results for vector-valued solutions derived in [13]. We also refer to [98, Thm. 2.1] for another estimate of the error term.

Theorem 3.21 *Let $\Lambda(n) = \mathrm{diag}\{\lambda_1(n), \ldots, \lambda_d(n)\}$ be invertible for $n \geq n_0$. Suppose that $\Lambda(n)$ satisfies the following discrete exponential/ordinary dichotomy condition: Assume there exist constants $K_1 > 0$ and $q \in (0,1)$ such that for each $1 \leq i,j \leq d$ either*

$$\prod_{N_1}^{N_2}\left|\frac{\lambda_j(k)}{\lambda_i(k)}\right| \leq K_1 q^{N_2-N_1} \qquad \forall\, n_0 \leq N_1 \leq N_2, \tag{3.86}$$

or that (3.9) *holds. Let $R(n)$ be a $d \times d$ matrix defined for all $n \geq n_0$ that satisfies*

$$\sum_{n=n_0}^{\infty}\left|\Lambda^{-1}(n)R(n)\right| < \infty. \tag{3.87}$$

Then (3.11) *has for n sufficiently large a fundamental matrix satisfying, as* $n \to \infty$

$$Y(n) = \left[I + \mathrm{O}\left(\sum_{k=n_2}^{n-1} q^{n-k} \left| \Lambda^{-1}(k) R(k) \right| \right) \right.$$
$$\left. + \mathrm{O}\left(\sum_{k=n}^{\infty} \left| \Lambda^{-1}(k) R(k) \right| \right) \right] \prod_{k=n_0}^{n-1} \Lambda(k). \tag{3.88}$$

Here n_2 *is determined in the proof of Theorem 3.8.*

Proof Note that the assumptions of Theorem 3.21 are stronger than the ones of Theorem 3.8, hence the proof of Theorem 3.8 remains valid. Here, we use the strengthened dichotomy condition (3.86) to estimate the entries in the $\hat{T}$-operator given in (3.22), which are precisely the error terms.

To that end, let $1 \le j, \nu \le d$. If $p_j = 1$, then $\left(W(n)[I-P]W^{-1}(k+1)\frac{R(k)}{\lambda_i(k)} \right)_{j\nu} = 0$ and from (3.23) and (3.86) follows that

$$\begin{aligned} \sum_{k=n_2}^{n-1} \left| \left(W(n) P W^{-1}(k+1) \frac{R(k)}{\lambda_i(k)} \right)_{j\nu} \right| &= \sum_{k=n_2}^{n-1} \left| \frac{r_{j\nu}(k)}{\lambda_j(k)} \right| \prod_{l=k}^{n-1} \left| \frac{\lambda_j(l)}{\lambda_i(l)} \right| \\ &\le K_1 \sum_{k=n_2}^{n-1} q^{n-k-1} \left| \frac{r_{j\nu}(k)}{\lambda_j(k)} \right| \\ &\le \hat{K} \sum_{k=n_2}^{n-1} q^{n-k-1} \left| \frac{r_{j\nu}(k)}{\lambda_j(k)} \right|. \end{aligned} \tag{3.89}$$

If $p_j = 0$, then $\left(W(n) P W^{-1}(k+1)\frac{R(k)}{\lambda_i(k)} \right)_{j\nu} = 0$ and, since the assumption (3.9) is the same as in Theorem 3.8, the estimate (3.24) remains valid.

From (3.22), (3.24), and (3.89) it follows that for $n \ge n_2$ that

$$\begin{aligned} \left| (\hat{T}z)(n) \right| &\le \max_j \sum_{\nu=1}^{d} \|z\| \hat{K} \left[\sum_{k=n_2}^{n-1} q^{n-k-1} \left| \frac{r_{j\nu}(k)}{\lambda_j(k)} \right| + \sum_{k=n}^{\infty} \left| \frac{r_{j\nu}(k)}{\lambda_j(k)} \right| \right] \\ &\le \|z\| \hat{K} \left(\sum_{k=n_2}^{n-1} q^{n-k-1} \left| \Lambda^{-1}(k) R(k) \right| + \sum_{k=n}^{\infty} \left| \Lambda^{-1}(k) R(k) \right| \right), \end{aligned} \tag{3.90}$$

and standard arguments now lead to (3.88). We note that $Y(n)$ in (3.88) is a fundamental solution matrix for n sufficiently large (see also Lemma 3.23). □

Remark 3.22 A similar result holds if we replace (3.87) by the column-wise assumption $R\Lambda^{-1} \in l^1$ (cf. Corollary 3.9).

To verify that the error estimates in (3.88) vanishes at infinity, we show

Lemma 3.23 *Under the hypotheses of Theorem 3.21*

$$\lim_{n\to\infty} \sum_{k=n_2}^{n-1} q^{n-k} \left|\Lambda^{-1}(k)R(k)\right| = 0.$$

Proof Let $\alpha := \sum_{k=n_2}^{\infty} \left|\Lambda^{-1}(k)R(k)\right|$. Given $\varepsilon > 0$, fix a number $N \in \mathbb{N}$ such that $q^N < \varepsilon/(2\alpha)$. Then it follows that

$$\begin{aligned}\sum_{k=n_2}^{n-1} q^{n-k} \left|\Lambda^{-1}(k)R(k)\right| &= \sum_{k=n_2}^{n-N-1} q^{n-k} \left|\Lambda^{-1}(k)R(k)\right| \\ &\quad + \sum_{k=n-N}^{n-1} q^{n-k} \left|\Lambda^{-1}(k)R(k)\right| \\ &< \varepsilon/2 + \sum_{k=n-N}^{n-1} \left|\Lambda^{-1}(k)R(k)\right| < \varepsilon,\end{aligned}$$

for all n sufficiently large. □

Remark 3.24 In Theorem 3.21, the dichotomy conditions (3.86) and (3.9) could also have been stated in the following equivalent way using discrete "exponential-ordinary dichotomies": For the fixed $1 \leq i \leq d$, let W denote a fundamental matrix of (3.15). Assume that there exists a projection P and constants $M \geq 1$ and $0 < q < 1$ such that

$$\begin{aligned}|W(n)PW^{-1}(k+1)| &\leq Mq^{n-k} \quad n > k \geq n_0 \\ |W(n)[I-P]W^{-1}(k+1)| &\leq M \quad k \geq n \geq n_0.\end{aligned}$$

In (3.88) we see that the remainder has two components. A more explicit result can be obtained by assuming a slow-decay condition on the perturbation in addition to the assumptions of Theorem 3.21.

Theorem 3.25 *Suppose that* $\Lambda(n)$ *and* $R(n)$ *satisfy all conditions of Theorem 3.21, and let* $q \in (0,1)$ *be defined as in Theorem 3.21. Suppose there exists non-negative* $\phi(n)$ *satisfying*

$$\phi(n) \geq \left|\Lambda^{-1}(n)R(n)\right| \text{ for } n \geq n_0$$

such that

$$\sum_{n=n_0}^{\infty} \phi(n) < \infty. \tag{3.91}$$

If (3.86) *is satisfied by at least one pair* (i,j), *assume that there exists* $1 \le b < 1/q$ *such that*

$$\phi(N_1)\, b^{N_1} \le \phi(N_2)\, b^{N_2} \quad \textit{for all } n_0 \le N_1 \le N_2. \tag{3.92}$$

Then (3.11) *has for n sufficiently large a fundamental matrix satisfying as* $n \to \infty$

$$Y(n) = \left[I + \mathrm{O}\left(\sum_{k=n}^{\infty} \phi(k) \right) \right] \prod^{n-1} \Lambda(k), \tag{3.93}$$

Proof The proof is a modification of that of Theorem 3.21. The main step is to use (3.92) to further estimate the $\hat{T}$-operator in (3.90). Observe that (3.90) implies that

$$\left| (\hat{T}z)(n) \right| \le \|z\| \hat{K} \left(\sum_{k=n_2}^{n-1} q^{n-k-1} \phi(k) + \sum_{k=n}^{\infty} \phi(k) \right).$$

Note that

$$\sum_{k=n_2}^{n-1} q^{n-k-1} \phi(k) = \sum_{k=n_2}^{n-1} q^{n-k-1} \frac{\phi(k) b^k}{b^k} \le \frac{1}{q} \sum_{k=n_2}^{n-1} q^{n-k} \frac{\phi(n) b^n}{b^k}$$
$$= \frac{\phi(n)}{q} \sum_{k=n_2}^{n-1} (qb)^{n-k} = \mathrm{O}(\phi(n)),$$

and therefore

$$\left| (\hat{T}z)(n) \right| = \mathrm{O}\left(\sum_{k=n}^{\infty} \phi(k) \right).$$

□

We emphasize that results in this section so far led to error estimates for a fundamental solution matrix provided that the growth condition (3.87) is satisfied. In [13] the following results were derived for vector-valued solution assuming that $\frac{|R(n)|}{|\lambda_i(n)|} \in l^1$ for a fixed value of i. We note that if this assumption is satisfied for *each*

$i \in \{1, \ldots, d\}$, then this would be a stronger assumption than (3.87). However, it is possible that (3.87) fails to hold but $\frac{|R(n)|}{|\lambda_i(n)|} \in l^1$ for some value of i.

Theorem 3.26 *In Theorem 3.21, assume that* (3.86) *or* (3.9) *hold for a fixed value of i, $1 \le i \le d$, and for all $1 \le j \le d$. Instead of* (3.87), *suppose that*

$$\sum_{n=n_0}^{\infty} \frac{|R(n)|}{|\lambda_i(n)|} < \infty.$$

Then (3.11) *has for n sufficiently large a vector-valued solution vector satisfying as $n \to \infty$*

$$y_i(n) = \left[e_i + O\left(\sum_{k=n_2}^{n-1} q^{n-k} \frac{|R(k)|}{|\lambda_i(k)|} \right) + O\left(\sum_{k=n}^{\infty} \frac{|R(k)|}{|\lambda_i(k)|} \right) \right] \prod^{n-1} \lambda_i(k),$$

for $1 \le i \le d$. Here e_i is the ith Euclidean unit vector and n_2 is determined in the proof of Theorem 3.8.

Moreover, if there exists $\phi_i(n) \ge |R(n)|/|\lambda_i(n)|$ for $n \ge n_0$ such that $\phi_i \in l^1$ and such that if (3.86) *is satisfied by at least one j, there exists $1 \le b < 1/q$ such that*

$$\phi_i(N_1)\, b^{N_1} \le \phi_i(N_2)\, b^{N_2} \quad \textit{for all } n_0 \le N_1 \le N_2,$$

then (3.11) *has for n sufficiently large a solution satisfying as $n \to \infty$*

$$y_i(n) = \left[e_i + O\left(\sum_{k=n}^{\infty} \phi_i(k) \right) \right] \prod^{n-1} \lambda_i(k) \qquad \textit{as} \quad n \to \infty.$$

3.8 Asymptotic Equivalence

We define two d-dimensional difference systems

$$x(n+1) = A(n)x(n) \qquad \text{and} \qquad y(n+1) = [A(n) + R(n)]y(n)$$

to be *(left) asymptotically equivalent* if there exist fundamental solution matrices satisfying

$$Y(n) = [I + E(n)]X(n), \qquad E(n) = o(1) \text{ as } n \to \infty. \tag{3.94}$$

Here we assume $A(n)$ and $A(n) + R(n)$ are both invertible for all $n \ge n_0$ and ask under what conditions on $A(n)$ and $R(n)$ does (3.94) hold. Theorems 3.4 and 3.8 answer this when $A(n) = \Lambda(n)$ is diagonal and satisfies the discrete Levinson's dichotomy conditions (3.8) and (3.9). In this section we are concerned with extending this to general $A(n)$. The following result (Theorem 3.27) is motivated by a paper of Naulin and Pinto [112] and we mainly follow their approach.

First observe that (3.94) holds if and only if $E(n)$ satisfies

$$E(n+1) = A(n)E(n)A^{-1}(n) + R(n)E(n)A^{-1}(n) + R(n)A^{-1}(n), \quad E(n) = o(1). \tag{3.95}$$

This two-sided nonhomogeneous matrix system can be treated as a d^2-dimensional vector/matrix equation in the following way: Introduce $\mathbf{q}(n) = \text{vec}(E(n)) \in \mathbb{C}^{d^2}$ to be the vector of columns of $E(n)$ stacked in order from top to bottom, starting with the first column of $Q(n)$. With the Kronecker product $A \otimes B$ defined in (2.129) and using the so-called Roth identity

$$\text{vec}\,(ABC) = (C^T \otimes A)\text{vec}\,B, \tag{3.96}$$

one can see that (3.95) becomes the nonhomogeneous system

$$\begin{aligned} \mathbf{q}(n+1) = &\left\{[A^{-1}(n)]^T \otimes A(n) + [A^{-1}(n)]^T \otimes R(n)\right\} \mathbf{q}(n) \\ &+ \left\{[A^{-1}(n)]^T \otimes R(n)\right\} \text{vec}\,(I_{d^2}). \end{aligned} \tag{3.97}$$

We consider (3.97) as a perturbation of the system

$$\mathbf{z}(n+1) = \left\{[A^{-1}(n)]^T \otimes A(n)\right\} \mathbf{z}(n), \tag{3.98}$$

and note that using the identity (2.130) one can check that

$$Z(n) = [X^{-1}(n)]^T \otimes X(n) \tag{3.99}$$

is a solution matrix for (3.98), where $X(n)$ is a fundamental solution of $x(n+1) = A(n)x(n)$. Note that using (2.131), one sees that $Z(n)$ is also fundamental.

Theorem 3.27 *Let $A(n)$ and $A(n) + R(n)$ be invertible for all $n \geq n_0$, and let $X(n)$ be a fundamental matrix for $x(n+1) = A(n)x(n)$. Define $Z(n)$ by (3.99). Assume $Z(n)$ satisfies the following dichotomy condition: there exist a projection P and a sequence $\beta(k) \geq 1$ such that all three of the following hold:*

$$\begin{cases} Z(n)P \to 0 & \text{as } n \to \infty \\ |Z(n)PZ^{-1}(k+1)| \leq \beta(k) & \text{for all } n_0 \leq k \leq n \\ |Z(n)[I-P]Z^{-1}(k+1)| \leq \beta(k) & \text{for all } n_0 \leq n \leq k \end{cases} \tag{3.100}$$

If $R(n)$ satisfies

$$\beta(n)|A^{-1}(n)||R(n)| \in l^1, \tag{3.101}$$

then there exists a fundamental solution matrix $Y(n)$ *of* $y(n+1) = [A(n)+R(n)]y(n)$ *satisfying* (3.94).

Proof With $Z(n)$ defined in (3.99) and hence satisfying (3.98), we fix $n_1 \geq n_0$ such that

$$\sum_{k=n_1}^{\infty} \beta(k)|A^{-1}(k)||R(k)| \leq \theta < 1.$$

We consider, for $n \geq n_1$,

$$(T\mathbf{q}(n)) = \sum_{k=n_1}^{n-1} Z(n)PZ^{-1}(k+1)\left[A^{-1}(k)\right]^T \otimes R(k)\right] [\mathbf{q}(k) + \mathbf{I}_{d^2}]$$

$$- \sum_{k=n}^{\infty} Z(n)[I-P]Z^{-1}(k+1)\left[A^{-1}(k)\right]^T \otimes R(k)\right] [\mathbf{q}(k) + \mathbf{I}_{d^2}].$$

Then one can show that

$$|(T\mathbf{q}(n))| \leq \{\|\mathbf{q}\| + 1\} \sum_{k=n_1}^{\infty} \beta(k)|A^{-1}(k)||R(k)| \leq \theta\{\|\mathbf{q}\| + 1\},$$

and

$$|(T\mathbf{q}_1)(n) - (T\mathbf{q}_2)(n))| \leq \theta\|\mathbf{q}_1 - \mathbf{q}_2\|.$$

Hence T is a contraction in the Banach space $\mathscr{B}$ of bounded sequences with the supremum-norm, and therefore $\mathbf{q}(n) = (T\mathbf{q}(n))$ has a unique solution in $\mathscr{B}$. Hence there exists a one-to-one correspondence between bounded solutions of (3.97) and (3.98). In particular, since $\mathbf{z}(n) \equiv 0$ is a bounded solution of (3.98), there exists a corresponding bounded solution $\mathbf{e}(n)$ of (3.97). Moreover, the requirement $Z(n)P \to 0$ implies for this solution that $\mathbf{e}(n) \to 0$ as $n \to \infty$. Therefore, (3.95) has a solution $E(n) \to 0$ as $n \to \infty$, which concludes the proof. □

3.9 (h, k)-Dichotomies for Difference Systems

As in the case for differential equations discussed in Sect. 2.8, the concepts of discrete ordinary, exponential, and exponential/ordinary dichotomies can be generalized in the following way:

Definition 3.28 (Pinto [125]) A linear system $x(n+1) = A(n)x(n)$ with $A(n)$ invertible for $n \geq n_0$ is said to possess an (h, k)-dichotomy if there exist a fundamental solution $X(n)$, a projection matrix P, two positive sequences $h(n)$ and

$k(n)$, and a positive constant c such that

$$\left|X(n)PX^{-1}(m)\right| \leq ch(n)/h(m) \quad \text{for all } n_0 \leq m \leq n, \tag{3.102}$$

and

$$\left|X(n)[I-P]X^{-1}(m)\right| \leq ck(m)/k(n) \quad \text{for all } n_0 \leq n \leq m. \tag{3.103}$$

In addition to these conditions, the sequences are usually assumed to also satisfy a "compensation law": there exists $\kappa > 0$ such that

$$\frac{h(n)k(n)}{h(m)k(m)} \leq \kappa \quad \text{for all } n_0 \leq m \leq n. \tag{3.104}$$

A special case occurs when $h(n) = k^{-1}(n)$ and the resulting dichotomy is then called an *h-dichotomy*. As Pinto observed, an (h, k) dichotomy with compensation law is automatically an h-dichotomy, but certain estimates might be better in the (h, k)-context.

Using these concepts, Pinto has obtained several results which extend Theorem 3.2, including estimates on error bounds. One example is

Theorem 3.29 (Pinto [125]) *If in addition to* (3.102), (3.103), *and* (3.104), *the perturbation* $R(n)$ *satisfies*

$$\frac{|R(n)|h(n-1)|}{|h(n)|} \in l^1, \tag{3.105}$$

then for any solution of $x(n+1) = A(n)x(n)$ *such that* $x(n)/h(n-1)$ *is bounded, there exists for large* n *a corresponding solution of* $y(n+1) = [A(n) + R(n)]y(n)$ *satisfying*

$$y(n+1) = x(n+1) + \mathrm{O}(h(n)) \qquad \text{as } n \to \infty.$$

Moreover, if $X(n)P/h(n) \to 0$ *as* $n \to \infty$, *then this can be improved to*

$$y(n+1) = x(n+1) + \mathrm{o}(h(n)) \qquad \text{as } n \to \infty.$$

Pinto employs operators acting on certain weighted Banach spaces which are similar to the ones used in the previous sections. We refer the reader to [125] for details.

More recently, Pinto [127] has also considered further extensions by including exponential factors in (3.102), (3.103), leading to so-called *expo-(h, k) dichotomy conditions*. See [127] for results and details.

Finally, we mention another application of (h, k) dichotomies which is related to an asymptotic equivalence considered in Sect. 3.8, but also concerns another extension of Theorem 3.2.

Theorem 3.30 (cf. Naulin/Pinto [112]) *Assume that* $x(n+1) = A(n)x(n)$ *has an* h*-dichotomy with* $h(n)$ *bounded and* $R(n)$ *satisfies*

$$\frac{1}{h(n+1)} \left|A^{-1}(n)\right| |R(n)| \in l^1. \tag{3.106}$$

Then $y(n+1) = [A(n)+R(n)]y(n)$ *has for* n *sufficiently large a fundamental solution matrix satisfying*

$$Y(n) = [I + \mathrm{o}(h(n))] \prod_{n_0}^{n-1} A(l) \qquad \text{as } n \to \infty. \tag{3.107}$$

The proof also employs operators related to the T-operator introduced in Sect. 3.2 acting on certain weighted Banach spaces and we refer to [112, Thm. 1 and Lemmas 1, 2] for details. Observe that in the special case when $A(n) = \Lambda(n)$ is diagonal and satisfies Levinson-type dichotomy conditions (3.8), (3.9) (in particular, $h(n) \equiv k(n) \equiv 1$), then (3.106) corresponds to

$$\left|\Lambda^{-1}(n)\right| |R(n)| \in l^1, \tag{3.108}$$

which is equivalent to the assumption (3.10). In this respect, one can appreciate Kooman's improvement of Theorem 3.4 (see Theorem 3.8) by realizing that it replaces (3.108) by the considerably weaker conditions $\left|\Lambda^{-1}(n)R(n)\right|$ or $\left|R(n)\Lambda^{-1}(n)\right| \in l^1$.

It is also interesting to compare Theorem 3.29 with Theorem 3.11 since the perturbation condition (3.36) in the situation of a weak dichotomy (3.34), (3.35) is similar to (3.106). In case that $h(n)$ is a bounded sequence, it is easy to show that h- or (h, k)-dichotomy conditions (3.102), (3.103) together with the compensation law (3.104) imply the weak dichotomy conditions (3.34), (3.35) and thus they are stronger (e.g., define $\beta(n) = M/h(n)$, where $h(n) \leq M$). Therefore, Theorem 3.29 does not imply Theorem 3.11 because in the main applications the function $\beta(n)$ is unbounded and this would violate the assumption that $h(n)$ is bounded. Also see Theorem 3.27, which was motivated by Theorem 3.29, but which is not a consequence of it in light of the reasons above.

3.10 Noninvertible Systems

The definition of a discrete ordinary dichotomy for linear systems of difference equations $x(n+1) = A(n)x(n)$ (see Definition 3.1) requires the existence of a fundamental matrix, i.e., the assumption that the coefficient matrices $A(n)$ are invertible for all n sufficiently large. If $A(n)$ is not invertible, then in place of using the inverse of a fundamental solution another device is required which substitutes for it both in the dichotomy condition as well as the T-operator. This situation was treated in some pioneering work by Henry in the framework of infinite dimensional Banach spaces [74, Defs. 7.6.1, 7.6.4], but here we will only consider finite dimensional spaces and follow in the large the treatment of Elaydi, Papschinopoulos, and Schinas [57].

We consider first the unperturbed system

$$x(n+1) = A(n)x(n) \qquad n \geq n_0, \tag{3.109}$$

where the $d \times d$ matrix $A(n)$ is not necessarily invertible.

The state transition matrix $\Phi(n, r)$ of (3.109) was defined in (3.44) for $n \geq r \geq n_0$ by

$$\Phi(n,r) = \begin{cases} A(n-1)A(n-2)\cdots A(r) & \text{if } n > r, \\ I & \text{if } n = r. \end{cases} \tag{3.110}$$

Thus $\Phi(n, r)$ is in general not invertible and exists only for $n \geq r$.

We continue with a definition of a dichotomy for noninvertible systems, which is a slight modification of the original definition by Henry [74, Def. 7.6.4].

Definition 3.31 (3.109) is said to have an H-ordinary dichotomy if there exist $d \times d$ projection matrices $P(n)$, $n \geq n_0$, and a positive constant K such that

(Hi)

$$P(n)\Phi(n,r) = \Phi(n,r)P(r), \qquad \forall\, n \geq r \geq n_0, \tag{3.111}$$

where $\Phi(n, r)$ was defined in (3.110) for $n \geq r$.

(Hii) For $n \geq r$, the restriction $\Phi(n,r)|_{\mathscr{R}(I-P(r))}$ is an isomorphism from $\mathscr{R}(I - P(r))$ onto $\mathscr{R}(I - P(n))$, and we define $\Phi(r, n)$ as the inverse map of $\Phi(n,r)|_{\mathscr{R}(I-P(r))}$ from $\mathscr{R}(I - P(n))$ to $\mathscr{R}(I - P(r))$;

(Hiii) $|\Phi(n,r)P(r)| \leq K$ for $n_0 \leq r \leq n$;

(Hiv) $|\Phi(n,r)\,[I - P(r)]\,| \leq K$ for $n_0 \leq n \leq r$.

We note that (3.111) is equivalent to the assumption

$$A(n)P(n) = P(n+1)A(n), \qquad \forall\, n \geq n_0.$$

For difference systems (3.109) with $A(n)$ being invertible for all $n \geq n_0$, Definition 3.31 can be seen to coincide with the usual definition of a discrete ordinary dichotomy given in Definition 3.1. This follows from letting $X(n)$ denote a fundamental matrix of (3.109) (hence $\Phi(n, r) = X(n)X^{-1}(r)$) and setting $P(n) = X(n)PX^{-1}(n)$ for all $n \geq n_0$.

Example 3.32

$$A(n) = \begin{pmatrix} \lambda_1(1+(-1)^n) & \\ & \lambda_2 \end{pmatrix}, \qquad \lambda_1, \lambda_2 \in \mathbb{C};\ |\lambda_2| > 1, \quad n \geq n_0,$$

has an H-ordinary dichotomy with $P(n) = \operatorname{diag}\{1,0\}$ and $K = \max\{2|\lambda_1|, 1\}$ for $n \geq n_0$.

In the following, we consider perturbations of system (3.109)

$$y(n+1) = [A(n) + R(n)]\, y(n), \qquad n \geq n_0, \tag{3.112}$$

with no assumptions on the invertibility of matrices. As in the study of perturbations of invertible difference systems, a first step is to establish a version of Coppel's result (Theorem 3.2).

Theorem 3.33 *Assume that the unperturbed system* (3.109) *possesses an H-ordinary dichotomy and that*

$$\sum_{n=n_0}^{\infty} |R(n)| < \infty. \tag{3.113}$$

Then for every bounded solutions of (3.109)*, there exists for n sufficiently large a bounded solution of* (3.112)*. Moreover, the difference between corresponding solutions of* (3.109) *and* (3.112) *tends to zero as $n \to \infty$ if $\Phi(n,m)P(m) \to 0$ as $n \to \infty$ for $n \geq m \geq n_0$.*

Proof The proof is based on a modification of the discrete T-operator of Theorem 3.2. As in this theorem, let l_∞ be the Banach space of bounded d-dimensional vector-valued sequences with the supremum norm

$$\|y\| = \sup_{n \geq n_1} |y(n)|,$$

where n_1 is picked sufficiently large such that

$$\theta = K \sum_{n=n_1}^{\infty} |R(n)| < 1.$$

For $n \geq n_1$, define an operator $\tilde{T}$ by

$$(\tilde{T}y)(n) = \sum_{k=n_1}^{n-1} \Phi(n,k+1)P(k+1)R(k)y(k) - \sum_{k=n}^{\infty} \Phi(n,k+1)[I - P(k+1)]R(k)y(k). \tag{3.114}$$

(In the invertible case, $(\tilde{T}y)(n)$ coincides with the T-operator given in (3.6) by setting $P(n) = X(n)PX^{-1}(n)$ and observing that $\Phi(n,r) = X(n)X^{-1}(r)$ where $X(n)$ denotes a fundamental matrix of (3.109).) Note that only the restriction of the state transition matrix $\Phi(n, k+1)|_{\mathscr{R}(I-P(k+1))}$ is assumed to be invertible which is given by Definition 3.31(Hii). By Definition 3.31(Hiii) and (Hiv),

$$|(\tilde{T}y)(n)| \le \sum_{k=n_1}^{\infty} K|R(k)y(k)| \le \|y\| \sum_{k=n_1}^{\infty} K|R(k)| \le \theta\|y\| < \|y\|,$$

i.e., $\tilde{T}$ maps l_∞ into l_∞. Similarly, for $y_i \in l_\infty$ $(i = 1, 2)$

$$\|\tilde{T}(y_2 - y_1)\| \le \theta\|y_2 - y_1\|,$$

hence $\tilde{T} : l_\infty \to l_\infty$ is a contraction. Therefore, given $x \in l_\infty$, the fixed point equation $y = x + \tilde{T}y$ has a unique solution $y \in l_\infty$. It can be checked that if $x(n)$ is a bounded solution of (3.109), then $y(n) = x(n) + (\tilde{T}y)(n)$ is a bounded solution of (3.112). Finally, given a solution y of (3.112) and $\varepsilon > 0$, fix $n_2 = n_2(\varepsilon) \ge n_1$ such that

$$K\sum_{n_2}^{\infty} |R(n)|\,|y(n)| \le K\|y\| \sum_{n_2}^{\infty} |R(n)| < \frac{\varepsilon}{2}.$$

Then for $n \ge n_2$

$$\begin{aligned}
|y(n) - x(n)| &= |(\tilde{T}y)(n)| \\
&\le \left| \sum_{k=n_0}^{n_2-1} \Phi(n, k+1)P(k+1)R(k)y(k) \right| + \frac{\varepsilon}{2} \\
&= \left| \sum_{k=n_0}^{n_2-1} \Phi(n, n_2)\Phi(n_2, k+1)P(k+1)R(k)y(k) \right| + \frac{\varepsilon}{2} \\
&\le |\Phi(n, n_2)P(n_2)| \sum_{k=n_0}^{n_2-1} |\Phi(n_2, k+1)R(k)y(k)| + \frac{\varepsilon}{2} < \varepsilon,
\end{aligned}$$

for all n sufficiently large if $\Phi(n, n_2)P(n_2) \to 0$ as $n \to \infty$ $(n \ge n_2)$. □

As in [57], the next goal is to establish a version of Theorem 3.4 for not necessarily invertible difference systems. For that purpose, consider

$$y(n+1) = [\Lambda(n) + R(n)]\,y(n) \qquad n \ge n_0, \tag{3.115}$$

where

$$\Lambda(n) = \text{diag}\{\lambda_1(n), \ldots, \lambda_d(n)\},$$

and $R(n)$ is a $d \times d$ matrix-valued perturbation for all $n \geq n_0$. For a fixed value of $i \in \{1, \ldots, d\}$, either $\lambda_i(n) \neq 0$ for all large n, say $n \geq n_1$, or $\lambda_i(n) = 0$ infinitely often.

The following theorem addresses the first case, where we assume without loss of generality that $\lambda_i(n) \neq 0$ for all $n \geq n_0$.

Theorem 3.34 *Assume that for a fixed $i \in \{1, \ldots, d\}$, $\lambda_i(n) \neq 0$ for all $n \geq n_0$, and that for all $1 \leq j \neq i \leq d$, the ordered pair (i,j) satisfies* (3.8) *or* (3.9). *Assume that the perturbation $R(n)$ satisfies $\sum_{n=0}^{\infty} \left|\frac{R(n)}{\lambda_i(n)}\right| < \infty$. Then* (3.115) *has for n sufficiently large a solution $y_i(n)$ satisfying*

$$y_i(n) = [e_i + o(1)] \prod_{r=n_0}^{n-1} \lambda_i(r) \qquad \text{as } n \to \infty. \tag{3.116}$$

Proof The proof follows closely the one of Theorem 3.4. However, without having the existence of a fundamental solution matrix of $w(n+1) = \Lambda(n)w(n)$ given, the proof relies on Theorem 3.33 rather than Theorem 3.2 and only establishes the existence of one vector-valued solution. For the fixed value of $i \in \{1, \ldots, d\}$, make the change of variables $y(n) = z(n) \prod_{k=n_0}^{n-1} \lambda_i(k)$, leading to

$$z(n+1) = \frac{1}{\lambda_i(n)} [\Lambda(n) + R(n)]\, z(n). \tag{3.117}$$

We also consider the unperturbed normalized system

$$w(n+1) = \frac{\Lambda(n)}{\lambda_i(n)} w(n), \tag{3.118}$$

and we will apply Theorem 3.33 to these normalized diagonal systems. For that purpose, let $P = \text{diag}\{p_1, \ldots, p_d\}$, where

$$p_j = \begin{cases} 1 & \text{if } (i,j) \text{ satisfies (3.8)}, \\ 0 & \text{otherwise}. \end{cases}$$

That is, P is a constant projection matrix in this diagonal case, (3.111) is (trivially) satisfied, and (3.15) possesses an H-ordinary dichotomy. Moreover, $\Phi(n,m)P \to 0$ as $n \to \infty$. Now, by Theorem 3.33 and since $w_i(n) = e_i$ is a bounded solution of (3.118), there exists a bounded solution $z_i(n)$ of (3.117) with $z_i(n) = e_i + o(1)$ as $n \to \infty$, which establishes (3.116). □

In case $\lambda_i(n) = 0$ infinitely often, the normalization leading to (3.117) is not possible. In such a case there are, of course, situations where solutions are identically zero (or else zero after a finite number of iterations) as well as possibly nontrivial solutions which tend to zero. However, it does not appear possible to quantitatively link the behavior of a sequence $\{\lambda_i(n)\}$ which is zero infinitely often with the existence of a nontrivial solution of (3.115) which is asymptotically equal to zero. Hence the most that one can say is the very weak statement that there is a solution $y_i(n)$ satisfying

$$y_i(n) = o(1) \qquad \text{as } n \to \infty.$$

Finally, we are concerned with a generalization of Theorem 3.34 to "blocked" systems of the form (3.42), where $\lambda_j(n) \neq 0$ for all n sufficiently large (without loss of generality for all $n \geq n_0$). Perturbations of such systems were considered in Sect. 3.5 under a strong "exponential dichotomy" assumption on the leading matrix and relative weak hypotheses on the perturbation. Here we weaken the assumptions on the leading matrix while strengthening the ones on the perturbation.

Theorem 3.35 *Suppose that the $d \times d$ matrix $A(n)$ is of the form* (3.42)*, where $A_{-1}(n)$ is a $(j-1)\times(j-1)$ matrix, $A_1(n)$ is a $(d-j)\times(d-j)$ invertible matrix, and $\lambda_j(n) \neq 0$ for all $n \geq n_0$. Suppose that the unperturbed reduced $(d-1)\times(d-1)$ system* (3.43) *possesses an H-ordinary dichotomy with projection matrix $P(n) = P$ given by*

$$P = \begin{pmatrix} I_{j-1} & \\ & 0_{d-j} \end{pmatrix}. \tag{3.119}$$

Suppose that $R(n)$ satisfies

$$\frac{|R(n)|}{|\lambda_j(n)|} \in l^1 \tag{3.120}$$

for this fixed value of j and that $\Phi(n,m)P \to 0$ as $n \to \infty$ for $n \geq m \geq n_0$. Then there exists for n sufficiently large a solution $y_j(n)$ of (3.46) *satisfying*

$$y_j(n) = [e_j + o(1)] \prod^{n-1} \lambda_j(k). \tag{3.121}$$

Proof The proof is a modification of the proof of Theorem 3.12, and we refer the reader to the proof of Theorem 3.12 for notation including $\sigma(n)$, $\hat{R}(n)$, $V(n, \sigma(n), \sigma(n+1))$, $\mathscr{B}$, and $\mathscr{W}$.

The beginning of the proof is the same as the proof of Theorem 3.12 up to (3.57) included. In what follows, we will also use (3.60) and (3.63) which still hold true. We now fix $n_1 \geq n_0$ sufficiently large such that

$$4K \sum_{k=n_1}^{\infty} \left| \frac{R(n)}{\lambda_j(n)} \right| \leq \theta < 1,$$

where K was given in Definition 3.31. To show that (3.56) has a solution $\sigma(n) = o(1)$ as $n \to \infty$, we define for $n \geq n_1$ the operator

$$(T\sigma)(n) = \sum_{k=n_1}^{n-1} \hat{\Phi}(n,k+1)P(k+1)V(k,\sigma(k),\sigma(k+1))$$
$$-\sum_{k=n}^{\infty} \hat{\Phi}(n,k+1)(I-P(k+1))V(k,\sigma(k),\sigma(k+1)), \quad (3.122)$$

where $\hat{\Phi}(n,r)$ is the state transition matrix of $w(n+1) = \hat{A}(n)w(n)$ and $\hat{A}(n)$ was defined in (3.43). Using (3.60), one can show for $\sigma(n) \in \mathscr{W}$ and $n \geq n_1$ that

$$|(T\sigma)(n)| \leq K\sum_{k=n_1}^{n-1} |V(k,\sigma(k),\sigma(k+1))| + K\sum_{k=n}^{\infty} |V(k,\sigma(k),\sigma(k+1))|$$
$$\leq 2K\left(\sum_{k=n_1}^{n-1}\left|\frac{R(k)}{\lambda_j(k)}\right| + \sum_{k=n}^{\infty}\left|\frac{R(k)}{\lambda_j(k)}\right|\right) = 2K\sum_{k=n_1}^{\infty}\left|\frac{R(k)}{\lambda_j(k)}\right| \leq \frac{1}{1}.$$

Also, for $n_1 < n_2 < n$,

$$|(T\sigma)(n)| \leq 2K\sum_{k=n_1}^{n_2-1}\left|\hat{\Phi}(n,k+1)P(k+1)\right|\left|\frac{R(k)}{\lambda_j(k)}\right| + 2K\sum_{k=n_2}^{\infty}\left|\frac{R(k)}{\lambda_j(k)}\right|.$$

Hence, given $\varepsilon > 0$, fix n_2 sufficiently large such that $2K\sum_{k=n_2}^{\infty}\left|\frac{R(k)}{\lambda_j(k)}\right| < \varepsilon/2$. By making n sufficiently large ($n > n_2$), $2K\sum_{k=n_1}^{n_2-1}\left|\hat{\Phi}(n,k+1)P(k+1)\right|\left|\frac{R(k)}{\lambda_j(k)}\right| < \varepsilon/2$. Hence $(T\sigma)(n)$ vanishes at infinity and we have shown that T maps $\mathscr{W}$ into $\mathscr{W}$.

To show that T is a contraction mapping, we use (3.63) to show that for $n \geq n_1$

$$|(T\sigma)(n) - (T\tilde{\sigma})(n)| \leq K\sum_{k=n_1}^{n-1} |V(k,\sigma(k),\sigma(k+1)) - V(k,\tilde{\sigma}(k),\tilde{\sigma}(k+1))|$$
$$+K\sum_{k=n}^{\infty} |V(k,\sigma(k),\sigma(k+1)) - V(k,\tilde{\sigma}(k),\tilde{\sigma}(k+1))|$$
$$\leq 4K\,\|\sigma-\tilde{\sigma}\|\sum_{k=n_1}^{\infty}\left|\frac{R(k)}{\lambda_j(k)}\right| \leq \theta\|\sigma-\tilde{\sigma}\|.$$

Hence,

$$\|T(\sigma-\tilde{\sigma})\| \leq \theta\|\sigma-\tilde{\sigma}\| < \|\sigma-\tilde{\sigma}\|.$$

Continuing the proof as the one in Theorem 3.12 establishes the existence of a solution $y_j(n)$ satisfying (3.50), with $l_j(k)$ given in (3.51) and $s_j(n) = o(1)$. Since $l_j(n) = O(R(n))$, one can rewrite this solution by (3.120) as

$$y_j(n) = [e_j + o(1)] \prod^{n-1} \lambda_j(l) \underbrace{\prod^{n-1} \left[1 + \frac{l_j(k)}{\lambda_j(k)}\right]}_{c+o(1)}$$

for some $c \neq 0$. Division by c implies (3.121), which concludes the proof. □

Chapter 4
Conditioning Transformations for Differential Systems

4.1 Chapter Overview

In this chapter we will consider linear systems of the form $x' = A(t)x$ and discuss various procedures which may be used for transforming such a system (if possible) into an L-diagonal form, so that the theorems in Chap. 2 could be used to obtain an asymptotic representation for solutions.

The transformations we will consider have the form

$$x = P(t)\,[I + Q(t)]\,y,$$

where $P(t)$ is an invertible, explicitly-constructed matrix (which we call a *preliminary transformation*) and $Q(t)$ is a matrix which tends to zero as $t \to \infty$. The goal of such transformations is to obtain an L-diagonal system $y' = [\Lambda + R]y$ to which Levinson's fundamental theorem (see Theorem 2.7) can be applied and yields an asymptotic solution $Y(t) = [I + E(t)] \exp\left[\int^t \Lambda(\tau)\,d\tau\right]$, where $E(t) \to 0$ as $t \to \infty$. Then the original system has a solution $X(t)$ with what we call an *asymptotic factorization* of the form

$$X(t) = P(t)\,[I + Q(t)]\,[I + E(t)] \exp\left[\int^t \Lambda(\tau)\,d\tau\right].$$

We emphasize that in such a factorization, the only terms which are not explicitly determined appear in $E(t)$. We know from Levinson's fundamental theorem that $E(t) = \mathrm{o}(1)$ as $t \to \infty$, but often Theorem 2.25 can be applied to yield more precise error estimates.

S. Bodine, D.A. Lutz, *Asymptotic Integration of Differential and Difference Equations*, Lecture Notes in Mathematics 2129,
DOI 10.1007/978-3-319-18248-3_4

What we call a preliminary transformation $P(t)$ is usually a product of matrices which takes advantage of particular structural properties of the asymptotically dominant terms in $A(t)$ and arrange that the influence of such terms is brought to the main diagonal, if possible, so that $x = P(t)w$ leads to a system of the form $w' = [\Lambda(t) + V(t)]w$, where $\Lambda(t)$ is diagonal. While there are no algorithms for doing this in general, we will discuss in Sect. 4.7 various options which are available. In Sect. 8.3 we will use such preliminary transformations as a first step in analyzing some classical second-order equations.

For most of this chapter we will focus on the factor (or factors) going into the $I+Q(t)$-terms. The role of a transformation $w = [I+Q(t)]y$ is to modify or *condition* a perturbed diagonal system $w' = [\Lambda+V]w$ (with V not in L^1) so that the transformed system satisfies the assumptions of Levinson's fundamental theorem. In this spirit, we refer to such transformations as *conditioning transformations*. Such a matrix $Q(t)$ may be either given in closed form, as an explicit integral, as the solution of an algebraic equation, or else as the solution of some elementary differential equation.

Beginning with Sect. 4.2 and continuing through Sect. 4.6, we will discuss various types of conditioning matrices and the role that they have played in applying Levinson's fundamental theorem to many other situations that were sometimes initially treated by quite different methods. This treatment of conditioning transformations is based on the concept of what we call *approximate equations* for Q. Depending upon special properties of Λ and V, we show how to construct various approximate equations which can lead to improvements in the perturbation terms. In measuring the size of such terms, it is frequently suitable to consider the standard L^p norms and remark that an improvement corresponds to decreasing p with the goal of achieving an L^1-perturbation for which Levinson's fundamental theorem applies.

While the presentation here will focus on several different situations which lead to corresponding asymptotic integration theorems, in applications it is important to realize that asymptotic integration of a given differential equation (or class of equations) is often a multi-step process involving repeated applications of several techniques. So a resulting asymptotic factorization can involve products of preliminary and conditioning transformations and we refer the reader to Sect. 8.4.2 for some examples of this.

It should be noted and emphasized that not all systems are amenable to this approach for several different reasons. One obstacle involves diagonalizability of the leading terms and whether that is possible or not. In Chap. 6 we will discuss Jordan systems and also some block-diagonalization techniques. Another obstacle is that even if the leading terms are diagonal, they might not satisfy Levinson's dichotomy conditions (recall the discussions in Sects. 2.3 and 2.4). However, it might be possible that conditioning transformations could still be used to modify the perturbation in case of a "weak dichotomy" (see Sect. 2.4), but there are no known general results along those lines and the techniques would probably involve ad hoc procedures.

4.2 Conditioning Transformations and Approximate Equations

We will now consider systems of the form

$$y' = [\Lambda(t) + V(t)]y \qquad t \geq t_0, \tag{4.1}$$

where $\Lambda(t)$ is a diagonal matrix and $V(t)$ is a "small" perturbation in a sense that will be explained later. If $V(t)$ does not satisfy the growth condition of Theorem 2.7, we attempt to use linear transformations of the form

$$y = [I + Q(t)]z, \tag{4.2}$$

where $Q(t) = o(1)$ as $t \to \infty$ in order to modify or *condition* the system into an equivalent one

$$z' = [\tilde{\Lambda}(t) + \tilde{V}(t)]z, \tag{4.3}$$

which either immediately satisfies the assumptions of Levinson's fundamental theorem or else comes "closer" to satisfying them in the sense that $\tilde{V}$ is smaller than V with respect to absolute integrability.

We mainly consider here cases when Λ or $\tilde{\Lambda}$ satisfy Levinson's dichotomy conditions and seek $\tilde{V} \in L^1$, but we remark that in light of Theorem 2.12, one could also try to use such conditioning matrices to modify V so that it satisfies a more restrictive growth condition in case Λ satisfies a weaker dichotomy condition.

Observe that one can always assume without loss in generality that diag $V(t) \equiv 0$ since the diagonal terms can be absorbed into Λ. Sometimes it is convenient to do this, but not always, so we reserve the option of making this additional assumption or not.

Letting

$$\tilde{\Lambda} = \Lambda + D,$$

where D is diagonal, one sees from

$$(I + Q)^{-1}(\Lambda + V)(I + Q) - (I + Q)^{-1}Q' = \tilde{\Lambda} + \tilde{V},$$

that D, Q, and $\tilde{V}$ must satisfy the so-called *transformation equation*

$$Q' = \Lambda Q - Q\Lambda + V - \tilde{V} + VQ - D - QD - Q\tilde{V}. \tag{4.4}$$

Observe that (2.4) is a special case of (4.4) in which $D \equiv 0$, $\Lambda = \tilde{\Lambda}$, and $\tilde{V} \equiv 0$. The point we want to emphasize is that if (2.4) would not have a solution

$E(t) = o(1)$, one might be able to find solutions of certain approximations to (4.4) which lead to a system (4.3) where Levinson's fundamental theorem would apply.

Here we think of Λ and V as given and we want to determine D, Q and $\tilde{V}$ so that (4.4) has a solution $Q = o(1)$ as $t \to \infty$ and $\tilde{V}$ is an improvement to V. In practice, one "solves" (4.4) by first selecting D (possibly depending on V and Q), then constructing a simpler approximate equation for Q, showing that this approximate equation has a solution $Q = o(1)$ as $t \to \infty$, and finally using the residual terms in (4.4) to calculate $\tilde{V}$ and show that it is an improvement to V.

As we shall see, the choices of D and a corresponding approximate equation depend very intimately on the structure of Λ and the properties of V with respect to its behavior when differentiated or integrated. While there is no comprehensive algorithm for this procedure, there are several principles that can be applied to conjecture what might be good choices. Whether the choices are successful then depends upon the quantitative behavior of Q and how that impacts on the resulting $\tilde{V}$.

We indicate now some possible choices for D and corresponding approximate equations which will be more fully developed and explained in the subsequent sections of this chapter.

We first consider choosing $D \equiv 0$. Then (4.4) simplifies to

$$Q' = \Lambda Q - Q\Lambda + V + VQ - [I + Q]\tilde{V}. \tag{4.5}$$

A simple choice for an approximate equation could be

$$Q' = V, \tag{4.6}$$

leading to $Q(t) = -\int_t^\infty V(\tau)\,d\tau$, provided that V is conditionally integrable so that Q is well-defined and tends to 0 as $t \to \infty$. This choice would yield

$$\tilde{V} = O\left(\|\Lambda Q - Q\Lambda\| + \|VQ\|\right), \qquad \text{as } t \to \infty,$$

which might or might not be an improvement to V. We will discuss situations related to conditional convergence in more detail in Sect. 4.6.

Another similar choice for an approximate equation in the case $D = 0$ is

$$Q' = \Lambda Q - Q\Lambda + V. \tag{4.7}$$

Provided there is a solution $Q = o(1)$ as $t \to \infty$, this would lead to

$$\tilde{V} = O\left(\|VQ\|\right), \qquad \text{as } t \to \infty.$$

We will consider situations involving this and other similar equations in Sects. 4.4 and 4.5. These cases involve some rather elementary differential (approximate) equations and showing that under appropriate conditions on Λ and V, there are solutions tending to zero.

Another alternative involves solving some *algebraic* approximate equations in the case $D = 0$ instead of differential equations as above. For example, one could think of choosing

$$\Lambda Q - Q\Lambda + V = 0. \tag{4.8}$$

Provided that (4.8) has a solution $Q = o(1)$ as $t \to \infty$, this leads to

$$\tilde{V} = O\left(\|Q'\| + \|VQ\|\right), \qquad \text{as } t \to \infty.$$

Again, depending upon properties of Λ and V, this may or may not be an improvement to V. In similar cases (which will be treated in Sect. 4.3) one sees that it is critical to be able to determine how Q inherits properties from V and what effect that has on the corresponding $\tilde{V}$.

If the choice $D \equiv 0$ is not possible, i.e., if this choice leads to approximate equations either having no solution $Q = o(1)$ as $t \to \infty$ or else the corresponding $\tilde{V}$ is not an improvement to V, then one could think of selecting D to depend on V and Q in some manner. As we shall see in the next section, a natural choice for D is

$$D = \text{diag}(VQ).$$

Putting $\tilde{V} \equiv 0$, the transformation equation (4.4) leads to the equation

$$Q' = \Lambda Q - Q\Lambda + V + VQ - \text{diag}\,(VQ) - Q\text{diag}\,(VQ). \tag{4.9}$$

Depending on properties of Λ and V, we shall then choose approximate equations and this will give rise to another set of asymptotic integration results that follow from Levinson's fundamental theorem. Such results in Sect. 4.3 are related to cases in which the eigenvalues of $\Lambda + V$ (or suitable approximations to them) are instrumental for determining the asymptotic behavior of solutions.

4.3 Reduction to Eigenvalues

In this section, we consider systems (4.1), where $\Lambda(t)$ is diagonal and $V(t)$ is a suitably small perturbation so that the equation

$$(I + Q)^{-1}[\Lambda + V](I + Q) = \tilde{\Lambda}, \tag{4.10}$$

is satisfied, where $\tilde{\Lambda}(t)$ is diagonal and $Q = o(1)$ as $t \to +\infty$. If this is the case, then the conditioning transformation (4.2) takes (4.1) into

$$z' = \left[\tilde{\Lambda} - (I + Q)^{-1}Q'\right] z, \tag{4.11}$$

to which we would seek to apply Theorem 2.7. If this is possible, it would yield a solution of (4.1) of the form

$$Y(t) = [I + \mathrm{o}(1)] \exp\left[\int^t \tilde{\Lambda}(\tau)\,d\tau\right] \qquad \text{as } t \to \infty, \tag{4.12}$$

and in this sense the eigenvalues of $\Lambda + V$ determine the asymptotic behavior of solutions.

This approach leads to several interesting questions. The first concerns finding conditions on Λ and V guaranteeing that (4.10) has such a solution. Next comes the question of finding $\tilde{\Lambda}$ or at least approximating it up to an L^1-perturbation so that the dichotomy conditions of Theorem 2.7 could be checked. Finally, the question arises of how Q' depends upon Λ and V so that $Q' \in L^1$ or that at least (4.11) is an improvement of (4.1) with respect to absolute integrability? Of course, we implicitly assume that V is not already in L^1, for otherwise Theorem 2.7 would apply immediately to (4.1) as long as its dichotomy conditions are satisfied.

From (4.10) observe that the columns of $I + Q$ are eigenvectors of $\Lambda + V$ and so the questions above all concern the asymptotic behavior of such eigenvectors and how they and the corresponding eigenvalues depend on Λ and V.

The simplest case to discuss occurs when $\Lambda(t) \to \Lambda_0$, a constant matrix, and $V(t) \to 0$. In this case one might as well just assume $\Lambda(t) = \Lambda_0$ without any loss in generality. Moreover, to keep the situation even simpler, assume that the eigenvalues of Λ_0 are distinct. This case was treated by Levinson ([100] or [53, Theorem 1.8.3]), who determined then the quantitative behavior of Q as a function of V as well as that of Q'. In Sect. 4.3.1 we will present two ways to determine this behavior, and also state Levinson's result based on it.

A much more general case was treated by Eastham (see [52], [53, Thm. 1.6.1]), in which the only assumption is that the eigenvalues $\lambda_i(t)$ are distinct for each $t_0 \leq t < \infty$, but might either have coalescing limiting values (at ∞) or have no limiting behavior at all as $t \to +\infty$. In the simpler, asymptotically constant case, however, the arguments are somewhat more transparent and so we will first treat this case in Sect. 4.3.1. In Sect. 4.3.2 we will present a modified treatment of the approach used by Eastham, resulting in what we call the "Levinson–Eastham lemma" which yields the quantitative behavior of Q and Q' in this case.

Finally, in Sect. 4.3.3 we will discuss some extensions to cases involving conditions on higher order derivatives, which are treated by iterating the transformations.

4.3.1 Asymptotically Constant Systems

Assuming that $\Lambda_0 = \operatorname{diag}\{\lambda_1, \ldots, \lambda_d\}$ has distinct eigenvalues and $V(t) \to 0$ as $t \to +\infty$, it follows that the characteristic polynomial $p(\lambda, t)$ of $\Lambda_0 + V(t)$ has, for all $t \geq t_1$ sufficiently large, distinct roots $\lambda_i(t) \to \lambda_i$ and corresponding linearly independent eigenvectors $x_i(t)$. Moreover, $x_i(t)$ can be normalized so that

$x_i(t) = e_i + q_i(t)$ with $q_i(t) \to 0$ as $t \to +\infty$. To determine more precisely how $q_i(t)$ depends upon $V(t)$, Levinson (see also [42, Ch. 3]) represented solutions of the equations

$$[\Lambda_0 + V(t)]\,[e_i + q_i(t)] = \lambda_i(t)[e_i + q_i(t)]$$

using cofactors of the matrix

$$\Lambda_0 + V(t) - \lambda_i(t)I,$$

from which it is easy to see that $q_i(t) = \mathrm{O}\left(|V(t)|\right)$ as $t \to +\infty$. Also, differentiating these formulas and using the fact that $\lambda_i'(t)$ can be calculated by differentiating $p(\lambda_i(t), t)$, one can show that $q_i'(t) = \mathrm{O}\left(|V'|\right)$ as $t \to +\infty$.

Another, more "analytic" way to obtain $I + Q(t)$ and deduce the dependence on $V(t)$ involves the use of projection matrices which map $\Lambda_0 + V(t)$ onto the eigenspaces. Using some classical formulas from operator theory, such projection matrices can be represented as Cauchy integrals. This technique was used by Coppel [43, pp. 111–113] and Harris–Lutz [69, Lemma 1]. By iterating this procedure, one can construct $I + Q(t)$ as an ordered product

$$I + Q(t) = \prod_{i=1}^{d} [I + Q_i(t)]\,,$$

where $Q_i(t)$ are defined inductively in the following manner:

$$Q_i(t) = [P_i(t) - P_i(\infty)]\,[2P_i(\infty) - I]\,,$$

$$P_i(t) = \frac{1}{2\pi i}\int_{\gamma_i} [zI - A_i(t)]^{-1}\,dz,$$

where

$$A_1(t) = \Lambda_0 + V(t), \qquad A_{i+1}(t) = [I + Q_i(t)]^{-1}\,A_i(t)\,[I + Q_i(t)]\,.$$

Here, γ_i is a positively-oriented circle not passing through any eigenvalue of Λ_0 and containing exactly one eigenvalue in its interior. Although this representation for $Q(t)$ is not practicable for calculations, it is convenient for determining how the regularity properties of $Q(t)$ are inherited from those of $A(t)$. For example,

$$Q'(t) = \sum_{k=1}^{d}\prod_{i=1}^{k-1} [I + Q_i(t)]\,Q_k'(t)\prod_{i=k+1}^{d} [I + Q_i(t)],$$

$$Q_i'(t) = \frac{1}{2\pi i}\int_{\gamma_i} [zI - A_i(t)]^{-1}\,A_i'(t)\,[zI - A_i(t)]^{-1}\,dz\;[2P_i(\infty) - I]$$

$$= \mathrm{O}\left(|A_i'(t)|\right),$$

and hence $Q'(t) = O(|V'(t)|)$ as $t \to \infty$. This method, when applicable, also provides immediate estimates of higher order derivatives (see Sect. 4.3.3). This approach is quite intimately tied to the assumption that Λ_0 is constant with distinct eigenvalues and does not easily extend to the situation of the more general results discussed in the the next section.

So applying Theorem 2.7 to $w' = \left[\Lambda(t) + (I+Q)^{-1}\{R(I+Q) - Q'\}\right] w$, one can obtain the following

Theorem 4.1 (Levinson) *Let A be a constant matrix with pairwise distinct eigenvalues and let $P^{-1}AP = \Lambda_0$. Let V' and R be continuous for $t \geq t_0$ and satisfy $V(t) \to 0$ as $t \to \infty$, V' and R both in L^1. Finally assume that the eigenvalues of $A + V(t)$, which exist and are continuous for $t \geq t_1 \geq t_0$ and are represented by the diagonal matrix $\Lambda(t) \to \Lambda_0$ as $t \to +\infty$, also satisfy Levinson's dichotomy conditions* (2.13), (2.14). *Then the system*

$$y' = [A + V(t) + R(t)]y$$

has a fundamental matrix $Y(t)$ satisfying

$$Y(t) = [P + o(1)] \exp\left[\int^t \Lambda(\tau)\, d\tau\right] \qquad \text{as } t \to \infty. \tag{4.13}$$

Remark 4.2 It is easy to construct examples showing that the assumption on A cannot be weakened to just assuming that A is diagonalizable without strengthening the conditions on V. Hence the assumption in Theorem 4.1 that the eigenvalues of A are "simple" (as roots of its characteristic polynomial) cannot in general be weakened to A just having simple elementary divisors (in spite of the fact that Theorem 2.7 does apply to $y' = [D + R(t)]y$ for any constant diagonal D as long as $R(t) \in L^1$).

Example 4.3 A simple example to which Theorem 4.1 can be applied is the two-dimensional system

$$x' = \begin{pmatrix} y \\ y' \end{pmatrix}' = \begin{pmatrix} 0 & 1 \\ -1 + \frac{\nu^2}{t^2} & -\frac{1}{t} \end{pmatrix} x \qquad t \geq 1,$$

which is easily seen to be equivalent to the classical Bessel equation $t^2y'' + ty' + (t^2 - \nu^2)y = 0$. Identifying

$$A = \begin{pmatrix} 0 & 1 \\ -1 & 0 \end{pmatrix}, \quad V(t) = \begin{pmatrix} 0 & 0 \\ 0 & -\frac{1}{t} \end{pmatrix}, \text{ and } R(t) = \begin{pmatrix} 0 & 0 \\ \frac{\nu^2}{t^2} & 0 \end{pmatrix},$$

and calculating the eigenvalues $\lambda_{1,2}(t) = -\frac{1}{2t} \pm i\sqrt{1 - \frac{1}{4t^2}}$ of $A + V(t)$, Theorem 4.1 and error estimates from Theorem 2.25 imply there exists an asymptotic solution of the form

$$X(t) = \left[\begin{pmatrix} 1 & 1 \\ i & -i \end{pmatrix} + \mathrm{O}\left(\frac{1}{t}\right)\right]\left[\begin{pmatrix} \exp\left(\int^t \lambda_1(\tau)\,d\tau\right) & 0 \\ 0 & \exp\left(\int^t \lambda_2(\tau)\,d\tau\right) \end{pmatrix}\right], \tag{4.14}$$

as $t \to \infty$. Of course, the system is also an example of a meromorphic differential equation, so applying that theory there is also a *formal* fundamental solution matrix of the (Hankel) form

$$\hat{X}(t) = \left[\begin{pmatrix} 1 & 1 \\ i & -i \end{pmatrix} + \sum_{k=1}^{\infty} \frac{E_k}{t^k}\right]\begin{pmatrix} t^{-1/2}e^{it} & 0 \\ 0 & t^{-1/2}e^{-it} \end{pmatrix},$$

which can be shown to be an *asymptotic expansion* for an actual solution in the following sense: For all $N \in \mathbb{N}$, there exists a function $E_N(t)$ such that

$$\left[\begin{pmatrix} 1 & 1 \\ i & -i \end{pmatrix} + \sum_{k=1}^{N} \frac{E_k}{t^k} + E_N(t)\right]\begin{pmatrix} t^{-1/2}e^{it} & 0 \\ 0 & t^{-1/2}e^{-it} \end{pmatrix}$$

is an actual solution and $E_N(t) = \mathrm{O}\left(t^{-(N+1)}\right)$ as $t \to +\infty$. The "solution" $\hat{X}$ is called formal because the power series is generally divergent for all t (see [3] or [160] for details, including a discussion of the interpretations in sectors of the complex t-plane).

To reconcile these two representations, observe that

$$\lambda_{1,2}(t) = -\frac{1}{2t} \pm i\left[1 - \frac{1}{8t^2} + \mathrm{O}\left(\frac{1}{t^4}\right)\right] \qquad \text{as } t \to \infty,$$

hence

$$\int^t \lambda_{1,2}(\tau)\,d\tau = -\frac{1}{2}\log t \pm it + \mathrm{O}(1/t) \qquad \text{as } t \to \infty.$$

Thus $X(t)$ in (4.14) can be re-written as

$$X(t) = \left[\begin{pmatrix} 1 & 1 \\ i & -i \end{pmatrix} + E(t)\right]\begin{pmatrix} t^{-1/2}e^{it} & 0 \\ 0 & t^{-1/2}e^{-it} \end{pmatrix},$$

where $E(t) = \mathrm{O}\left(\frac{1}{t}\right)$ as $t \to \infty$. Thus the formal power series in $\hat{X}(t)$ can be seen to be an asymptotic expansion for $E(t)$ as $t \to \infty$ in the above sense.

The example above indicates that the eigenvalues of $A + V(t)$ can sometimes be more simply approximated up to L^1-perturbations and such approximations can equally and more effectively be used in Theorem 4.1. We now wish to show this is the case whenever the (limiting) eigenvalues of A are explicitly known and (in addition to $V(t) = o(1)$) also $V(t) \in L^p$ for some $p < \infty$.

As an inductive step in these calculations, we first prove

Lemma 4.4 *Let $\Lambda(t) = \operatorname{diag}\{\lambda_1(t), \ldots, \lambda_d(t)\} = \Lambda_0 + o(1)$ as $t \to +\infty$, where Λ_0 has distinct entries. Assume that $V(t)$ is defined for $t \geq t_0$ and satisfies $V(t) = o(1)$ as $t \to +\infty$. If $T(t) = (t_{ij}(t))$ is defined for $i \neq j$ by*

$$t_{ij}(t) = \frac{v_{ij}(t)}{\lambda_j(t) - \lambda_i(t)}, \tag{4.15}$$

which exists for $t \geq t_1$ sufficiently large, and $t_{ii} = 0$, then T satisfies

$$\Lambda(t)T(t) - T(t)\Lambda(t) + V(t) - \operatorname{diag} V(t) = 0. \tag{4.16}$$

If t_1 is also so large that $I + T(t)$ is invertible for $t \geq t_1$, then

$$[I + T(t)]^{-1}[\Lambda(t) + V(t)][I + T(t)] = \Lambda(t) + \operatorname{diag} V(t) + U(t), \tag{4.17}$$

where $U(t) = O\left(|V(t)|^2\right)$ as $t \to \infty$.

Proof Using the definition of T, it is easy to show that (4.16) holds. Then (4.17) is equivalent to U satisfying

$$U(t) = [I + T(t)]^{-1}[V(t)T(t) - T(t)\operatorname{diag}V(t)].$$

Now (4.15) implies that $T(t) = O(|V(t)|)$ and hence

$$U(t) = O\left(|V(t)|^2\right) \qquad \text{as } t \to \infty.$$

□

Now consider

$$\Lambda_0 + V(t),$$

where Λ_0 is a constant matrix with distinct entries, $V(t) \in L^p$ for some $p > 1$ and $V(t) = o(1)$ as $t \to +\infty$. We are interested in constructing a sequence of $I + T_n(t)$ transformations that diagonalize $\Lambda_0 + V(t)$ up to an L^1-perturbation. From Lemma 4.4 with $\Lambda(t) = \Lambda_0$, there exists, for $t \geq t_1$, $T(t) = T_1(t) = O(|V(t)|)$ and $U_1(t) = O(|V(t)|^2) \in L^{p/2}$. If $p \leq 2$, we are finished with $\Lambda_1(t) = \Lambda_0 + \operatorname{diag} V(t)$. If not, we note from (4.15) follows that T_1 is a polynomial in the elements of $V(t)$ and, moreover, $[I+T(t)]^{-1} = I - T + T^2 - \ldots T^k + L^1$ for some appropriately chosen positive integer k. Then $U(t)$ in (4.17) is also a polynomial in v_{ij} with coefficients that are rational in $\lambda_i - \lambda_j$. Now apply Lemma 4.4 to $\Lambda_1(t) + U_1(t)$. Then there exists, for $t \geq t_2$, $T_2(t) = O(|U_1(t)|)$ such that (4.17) holds, i.e.,

$$[I + T_2(t)]^{-1}[\Lambda_1(t) + U_1(t)][I + T_2(t)] = \Lambda_2(t) + U_2(t), \tag{4.18}$$

with

$$\Lambda_2(t) = \Lambda_1(t) + \operatorname{diag} U_1(t),$$

and

$$U_2(t) = O(|U_1(t)|^2) = O\left(|V(t)|^4\right).$$

Since

$$(T_2)_{ij} = \frac{u_{ij}}{(\Lambda_1)_j - (\Lambda_1)_i} = \frac{u_{ij}}{(\Lambda_0)_j - (\Lambda_0)_i} \frac{1}{1 + \frac{v_{ii} - v_{jj}}{(\Lambda_0)_j - (\Lambda_0)_i}},$$

an expansion of the last fraction as a finite geometric sum in powers of $V(t)$ plus an L^1-perturbation shows that $T_2(t)$ is (up to an L^1-perturbation) a polynomial in entries of $V(t)$ with coefficients that are rational functions in Λ_0. An argument analogous to (4.18) shows that $[I + T_2(t)]^{-1}$ is also a polynomial in entries of $V(t)$ with coefficients that are rational functions in Λ_0.

Then by induction (after a finite number of steps $N \leq \log_2 p$) for $t \geq t^*$ sufficiently large, there exists a finite product of elementary factors

$$I + T(t) = [I + T_1(t)]\,[I + T_2(t)]\,\ldots\,[I + T_N(t)]$$

which diagonalizes $\Lambda_0 + V(t)$ up to an L^1-perturbation. It follows from the construction above that since all the matrices T_i, $(I + T_i)^{-1}$, U_i, and $\Lambda_i(t)$ are polynomials in the elements of V, then so is the resulting diagonal part. We formulate this result as

Proposition 4.5 *Let Λ_0 be diagonal with distinct eigenvalues. Assume that $V(t) = o(1)$ as $t \to \infty$ and that $V \in L^p[t_0, \infty)$ for some $p < \infty$. Then there exists for $t \geq t^*$ sufficiently large a matrix $T(t) = O(|V(t))|$ as $t \to \infty$ such that*

$$[I + T(t)]^{-1}[\Lambda_0 + V(t)][I + T(t)] = \tilde{\Lambda}(t) + \tilde{U}(t),$$

where $\tilde{\Lambda}(t) = \Lambda_0 + D(t)$, $\tilde{U}(t) \in L^1[t^*, \infty)$, and $D(t)$ is diagonal with entries that are polynomials in the elements of V with coefficients which are rational functions in Λ_0, all of which can be explicitly calculated.

Also see a similar discussion in Eastham's book [53, pp. 26–27]. Instead of solving the linear system (4.16) resulting in $U(t) = O\left(|V|^2\right)$, it was pointed out in [71, p. 574] that solving the somewhat more complicated (but still linear) system

$$\Lambda\hat{Q} - \hat{Q}\Lambda + V + V\hat{Q} - \operatorname{diag}\{V + V\hat{Q}\} - \hat{Q}\operatorname{diag}V = 0$$

would result in a corresponding residual term $\hat{U}(t) = O\left(|V|^3\right)$. This would yield an L^1-approximation to the eigenvalues of $A + V(t)$ in fewer steps, but each more complicated.

4.3.2 Eastham's Generalization

Eastham considered the case of coalescing eigenvalues or more general situations of perturbed diagonal systems (4.1), where $\Lambda(t)$ may have no limit as $t \to +\infty$. He proved a far-reaching extension of Theorem 4.1 (see [53, Theorem 1.6.1]), where the basic idea was to generalize Levinson's result on Q and then apply the Theorem 2.7. We present here a somewhat modified treatment of this result which is in the same spirit as that of Levinson and Eastham, so we will refer to it as the "Levinson–Eastham lemma." We will then apply it to obtain a modification of Eastham's theorem.

Lemma 4.6 (Levinson–Eastham) *Assume that* $\Lambda(t) = \operatorname{diag}\{\lambda_1(t), \ldots, \lambda_d(t)\}$ *and* $V(t)$ *are continuous* $d \times d$ *matrices for* $t \geq t_0$ *satisfying*

$$\lambda_i(t) \neq \lambda_j(t) \text{ for all } i \neq j, \tag{4.19}$$

$$\operatorname{diag} V(t) \equiv 0, \tag{4.20}$$

and

$$\frac{v_{ik}(t)}{\lambda_i(t) - \lambda_j(t)} \to 0 \qquad \text{as } t \to \infty \tag{4.21}$$

for all i, j, k *such that* $i \neq k$ *and* $i \neq j$. *Then there exists for* $t \geq t_1$ *sufficiently large a* $d \times d$ *matrix* $Q(t)$, $\operatorname{diag} Q \equiv 0$, *satisfying* (4.10) *and*

$$\|Q(t)\| = O\left(\max_{i \neq j}\left|\frac{v_{ij}(t)}{\lambda_j(t) - \lambda_i(t)}\right|\right) \qquad \text{as } t \to \infty. \tag{4.22}$$

Moreover, if $v_{ik}/(\lambda_i - \lambda_j)$ *are continuously differentiable for all* i, j, k *such that* $i \neq j$, *then* Q *is also continuously differentiable and satisfies*

$$\begin{aligned} \|Q'(t)\| &= O\left(\max_{i \neq j}\left|\left(\frac{v_{ij}}{\lambda_i - \lambda_j}\right)'\right|\right) \\ &\quad + O\left(\max_{j \neq i \neq k}\left|\left(\frac{v_{ik}}{\lambda_i - \lambda_j}\right)'\right| \|Q\|\right) \qquad \text{as } t \to \infty. \end{aligned} \tag{4.23}$$

Proof Letting $\tilde{\Lambda} = \Lambda + D$, where D is diagonal, it follows that (4.10) is equivalent to

$$Q\Lambda - \Lambda Q = V + VQ - D - QD. \tag{4.24}$$

Recalling (4.20) and also normalizing Q so that $\operatorname{diag} Q \equiv 0$, taking the diagonal terms on both sides in (4.24) leads to

$$D = \operatorname{diag}(VQ), \tag{4.25}$$

and then (4.24) can be re-written as

$$Q\Lambda - \Lambda Q = V + VQ - \operatorname{diag}(VQ) - Q\operatorname{diag}(VQ). \tag{4.26}$$

It remains to first determine conditions on (Λ, V) so that (4.26) has a solution $Q = o(1)$ as $t \to \infty$ and then to estimate Q and Q'.

With the diagonal terms on both sides of (4.26) being identically zero, it turns out to be convenient to re-write (4.26) as a $(d-1)d$-dimensional vector/matrix system for the off-diagonal elements. To that end, for a given $d \times d$ matrix M with $\operatorname{diag} M = 0$, we introduce the notation

$$M \hat{=} \mathbf{m}$$

to mean that the off-diagonal elements in M are re-written as the $d(d-1)$-dimensional column vector $\hat{m}$ according to the enumeration of the columns, left to right. Putting $Q \hat{=} \mathbf{q}$, easy calculations show that

$$Q\Lambda - \Lambda Q \hat{=} \hat{\Lambda}\mathbf{q},$$

where $\hat{\Lambda}$ is the $d(d-1) \times d(d-1)$ diagonal matrix

$$\hat{\Lambda} = \operatorname{diag}\{\lambda_1 - \lambda_2,\ \lambda_1 - \lambda_3,\ \ldots\ ,\lambda_1 - \lambda_d;\ \lambda_2 - \lambda_1, \lambda_2 - \lambda_3,\ \ldots\ ,\lambda_2 - \lambda_d\ ;\ \ldots\},$$

and

$$VQ - \operatorname{diag}(VQ) \hat{=} \hat{V}\mathbf{q},$$

where $\hat{V}$ is the $d(d-1) \times d(d-1)$ block-diagonal matrix

$$\hat{V} = \operatorname{diag}\{V_1,\ \ldots\ , V_d\}$$

and V_i is the $(d-1) \times (d-1)$-dimensional principal sub-matrix formed from V by deleting the ith row and ith column. Finally,

$$Q\operatorname{diag}(VQ) \hat{=} \hat{D}(V, Q)\mathbf{q},$$

where $\hat{D}(V,Q)$ is the $d(d-1) \times d(d-1)$ diagonal matrix

$$\hat{D}(V,Q) = \operatorname{diag}\{D_1, \ldots, D_d\} \tag{4.27}$$

and

$$D_i = \left(\sum_{j \neq i} v_{ij} q_{ji} \right) I_{d-1}. \tag{4.28}$$

With this notation, (4.26) is equivalent to

$$\hat{\Lambda}\mathbf{q} = \mathbf{v} + \hat{V}\mathbf{q} - \hat{D}(V,Q)\mathbf{q}, \tag{4.29}$$

and since by assumption (4.19) $\hat{\Lambda}$ is invertible for all $t \geq t_0$, (4.29) is equivalent to

$$\mathbf{q} = \hat{\Lambda}^{-1}\mathbf{v} + \hat{\Lambda}^{-1}\hat{V}\mathbf{q} - \hat{\Lambda}^{-1}\hat{D}(V,Q)\mathbf{q}. \tag{4.30}$$

For future reference, we remark that $\hat{\Lambda}^{-1}\mathbf{v}$ is the $d(d-1)$ column vector whose transpose is given by

$$\left[\hat{\Lambda}^{-1}\mathbf{v}\right]^T = \left\{ \frac{v_{21}}{\lambda_1 - \lambda_2}, \frac{v_{31}}{\lambda_1 - \lambda_3}, \ldots, \frac{v_{d1}}{\lambda_1 - \lambda_d}; \ \ldots \ ; \right.$$
$$\left. \frac{v_{1d}}{\lambda_d - \lambda_1}, \ldots, \frac{v_{d-1,d}}{\lambda_d - \lambda_{d-1}} \right\}, \tag{4.31}$$

$\hat{\Lambda}^{-1}\hat{V}$ is a $d(d-1) \times d(d-1)$ block-diagonal matrix whose elements in the ith block have the form

$$\frac{v_{jk}}{\lambda_i - \lambda_j} \qquad j \neq i, \quad j \neq k \tag{4.32}$$

and $\hat{\Lambda}^{-1}\hat{D}(V,Q)$ is a $d(d-1) \times d(d-1)$ diagonal matrix with elements of the form

$$\sum_{k \neq i} \frac{v_{ik} q_{ki}}{\lambda_i - \lambda_j}, \qquad (i \neq j). \tag{4.33}$$

Let $T[\mathbf{q}]$ denote the (quadratic) operator defined by

$$T[\mathbf{q}] = \hat{\Lambda}^{-1}\hat{V}\mathbf{q} - \hat{\Lambda}^{-1}\hat{D}(V,Q)\mathbf{q} \tag{4.34}$$

on the Banach space $\mathscr{B}_0$ of $d(d-1)$-dimensional vector-valued functions tending to zero as $t \to \infty$ equipped with the sup-norm $\|\mathbf{q}\| = \sup_{t \geq t_1} |\mathbf{q}(t)|$ for some t_1 specified below ($t_1 \geq t_0$). It is important to remember that (4.27) and (4.28) imply that the elements of $\hat{D}(V,Q)$ are linear functions of Q, hence linearly depend on the elements in $\mathbf{q}$. We claim that T is a contraction mapping from $\mathscr{B}_0$ into itself. To substantiate this claim, we consider for $Q_i \hat{=} \mathbf{q}_i$ $(i=1,2)$

$$\begin{aligned} T[\mathbf{q}_1]-T[\mathbf{q}_2] &= \hat{\Lambda}^{-1}\hat{V}\,[\mathbf{q}_1-\mathbf{q}_2] - \hat{\Lambda}^{-1}\left[\hat{D}(V,Q_1)\mathbf{q}_1 - \hat{D}(V,Q_2)\mathbf{q}_2\right] \\ &= \hat{\Lambda}^{-1}\hat{V}\,[\mathbf{q}_1-\mathbf{q}_2] - \hat{\Lambda}^{-1}\left[\hat{D}(V,Q_1) - \hat{D}(V,Q_2)\right]\mathbf{q}_1 \\ &\quad - \hat{\Lambda}^{-1}\hat{D}(V,Q_2)\,[\mathbf{q}_1-\mathbf{q}_2]\,, \end{aligned}$$

and therefore

$$\begin{aligned} \|T[\mathbf{q}_1]-T[\mathbf{q}_2]\| \le \|\hat{\Lambda}^{-1}\hat{V}\|\;\|\mathbf{q}_1-\mathbf{q}_2\| \\ + \left\|\hat{\Lambda}^{-1}\left[\hat{D}(V,Q_1)-\hat{D}(V,Q_2)\right]\right\|\;\|\mathbf{q}_1\| \\ + \;\; \|\hat{\Lambda}^{-1}\hat{D}(V,Q_2)\|\;\|\mathbf{q}_1-\mathbf{q}_2\|. \end{aligned} \tag{4.35}$$

By the descriptions of the elements of $\hat{\Lambda}^{-1}\hat{V}$ and $\hat{\Lambda}^{-1}\hat{D}(V,Q)$ given in (4.32) and (4.33), respectively, and (4.21), it follows that

$$\|T[\mathbf{q}_1]-T[\mathbf{q}_2]\| \le \left\|\hat{\Lambda}^{-1}\left[\hat{D}(V,Q_1)-\hat{D}(V,Q_2)\right]\right\|\;\|\mathbf{q}_1\| + (k_1+k_3)\|\mathbf{q}_1-\mathbf{q}_2\|,$$

where k_1 and k_3 are positive constants that can be taken arbitrarily small for $t \ge t_1$ sufficiently large. From the definition of $\hat{D}(V,Q)$, it follows that the ith diagonal block in $\hat{D}(V,Q_1)-\hat{D}(V,Q_2)$ equals $\sum_{j\neq i} v_{ij}\left(q_{ji}^{(1)} - q_{ji}^{(2)}\right) I_{d-1}$, where $q_{ji}^{(1)}$ resp. $q_{ji}^{(2)}$ denote the entries of $\mathbf{q}_1$, resp. $\mathbf{q}_2$. So the elements in the ith diagonal block of $\hat{\Lambda}^{-1}\left[\hat{D}(V,Q_1)-\hat{D}(V,Q_2)\right]$ equal

$$\sum_{j\neq i} v_{ij}\left(q_{ji}^{(1)} - q_{ji}^{(2)}\right) \operatorname{diag}\left\{\frac{1}{\lambda_i-\lambda_1},\ldots,\frac{1}{\lambda_i-\lambda_d}\right\},$$

and therefore

$$\left\|\hat{\Lambda}^{-1}\left[\hat{D}(V,Q_1)-\hat{D}(V,Q_2)\right]\right\| \le k_2\|\mathbf{q}_1-\mathbf{q}_2\|$$

from assumption (4.21) again, where the positive constant k_2 can also be made arbitrarily small by taking $t \ge t_1$ sufficiently large. Therefore, recalling that $\|\mathbf{q}_1\|$ is bounded, T is a contraction for such $t \ge t_1$ sufficiently large. To see that T maps $\mathscr{B}_0$ into itself, recall that $\hat{\Lambda}^{-1}\hat{V}$ and $\hat{\Lambda}^{-1}\hat{D}(V,Q)$ both tend to zero as $t \to \infty$.

Finally, since $\hat{\Lambda}^{-1}\mathbf{v}$ defined in (4.31) also vanishes at infinity, it follows from the Contraction Mapping Principle that (4.30), re-written using (4.34) as

$$\mathbf{q} = \hat{\Lambda}^{-1}\mathbf{v} + T[\mathbf{q}],$$

has a unique solution $\mathbf{q} \in \mathscr{B}_0$, which is a solution of (4.30). Rewriting (4.30) as

$$\left[I - \hat{\Lambda}^{-1}\hat{V} - \hat{\Lambda}^{-1}\hat{D}(V,Q)\right]\mathbf{q} = \hat{\Lambda}^{-1}\mathbf{v},$$

and noting that the elements of $\hat{\Lambda}^{-1}\mathbf{v}$ are of the form $v_{ij}/(\lambda_j - \lambda_i)$, we see that (4.22) follows by (4.31) since $\hat{\Lambda}^{-1}\hat{V}$ and $\hat{\Lambda}^{-1}\hat{D}(V,Q)$ go to zero as $t \to \infty$.

Now to prove (4.23), we first use the implicit function theorem to establish the differentiability of $\mathbf{q}(t)$ (i.e., also $Q(t)$) provided that $v_{jk}/(\lambda_i - \lambda_j)$ are continuously differentiable for all i, j, k such that $i \neq j$. To that end, note that (4.30) can be thought of as $F(t, \mathbf{q}) = 0$ with $F(\infty, 0) = 0$ and $F_{\mathbf{q}}(\infty, 0) = I$. For F to be in C^1, as assumed in the implicit function theorem, the partial derivatives of F have to be continuous. Since $\frac{\partial F}{\partial q_l}$ consists of terms of the form $v_{jk}/(\lambda_i - \lambda_j)$ and $v_{ik}/(\lambda_i - \lambda_j)$, these partial derivatives are continuous by the assumption that V and Λ are continuous and by (4.21). Now $\frac{\partial F}{\partial t}$ consists of terms of the form $\pm[v_{jk}/(\lambda_i - \lambda_j)]'$ which we require to be continuous. So the implicit function theorem implies that $Q = Q(t)$ is continuously differentiable for all t sufficiently large. Hence, we can differentiate (4.30) to obtain

$$\begin{aligned} \mathbf{q}' = \left(\hat{\Lambda}^{-1}\mathbf{v}\right)' + \left[\left(\hat{\Lambda}^{-1}\hat{V}\right)' + \left(\hat{\Lambda}^{-1}\hat{D}(V,Q)\right)'\right]\mathbf{q} \\ + \left[\hat{\Lambda}^{-1}\hat{V} + \hat{\Lambda}^{-1}\hat{D}(V,Q)\right]\mathbf{q}'. \end{aligned} \tag{4.36}$$

From the definition of $\hat{D}(V,Q)$ in (4.27) and (4.28), we obtain in the ith block

$$\begin{aligned} \left[\hat{\Lambda}^{-1}\hat{D}(V,Q)\right]_i' &= \sum_{k\neq i} \operatorname{diag}\left\{\left(\frac{v_{ik}}{\lambda_i - \lambda_1}\right)' q_{ki}, \cdots, \left(\frac{v_{ik}}{\lambda_i - \lambda_d}\right)' q_{ki}\right\} \\ &\quad + \sum_{k\neq i} \operatorname{diag}\left\{\left(\frac{v_{ik}}{\lambda_i - \lambda_1}\right) q'_{ki}, \cdots, \left(\frac{v_{ik}}{\lambda_i - \lambda_d}\right) q'_{ki}\right\} \\ &=: G_i + H_i. \end{aligned}$$

Putting $G = \operatorname{diag}\{G_1, \ldots, G_d\}$, and $H = \operatorname{diag}\{H_1, \ldots, H_d\}$, it follows that

$$\|G\| = \mathrm{O}\left(\max_{k\neq i\neq j}\left|\left(\frac{v_{ik}}{\lambda_i - \lambda_j}\right)'\right| \|Q\|\right)$$

and that $H\mathbf{q}$ can be re-written as $H\mathbf{q} = H_0\mathbf{q}'$, where H_0 is the $d(d-1)$ matrix whose entries are of the form $q_{mn}\sum_{k\neq i}\frac{v_{ik}}{\lambda_i-\lambda_j}$. Therefore

$$\left[\hat{\Lambda}^{-1}\hat{D}(V,Q)\right]'\mathbf{q} = (G+H)\mathbf{q} = G\mathbf{q} + H_0\mathbf{q}',$$

and

$$\|H_0\| = \mathrm{O}\left(\max_{k\neq i\neq j}\left|\frac{v_{ik}}{\lambda_i-\lambda_j}\right| \|Q\|\right) = \mathrm{o}(1) \quad \text{as } t\to\infty.$$

As shown above, $Q(t)\to 0$ as $t\to\infty$, thus (4.36) can be expressed for $t\to\infty$ as

$$\left[I - \hat{\Lambda}^{-1}\hat{V} - \hat{\Lambda}^{-1}\hat{D}(V,Q) - H_0\right]\mathbf{q}' = (\hat{\Lambda}^{-1}\mathbf{v})' + \left[(\hat{\Lambda}^{-1}\hat{V})' + G\right]\mathbf{q}.$$

Since the coefficient matrix in front of $\mathbf{q}'$ tends to the identity matrix I as $t\to\infty$, we obtain

$$\|\mathbf{q}'\| = \mathrm{O}\left(\left|(\hat{\Lambda}^{-1}\mathbf{v})'\right|\right) + \mathrm{O}\left(\left|(\hat{\Lambda}^{-1}\hat{V})'\right| \|Q\|\right) + \mathrm{O}\left(\max\left|\left(\frac{v_{ik}}{\lambda_i-\lambda_j}\right)'\right| \|Q\|^2\right),$$

as $t\to\infty$, which yields (cp. (4.31) and (4.32))

$$\begin{aligned}\|Q'\| &= \mathrm{O}\left(\max\left|\left(\frac{v_{ij}}{\lambda_j-\lambda_i}\right)'\right|\right) + \mathrm{O}\left(\max\left|\left(\frac{v_{jk}}{\lambda_i-\lambda_j}\right)' \|Q\|\right|\right) \\ &\quad + \mathrm{O}\left(\max\left|\left(\frac{v_{ik}}{\lambda_i-\lambda_j}\right)'\right| \|Q\|^2\right) \\ &= \mathrm{O}\left(\max\left|\left(\frac{v_{ij}}{\lambda_j-\lambda_i}\right)'\right|\right) + \mathrm{O}\left(\max\left|\left(\frac{v_{ik}}{\lambda_i-\lambda_j}\right)'\right| \|Q\|\right) \qquad \text{as } t\to\infty.\end{aligned}$$

□

Remark 4.7 In his proof, Eastham [53, Theorem 1.6.1] did not choose to normalize diag $V \equiv 0$ as we do, but his equations (1.6.17) are equivalent to Eqs. (4.29) and (4.30). We found it convenient for the arguments used here to express the equations in system form using matrices and quadratic forms and use an operator approach for their solution as an alternative to Eastham's successive approximations. Note we have assumed that $v_{ik}/(\lambda_i-\lambda_j)$ $(i\neq j,k)$ are continuously differentiable in order to apply the standard version of the implicit function theorem to $\mathbf{q} = F(t,\mathbf{q})$. Eastham instead assumes that $v_{ik}/(\lambda_i-\lambda_j)$ are in L^1 and uses this assumption together with the dominated convergence theorem to obtain the estimate for Q'.

Now applying Lemma 4.6, we obtain the following extension of Theorem 4.1, which is comparable to Eastham's theorem [53, Theorem 1.6.1].

Theorem 4.8 (Levinson–Eastham) *Consider*

$$y' = [\Lambda(t) + V(t) + R(t)]\,y, \qquad t \geq t_0, \tag{4.37}$$

where $\Lambda(t) = \operatorname{diag}\{\lambda_1(t), \ldots, \lambda_d(t)\}$ *and* $V(t)$ *are continuous* $d \times d$ *matrices for* $t \geq t_0$ *satisfying* (4.19)–(4.21) *and where the continuous* $d \times d$ *matrix* $R(t) \in L^1[t_0, \infty)$. *Assume that* $v_{ik}/(\lambda_j - \lambda_i)$ *are continuously differentiable (a.e.) for all* i, j, k *such that* $i \neq j$ *and satisfy*

$$\left(\frac{v_{ij}}{\lambda_i - \lambda_j}\right)' \in L^1[t_0, \infty) \qquad 1 \leq i \neq j \leq d, \tag{4.38}$$

and for all $j \neq i \neq k$

$$\left(\frac{v_{ik}}{\lambda_i - \lambda_j}\right)' \max_{i \neq j} \left|\frac{v_{ij}}{\lambda_i - \lambda_j}\right| \in L^1[t_0, \infty). \tag{4.39}$$

If the eigenvalues of $\Lambda(t) + V(t)$ *also satisfy Levinson's dichotomy conditions* (2.13), (2.14), *then* (4.37) *has a fundamental matrix satisfying*

$$Y(t) = [I + o(1)] \exp\left[\int^t \hat{\Lambda}(\tau)\, d\tau\right], \qquad \text{as } t \to \infty,$$

where $\hat{\Lambda}$ *is a diagonal matrix consisting of the eigenvalues of* $\Lambda + V$.

Proof In (4.37), we make the conditioning transformation $y = [I + Q(t)]z$, where $Q(t)$ satisfies (4.10) and (4.22) by Lemma 4.6. Then it follows from (4.37) that

$$z' = \left[\hat{\Lambda}(t) + R_1(t) - [I + Q(t)]^{-1} Q'(t)\right] z,$$

where $R_1(t) = (I + Q)^{-1} R (I + Q)$. Since $Q(t) \to 0$ as $t \to \infty$ by (4.22) and (4.21), $R_1 \in L^1[t_0, \infty)$ and it suffices to show that $Q' \in L^1[t_0, \infty)$ in order to apply Theorem 2.7. But this follows from (4.22), (4.38) and (4.39), recalling that that Q' was estimated in (4.23). Since the diagonal matrix $\hat{\Lambda}(t)$ satisfies Levinson's dichotomy condition, the conclusion follows from Theorem 2.7. □

Remark 4.9 In his statement of Theorem 4.8 (see [53, Theorem 1.6.1]), Eastham assumes that

$$\frac{V(t)}{\lambda_i(t) - \lambda_j(t)} \to 0 \qquad \text{as } t \to \infty,$$

for all $i \neq j$, which means

$$\frac{v_{kl}(t)}{\lambda_i(t) - \lambda_j(t)} \to 0 \qquad \text{as } t \to \infty, \tag{4.40}$$

for all $k \neq l$, $i \neq j$. He also assumes that

$$\left(\frac{V(t)}{\lambda_i(t) - \lambda_j(t)}\right)' \in L^1[t_0, \infty). \tag{4.41}$$

In his proof, however, it is seen that conditions (4.21) are sufficient for the existence of $Q(t) = \mathrm{o}(1)$ satisfying (4.10). (Likewise, his condition (1.6.3) can be weakened to $\left(v_{ik}/(\lambda_j - \lambda_i)\right)' \in L^1$ for all i, j, k with $i \neq j$.) The example with

$$\Lambda(t) + V(t) = \begin{pmatrix} 0 & & \\ & \frac{1}{\sqrt{t}} & \\ & & \frac{1}{t} \end{pmatrix} + \begin{pmatrix} 0 & \mathrm{O}\left(\frac{1}{t^{1+\varepsilon}}\right) & \mathrm{O}\left(\frac{1}{t^{1+\varepsilon}}\right) \\ \mathrm{O}\left(\frac{1}{t^{1/2+\varepsilon}}\right) & 0 & \mathrm{O}\left(\frac{1}{t^{1/2+\varepsilon}}\right) \\ \mathrm{O}\left(\frac{1}{t^{1+\varepsilon}}\right) & \mathrm{O}\left(\frac{1}{t^{1+\varepsilon}}\right) & 0 \end{pmatrix} \qquad (\varepsilon > 0)$$

satisfies (4.21), but not the more restrictive (4.40).

Observe that both the conditions (4.21) as well as the above example display a certain kind of asymmetry. This is due to the fact that in expressing (4.26) as a nonlinear vector system (4.29), we chose $\mathbf{q}$ corresponding to a concatenation of the non-diagonal elements in the *columns* of Q. If we had chosen *rows* instead, the *dual* conditions

$$\frac{v_{kj}}{\lambda_i(t) - \lambda_j(t)} \to 0 \qquad \text{as } t \to \infty, \tag{4.42}$$

can be seen to be sufficient and could be used in place of (4.21). The corresponding statements (4.23) and assumption (4.39) should then also be modified to reflect the dual assumption (4.42).

Finally, we note that conditions (4.38) and (4.39) are weaker than Eastham's original conditions (4.40) and (4.41). But on the other hand the assumption of continuous differentiability is more restrictive.

As an immediate application of Theorem 4.8, consider the meromorphic system

$$y' = A(z)y, \qquad A(z) = z^r \sum_{k=0}^{\infty} \frac{A_k}{z^k}. \tag{4.43}$$

Here $r \geq 0$ is an integer and $A(z)$ converges for $z \in \mathbb{C}$, $|z| > a$. For the moment, we assume without loss of generality that $z \in \mathbb{R}$, $z > a$, but note that the discussion remains the same on every ray $z = \rho e^{i\theta}$ with $\rho > a$. We also suppose that the leading matrix is diagonal, i.e., $A_0 = \mathrm{diag}\{\lambda_1, \ldots, \lambda_d\}$, where $\lambda_i \neq \lambda_j$ for $1 \leq i \neq j \leq d$. We re-write

$$A(z) = \Lambda(z) + V(z) + R(z),$$

where

$$\Lambda(z) = z^r A_0 + \operatorname{diag}\left(z^r \sum_{k=1}^{r+1} \frac{A_k}{z^k}\right),$$

$$V(z) = z^r \sum_{k=1}^{r+1} \frac{A_k}{z^k} - \operatorname{diag}\left(z^r \sum_{k=1}^{r+1} \frac{A_k}{z^k}\right),$$

$$R(z) = z^r \sum_{k=r+2}^{\infty} \frac{A_k}{z^k}.$$

Noting that the eigenvalues of $\Lambda(z) + V(z) = z^r \sum_{k=0}^{r+1} \frac{A_k}{z^k}$ automatically satisfy Levinson's dichotomy conditions, one finds that $\Lambda(z)$, $V(z)$ and $R(z)$ satisfy (for z sufficiently large) the hypotheses of Theorem 4.8. Therefore (4.43) has for z sufficiently large a fundamental matrix of the form

$$Y(z) = [I + E(z)] \exp\left[\int^z \tilde{\Lambda}(\zeta)\, d\zeta\right],$$

where $\tilde{\Lambda}(z)$ is a diagonal matrix consisting of eigenvalues of $\Lambda(z)+V(z)$ and $E(z) = o(1)$ as $z \to \infty$.

We refer the reader to Balser [3] for a modern discussion of the meromorphic theory and in particular how so-called *formal fundamental solutions* can be constructed not only in this case, but also when A_0 is a general constant matrix.

In this book we are, however, interested primarily in treating differential equations which are not in this meromorphic class. For example the scalar, second order differential equation $y'' + \left[\lambda + \frac{\sin t^\alpha}{t^\beta}\right] y = 0$, which will be considered in Chap. 8, is not in the meromorphic class, yet it still can be asymptotically integrated for a wide range of parameter values of α and β. Even though the meromorphic theory does not apply, it is a motivating force behind some of the techniques we use, which are modifications and adaptations of ones which have been useful there.

We conclude this section by mentioning another extension of Theorem 4.1. Gingold [66] considered a system equivalent to

$$y' = [\psi(t)\hat{\Lambda}(t) + V(t) + R(t)]y, \qquad t \geq t_0 \tag{4.44}$$

where $\psi(t)$ is a nonzero scalar function, $\hat{\Lambda}(t) = \operatorname{diag}\{\hat{\lambda}_1(t), \ldots, \hat{\lambda}_d(t)\}$ such that

$$\lim_{t\to\infty} [\hat{\lambda}_i(t) - \hat{\lambda}_j(t)] := \hat{\lambda}_{ij} \text{ exists and } \hat{\lambda}_{ij} \neq 0 \text{ for } i \neq j, \tag{4.45}$$

$V(t) = o(\psi(t))$ as $t \to \infty$, diag $V \equiv 0$, and $R \in L^1[t_0, \infty)$. Furthermore, $\psi \hat{\Lambda}$ was assumed to satisfy Levinson's dichotomy conditions (2.13), (2.14), and

$$|(\psi^{-1}V)'| \text{ and } |\hat{\Lambda}'||\psi^{-1}V| \in L^1[t_0, \infty) \tag{4.46}$$

and

$$\psi^{-1} \sum |v_{ik}v_{lm}| \in L^1[t_0, \infty), \tag{4.47}$$

where the summation is taken over the indices i, k, l, m all in $\{1, \dots d\}$ such that $i \neq k, l \neq m, i \neq l, k \neq m$. Finally, ψ, $\hat{\Lambda}$, and V are all in $C^1[t_0, \infty)$. Under these assumptions, Gingold showed that (4.44) has a fundamental solution matrix of the form

$$Y(t) = [I + o(1)] \exp\left[\int^t \psi(\tau)\hat{\Lambda}(\tau)\, d\tau\right] \qquad \text{as } t \to \infty. \tag{4.48}$$

We first note that this result is a special case of Theorem 4.8. It is straightforward to show that the assumptions above ensure that the hypotheses of Theorem 4.8 are satisfied for $\Lambda(t) := \psi(t)\hat{\Lambda}(t)$, which then implies the existence of a fundamental matrix of (4.44) of the form

$$Y(t) = [I + o(1)] \exp\left[\int^t \tilde{\Lambda}(\tau)\, d\tau\right],$$

where $\tilde{\Lambda}(t)$ consists of the eigenvalues of $\psi(t)\hat{\Lambda}(t) + V(t)$. It follows from (4.10) and (4.25) that $\tilde{\Lambda} = \psi\hat{\Lambda} + \text{diag}\,(VQ)$. Finally, (4.22) and (4.47) imply that $VQ \in L^1[t_0, \infty)$ which establishes (4.48).

Secondly, one can show Gingold's result directly (and shorten the original proof significantly), by making for t sufficiently large in (4.44) a conditioning transformation $y = [I + Q(t)]z$, where

$$q_{ij} = \frac{-v_{ij}}{\psi(\hat{\lambda}_i - \hat{\lambda}_j)} \qquad i \neq j,$$

and $q_{ii} = 0$. Note that the assumptions above imply that $Q(t)$ is well-defined for t sufficiently large and that $Q(t) \to 0$ as $t \to \infty$. It follows that $\psi(\hat{\Lambda}Q - Q\hat{\Lambda}) + V = 0$, and the transformation $y = [I + Q(t)]z$ takes (4.44) into

$$z' = \left(\psi\hat{\Lambda} + [I + Q]^{-1}\left[VQ - Q' + R(I + Q)\right]\right) z.$$

Now $Q' \in L^1[t_0, \infty)$ by (4.45) and (4.46), and the (i, m)-element of VQ is given by

$$(VQ)_{im} = -\sum_{\mu=1}^{d} \frac{v_{i\mu}v_{\mu m}}{\psi(\hat{\lambda}_\mu - \hat{\lambda}_m)},$$

which can be seen to be absolutely integrable by setting $k = l$ in (4.47). Hence

$$z' = \left(\psi \hat{\Lambda} + R_1\right) z \qquad \text{where } R_1 \in L^1[t_0, \infty),$$

and $\psi \hat{\Lambda}$ satisfies Levinson's dichotomy conditions, hence (4.48) follows from Theorem (2.7).

4.3.3 *Higher Derivatives*

Now we turn to the question of weakening Levinson's condition $V' \in L^1$ (or the more general conditions (4.38), (4.39)). To our knowledge, the first such case of this is due to A. Devinatz [48], who assumed (see the result below for more details, including other hypotheses) that $V'' \in L^1$. This was later extended to conditions on higher order derivatives (see [69] and [53]). In these extensions, the idea is to make further conditioning transformations so that in the end Levinson's fundamental theorem may still be applied. In doing so, it could happen that extra terms enter the diagonal, perturbing it from just the eigenvalues of $A+V(t)$, so additional conditions are required to insure that the extra term are in L^1. Or, without such conditions, the resulting asymptotic integration might contain some not absolutely integrable, but computable perturbations of the eigenvalues. These possibilities then give rise to many different kinds of results, but all following the same basic theme of using suitable conditioning transformations followed by an application of Theorem 2.7.

Theorem 4.10 (Devinatz) *Let A be a constant matrix with distinct eigenvalues, $V''(t)$ and $R(t)$ be continuous for $t \geq t_0$, $V(t) \to 0$ as $t \to \infty$,*

$$(V')^2,\ VV',\ V'',\ R \in L^1[t_0, \infty), \tag{4.49}$$

and assume that the eigenvalues of $A + V(t)$ satisfy Levinson's dichotomy condition (see Definition 2.6). Then

$$y' = [A + V(t) + R(t)]y \tag{4.50}$$

has a fundamental matrix of the form (4.13) as $t \to \infty$, where $P^{-1}AP$ is a diagonal matrix, and where $\Lambda(t)$ in (4.13) is a diagonal matrix with components the eigenvalues of $A + V(t)$.

Proof In (4.50), put $y = Pz$, where $P^{-1}AP = \Lambda = \mathrm{diag}\{\lambda_1, \dots, \lambda_d\}$. Then

$$z' = \left[\Lambda + \hat{V}(t) + \hat{R}(t)\right] z,$$

where $\hat{V}(t) = P^{-1}V(t)P$ and $\hat{R}(t) = P^{-1}R(t)P$, and $\hat{V}$, $\hat{R}$ satisfy (4.49). Put $z = [I + Q(t)]\, w$ for $t \geq t_1 \geq t_0$, where $Q(t) \to 0$, diag $Q(t) \equiv 0$, $I + Q(t)$ is invertible for $t \geq t_1$ and

$$[I + Q(t)]^{-1}\, [\Lambda + \hat{V}(t)]\, [I + Q(t)] = \Lambda(t). \tag{4.51}$$

Hence $\Lambda(t)$ consists of eigenvalues of $A + V(t)$. In particular, (see [69, Lemma 1])

$$Q^{(k)} = \mathrm{O}\left(|V^{(k)}(t)|\right) \qquad \text{for } k = 0, 1, 2. \tag{4.52}$$

It follows for $t \geq t_1$ that

$$w' = \left[\Lambda(t) - [I + Q(t)]^{-1} Q'(t) + [I + Q(t)]^{-1}\, \hat{R}(t)\, [I + Q(t)]\right] w. \tag{4.53}$$

Using the identity

$$[I + Q(t)]^{-1} = I - Q(t) + Q^2(t)[I + Q(t)]^{-1},$$

and noting that $QQ' \in L^1[t_1, \infty)$ by (4.49) and (4.52), (4.53) can be re-written as

$$w' = \left[\Lambda(t) - Q'(t) + \tilde{R}(t)\right], \qquad \tilde{R} \in L^1[t_1, \infty). \tag{4.54}$$

We next claim that Q' does not have an impact of the exponential term in the solution matrix. To that end, write

$$\Lambda(t) = \Lambda + D(t)$$

for some diagonal matrix $D(t) \to 0$ as $t \to \infty$. Using (4.51) and (4.52), it follows that $D(t) = \mathrm{O}\left(|V(t)|\right)$. Define $T(t)$ by

$$\Lambda T(t) - T(t)\Lambda + Q'(t) = 0,$$

or $t_{ij}(t) = q'_{ij}(t)/(\lambda_j - \lambda_i)$ $(i \neq j)$ and $t_{ii} \equiv 0$. Thus $T(t) = \mathrm{O}\left(|V'(t)|\right)$. Note the absolute integrability of V'' implies that $V'(t) = C - \int_t^\infty V''(\tau)\, d\tau$ for some constant matrix C. Moreover, since $V' \in L^2$, it follows that $C = 0$. Hence $V'(t)$ and $T(t)$ vanish at infinity, and $I + T(t)$ is invertible for sufficiently large t.

Put

$$w = [I + T(t)]v.$$

Then (4.54) implies that for all t sufficiently large

$$\begin{aligned} v' &= \{\Lambda(t) + [I + T(t)]^{-1} [D(t)T(t) - T(t)D(t) \\ &\quad -Q'(t)T(t) - T'(t) + \tilde{R}(I + T)]\} v \\ &= \{\Lambda(t) + [\mathrm{O}(|V||V'|) + \mathrm{O}(|V'|^2) + \mathrm{O}(|V''|) + \mathrm{O}(|R|)]\} v. \end{aligned} \tag{4.55}$$

Recalling (4.49), (4.55) can be seen to be an absolutely integrable perturbation of the diagonal system $z' = \Lambda(t)z$ which satisfies Levinson's dichotomy conditions. Therefore Theorem 2.7 implies that (4.55) has a fundamental matrix satisfying, as $t \to \infty$,

$$V(t) = [I + \mathrm{o}(1)] \exp\left[\int^t \Lambda(\tau)\, d\tau\right].$$

Hence (4.50) has a fundamental matrix for the form

$$Y(t) = P[I + Q(t)][I + T(t)]V(t) = [P + \mathrm{o}(1)] \exp\left[\int^t \Lambda(\tau)\, d\tau\right],$$

as $t \to \infty$. □

We end this section with several observations.

Devinatz [48, p. 362] considered a scalar example with $V(t)$ given by

$$V(t) = -\int_{t^\alpha}^{\infty} \frac{\sin \tau}{\tau}\, d\tau, \qquad \frac{1}{2} < \alpha < 1.$$

Using integration by parts, he showed that V satisfies the assumptions of Theorem 4.10, but not of Theorem 4.1.

We should note that Devinatz formulated his result in a more general way (see [48, p. 354]). However, the main idea was a perturbation V satisfying the hypotheses of Theorem 4.10, and we present his idea in this somewhat simplified form. We refer to [69, Remark 3.1] which addresses the original Devinatz result in this simplified framework.

Harris and Lutz [69, Theorem 1] gave a generalization of Theorem 4.10 to higher order derivatives of perturbations of constant diagonal systems, and we refer the reader to this publication for details.

In [53, Theorem 1.7.1], Eastham gives another type of extension to perturbations of general diagonal systems $y' = [\Lambda(t) + V(t) + R(t)]y$ with conditions involving higher order derivatives of $\Lambda(t)$ and $V(t)$. Using his approach described above in Theorem 4.8, he gives sufficient conditions for an asymptotic integration result of the form

$$Y(t) = [I + \mathrm{o}(1)] \exp\left[\int^t \{\Lambda(\tau) + \operatorname{diag} V(\tau)\}\, d\tau\right].$$

These conditions concern an inductively-defined sequence of matrices formed from $V(t)$ and $\Lambda(t)$.

In case some of the absolute-integrability conditions might not hold (as in the theorems of Devinatz and Harris–Lutz), we note that a sequence of conditioning transformations might still lead to an asymptotic integration result, but with appropriate modification to $\Lambda(t)+\text{diag } V(t)$ arising from a sequence of conditioning transformations. Since there are many options for such transformations, we just emphasize that the procedures used by Eastham and Harris–Lutz might still apply and yield some asymptotic integration results even though the conditions in the theorems are not all satisfied.

4.4 L^p-Perturbations with $p > 1$

In this section we consider perturbed diagonal systems $y' = [\Lambda(t) + V(t)]y$, where V is in some L^p-class with $1 < p < \infty$. Easy examples already show that generally the asymptotic behavior of solutions can be markedly different than when $V \in L^1$ even if Levinson's dichotomy conditions hold (see Sect. 2.3).

Results of this type for $V \in L^p$ were first obtained by Hartman and Wintner [73, Part III] in the case $1 < p \leq 2$. Their methods, involving solutions of Riccati differential inequalities and estimates, appear to be quite different from those which we will employ. We will follow here the approach in Harris and Lutz [71], who showed that the Hartman–Wintner result for fundamental matrices can be more easily proven using a certain conditioning transformation and then applying Levinson's fundamental theorem. We note, however, that in the case $1 < p \leq 2$, Theorem 2.19 yields a proof for a vector-valued solution using direct estimates of the T-operator. However, the proof-technique using a conditioning transformation has the advantage to lead to extensions of those results for $2 < p < \infty$ (see Corollary 4.12).

The dichotomy conditions used by Hartman–Wintner and Harris–Lutz involve "pointwise" assumptions that

$$\text{either Re}\,[\lambda_i(t) - \lambda_j(t)] \geq \delta > 0 \qquad \text{or} \qquad \text{Re}\,[\lambda_i(t) - \lambda_j(t)] \leq -\delta < 0, \tag{4.56}$$

for $t \geq t_0$ and $i \neq j$. These imply an "exponential dichotomy" for solutions of the unperturbed equation. Hsieh and Xie [80] later showed that such pointwise conditions can be weakened to "averaged exponential dichotomy conditions" (see (4.57) and (4.58) below) in case $1 < p \leq 2$ and obtained the following result. While their proof was based on a block-diagonalization result by Sibuya [136], we will present the proof in the spirit of Harris and Lutz [71].

Theorem 4.11 (Hartman–Wintner) *Let $\Lambda(t)$ and $V(t)$ be continuous $d \times d$-matrices for all $t \geq t_0$. Assume that $\Lambda(t) = \text{diag}\,\{\lambda_1(t), \ldots, \lambda_d(t)\}$ and assume its* *entries satisfy for all* $1 \leq i \neq j \leq d$ *the exponential dichotomy condition*

$$\text{either } e^{\mathrm{Re}\int_s^t \{\lambda_i(\tau)-\lambda_j(\tau)\}\,d\tau} \leq Ke^{-\delta(t-s)}, \quad t_0 \leq s \leq t \tag{4.57}$$

$$\text{or } \quad e^{\mathrm{Re}\int_s^t \{\lambda_i(\tau)-\lambda_j(\tau)\}\,d\tau} \leq Ke^{-\delta(s-t)}, \quad t_0 \leq t \leq s, \tag{4.58}$$

where $K \geq 1$ *and* $\delta > 0$ *are constants. Assume that there exists a constant* p, $1 < p \leq 2$ *such that* $V \in L^p[t_0, \infty)$, *i.e.,*

$$\int_{t_0}^{\infty} |V(t)|^p\,dt < \infty. \tag{4.59}$$

Then

$$y'(t) = [\Lambda(t) + V(t)]\,y \tag{4.60}$$

has a fundamental matrix satisfying

$$Y(t) = [I + o(1)]\exp\left[\int^t \{\Lambda(\tau) + \operatorname{diag} V(\tau)\}\,d\tau\right] \qquad \text{as } t \to \infty. \tag{4.61}$$

Proof In (4.60), we set

$$y = [I + Q(t)]z, \tag{4.62}$$

where $\operatorname{diag} Q(t) \equiv 0$ and the off-diagonal elements of Q will be chosen as appropriate solutions of

$$Q' = \Lambda Q - Q\Lambda + V - \operatorname{diag} V. \tag{4.63}$$

In coordinate representation, this is equivalent to

$$q'_{ij} = (\lambda_i - \lambda_j)q_{ij} + v_{ij} \qquad 1 \leq i \neq j \leq d.$$

If the ordered pair (i,j), $i \neq j$, satisfies (4.57), choose the solution

$$q_{ij}(t) = \int_{t_0}^{t} \exp\left[\int_s^t [\lambda_i(\tau) - \lambda_j(\tau)]\,d\tau\right] v_{ij}(s)\,ds. \tag{4.64}$$

If the ordered pair (i,j) satisfies (4.58), choose the solution

$$q_{ij}(t) = -\int_{t}^{\infty} \exp\left[\int_s^t [\lambda_i(\tau) - \lambda_j(\tau)]\,d\tau\right] v_{ij}(s)\,ds. \tag{4.65}$$

In the following, we will show that the matrix $Q(t)$ with off-diagonal elements given by (4.64) and (4.65) is well defined and that it has the properties $Q(t) \to 0$ as $t \to \infty$ and $Q \in L^p[t_0, \infty)$. It is then straightforward to show that (4.64) and (4.65) are solutions of (4.63).

To show that (4.65) is well defined, note that Hölder's inequality implies for $1/p + 1/p' = 1$ that

$$\begin{aligned}|q_{ij}(t)| &\leq K\left[\int_t^\infty e^{-p'\delta(s-t)}\,ds\right]^{1/p'}\left[\int_t^\infty |v_{ij}(s)|^p\,ds\right]^{1/p}\\ &= K\left[1/(\delta p')\right]^{1/p'}\left[\int_t^\infty |v_{ij}(s)|^p\,ds\right]^{1/p},\end{aligned}$$

hence $q_{ij}(t)$ given in (4.65) is well-defined and satisfies that $q_{ij}(t) \to 0$ as $t \to \infty$.

Similarly, if q_{ij} is defined in (4.64), then it follows by (4.57) that

$$|q_{ij}(t)| \leq K\int_{t_0}^t e^{-\delta(t-s)}|v_{ij}(s)|\,ds.$$

Given $\varepsilon > 0$, fix $t_1 \geq t_0$ sufficiently large such that

$$\left[\int_{t_1}^\infty |v_{ij}(s)|^p\,ds\right]^{1/p} < \frac{\varepsilon}{2K}\left(p'\delta\right)^{1/p'}.$$

Hence applying Hölder's inequality once more for $t \geq t_1$ shows that

$$\begin{aligned}|q_{ij}(t)| &\leq e^{-\delta t}K\int_{t_0}^{t_1} e^{\delta s}|v_{ij}(s)|\,ds\\ &\quad + K\left(\int_{t_1}^t e^{-p'\delta(t-s)}\,ds\right)^{1/p'}\left(\int_{t_1}^\infty |v_{ij}(s)|^p\,ds\right)^{1/p}\\ &= e^{-\delta t}K\int_{t_0}^{t_1} e^{\delta s}|v_{ij}(s)|\,ds + \varepsilon/2,\end{aligned}$$

hence $|q_{ij}(t)| < \varepsilon$ for all t sufficiently large. Therefore we have shown that $Q(t) \to 0$ as $t \to \infty$.

To show that $Q \in L^p[t_0, \infty)$, note that by extending the functions in the integrand of (4.64) and (4.65) to the whole real line so that they are zero outside of their natural domains, both integrals can be expressed in the form

$$g(t) = \int_{-\infty}^\infty h(t-s)v(s)\,ds,$$

where $h(u) \in L^1(\mathbb{R})$ and $v(s) \in L^p(\mathbb{R})$. Now a standard result from real analysis, e.g., Rudin [135, pp. 146–148], implies that $g(t) \in L^p(\mathbb{R})$. Hence $Q(t) \in L^p[t_0, \infty)$.

Since $Q(t) \to 0$ as $t \to \infty$, $I + Q(t)$ is invertible for t sufficiently large, say $t \geq t_2 \geq t_1 \geq t_0$, and then (4.60), (4.62) and (4.63) imply that

$$z' = [\Lambda(t) + \text{diag}\, V(t) + R(t)]\, z, \qquad t \geq t_2, \tag{4.66}$$

where

$$R(t) = (I + Q)^{-1}\, [VQ - Q\, \text{diag}\, V] \tag{4.67}$$

is well-defined for sufficiently large t. Moreover, R is absolutely integrable as can be seen by using Hölder's inequality and noting that $Q \in L^{p'}[t_0, \infty)$ $(1/p + 1/p' = 1)$ as it is in $L^p[t_0, \infty)$ and bounded.

We finally claim that $\Lambda + \text{diag}\, V$ satisfies Levinson's dichotomy conditions (see Definition 2.6). For that purpose, assume first that an index pair (i, j) satisfies (4.57). Then for $V(t) \in L^p[t_0, \infty)$, it follows for $t \geq s \geq t_0$

$$\begin{aligned} \int_s^t \text{Re}\{\lambda_i + v_{ii} - \lambda_j - v_{jj}\} &\leq c - \delta(t - s) + \int_s^t (|v_{ii}| + |v_{jj}|)\, d\tau \\ &\leq c - \delta(t - s) + 2\|V\|_p (t - s)^{1/p'} \leq K_1, \end{aligned}$$

for some constant K_1, and hence (2.22) holds. It can be shown similarly, that if an index pair (i, j) satisfies (4.58), then

$$\int_s^t \text{Re}\{\lambda_i + v_{ii} - \lambda_j - v_{jj}\} \geq K_2, \qquad t_0 \leq s \leq t$$

i.e., (2.23) holds. Thus $\Lambda(t) + \text{diag}\, V(t)$ satisfy Levinson's dichotomy conditions.

By Theorem 2.7, (4.66) has for $t \geq t_2$ a fundamental matrix satisfying

$$Z(t) = [I + o(1)] \exp\left[\int^t \{\Lambda(\tau) + \text{diag}\, V(\tau)\}\, d\tau\right].$$

Extending this solution to $t \geq t_0$, recalling (4.62) and the fact that $Q(t) \to 0$ as $t \to \infty$ completes the proof. □

As mentioned above, the original work by Hartman and Wintner [73] as well as in the simplified proof of this result by Harris and Lutz [71], the dichotomy condition in Theorem 4.11 was stated in the more restrictive way (4.56) instead of (4.57), (4.58). For example, if $\lambda_1(t) = \sin t$ and $\lambda_2(t) = 1/2$, then the index pair $(1, 2)$ satisfies the averaged condition (4.57), (4.58), but not the pointwise condition (4.56). However, the Harris–Lutz proof using the $I + Q$-transformation remains unchanged when relaxing this pointwise dichotomy condition to (4.57), (4.58).

We want to emphasize that the condition $p \leq 2$ was not used in the proof of Theorem 4.11 until the sentence following (4.67). This observation leads to the following

Corollary 4.12 *Let $\Lambda(t)$ and $V(t)$ be continuous $d \times d$-matrices for all $t \geq t_0$. Assume that $\Lambda(t) = \text{diag}\{\lambda_1(t), \dots, \lambda_d(t)\}$ and that its entries satisfy for $1 \leq i \neq j \leq d$ the exponential dichotomy condition* (4.57), (4.58). *Suppose V satisfies* (4.59) *for some $p > 1$. Then there exists for t sufficiently large matrix $Q(t)$,* $\text{diag}\, Q(t) = 0$, *$Q(t) \to 0$ as $t \to \infty$ such that $y = [I + Q(t)]\tilde{y}$ leads in* (4.60) *to*

$$\tilde{y}' = [\Lambda(t) + \text{diag}\, V(t) + V_1(t)]\, \tilde{y}, \tag{4.68}$$

where $\tilde{V}_1(t) \in L^{p/2}$.

This follows immediately from the proof of Theorem 4.11, in particular from (4.66) and (4.67), noting that now R in (4.67) is by Hölder's inequality in $L^{p/2}[t_1, \infty)$ for some t_1 sufficiently large.

Moreover, $\Lambda_1 := \Lambda(t) + \text{diag}\, V(t)$ satisfies again a dichotomy condition (4.57), (4.58) with positive constants K_1 and δ_1, where $0 < \delta_1 < \delta$. This follows from Hölder's inequality and noting that (4.57) implies for $t_0 \leq s \leq t$

$$e^{\int_s^t \{\lambda_i(\tau) - \lambda_j(\tau) + v_{ii}(\tau) - v_{jj}(\tau)\}\, d\tau} \leq K e^{-\delta(t-s)} e^{\kappa(t-s)^{1/p'}},$$

where κ is a positive constant. A similar argument holds in the case (4.58). Hence one can do another $I + Q_1(t)$ transformation in (4.68) and an inductive argument shows that the system can eventually be reduced to one in L-diagonal form. We formalize this in the following theorem (see Harris and Lutz [71, p. 578] and Eastham [53, Theorem 1.5.2]).

Theorem 4.13 *Let $\Lambda_1(t) = \Lambda(t) = \text{diag}\{\lambda_1(t), \dots, \lambda_d(t)\}$ be continuous and satisfy* (4.57), (4.58) *for $t \geq t_0$. Let the $d \times d$ matrix $V_1(t) = V(t)$ be continuous and in $L^p[t_0, \infty)$ for some p, $1 < p < \infty$. Let $M \in \mathbb{N}$ such that*

$$2^{M-1} < p \leq 2^M.$$

Define recursively for $1 \leq i \leq M - 1$

$$\begin{aligned} \Lambda_{i+1} &= \Lambda_i + \text{diag}\, V_i, \\ Q_i' &= \Lambda_i Q_i - Q_i \Lambda + V_i - \text{diag}\, V_i, \\ V_{i+1} &= [I + Q_i]^{-1}[V_i Q_i - Q_i \text{diag}\, V_i]. \end{aligned}$$

Then the iterated transformation

$$y = [I + Q_1(t)][I + Q_2(t)] \dots [I + Q_M(t)] y_M$$

takes (4.66) *for some* $t_M \geq t_0$ *into*

$$y_M' = [\Lambda_M(t) + V_M(t)]y_M \qquad t \geq t_M \geq t_0,$$

which has a fundamental matrix

$$Y_M(t) = [I + o(1)] \exp\left[\int^t \Lambda_M(\tau)\, d\tau\right] \qquad as \quad t \to \infty,$$

where $\Lambda_M(t)$ *is explicitly computable.*

4.5 Special Dichotomies and Triangular Perturbations

In this section, we are interested in treating differential systems of the form

$$y' = [\Lambda(t) + V(t) + R(t)]\, y, \qquad t \geq t_0, \tag{4.69}$$

where Λ is a diagonal matrix satisfying some dichotomy condition, V is (up to a permutation) strictly triangular, and R is an L^1-perturbation.

In [78, 79] Hsieh and Xie investigated such triangular perturbations. They considered two classes of dichotomy conditions and investigated admissible perturbations $V(t)$ to assure that (4.69) has a fundamental matrix of the form

$$Y(t) = [I + o(1)] \exp\left[\int^t \Lambda(s)\, ds\right] \qquad \text{as } t \to \infty. \tag{4.70}$$

In these two cases, given that Λ satisfies a certain dichotomy condition, they showed that the size of a strictly triangular form of V can be significantly "larger" than that of a "full-size" perturbations in order for (4.70) to hold. For example, if Λ satisfies the exponential dichotomy condition (4.57), (4.58), they showed that all that is required of V is that it satisfies $V(t) = o(1)$ as $t \to \infty$ for (4.70) to hold. Introducing somewhat weaker exponential type dichotomies parameterized by a constant α ($0 \leq \alpha < 1$), they also showed that if $V(t)$ approaches zero at a faster rate (depending on α), (4.70) still holds. Bodine and Lutz [15] later on extended their results to a general "balance" between the dichotomy condition on the diagonal and the growth condition of the triangular perturbation. This newer approach was based on a finite iteration of conditioning transformations, leading to a more transparent treatment than that in Hsieh and Xie, who used successive approximations and detailed estimates, and also allowing for weaker assumptions. We follow here the approach in [15] and present results in the following two theorems. In Theorem 4.14, we neglect the absolutely integrable perturbation R. In Theorem 4.15, we also allow such an R and then need to assume an extra dichotomy condition for Λ.

Theorem 4.14 *Let* $\Lambda = \mathrm{diag}\,\{\lambda_1(t), \ldots, \lambda_d(t)\}$ *be continuous for* $t \geq t_0$. *Suppose that* $V(t)$ *is a continuous and strictly upper triangular* $d \times d$ *matrix for* $t \geq t_0$. *Assume that there exists a continuous scalar function* $\varphi(t)$ *defined for* $t \geq t_0$ *satisfying*

$$\left|v_{ij}(t)\right| \leq \varphi(t) \qquad 1 \leq i < j \leq d, \; t \geq t_0 \tag{4.71}$$

and such that the following hold: for all $1 \leq i < j \leq d$

$$\text{either} \qquad \lim_{t\to\infty} \int_{t_0}^{t} \exp\left[\mathrm{Re} \int_s^t \{\lambda_i(\tau) - \lambda_j(\tau)\}\, d\tau\right] \varphi(s)\, ds = 0, \tag{4.72}$$

or (for $t \geq t_0$*)*

$$\begin{cases} \displaystyle\int_t^{\infty} \exp\left[\mathrm{Re} \int_s^t \{\lambda_i(\tau) - \lambda_j(\tau)\}\, d\tau\right] \varphi(s)\, ds \text{ is well-defined } \forall \; t \geq t_0 \\ \displaystyle\text{and } \lim_{t\to\infty} \int_t^{\infty} \exp\left[\mathrm{Re} \int_s^t \{\lambda_i(\tau) - \lambda_j(\tau)\}\, d\tau\right] \varphi(s)\, ds = 0. \end{cases} \tag{4.73}$$

Then (4.60) *has a fundamental matrix of the form* (4.70).

Proof The proof uses the same conditioning transformation as the one of Theorem 4.11. However, it relies on the algebraic structure of the perturbation instead of properties of L^p-functions. More specifically, we make in (4.60) the transformation $y(t) = [I + Q(t)]\, z(t)$, where Q is a strictly upper triangular matrix satisfying the differential equation (4.63). In coordinate representation, this is equivalent to

$$q_{ij}' = (\lambda_i - \lambda_j) q_{ij} + v_{ij} \qquad 1 \leq i < j \leq d.$$

(Note that V being upper triangular implies that Q is upper triangular.) If the ordered pair (ij) satisfies (4.72), choose a solution $q_{ij}(t)$ of the form (4.64), and if it satisfies (4.73), choose a solution $q_{ij}(t)$ of the form (4.65). By hypotheses (4.72) and (4.73), Q is well-defined and

$$Q(t) = o(1) \qquad \text{as } t \to \infty.$$

By the upper triangular nature of Q, $I + Q(t)$ is invertible for all $t \geq t_0$. Then it follows that

$$z' = [\Lambda + V_1]\, z, \quad t \geq t_0,$$

where

$$V_1 = (I + Q)^{-1} V Q.$$

Since both V and Q are strictly upper triangular, the product VQ is also a strictly upper triangular matrix and, moreover, it has additional zeros on the first super-diagonal, and the same holds true for V_1. From (4.71) and the boundedness of Q it follows that there exists a positive constant c_1 such that

$$|V_1(t)| \leq c_1\varphi(t) =: \varphi_1(t) \qquad \forall\; t \geq t_0, \quad 1 \leq i < j \leq d.$$

Note that (4.72) and (4.73) still hold with φ replaced by φ_1. Emphasizing that the matrix V_1 contains an additional super-diagonal of zeros, one continues by making the transformation $z(t) = [I + Q_1(t)]\, z_1(t)$ for $t \geq t_0$, where $Q_1' = \Lambda Q_1 - Q_1\Lambda + V_1$ and Q_1 is therefore strictly upper triangular with additional zeros on the first super-diagonal. Moreover Q_1 vanishes at infinity and $I + Q_1(t)$ is invertible for all $t \geq t_0$. Proceeding as before, this leads to a system

$$z_1' = [\Lambda(t) + V_2(t)]\, z_1,$$

where $V_2 = (I + Q_1)^{-1} V_1 Q_1$. Observe that $V_1 Q_1$ is strictly upper triangular with two more additional super-diagonals of zeros and so is V_2. Continuing by induction, one finds that

$$V_{d-1} \equiv 0,$$

and thus

$$z_{d-1}' = \Lambda(t) z_{d-1}.$$

Hence $Z_{d-1} = \exp\left[\int^t \Lambda(\tau)\, d\tau\right]$, and recalling the asymptotic behavior of the conditioning transformation establishes (4.70). □

It is straightforward to show that the strictly upper triangular perturbation V in Theorem 4.14 can be replaced by a strictly lower triangular perturbation V. We continue by extending this result to allow additional absolutely integrable perturbations.

Theorem 4.15 *Assume that Λ and V satisfy the assumptions of Theorem 4.14. Let R be a continuous $d \times d$-matrix satisfying*

$$\int_{t_0}^{\infty} |R(t)|\, dt < \infty. \tag{4.74}$$

Assume that, in addition to the assumptions of Theorem 4.14, $\Lambda(t)$ satisfies Levinson's dichotomy conditions (2.13), (2.14). *Then* (4.69) *has a fundamental matrix satisfying* (4.70).

Proof The proof employs exactly the same sequence of conditioning transformations as the one of Theorem 4.14. Note that the first transformation $y = [I + Q]z$

applied to (4.69) leads to

$$z' = \left[\Lambda + (I+Q)^{-1}VQ + (I+Q)^{-1}R(I+Q)\right]z,$$

and, by the boundedness of Q, $R_1 = (I+Q)^{-1}R(I+Q)$ satisfies (4.74). Continuing as in the proof of Theorem 4.14, one finds that

$$z_{d-1}' = [\Lambda(t) + R_{d-1}(t)]\, z_L,$$

where $R_{d-1} \in L^1[t_0, \infty)$. Applying Theorem 2.7 establishes the result of Theorem 4.15. □

In [15], there were also sufficient conditions given for the hypotheses (4.72) and (4.73) as well as Levinson's dichotomy conditions (2.13) and (2.14) to hold. In what follows, we will just state this result and one corollary and refer to [15] for proofs, further corollaries, and examples.

Theorem 4.16 *Let Λ, V, and R be continuous $d \times d$-matrices for all $t \geq t_0$. Assume*

$$\Lambda(t) = \operatorname{diag}\{\lambda_1(t), \ldots, \lambda_d(t)\} \qquad \forall\, t \geq t_0.$$

Suppose that there exist a positive constant κ and a C^1 function $\psi : [t_0, \infty) \to \mathbb{R}$ satisfying

$$\psi'(t) > 0 \quad \textit{for } t \geq t_0, \qquad \lim_{t\to\infty} \psi(t) = \infty, \tag{4.75}$$

such that for each pair of integers $1 \leq i \neq j \leq d$

$$\textit{either} \quad \operatorname{Re} \int_s^t \{\lambda_i(\tau) - \lambda_j(\tau)\}\, d\tau \leq \kappa - \psi(t) + \psi(s), \qquad \forall\ t_0 \leq s \leq t, \tag{4.76}$$

$$\textit{or} \quad \operatorname{Re} \int_s^t \{\lambda_i(\tau) - \lambda_j(\tau)\}\, d\tau \leq \kappa - \psi(s) + \psi(t) \qquad \forall\ t_0 \leq t \leq s. \tag{4.77}$$

Let $V(t)$ be a strictly upper triangular matrix and suppose that there exists a continuous scalar function

$$\varphi(t) \geq \left|v_{ij}(t)\right| \qquad 1 \leq i < j \leq d, \ \ t \geq t_0,$$

satisfying

$$\varphi(t) = \mathrm{o}\left(\psi'(t)\right) \qquad \textit{as } t \to \infty. \tag{4.78}$$

Finally, assume that $R(t)$ satisfies (4.74)*. Then* (4.69) *has a fundamental matrix of the form* (4.70)*.*

Choosing $\psi(t) = \beta \ln t$ leads to the following

Corollary 4.17 *In Theorem 4.16 , assume that there exists $\beta > 0$ and $K \geq 1$ such that for each pair of indices (i,j) with $1 \leq i \neq j \leq d$*

$$\text{either} \qquad \exp\left[\operatorname{Re} \int_s^t \{\lambda_i(\tau) - \lambda_j(\tau)\}\, d\tau\right] \leq K \left(\frac{s}{t}\right)^\beta, \qquad t_0 \leq s \leq t$$

$$\text{or} \qquad \exp\left[\operatorname{Re} \int_s^t \{\lambda_i(\tau) - \lambda_j(\tau)\}\, d\tau\right] \leq K \left(\frac{t}{s}\right)^\beta, \qquad t_0 \leq t \leq s,$$

and that the majorant satisfies

$$\varphi(t) = \mathrm{o}\left(\frac{1}{t}\right) \qquad \text{as } t \to \infty.$$

Then (4.69) *has a fundamental matrix of the form* (4.70).

4.6 Conditionally Integrable Perturbations

4.6.1 Some Initial Remarks

In this section, we consider various systems containing perturbations V which are conditionally integrable on $[t_0, \infty)$. This means that

$$\lim_{T\to\infty} \int_{t_0}^T V(\tau)\, d\tau$$

exists, and it follows that if one defines for $t \geq t_0$

$$Q(t) = -\int_t^\infty V(\tau)\, d\tau \tag{4.79}$$

then $Q(t) = \mathrm{o}(1)$ as $t \to \infty$. Therefore for $t \geq t_1$ sufficiently large, $I + Q(t)$ is invertible and $[I + Q(t)]^{-1}$ is bounded. In addition, $Q' = V$, hence the linear transformation $y = [I + Q(t)]z$ takes

$$y' = [\Lambda(t) + V(t) + R(t)], \qquad t \geq t_0 \tag{4.80}$$

into

$$\begin{aligned} z' &= [I + Q]^{-1} [\Lambda + \Lambda Q + V + VQ + R(I + Q) - V]\, z \\ &= \left[\Lambda(t) + \tilde{R}(t)\right] z, \end{aligned}$$

where $\tilde{R}(t) = (I+Q)^{-1}[\Lambda Q - Q\Lambda + VQ + R(I+Q)]$. It follows that if

(a) $\Lambda(t)$ satisfies Levinson's dichotomy conditions (2.13), (2.14),
(b) $\Lambda Q - Q\Lambda$, VQ, and R are in $L^1[t_0,\infty)$,

then $\tilde{R} \in L^1[t_1,\infty)$ and one may apply Theorem 2.7 to obtain

Theorem 4.18 *Let $\Lambda(t)$, $V(t)$, and $R(t)$ be $d \times d$ continuous matrices for $t \geq t_0$. Assume that $V(t)$ is conditionally integrable on $[t_0,\infty)$ and let $Q(t)$ be defined in* (4.79)*. If $\Lambda(t)$, $V(t)$, and $R(t)$ satisfy (a) and (b), then the linear system* (4.80) *has a fundamental solution satisfying*

$$Y(t) = [I + o(1)]\exp\left(\int^t \Lambda(\tau)\,d\tau\right) \qquad as\ t \to +\infty.$$

This situation has been considered by Harris and Lutz [70, Theorem 3.1] and also by Medina and Pinto [107, Theorem 2.1]. The example

$$y' = \left[\frac{1}{\sqrt{t}}\begin{bmatrix}1 & 0\\ 0 & -1\end{bmatrix} + t^{-p}\sin t\, V_0\right] y, \qquad t \geq 1,$$

where $\frac{1}{2} < p < 1$ and V_0 is any constant matrix was presented in [107] and illustrates the applicability of Theorem 4.18, while at the same time none of the Theorems 2.7, 4.8, nor 4.11 appear to apply. In this example, it is not particularly relevant that the leading diagonal matrix is real or that the eigenvalues are distinct. The only essential feature is that $Q(t) = O(t^{-p})$ and so $p > 1/2$ implies $\Lambda Q - Q\Lambda$ and VQ are in L^1. Theorem 2.21 applies, however, for $p > 3/4$.

In situations where $\Lambda(t)$ does not tend to zero, the applicability of Theorem 4.18 is considerably restricted since $\Lambda Q - Q\Lambda$ might then be of the same order of magnitude as V, and then $\tilde{R}$ would not be an improvement. Such a situation is encountered when treating systems of the form

$$y' = \left[\begin{pmatrix}\lambda i & 0\\ 0 & -\lambda i\end{pmatrix} + \frac{it^{-p}\sin\theta t}{2\lambda}\begin{pmatrix}1 & 1\\ -1 & -1\end{pmatrix}\right] y, \qquad t \geq t_0 > 0, \tag{4.81}$$

which come about when studying perturbations of the harmonic linear oscillator $x'' + \left[\lambda^2 + t^{-p}\sin\theta t\right] = 0$, where $\lambda, \theta \in \mathbb{R}$, $\lambda \neq 0$, and $p > 0$ (see Chap. 8). Even though Theorem 4.18 is not applicable, one can consider first applying the diagonal transformation $y = \operatorname{diag}\left\{e^{i\lambda t}, e^{-i\lambda t}\right\}\tilde{y}$, which yields

$$\begin{aligned}\tilde{y}' &= \frac{it^{-p}\sin\theta t}{2\lambda}\begin{pmatrix}1 & \exp(-2i\lambda t)\\ -\exp(2i\lambda t) & -1\end{pmatrix}\tilde{y}\\ &= \left[0 + \tilde{V}(t)\right]\tilde{y}.\end{aligned}$$

Observe now that $\tilde{V}$ is conditionally integrable on $[t_0, \infty)$ provided $\theta \neq \pm 2\lambda$ and in this case

$$Q(t) = -\int_t^\infty \tilde{V}(\tau)\, d\tau = O\left(t^{-p}\right).$$

So if $\theta \neq \pm 2\lambda$ and $p > 1/2$, Theorem 4.18 can be applied and leads to an asymptotic representation for solutions. On the other hand, if $\theta = \pm 2\lambda$, then

$$\sin(2\lambda t)\exp(-2i\lambda t) = \frac{1}{2i}\left[1 - \exp(-4i\lambda t)\right),$$

hence $\tilde{V}(t)$ is not conditionally integrable when $p \leq 1$. See [2, 32, 70, 71] where such examples are discussed in several cases when $1/2 \leq p \leq 1$ and θ assumes the values 2λ or λ. In particular, for these values of θ, solutions of (4.81) can become unbounded or tend to 0 as $t \to \infty$. Eastham [52] uses the term *resonance* to describe the phenomenon created when certain frequencies in the perturbation force solutions to strongly deviate from the periodic solutions of the unperturbed equation. Even when $\theta \neq 2\lambda, \lambda$ and $p = 1/2$ (see [70, Example 3]) solutions may mildly deviate from the periodic unperturbed ones, but not as strongly as in the resonant cases and they do remain bounded.

In the remainder of this section, we will discuss three situations involving conditionally integrable perturbations. The first concerns Theorem 4.18 in case $\Lambda \equiv 0$, which is a classical result attributed to Bôcher and related to the concept of linear asymptotic equilibria. Wintner extended Bôcher's result using an iteration of transformations and we discuss this in Sect. 4.6.2. Next, we consider several results due to Trench [154] in Sect. 4.6.3 concerning systems containing conditionally integrable perturbations, where conditioning transformations are used which combine ones as above with ones of the Hartman–Wintner type (Sect. 4.4) and also error estimates.

Next, in Sect. 4.6.4 we discuss some sufficient conditions due to Cassel [33] which can be applied to imply non-resonance for a broad class of systems $y' = [\Lambda_0 + R(t)]y$, including examples such as (4.81) (for all $p > 0$) as well as many other interesting classes of perturbations. Finally, in Sect. 4.6.5 we briefly mention some other results involving conditionally convergent perturbations related to the method of averaging.

4.6.2 A Theorem of Wintner and Extensions

A system

$$y' = A(t)y \tag{4.82}$$

($t \geq t_0$) is said to have a *linear asymptotic equilibrium* if for each constant vector c there is a solution $y(t)$ of (4.82) such that $y(t) \to c$ as $t \to \infty$. An equivalent definition would be that (4.82) has a fundamental matrix satisfying $Y(t) = I + o(1)$ as $t \to \infty$. Wintner [162] established some sufficient conditions for (4.82) having a linear asymptotic equilibrium in the following result:

Theorem 4.19 (Wintner) *Let $A(t)$ be defined and continuous for $t \geq t_0$ and assume that the following inductively defined matrices $A_j(t)$ exist:*

$$A_j(t) = -\int_t^{\infty} A(s)A_{j-1}(s)\,ds \qquad \text{for } j = 1, \ldots, k; \qquad A_0(t) \equiv I. \tag{4.83}$$

Furthermore, assume that

$$A(t)A_k(t) \in L^1[t_0, \infty). \tag{4.84}$$

Then

$$y' = A(t)y, \qquad t \geq t_0, \tag{4.85}$$

has a fundamental matrix satisfying

$$Y(t) = I + o(1) \qquad \text{as } t \to \infty. \tag{4.86}$$

Proof Let

$$y = S_k(t)\tilde{y}, \tag{4.87}$$

where

$$S_k(t) = I + A_1(t) + \ldots + A_k(t). \tag{4.88}$$

Since $A_j(t) = o(1)$ as $t \to \infty$ for $1 \leq j \leq k$, there exists $t_1 \geq t_0$ such that $S_k(t)$ is invertible for $t \geq t_1$. Then for each such t,

$$\tilde{y}' = [S_k(t)]^{-1}\left\{A(t)S_k(t) - S_k'(t)\right\}\tilde{y}.$$

Since $A_j'(t) = A(t)A_{j-1}(t)$ for all $1 \leq j \leq k$, it follows that

$$\tilde{y}' = [S_k(t)]^{-1}A(t)A_k(t)\tilde{y} =: R(t)\tilde{y}, \tag{4.89}$$

where by (4.84)

$$R(t) \in L^1[t_1, \infty).$$

Interpreting this as a perturbation of the trivial system $x' = 0$, Theorem 2.7 implies there exists a solution $\tilde{Y}(t) = I + o(1)$ as $t \to \infty$, hence also

$$Y(t) = [I + A_1(t) + \dots + A_k(t)][I + o(1)] = I + o(1).$$

□

Remark 4.20 Wintner's proof [162, pp. 186–187] does not explicitly concern a transformation $I + A_1(t) + \dots + A_k(t)$, but his inductive argument involving fundamental matrices and application of Bôcher's theorem for $\tilde{y}'(t) = R(t)\tilde{y}$ is essentially equivalent to the approach we have used.

Remark 4.21 Wintner also obtained an equivalent result concerned with the inductively defined sequence

$$\hat{A}_j(t) = \int_t^\infty \hat{A}_{j-1}(s)A(s)\, ds, \qquad \hat{A}_0(t) \equiv I. \tag{4.90}$$

That result states that the existence of the matrices $\hat{A}_j(t)$ for $j = 1, \dots, k$ and the assumption $\hat{A}_k(t)A(t) \in L^1$ is also sufficient for (4.85) having a fundamental matrix $Y = I + o(1)$. As Wintner also points out, it is not obvious that these conditions are equivalent to the ones involving $A_j(t)$ because of non-commutativity of matrix multiplication. However, using the fact that $(Y^{-1})^T(t)$ is a fundamental solution matrix for the adjoint system $y' = -A^T(t)y$, and $X(t) = I + o(1)$ if and only if $(X^{-1})^T(t) = I + o(1)$, one can check the equivalence of the two sets of conditions.

Remark 4.22 Wintner also presents (see [162, pp. 187–188]) an example of a 2×2 triangular system which has a fundamental matrix satisfying (4.86), but the coefficient matrix $A(t)$ is not conditionally integrable. Hence the existence of $\int_{t_0}^\infty A(s)\, ds$ is not necessary for the conclusion $Y(t) = I + o(1)$.

Trench [152, Corollary 3] considered an improvement of Theorem 4.19, where the error term $o(1)$ is quantified. Essentially, his result is contained in the following

Theorem 4.23 (Trench) *Let $w(t)$ be a positive, continuous and non-increasing function on $[t_0, \infty)$ and suppose that $\lim_{t\to\infty} w(t) = 0$. Suppose that integrals $\hat{A}_j(t)$ given in (4.90) converge for $1 \leq j \leq k+1$ and that*

$$|\hat{A}_j(t)| = O(w(t)) \qquad 1 \leq j \leq k+1. \tag{4.91}$$

Assume that

$$\limsup_{t\to\infty} \frac{1}{w(t)} \int_t^\infty |\hat{A}_k(s)A(s)| w(s)\, ds = \theta < 1 \tag{4.92}$$

Then (4.85) has a fundamental matrix satisfying

$$Y(t) = I + O(w(t)) \qquad \text{as } t \to \infty.$$

While Trench's proof was based on applying the Schauder–Tychonov theorem to a certain operator equation, we will give here a different proof using preliminary and conditioning transformations which allows us to weaken his assumption (4.92) significantly by replacing it with (4.93). To be more specific, we will prove the following

Theorem 4.24 (Bodine/Lutz) *Let w be a positive, continuous, and non-increasing function on $[t_0, \infty)$ and suppose that $\lim_{t\to\infty} w(t) = 0$. Suppose that integrals $\hat{A}_j(t)$ given in* (4.90) *converge for $1 \le j \le k+1$, and satisfy* (4.91) *for $1 \le j \le k+1$. Assume that*

$$\int_t^\infty |\hat{A}_k(s)A(s)|w(s)\,ds = O(w(t)) \qquad \text{as } t \to \infty. \tag{4.93}$$

Then (4.85) *has a fundamental matrix satisfying*

$$Y(t) = I + O(w(t)) \qquad \text{as } t \to \infty. \tag{4.94}$$

Proof As in Theorem 4.19 above, define

$$\hat{S}_k(t) := I + \hat{A}_1(t) + \ldots + \hat{A}_k(t),$$

hence $\hat{S}_k(t) = I + O(w(t))$ as $t \to \infty$ by (4.91). In particular, $\hat{S}_k(t)$ is invertible for all t sufficiently large, say $t \ge t_1$. Then, for $t \ge t_1$, the transformation

$$y_1(t) = \hat{S}_k(t)y, \tag{4.95}$$

leads in (4.85) to

$$y_1' = \hat{A}_k(t)A(t)\hat{S}_k^{-1}(t)y_1.$$

Put

$$\hat{S}_k^{-1}(t) = I + H(t),$$

then

$$H(t) = O(w(t)) \qquad \text{as } t \to \infty.$$

To see this, observe that by (4.91) $\hat{S}_k^{-1}(t) = [I + U(t)]^{-1}$, where $U(t) = O(w(t))$. Choose $t \ge t_2(\ge t_1)$ sufficiently large such that $|U(t)| \le 1/2$. Then $I + H(t) = [I + U(t)]^{-1} = I - U(t) + U^2(t) - U^3(t) + \ldots$, from which follows that $H(t) = O(w(t))$. Therefore

$$y_1' = \left[\hat{A}_k(t)A(t) + \hat{A}_k(t)A(t)H(t)\right]y_1. \tag{4.96}$$

To remove the term $\hat{A}_k(t)A(t)$ in (4.96), we put

$$y_1 = [I - \hat{A}_{k+1}(t)]y_2. \tag{4.97}$$

Recall that $\hat{A}'_{k+1} = -\hat{A}_k A$ and $\hat{A}_{k+1} = O(w)$. Then

$$y_2' = [0 + R(t)]y_2, \tag{4.98}$$

where

$$R := [I - \hat{A}_{k+1}(t)]^{-1}\left[\hat{A}_k(t)A(t)O(w(t))\right] = O\left(\left|\hat{A}_k(t)A(t)\right| w(t)\right) \in L^1.$$

By Theorem 2.7, (4.98) has a fundamental matrix satisfying $Y_2(t) = I + o(1)$ as $t \to \infty$. A sharper estimate, however, can be derived from applying Theorem 2.25 which implies that (4.98) has a fundamental matrix satisfying

$$Y_2(t) = I + O\left(\int_t^\infty \left(|\hat{A}_k(s)A(s)|w(s)\right) ds\right) = I + O\left(w(t)\right),$$

where the last equality follows from (4.93). From (4.95) and (4.97) follows that (4.85) has for t sufficiently large a fundamental system satisfying

$$\begin{aligned} Y(t) &= S_k^{-1}(t)[I - \hat{A}_{k+1}(t)]\left[I + O\left(w(t)\right)\right] \\ &= [I + O\left(w(t)\right)]^3 = I + O\left(w(t)\right), \end{aligned}$$

and the proof is complete. □

We continue with a similar extension of Theorem 4.19, in terms of matrices $A_j(t)$ instead of $\hat{A}_j(t)$.

Theorem 4.25 *Let $w(t)$ be a positive, continuous and non-increasing function on $[t_0, \infty)$ and suppose that $\lim_{t\to\infty} w(t) = 0$. Suppose that integrals $A_j(t)$ given in* (4.83) *converge for $1 \le j \le k+1$, and satisfy*

$$|A_j(t)| = O(w(t)) \qquad 1 \le j \le k+1.$$

Assume that

$$\int_t^\infty |A(s)A_k(s)|w(s)\, ds = O(w(t)) \qquad \text{as } t \to \infty. \tag{4.99}$$

Then (4.85) *has a fundamental matrix of the form* (4.94).

Proof As in the proof of Theorem 4.19, we first make the change of variables (4.87), with $S_k(t)$ defined in (4.88). Note that $S_k(t) = I + O(w(t))$ by (4.91). It follows that

$$\tilde{y}' = [S_k(t)]^{-1}A(t)A_k(t)\tilde{y}$$

Writing $[S_k(t)]^{-1} = I + H(t)$, it follows that

$$\tilde{y}' = [A(t)A_k(t) + H(t)A(t)A_k(t)]\,\tilde{y},$$

and an argument parallel to that used in Theorem 4.24 shows that $H(t) = O(w(t))$. To remove the leading conditionally integrable term $A(t)A_k(t)$, we put

$$\tilde{y} = [I + A_{k+1}(t)]\hat{y},$$

which implies that

$$\hat{y}' = [I + A_{k+1}]^{-1}\,[HAA_k + (I + H)AA_kA_{k+1}]\,\hat{y} = O\left(|A(t)A_k(t)|w(t)\right)\hat{y},$$

which we consider a perturbation of the trivial system $z' = 0$. In the same way as above, we obtain

$$\hat{Y}(t) = I + O\left(\int_t^\infty (|A(s)A_k(s)|w(s))\;ds\right) = I + O\left(w(t)\right),$$

as $t \to +\infty$ and the rest of the proof is parallel to that of Theorem 4.24. □

4.6.3 Conditionally Integrable Perturbations of Constant Systems

One interpretation of Theorem 4.23 and its improvement Theorem 4.24 is that the use of a conditioning transformation (as in the proof of Theorem 4.24) instead of a fixed point argument (as in the proof of Theorem 4.23) allows one to replace assumption (4.92) by the weaker assumption (4.99).

This approach can also be applied to systems $y' = [\Lambda + R(t)]y$, where $\Lambda \neq 0$ and R is conditionally integrable. This augments some of the results in Sect. 2.6, where other assumptions on Λ and R were made and the asymptotic results followed directly using the T-operator approach. Here, we will instead first apply a conditioning transformation followed by applying Levinson's fundamental theorem.

Theorem 4.26 *Suppose $\Lambda = \operatorname{diag}\{\lambda_1, \ldots, \lambda_d\}$ and put $\mu_i = \operatorname{Re}\lambda_i$ $(1 \leq i \leq d)$. Let $R(t) = \{r_{ij}(t)\}$ be a continuous $d \times d$ matrix on $[t_0, \infty)$ satisfying*

$$\int_t^\infty R(s)\,ds = O(\varphi(t)), \tag{4.100}$$

and

$$\int_t^\infty |R(s)|\varphi(s)\, ds = \mathrm{O}(\varphi(t)), \tag{4.101}$$

where $\varphi(t)$ is non-increasing and $\lim_{t\to\infty}\varphi(t) = 0$. *Let*

$$\alpha := \min_{1\le i,j\le d}\left\{|\mu_i - \mu_j| : \mu_i \neq \mu_j\right\}.$$

(In case that $\mu_i = \mu_j$ for all $1 \le i,j \le d$, put $\alpha = 0$.) Whenever $\mu_i = \mu_j$, suppose that

$$\int_t^\infty e^{(\lambda_i-\lambda_j)s} r_{ij}(s)\, ds = \mathrm{O}(\varphi(t)). \tag{4.102}$$

If $\alpha > 0$, suppose that there exists $\rho \in [0,\alpha)$ such that

$$e^{\rho t}\varphi(t) \textit{ is nondecreasing for } t \ge t_0. \tag{4.103}$$

Then

$$y' = [\Lambda + R(t)]y, \qquad t \ge t_0 \tag{4.104}$$

has a fundamental matrix $Y(t)$ satisfying

$$Y(t) = [I + \mathrm{O}(\varphi(t))]\, e^{\Lambda t}, \qquad \textit{as } t \to \infty. \tag{4.105}$$

Proof For fixed $r \in \{1, \ldots, d\}$, put $y = e^{\lambda_r t}\tilde{y}$. Hence

$$\tilde{y}' = [\Lambda - \lambda_r I + R(t)]\tilde{y} =: [\tilde{\Lambda} + R(t)]\tilde{y}.$$

Make the conditioning transformation $\tilde{y} = [I + Q(t)]z$, where

$$\tilde{\Lambda}Q - Q\tilde{\Lambda} + R = Q'.$$

Written in coordinates, this is equivalent to

$$q_{ij}' = [\tilde{\lambda}_i - \tilde{\lambda}_j]q_{ij} + r_{ij} = [\lambda_i - \lambda_j]q_{ij} + r_{ij}.$$

We remark that this is closely related to the conditioning transformation used in the proof of Theorem 4.11, but we do not require $\operatorname{diag} Q(t) \equiv 0$ here. For each index pair (i,j), $1 \le i,j \le d$, we define

$$p_{ij}(t) = -\int_t^\infty r_{ij}(s)\, ds,$$

hence by (4.100)

$$p_{ij}(t) = \mathrm{O}(\varphi(t)), \qquad 1 \leq i,j \leq d. \tag{4.106}$$

1. If Re $(\lambda_i - \lambda_j) > 0$, we define

$$q_{ij}(t) = -\int_t^\infty e^{(\lambda_i-\lambda_j)(t-s)} r_{ij}(s)\, ds, \tag{4.107}$$

   and integration by parts shows that

$$q_{ij}(t) = p_{ij}(t) - (\lambda_i - \lambda_j) \int_t^\infty e^{(\lambda_i-\lambda_j)(t-s)} p_{ij}(s)\, ds,$$

   thus, using the monotonicity of $\varphi(t)$,

$$\begin{aligned} |q_{ij}(t)| &\leq \mathrm{O}(\varphi(t)) + \mathrm{O}\left(\int_t^\infty e^{\alpha(t-s)}\varphi(s)\, ds\right) \\ &\leq \mathrm{O}(\varphi(t))\left[1 + \int_t^\infty e^{\alpha(t-s)}\, ds\right] = \mathrm{O}(\varphi(t)). \end{aligned}$$

2. If Re $(\lambda_i - \lambda_j) = 0$, we define $q_{ij}(t)$ again by (4.107) and note that from (4.102) follows directly that

$$\int_t^\infty e^{(\lambda_i-\lambda_j)(t-s)} r_{ij}(s)\, ds = \mathrm{O}(\varphi(t)).$$

3. If Re $(\lambda_i - \lambda_j) < 0$, we define

$$q_{ij}(t) = \int_{t_0}^t e^{(\lambda_i-\lambda_j)(t-s)} r_{ij}(s)\, ds,$$

   and observe that from integration by parts and the slow-decay condition (4.103) follows that

$$\begin{aligned} q_{ij}(t) &= p_{ij}(t) - e^{(\lambda_i-\lambda_j)(t-t_0)} p_{ij}(t_0) + \int_{t_0}^t [\lambda_i - \lambda_j] e^{(\lambda_i-\lambda_j)(t-s)} p_{ij}(s)\, ds \\ &= p_{ij}(t) + \mathrm{O}\left(e^{-\alpha(t-t_0)}\varphi(t_0)\right) + \mathrm{O}\left(\int_{t_0}^t e^{-\alpha(t-s)}\varphi(s)\, ds\right), \\ |q_{ij}(t)| &\leq |p_{ij}(t)| + \mathrm{O}\left(e^{(\rho-\alpha)(t-t_0)}\varphi(t)\right) + \mathrm{O}\left(\varphi(t)\int_{t_0}^t e^{(\rho-\alpha)(t-s)}\, ds\right) \\ &= \mathrm{O}\left(\varphi(t)\right). \end{aligned}$$

Hence

$$Q(t) = \mathrm{O}\left(\varphi(t)\right), \tag{4.108}$$

thus $I + Q(t)$ is invertible for t sufficiently large, and

$$z' = \left[\tilde{\Lambda} + [I + Q(t)]^{-1} R(t) Q(t)\right] z = \left[\tilde{\Lambda} + \mathrm{O}\left(|R(t)|\varphi(t)\right)\right] z. \tag{4.109}$$

While $|R(t)|\varphi(t) \in L^1[t_0, \infty)$ by (4.101) and hence Theorem 2.7 implies that (4.109) has a fundamental matrix of the form $Z(t) = [I + \mathrm{o}(1)] \exp[\tilde{\Lambda} t]$, the stronger result (4.105) requires additional work. In particular, we make one further conditioning transformation to replace $\mathrm{O}(|R(t)|\varphi(t))$ by $\mathrm{O}\left(|R(t)|\varphi^2(t)\right)$. For that purpose, re-write (4.109) as

$$z' = \left[\tilde{\Lambda} + \hat{R}(t)\right] z, \qquad \text{where } \hat{R}(t) = \mathrm{O}\left(|R(t)|\varphi(t)\right),$$

and put $z = [I + \hat{Q}(t)]\hat{z}$, where

$$\tilde{\Lambda}\hat{Q} - \hat{Q}\tilde{\Lambda} + \hat{R} = \hat{Q}'.$$

Then one can repeat the line of arguments involving integration by parts used in the first conditioning transformation (with R and Q replaced by $\hat{R}$ and $\hat{Q}$, respectively, and noting that $\tilde{\lambda}_i - \tilde{\lambda}_j = \lambda_i - \lambda_j$) to show that if Re $(\lambda_i - \lambda_j) > 0$ or Re $(\lambda_i - \lambda_j) < 0$, then by (4.108) $\hat{q}_{ij}(t) = \mathrm{O}\left(\varphi(t)\right)$. Moreover, if Re $(\lambda_i - \lambda_j) = 0$, it follows from (4.101) that

$$\left|\hat{q}_{ij}(t)\right| = \left|\int_t^\infty e^{(\lambda_i - \lambda_j)(t-s)} \hat{r}_{ij}(s)\, ds\right| = \mathrm{O}\left(\int_t^\infty |R(s)|\varphi(s)\, ds\right) = \mathrm{O}(\varphi(t)).$$

Hence $\hat{Q}(t) = \mathrm{O}\left(\varphi(t)\right)$, thus $I + \hat{Q}(t)$ is invertible for t sufficiently large, and

$$\hat{z}' = \left[\tilde{\Lambda} + [I + \hat{Q}(t)]^{-1} \hat{R}(t) \hat{Q}(t)\right] \hat{z} = \left[\tilde{\Lambda} + \tilde{R}(t)\right] \hat{z}, \tag{4.110}$$

where

$$\tilde{R}(t) = \mathrm{O}\left(|R(t)|\varphi^2(t)\right).$$

While Theorem 2.7 could be applied to (4.110), the goal here is a more precise error estimate. We note that we cannot just apply Theorem 2.25, since the matrix $\tilde{R}(t)$ as defined does not satisfy the slow-decay condition required in this theorem. Instead we use a slightly modified argument in what follows.

For the index r fixed above, let $P = \operatorname{diag}\{p_1, \ldots, p_d\}$, where $p_i = 1$ if Re $\{\lambda_i - \lambda_r\} < 0$ and $p_i = 0$ else. We claim that (4.110) has a solution for $t \geq t_1$, t_1 provided that

$$\begin{aligned}\hat{z}(t) &= e_r + \int_{t_1}^{t} e^{\tilde{A}(t-s)} P\tilde{R}(s)\tilde{z}(s)\, ds - \int_{t}^{\infty} e^{\tilde{A}(t-s)} [I - P]\tilde{R}(s)\hat{z}(s)\, ds \\ &:= e_r + (T\hat{z})(t).\end{aligned}$$

Let $\mathscr{B} = \{z(t) : \|z\| := \sup_{t \geq t_1} |z(t)| < \infty$ and let $\hat{z} \in \mathscr{B}$. Then there exists $K > 0$ such that

$$\begin{aligned}|(T\hat{z})(t)| &\leq K\|\hat{z}\| \left[\int_{t_1}^{t} e^{-\alpha(t-s)} \varphi(s)|R(s)|\varphi(s)\, ds + \int_{t}^{\infty} |R(s)|\varphi^2(s)\, ds \right] \\ &\leq K\|\hat{z}\|\varphi(t) \left[\int_{t_1}^{t} e^{(\rho-\alpha)(t-s)} |R(s)|\varphi(s)\, ds + \int_{t}^{\infty} |R(s)|\varphi(s)\, ds \right],\end{aligned}$$

using once more the monotonicity of φ and the slow-decay condition (4.103). In particular,

$$|(T\hat{z})(t)| \leq K\|\hat{z}\|\varphi(t) \int_{t_1}^{\infty} |R(s)|\varphi(s)\, ds,$$

which shows that T maps $\mathscr{B}$ into $\mathscr{B}$ and similar arguments show that it is a contraction for $t \geq t_1$, t_1 sufficiently large. Hence there exists a unique fixed point solution $\hat{z}_0$. Put $\hat{K} = K\|\hat{z}_0\|$. Moreover, given $\varepsilon > 0$, fix $t_2 \geq t_1$ such that $\int_{t_2}^{\infty} |R(s)|\varphi(s)\, ds < \varepsilon/(2\hat{K})$. Then, for $t \geq t_2$,

$$\begin{aligned}|(T\hat{z}_0)(t)| &\leq \hat{K}\varphi(t) \left[\int_{t_1}^{t_2} e^{(\rho-\alpha)(t-s)} |R(s)|\varphi(s)\, ds + \int_{t_2}^{t} e^{(\rho-\alpha)(t-s)} |R(s)|\varphi(s)\, ds \right. \\ &\quad \left. + \int_{t}^{\infty} |R(s)|\varphi(s)\, ds \right] \\ &\leq \hat{K}\varphi(t) \left[\int_{t_1}^{t_2} e^{(\rho-\alpha)(t-s)} |R(s)|\varphi(s)\, ds + \int_{t_2}^{\infty} |R(s)|\varphi(s)\, ds \right] \\ &\leq \varphi(t) \left[\hat{K} \int_{t_1}^{t_2} e^{(\rho-\alpha)(t-s)} |R(s)|\varphi(s)\, ds + \varepsilon/2 \right].\end{aligned}$$

Note that for t sufficiently large, $\left| \hat{K} \int_{t_1}^{t_2} e^{(\rho-\alpha)(t-s)} |R(s)|\varphi(s)\, ds \right| < \frac{\varepsilon}{2}$, hence (4.110) has a solution

$$\hat{z}_0(t) = e_r + o(\varphi(t)) \qquad \text{as } t \to \infty.$$

Therefore (4.104) has a solution satisfying

$$y_r(t) = [I + \mathrm{O}(\varphi(t))]^2[e_r + \mathrm{o}(\varphi(t))]e^{\lambda_r t} = [e_r + \mathrm{O}(\varphi(t))]e^{\lambda_r t}.$$

□

Remark 4.27 Theorem 4.26 is an improvement of a result of Trench [155, Theorem 2], which assumed

$$\int_t^\infty |R(s)|\varphi(s)\,ds = \mathrm{o}(\varphi(t)), \tag{4.111}$$

instead of the weaker (4.101). As in the previous section, using a conditioning transformation instead of a fixed point argument allows one to relax a suitable $\mathrm{o}(\cdot)$ condition by replacing it by a $\mathrm{O}(\cdot)$ condition. Using a fixed point argument, Trench naturally derived the existence of one vector-valued solution instead of an entire fundamental matrix under conditions adjusted to the existence of just one single solution. Trench also considered certain exponential weight factors that we will not pursue here.

We conclude this section with the following example: Consider

$$y'' + \left[1 + \frac{\sin t}{t}\right]y = 0, \qquad t \geq 1. \tag{4.112}$$

This can be re-written as the system

$$\mathbf{y}' = \begin{pmatrix} 0 & 1 \\ -1 - \frac{\sin t}{t} & 0 \end{pmatrix}\mathbf{y}.$$

Setting $\mathbf{y} = \begin{pmatrix} 1 & 1 \\ i & -i \end{pmatrix}\tilde{y}$ yields

$$\tilde{y}' = \left[\begin{pmatrix} i & \\ & -i \end{pmatrix} + \frac{i \sin t}{2t}\begin{pmatrix} 1 & 1 \\ -1 & -1 \end{pmatrix}\right]\tilde{y} = [\Lambda + R(t)]\tilde{y}. \tag{4.113}$$

Then the hypotheses of Theorem 4.26 are satisfied with $\varphi(t) = 1/t$ and $\alpha = 0$. Hence (4.112) has two linearly independent solutions satisfying

$$y_\pm(t) = \left[1 + \mathrm{O}\left(\frac{1}{t}\right)\right]e^{\pm it}, \qquad (t \to \infty).$$

However, Trench's result [155, Theorem 2] does not apply since it assumes that (4.111) holds, which is not satisfied in this example.

4.6.4 Conditions for Non-resonance

As observed in Sect. 4.6.1, resonance for certain perturbations of the harmonic linear oscillator such as (4.81) corresponds to situations when solutions become either unbounded or can tend to zero as $t \to +\infty$. In examples, this usually occurs for certain frequencies θ, and in this section we are interested in conditions when such resonance does not occur. This subject has an extensive literature, and an excellent treatment can be found in [53, Chap. 4] containing many results up through the 1980s.

One of these results, particularly important for certain examples and applications, concerns equations of the form

$$y' = [i\Lambda_0 + \xi(t)P(t)]\, y \qquad t \geq t_0, \tag{4.114}$$

where Λ_0 , $\xi(t)$, and $P(t)$ satisfy the following properties:

(a) $\Lambda_0 = \operatorname{diag}\{\lambda_1, \ldots, \lambda_d\}$, $\lambda_j \in \mathbb{R}$, $\lambda_j \neq \lambda_k$ for $j \neq k$;
(b) the scalar-valued function ξ is continuously differentiable on $[t_0, \infty)$, $\xi(t) = o(1)$ as $t \to \infty$, $\xi' \in L^1[t_0, \infty)$, and $\xi \in L^q[t_0, \infty)$ for some q, $1 < q < +\infty$;
(c) the matrix $P(t)$ is continuous and periodic of period τ;
(d) the eigenvalues λ_j of Λ_0 and τ satisfy the *non-resonance assumptions*

$$\frac{(\lambda_j - \lambda_k)\tau}{2\pi} \notin \mathbb{Z}, \qquad \forall\ 1 \leq j \neq k \leq d. \tag{4.115}$$

Under these conditions, it is shown (see [53, Lemma 4.1.1]) that there exists a matrix $\Phi \in C^1[t_0, \infty)$, which is τ-periodic and satisfies the equation

$$\Phi' + i\Phi\Lambda_0 - i\Lambda_0\Phi = P - \operatorname{diag} P, \tag{4.116}$$

with $\operatorname{diag}\Phi \equiv 0$. As a consequence, it follows that (see [53, Lemma 4.6.1]) that the conditioning transformation

$$y = [I + \xi(t)\Phi(t)]\, \tilde{y}, \tag{4.117}$$

is invertible for all $t \geq t_1$ sufficiently large and takes (4.114) into a system of the form

$$\tilde{y}' = \left[i\Lambda_0 + \xi \operatorname{diag} P + \tilde{R}_1 + \tilde{R}_2\right] \tilde{y}, \tag{4.118}$$

where $\tilde{R}_1 = \sum_{m=2}^{\lfloor q \rfloor + 1} \xi^m P_m$, with P_m continuous and τ-periodic, and $\tilde{R}_2 \in L^1[t_1, \infty)$. This leads, after $\lfloor q \rfloor + 1$ such transformations, to a complete diagonalization of (4.114) up to an L^1-perturbation. Letting $E(t) = \exp(i\Lambda_0 t)$, the proof consists of first showing that

$$\Phi(t) = E(t)\left[\int_{t_0}^{t} E(-s)\,\{P(s) - \operatorname{diag} P(s)\}\, E(s)\, ds + C\right] E(-t)$$

is a solution of (4.116) for $t \geq t_0$ and any constant matrix C. Then using the non-resonance assumption (d), one can explicitly construct C such that Φ is τ−periodic. Finally (4.118) follows immediately by substituting (4.117) into (4.114) and using (4.116).

Observe that the choice $\xi(t) = t^{-p}$, $t \geq t_0 > 0$ satisfies (b) for any $p > 0$ and consequently this result applies to (4.81) provided θ is suitably restricted so that $\tau = 2\pi/\theta$ satisfies (d). See more about such examples in Sect. 8.6.1.

Comparing this construction of a conditioning transformation with the alternative procedure of first applying $y = E(t)\hat{y}$ to reduce (4.114) to

$$\hat{y}' = \left[\xi(t)E^{-1}(t)P(t)E(t)\right]\hat{y}(t) =: V(t)\hat{y},$$

one sees how assumptions (d) are related to the conditional integrability of V. Namely, associating $P(t)$ with its formal Fourier expansion $\sum_{-\infty}^{+\infty} c_n e^{i(n2\pi t)/\tau}$, one sees that (4.115) guarantees that V is conditionally integrable on $[t_0, \infty)$. However, in applications such as to (4.81), the only integers n which matter in (4.115) are the ones for which $c_n \neq 0$, so for trigonometric polynomials, this is always just a finite set. An advantage of (4.117) is that it also can be used in cases where $Q(t) = -\int_t^\infty V(s)\,ds$ cannot be suitably approximated using integration by parts. For examples, see Sect. 8.6.1.

In a significant extension of the above non-resonance results, Cassell [33] substantially weakened conditions (a)–(c) by replacing them with:

There exists $\tau > 0$ such that

$$\Delta R(t) := R(t + \tau) - R(t) \in L^1[t_0, \infty), \tag{4.119}$$

and

$$J(|R|)(t) := \int_t^{t+\tau} |R(s)|\,ds = o(1) \qquad \text{as } t \to \infty. \tag{4.120}$$

It is straightforward to check for $R(t) = \xi(t)P(t)$ satisfying (a)–(c) that (4.119) and (4.120) hold. Moreover, (4.119) and (4.120) do not seem to imply that R is even conditionally integrable on $[t_0, \infty)$. However, we include Cassell's result in this section because conditionally integrable perturbations R are a major class for which (4.119) and (4.120) might apply and the approximate equation for (4.122) below is a natural extension of (4.116).

We are first interested in the question of whether under these conditions there exists a conditioning transformation of the form $\mathcal{Q}(t) = I + o(1)$ with the normalization $\operatorname{diag}\mathcal{Q}(t) \equiv I$ that diagonalizes

$$y' = [\Lambda_0 + R(t)]y \tag{4.121}$$

up to a L^1-perturbation. That is, putting $y = \mathcal{Q}(t)\tilde{y}$, Cassell raised the question if there exists a diagonal matrix $D(t)$ such that

$$\tilde{y}' = [\Lambda_0 + D(t) + \tilde{R}(t)]\tilde{y},$$

where $\tilde{R}(t) \in L^1$. Neglecting the L^1-perturbation $\tilde{R}(t)$, this amounts to asking if there exists a matrix $\mathcal{Q}(t) = I + o(1)$ such that

$$\mathcal{Q}'(t) = \Lambda_0\mathcal{Q}(t) - \mathcal{Q}(t)\Lambda_0 + R(t)\mathcal{Q}(t) - \mathcal{Q}(t)D(t). \tag{4.122}$$

Note that with the normalization diag $\mathcal{Q}(t) \equiv I$, looking at the elements on the diagonal it follows that D satisfies

$$D = \operatorname{diag}(R\mathcal{Q}).$$

As usual, one shows the existence of such $\mathcal{Q}(t)$ by considering an appropriate integral equation. While Cassell assumed that coefficient matrices are Lebesgue integrable, we assume for simplicity that they are continuous. We follow his exposition and begin with a preliminary lemma.

Throughout this section, we take the norm of a matrix as

$$|A| = \max_i \sum_{j=1}^{d} |a_{ij}| .$$

Finally, for matrices $A = A(t)$, Cassell defined an operator $\mathcal{K}$ by

$$(\mathcal{K}A)_{ij} = \begin{cases} 0 & \text{if } i = j \\ \frac{a_{ij}}{1-e^{(\lambda_i-\lambda_j)\tau}} & \text{if } i \neq j. \end{cases} \tag{4.123}$$

Lemma 4.28 *Let τ be a fixed positive number, and let $\Lambda_0 = \{\lambda_1, \ldots, \lambda_d\}$ be a constant diagonal matrix satisfying* (4.115). *Let $R(t)$ be a continuous $d \times d$ matrix satisfying* (4.119) *and* (4.120).

Then the matrix equation

$$\mathcal{Q}(t) = I + \mathcal{K} \int_t^{t+\tau} e^{\Lambda_0(t+\tau-u)} F(u) e^{-\Lambda_0(t+\tau-u)}\, du, \tag{4.124}$$

where

$$F = R\mathcal{Q} - \mathcal{Q}\operatorname{diag}(R\mathcal{Q}), \tag{4.125}$$

has a bounded and differentiable solution $\mathcal{Q}(t)$ on $[t_1, \infty)$ for t_1 sufficiently large ($t_1 \geq t_0$). Moreover, $\operatorname{diag} \mathcal{Q} = I$, *and $\mathcal{Q}(t) \to I$ as $t \to \infty$, hence $\mathcal{Q}(t)$ is nonsingular and $\mathcal{Q}(t)^{-1}$ is bounded for t sufficiently large.*

Proof Fix $\tau > 0$. Put

$$M_0 := \max_{1 \leq i \neq j \leq d} \max_{t \leq u \leq t+\tau} \left| \frac{e^{(\lambda_i - \lambda_j)(t+\tau-u)}}{1 - e^{(\lambda_i - \lambda_j)\tau}} \right|, \tag{4.126}$$

hence M_0 can be re-written as

$$M_0 = \max_{1 \leq i \neq j \leq d} \max_{0 \leq \sigma \leq \tau} \left| \frac{e^{(\lambda_i - \lambda_j)\sigma}}{1 - e^{(\lambda_i - \lambda_j)\tau}} \right|,$$

hence M_0 is independent of t. We re-write (4.124) as

$$\mathcal{Q}(t) = I + (T\mathcal{Q})(t).$$

As a preliminary observation, we note that from (4.123) and (4.124) follows that

$$\begin{aligned}
|(T\mathcal{Q})(t)| &\leq \max_i \sum_{j \neq i} \left| \int_t^{t+\tau} \frac{e^{(\lambda_i - \lambda_j)(t+\tau-u)} F_{ij}(u)}{1 - e^{(\lambda_i - \lambda_j)\tau}}\, du \right| \\
&\leq M_0 \int_t^{t+\tau} \max_i \sum_{j \neq i} \left| F_{ij}(u) \right| du \\
&= M_0 \int_t^{t+\tau} |F(u)|\, du = M_0 \left(J|F| \right)(t).
\end{aligned} \tag{4.127}$$

Also observe that (4.125) implies that

$$|F| \leq \left(|\mathcal{Q}| + |\mathcal{Q}|^2 \right) |R|. \tag{4.128}$$

With $J(|R|)(t)$ defined in (4.120), fix $t_1 \geq t_0$ sufficiently large such that

$$J(|R|)(t) \leq \frac{1}{4M_0}, \qquad \text{for } t \geq t_1. \tag{4.129}$$

We choose $\|A\| = \sup_{t \geq t_1} |A(t)|$, $\mathcal{B} = \{A(t) \text{ continuous on } [t_1, \infty) : \|A\| < \infty\}$ and $\mathcal{W} = \{A(t) \in \mathcal{B} : \|A\| \leq 1\}$. We claim that for $t \geq t_1$, T maps $\mathcal{W}$ into $\mathcal{W}$ and is a contraction. To substantiate the first claim, let $\mathcal{Q} \in \mathcal{W}$. Then it follows from (4.127), (4.128), and (4.129) that for $t \geq t_1$

$$|(T\mathcal{Q})(t)| \leq M_0 \left(\|\mathcal{Q}\| + \|\mathcal{Q}\|^2 \right) J(|R|)(t) \leq 2M_0 J(|R|)(t) \leq \frac{1}{2}, \tag{4.130}$$

hence T maps $\mathscr{W}$ into $\mathscr{W}$. To show that T is a contraction on $\mathscr{W}$, let $\mathscr{Q}_i \in \mathscr{W}$ for $i = 1, 2$. With $F_i = R\mathscr{Q}_i - \mathscr{Q}_i \text{diag}(R\mathscr{Q}_i)$, one can use the triangle inequality to show that

$$|F_1(t) - F_2(t)| \leq 3|R(t)| \|\mathscr{Q}_1 - \mathscr{Q}_2\|,$$

hence

$$\begin{aligned} |(T\mathscr{Q}_1(t) - (T\mathscr{Q}_2(t)| &\leq 3M_0 \|\mathscr{Q}_1 - \mathscr{Q}_2\| \int_t^{t+\tau} |R(u)| \, du \\ &= 3M_0 \|\mathscr{Q}_1 - \mathscr{Q}_2\| J(|R|)(t) \\ &\leq \frac{3}{4} \|\mathscr{Q}_1 - \mathscr{Q}_2\|. \end{aligned}$$

The Banach Fixed Point principle implies the existence of a unique continuous solution $\mathscr{Q}(t)$ of (4.124) with $\mathscr{Q} \in \mathscr{W}$. It is clear from (4.124) that $\mathscr{Q}$ is also differentiable. Moreover, $(T\mathscr{Q})(t) \to 0$ as $t \to \infty$ by (4.120) and (4.130). Therefore $\mathscr{Q}(t) \to I$ as $t \to \infty$ and hence $\mathscr{Q}(t)$ is invertible with a bounded inverse for t sufficiently large. □

Now differentiating (4.124) yields that

$$\begin{aligned} \mathscr{Q}'(t) &= \Lambda_0\{\mathscr{Q}(t) - I\} - \{\mathscr{Q}(t) - I\}\Lambda_0 + \mathscr{K}\left(F(t+\tau) - e^{\Lambda_0 \tau} F(t) e^{-\Lambda_0 \tau}\right) \\ &= \Lambda_0 \mathscr{Q}(t) - \mathscr{Q}(t)\Lambda_0 + \mathscr{K}\left(F(t) - e^{\Lambda_0 \tau} F(t) e^{-\Lambda_0 \tau}\right) + \mathscr{K}\left(F(t+\tau) - F(t)\right) \\ &= \Lambda_0 \mathscr{Q}(t) - \mathscr{Q}(t)\Lambda_0 + F(t) + \mathscr{K}\left(\Delta F(t)\right). \end{aligned} \tag{4.131}$$

Theorem 4.29 (Cassell) *Let τ be a fixed positive number, and suppose that Λ_0 and $R(t)$ satisfy* (4.115), (4.119), *and* (4.120). *Fix $t_1 \geq t_0$ as in Lemma 4.28. For $t \geq t_1$, let*

$$\Lambda(t) = \Lambda_0 + \text{diag}\,(R(t)\mathscr{Q}(t)),$$

where $\mathscr{Q}(t)$ is a solution of (4.124) *defined in Lemma 4.28 for $t \geq t_1$ ($\mathscr{Q} \in \mathscr{W}$). If $\Lambda(t)$ satisfies Levinson's dichotomy conditions* (2.13), (2.14), *then*

$$y' = [\Lambda(t) + R(t)]y, \qquad t \geq t_0, \tag{4.132}$$

has a fundamental matrix satisfying

$$Y(t) = [I + o(1)] \exp\left[\int^t \Lambda(s) \, ds\right].$$

Proof In (4.132), put $y = \mathscr{Q}(t)\tilde{y}$, where $\mathscr{Q}(t)$ is the solution of (4.124) established in Lemma 4.28. Then it follows from (4.125) and (4.131) that

$$\begin{aligned}\tilde{y}' &= \left[\mathscr{Q}^{-1}\left(\Lambda_0\mathscr{Q} + R\mathscr{Q} - \mathscr{Q}'\right)\right]\tilde{y} \\ &= \left[\Lambda_0 + \operatorname{diag}(R\mathscr{Q}) - \mathscr{Q}^{-1}\mathscr{K}(\Delta F(t))\right]\tilde{y}.\end{aligned}$$

Recall that $\lim_{t\to\infty}\mathscr{Q}(t) = I$ and, in particular, $\mathscr{Q}^{-1}$ is bounded for t sufficiently large. Then the result will follow from Theorem 2.7 if we show that $\Delta F \in L^1$ and therefore also $\mathscr{K}(\Delta F(t)) \in L^1$.

To substantiate this claim, note that from (4.124) and (4.125) follows that

$$\Delta\mathscr{Q}(t) = \mathscr{K}\int_t^{t+\tau} e^{\Lambda_0(t+\tau-u)}\Delta F(u)e^{-\Lambda_0(t+\tau-u)}\,du,$$

and

$$\Delta F = (\Delta R)E\mathscr{Q} + R\Delta\mathscr{Q} - (E\mathscr{Q})\operatorname{diag}\{(\Delta R)E\mathscr{Q} + R\Delta\mathscr{Q}\} - (\Delta\mathscr{Q})\operatorname{diag}(R\mathscr{Q}),$$

where E is the shift operator defined by $Er(t) = r(t+\tau)$. Hence

$$|\Delta\mathscr{Q}(t)| \le M_0\int_t^{t+\tau}|\Delta F(u)|\,du = M_0J(|\Delta F|)(t),$$

and

$$|\Delta F(t)| \le (\|\mathscr{Q}\| + \|\mathscr{Q}\|^2)|\Delta R(t)| + (2\|\mathscr{Q}\| + 1)|R(t)|\Delta\mathscr{Q}|. \tag{4.133}$$

It follows for $t_2 > t_1 + \tau$ that

$$\begin{aligned}\int_{t_1+\tau}^{t_2}|\Delta F(t)|\,dt &\le (\|\mathscr{Q}\| + \|\mathscr{Q}\|^2)\int_{t_1+\tau}^{\infty}|\Delta R(t)|\,dt \\ &\quad + (2\|\mathscr{Q}\| + 1)M_0\int_{t_1+\tau}^{t_2}|R(t)|J(|\Delta F(t)|)\,dt.\end{aligned}$$

A change in the order of integration implies that

$$\begin{aligned}\int_{t_1+\tau}^{t_2}|\Delta F(t)|\,dt &\le (\|\mathscr{Q}\| + \|\mathscr{Q}\|^2)\int_{t_1+\tau}^{\infty}|\Delta R(t)|\,dt \\ &\quad + (2\|\mathscr{Q}\| + 1)M_0\int_{t_1+\tau}^{t_2+\tau}|\Delta F(u)|J(|R(u-\tau)|\,du.\end{aligned}$$

Now making t_1 sufficiently large, such that $(2\|\mathcal{Q}\| + 1)M_0 J(|R(u-\tau)| \leq \theta < 1$ for all $u \geq t_1 + \tau$, it yields that

$$[1-\theta]\int_{t_1+\tau}^{t_2} |\Delta F(t)|\,dt \leq (\|\mathcal{Q}\| + \|\mathcal{Q}\|^2)\int_{t_1+\tau}^{\infty} |\Delta R(t)|\,dt + \theta J(|\Delta F|)(t_2).$$

By (4.119), (4.120), and (4.133), $J(|\Delta F|)$ is bounded, so letting t_2 go to infinity shows that $\Delta F \in L^1$, which completes the proof. □

As remarked earlier, Theorem 4.29 is an impressive extension of earlier results concerning non-resonance due to the generality of conditions (4.119) and (4.120) which do not explicitly involve conditional convergence of R nor that $R \in L^p$, $p < \infty$. However, this generality comes at a price and concerns the fact that solving (4.124) for Q and checking the dichotomy conditions is not necessarily an easy matter because of the non-linearity of F. But in special cases such as $R(t) = \xi(t)P(t)$ mentioned earlier (see [53, Chap. 4]), the solution of (4.124) can be suitably approximated (up to L^1-terms) by an explicit iteration of $I+Q$-transformations.

4.6.5 Some Further Results Involving Averaging

In this section, we are concerned with some results leading to the asymptotic integration of systems

$$y' = A(t)y, \tag{4.134}$$

where resonance may occur. We consider classes of matrices $A(t)$, $t \geq t_0$, where $A(t)$ is oscillatory and tends to zero as $t \to \infty$, however the methods and results in the foregoing sections to do not apply. One reason is that either $A(t)$ is not conditionally integrable on $[t_0, \infty)$ or even if it would be so, after one or more conditioning transformations, the new system contains terms which are not conditionally integrable. Such a situation arises, for example in (4.81) when $\lambda = \pm 1$ or ± 2 and resonance may occur.

In this case it is necessary to modify the reductive procedure in order to obtain a reduced system to which Theorem 2.7 may be applied.

The types of coefficient matrices we will discuss here contain entries which are finite linear combinations of functions of the form $f(t) = v(t)p(t)$, where

$$v(t) = o(1) \text{ as } t \to +\infty, \qquad v'(t) \in L^1[t_0, \infty). \tag{4.135}$$

and $p(t)$ is periodic of period θ.

If it happens that $\int_0^\theta p(t)\,dt = 0$, then integration by parts yields

$$\int_{t_0}^{T} v(t)p(t)\,dt = v(T)P(T) - v(t_0)P(t_0) - \int_{t_0}^{T} v'(t)P(t)\,dt,$$

where $P(t) = \int_{t_0}^{t} p(\tau)\,d\tau$ is easily seen to be periodic of period θ and hence bounded for $t \geq t_0$. Therefore $\lim_{T\to\infty} \int_{t_0}^{T} v(t)p(t)\,dt$ exists and $v(t)p(t)$ is conditionally integrable. Similarly,

$$q(t) := -\int_t^\infty v(\tau)p(\tau)\,d\tau = v(t)P(t) + \int_t^\infty v'(\tau)P(\tau)\,d\tau,$$

is well-defined and might be a suitable candidate for entries in an $I + Q(t)$-transformation.

If $\int_0^\theta p(t)\,dt \neq 0$, let

$$M[p] = \frac{1}{\theta}\int_0^\theta p(t)\,dt$$

denote the mean or average value of p. Then defining

$$\tilde{p}(t) = p(t) - M[p]$$

it follows that $\tilde{p}$ is periodic of mean zero, hence as above $v(t)\tilde{p}(t)$ is conditionally integrable and

$$v(t)p(t) = v(t)M[p] + v(t)\tilde{p}(t)$$

decomposes such functions into a residual term (which will contribute to resonance in the system) and a conditionally integrable part (which can be affected using conditioning transformations). These observations form the basis of a procedure knows as the *method of averaging*, and we now briefly discuss two situations where this method has been applied. We note that this method is also used for equations such as

$$y'' = [\lambda^2 + \varepsilon p(t)]y = 0,$$

where ε is a small parameter instead of an o(1) perturbation (see [163]).

One case treated by Burd, Nesterov, and Karakulin (see [29–31]) involves systems (4.134) with

$$A(t) = \sum_{i=1}^{n} v_i(t)A_i(t) + R(t),$$

where the $v_i(t)$ are scalar functions satisfying (4.135) above, the $A_i(t)$ are matrices consisting of trigonometric polynomials, and $R(t) \in L^1[t_0,\infty)$. In addition, the $v_i(t)$ are assumed to satisfy the property that there exists $K \in \mathbb{N}$ such that *all* products of $K+1$ factors from $\{v_i(t)\}_1^n$ are in L^1. Then it is shown (see [31, Theorem 2.1]) that the system (4.134) can be reduced to a system of the form

$$\tilde{y}' = \left(\sum_{i=1}^{n} v_i(t) A_i + \sum_{1 \le i_1 \le i_2 \le n} v_{i1}(t) v_{i2}(t) A_{i1,i2} \right.$$
$$\left. + \cdots + \sum_{1 \le i_1 \le \ldots \le i_K \le n} v_{i1}(t) \cdots v_{iK}(t) A_{i1,\cdots,iK} + \tilde{R}(t) \right) \tilde{y},$$

where the matrices A_i, $A_{i1,2}$ etc. are constant and $\tilde{R} \in L^1$. This is accomplished by a conditioning transformation of the form

$$y = \left[I + \sum_{i=1}^{n} v_i(t) P_i(t) + \sum_{1 \le i_1 \le i_2 \le n} v_{i1}(t) v_{i2}(t) P_{i1,i2}(t) \right.$$
$$\left. + \cdots + \sum_{1 \le i_1 \le \ldots \le i_K \le n} v_{i1}(t) \cdots v_{iK}(t) P_{i1,\cdots,iK}(t) \right] \tilde{y},$$

where the entries of the matrices $P_i(t), P_{i1,i2}$ etc. are trigonometric polynomials of mean zero. In treating sums of periodic functions with different periods, Burd and Nesterov define

$$M[p] = \lim_{T \to \infty} \frac{1}{T} \int_0^T p(t)\, dt,$$

which can be seen to be equivalent to that above. In fact, for the case of trigonometric polynomials which can be expressed as

$$p(t) = \sum_{j=1}^{N} c_j e^{ib_j t} = \sum_{j=1}^{N} p_j(t), \qquad b_j \in \mathbb{R}, \quad b_j \neq b_k \text{ for } j \neq k,$$

it follows that $M[p_j] = c_j$ if $b_j = 0$ and $M[p_j] = 0$ otherwise. In order to apply Levinson's fundamental theorem to obtain an asymptotic representation for solutions, one requires more information about the $v_i(t)$ and the coefficient matrices in the reduced system. See [29] and [31] for details of the proof and applications to systems such as (4.81).

A second application of the method of averaging concerns matrices of the form

$$A(t) = \frac{A}{t} + \sum_{k=1}^{\infty} \frac{B_k(t)}{t^k},$$

where A is constant and $B_k(t)$ are periodic matrices of normalized period 1. If the eigenvalues of $A_1 = A + M[B_1(t)]$ do not differ by non-zero integers, then it can be shown that a formal transformation

$$y = \left[I + \sum_{k=1}^{\infty} \frac{P_k(t)}{t^k}\right] \tilde{y}$$

reduces the system (4.134) to $\tilde{y}'(t) = \frac{A_1}{t}\tilde{y}$. Hence $y' = A(t)y$ has a formal fundamental solution matrix

$$\hat{Y}(t) = \left[I + \sum_{k=1}^{\infty} \frac{P_k(t)}{t^k}\right] t^{A_1},$$

which can be shown (e.g., by applying Levinson's fundamental theorem) to be an asymptotic expansion (in the sense of Poincaré) for an actual solution. This result can be found in [14] along with similar ones in case the eigenvalue assumption on A_1 is not satisfied or when the expansion of $A(t)$ contains other powers of $1/t$. Such results as these can be thought of as extending the classical Poincaré-Frobenius method for solving differential equations having regular/irregular singularities (at ∞) to equations where the coefficients are periodic functions instead of constants.

4.7 Preliminary Transformations

The goal in this section is to indicate several types of explicit linear transformations $x = P(t)y$, which might be used to bring a given system

$$x' = A(t)x \tag{4.136}$$

into what could be called a "pre-L-diagonal form"

$$y' = [\Lambda(t) + V(t)]y,$$

were Λ is diagonal and V is small, but not necessarily in L^1.

1. First, we consider constant, invertible transformations

$$x = Py,$$

which may be used to bring the coefficient matrix of the largest order in $A(t)$ into Jordan canonical form. For example, if in (4.136)

$$A(t) = \alpha(t)A + B(t) \tag{4.137}$$

where $B(t) = o(|\alpha(t)|)$ as $t \to \infty$, then $x = Py$ transforms (4.136) into

$$y' = [\alpha(t)P^{-1}AP + \tilde{B}(t)]y,$$

with $\tilde{B}(t) = o(|\alpha(t)|)$ as $t \to \infty$.

2. If $P^{-1}AP$ would be diagonal, then conditioning transformations $y = [I + Q(t)]\tilde{y}$ as in the previous sections may be used to try to further diagonalize $\tilde{B}(t)$ up to L^1-terms. If A is not diagonalizable, blocked conditioning transformations may be used to create a diagonally-blocked matrix plus a blocked perturbation with suitably small entries in the off-diagonal blocks. This works especially well for diagonal blocks whose leading coefficient matrices have different eigenvalues, but it is also possible to effect some entries within a single block having just one eigenvalue, but several Jordan blocks.
3. Two types of scaling transformations can be applied to normalize a system (4.136). One is a scalar transformation $y = a(t)\tilde{y}$ which takes (4.136) into $\tilde{y}' = \left[A(t) - \frac{a'(t)}{a(t)}I\right]\tilde{y}$ and is usually used to normalize the diagonal elements of $A(t)$. The other is a change of independent variable $t = f(\tau)$, $y(f(\tau)) = \hat{y}(\tau)$ which yields $\frac{d\hat{y}}{d\tau} = f'(\tau)A(f(\tau))\hat{y}$ and is used to normalize the growth of the coefficient matrix.
4. In case $A(t)$ contains terms above and below the main diagonal of different orders of magnitude and the above-mentioned transformations are not effective, certain kinds of diagonal matrices called *shearing transformations* can sometimes be applied. Their goal is to equalize or balance such terms and create a new diagonally-dominant coefficient matrix. A typical situation calling for a shearing transformation is if $A(t) = N + R(t)$, where N is the full nilpotent block in upper-triangular Jordan form and $R(t) = o(1)$. A shearing transformation $y = \text{diag}\{1, \beta(t), \beta^2(t), \ldots, \beta^{d-1}(t)\}\tilde{y}$ then multiplies each element in N by $\beta(t)$ and elements below the diagonal by various powers of $1/\beta(t)$. A good choice for $\beta(t)$ can sometimes lead to a system whose leading coefficient is non-nilpotent and then transformations such as in 1.,2., or 3. above can be applied to diagonalize or block-diagonalize the system.

While there is no comprehensive algorithm to treat general systems, in the case of meromorphic coefficients $A(t)$, transformations of the above types can be shown to always yield a so-called formal fundamental solution matrix that establishes the asymptotic behavior. (See [3] for a modern discussion of an algorithmic treatment as well as references to symbolic computational tools for implementing it.)

In Chap. 8 we will discuss a variety of second-order scalar differential equations and show how transformations of the above type can lead to normalized systems for which Levinson's fundamental theorem can be applied. We conclude this section with an example which also demonstrates how such transformations can be applied.

Consider the two-dimensional system

$$y' = \begin{bmatrix} 0 & 1 \\ \frac{1}{\log t} & 0 \end{bmatrix} y \qquad \text{for } t \geq e,$$

where the leading coefficient is already in Jordan form and nilpotent. The shearing transformation

$$y = \operatorname{diag}\{1, 1/\sqrt{\log t}\}\tilde{y}$$

yields

$$\tilde{y}' = \begin{bmatrix} 0 & \dfrac{1}{\sqrt{\log t}} \\ \dfrac{1}{\sqrt{\log t}} & \dfrac{1}{(2t\log t)} \end{bmatrix} \tilde{y}$$

and the constant transformation

$$\tilde{y} = \begin{bmatrix} 1 & 1 \\ -1 & 1 \end{bmatrix} \hat{y}$$

yields $\hat{y}' = [\Lambda(t) + V(t)]\hat{y}$ with

$$\Lambda(t) = \frac{1}{\sqrt{\log t}} \begin{bmatrix} -1 & 0 \\ 0 & 1 \end{bmatrix} \quad \text{and} \quad V(t) = \frac{1}{4t\log t} \begin{bmatrix} 1 & -1 \\ -1 & 1 \end{bmatrix}.$$

Since V is not in L^1, Theorem 2.7 cannot be applied. However, one can easily check that $|V|/(\lambda_1 - \lambda_2) \to 0$ as $t \to \infty$ and that $(V/(\lambda_1 - \lambda_2))' \in L^1$, and that the eigenvalues of $\lambda + V$ satisfy Levinson's dichotomy conditions, so Theorem 4.8 may be used to obtain an asymptotic integration.

But in this case, one can also use the conditioning transformation

$$\hat{y} = \begin{bmatrix} 1 & -\frac{1}{8t\sqrt{\log t}} \\ \frac{1}{8t\sqrt{\log t}} & 1 \end{bmatrix} \breve{y},$$

to obtain

$$\breve{y}' = \left(\begin{bmatrix} \frac{-1}{\sqrt{\log t}} + \frac{1}{4t\log t} & 0 \\ 0 & \frac{1}{\sqrt{\log t}} + \frac{1}{4t\log t} \end{bmatrix} + R(t) \right) \breve{y},$$

with $R \in L^1$. Therefore Theorem 2.7 applies and leads to an asymptotic integration. To analyze the behavior of the main terms on the diagonal, one can use integration by parts to see that

$$\int^t \frac{ds}{\sqrt{\log s}} = \frac{t}{\sqrt{\log t}} \left(1 + O\left(\frac{1}{\log t}\right)\right) \qquad \text{as } t \to \infty.$$

This example also illustrates the close relationship that preliminary transformations may have with eigenvectors of $A(t)$. Observe that

$$A(t) = \begin{bmatrix} 0 & 1 \\ \frac{1}{\log t} & 0 \end{bmatrix}$$

has eigenvectors

$$x_1 = \begin{bmatrix} 1 \\ \frac{-1}{\sqrt{\log t}} \end{bmatrix} \quad \text{and} \quad x_2 = \begin{bmatrix} 1 \\ \frac{1}{\sqrt{\log t}} \end{bmatrix},$$

so the similarity transformation $P^{-1}AP$ with $P(t) = [x_1, x_2]$ diagonalizes A. But

$$P(t) = \begin{bmatrix} 1 & 0 \\ 0 & \frac{1}{\sqrt{\log t}} \end{bmatrix} \begin{bmatrix} 1 & 1 \\ -1 & 1 \end{bmatrix}$$

is just the product of the shearing transformation and constant transformation we used above. Moreover, the logarithmic derivative $P^{-1}P'$ which comes from making the linear transformation contains the terms $-\frac{1}{4t \log t}$ which must be used to modify the eigenvalues of A in the asymptotic integration formula. One could also have calculated the eigenvalues of $\Lambda + V$ above and used them, but that would yield a more complicated integrand. The difference between the two asymptotic representations is of course just an L^1 function which can be absorbed into a factor $c + o(1)$.

Chapter 5
Conditioning Transformations for Difference Systems

5.1 Chapter Overview

In this chapter we will consider linear difference systems of the form $x(n+1) = A(n)x(n)$, where $\det A(n) \neq 0$ for all $n \geq n_0$. Various procedures will be discussed (similar to those in the preceding chapter) for bringing such a system (if possible) into what we have called an L-diagonal form, so that the results of Chap. 3 may be used.

The transformations we will consider have the form

$$x(n) = P(n)[I + Q(n)]y(n),$$

where $P(n)$ (called a preliminary transformation) will be explicit and invertible for sufficiently large n, and $Q(n)$ is a matrix tending to zero as $n \to \infty$ or having a special structure to ensure invertibility (e.g., strictly upper or lower triangular). Such a $Q(n)$ is either explicitly known or implicitly constructed as the solution of some simpler algebraic or difference equations. Analogously to the case of differential equations, we also use the term *conditioning transformation* for

$$y(n) = [I + Q(n)]z(n),$$

because its purpose is to modify a system of the form

$$y(n+1) = [\Lambda(n) + V(n)]y(n)$$

into a system which will be in L-diagonal form so that the results of Chap. 3 may be applied.

Beginning with Sect. 5.2 and continuing through Sect. 5.6, we will discuss various types of conditioning matrices and the roles that they have played in

S. Bodine, D.A. Lutz, *Asymptotic Integration of Differential and Difference Equations*, Lecture Notes in Mathematics 2129,
DOI 10.1007/978-3-319-18248-3_5

applying the difference equations analogue of Levinson's fundamental theorem (Theorem 3.4) to many other situations, some of which were treated initially by quite different methods. After that, we will discuss some techniques that have been used to construct preliminary transformations to bring, if possible, a system into a form where the main (largest) terms lie on the diagonal. As with differential equations, such transformations involve the most "ad hoc" part of the treatment here and some of them are related to similar ones used for differential equations.

In Chap. 9 we will see how some of the preliminary transformations are applied, especially for scalar, linear difference equations and, in particular, those of second order.

As with differential equations, we point out that no general theory exists for the wide range of difference equations to which the methods are potentially applicable. The same obstacles involve being able to isolate the main growth terms on the diagonal, and then reduce the growth of the off-diagonal terms with conditioning transformations, and finally check that appropriate dichotomy conditions hold.

Note that in the case of so-called meromorphic linear difference systems $x(z+1) = A(z)x(z)$, where

$$A(z) = \sum_{k=-m}^{\infty} A_k z^{-k}$$

and $\det A(z) \neq 0$ for $r < |z| < \infty$, there is a complete theory which produces so-called *formal fundamental solution matrices* $\hat{X}(z)$ and then discusses their asymptotic meaning as $|z| \to \infty$ in suitable sectors of the complex plane. See [82] for details, especially how such formal solutions are constructed, since many of these techniques can be adapted to the more general (non-meromorphic) equations and motivate some of the transformations we will use.

5.2 Conditioning Transformations and Approximate Equations

We will consider in this section and following ones, systems of the form

$$y(n+1) = [\Lambda(n) + V(n)]y(n), \tag{5.1}$$

where $\Lambda(n)$ is diagonal with elements that are non-zero for n sufficiently large, and $V(n)$ is small in some sense, but not absolutely summable as required by Theorem 3.4.

The conditioning transformations have the form

$$y(n) = [I + Q(n)]z(n), \tag{5.2}$$

with the goal of transforming (5.1) into

$$z(n+1) = [\tilde{\Lambda}(n) + \tilde{V}(n)]z(n),$$

where $\tilde{\Lambda}(n)$ is again diagonal and invertible for large n and $\frac{\tilde{V}(n)}{\tilde{\lambda}_i(n)}$ is either absolutely summable or else an "improvement" to (5.1) in the sense that $V(n)$ is smaller by an order of magnitude and a repeat of similar transformations to further reduce the perturbation is possible.

In (5.1) one may assume that $\operatorname{diag} V(n) \equiv 0$ and we often find it convenient to make that assumption. If we let $\tilde{\Lambda}(n) = \Lambda(n) + D(n)$, where D is a diagonal, yet undetermined matrix, then D, Q, and $\tilde{V}$ must satisfy the transformation equation

$$\begin{aligned} V(n) + \Lambda(n)Q(n) + V(n)Q(n) &= D(n) + \tilde{V}(n) + Q(n+1)\Lambda(n) \\ &\quad + Q(n+1)D(n) + Q(n+1)\tilde{V}(n). \end{aligned} \tag{5.3}$$

Here we think of Λ and V as given and want to find D and $\tilde{V}$ so that there exists a solution $Q = o(1)$ as $n \to \infty$.

As for differential equations, the simpler case occurs when one can take $D \equiv 0$. Then (5.3) becomes

$$V(n) + \Lambda(n)Q(n) + V(n)Q(n) = Q(n+1)\Lambda(n) + [I + Q(n+1)]\tilde{V}(n). \tag{5.4}$$

Depending upon different additional assumptions on Λ and V, one can construct matrices $\tilde{Q}$ which are solutions of so-called "approximate equations" which are easier to solve.

For example, if $\tilde{Q}$ satisfies the approximate equation

$$V(n) + \Lambda(n)\tilde{Q}(n) - \tilde{Q}(n+1)\Lambda(n) = 0,$$

and $\tilde{Q}(n) = o(1)$ as $n \to \infty$, then the transformation $y(n) = [I + \tilde{Q}(n)]\tilde{y}(n)$ takes (5.1) into

$$\tilde{y}(n+1) = [\Lambda(n) + \tilde{V}(n)]\tilde{y}(n),$$

where $\tilde{V}(n) = [I + \tilde{Q}(n+1)]^{-1} V(n)\tilde{Q}(n)$, which is always an improvement to (5.1) and is particularly significant if $\tilde{Q}(n) = O(|V(n)|)$ as $n \to \infty$. This situation will be considered in Sect. 5.4.

If choosing $D \equiv 0$ as above does not lead to an improvement, one could return to (5.3) and assume $\operatorname{diag} V(n) = \operatorname{diag} \tilde{V}(n) = \operatorname{diag} Q(n) \equiv 0$. Then one obtains by taking diagonal parts

$$D(n) = \operatorname{diag}[V(n)Q(n)] - \operatorname{diag}[Q(n+1)\tilde{V}(n)], \tag{5.5}$$

therefore (5.3) reduces to the non-linear two sided matrix equation

$$\begin{aligned} &V(n) + \Lambda(n)Q(n) + V(n)Q(n) - Q(n+1)\Lambda(n) \\ &\quad = [I + Q(n+1)]\left\{\operatorname{diag}\left(V(n)Q(n) - Q(n+1)\tilde{V}(n)\right) + \tilde{V}(n)\right\}. \end{aligned}$$

If there would exist a matrix $\tilde{Q}(n) = o(1)$ as $n \to \infty$ satisfying the approximate equation

$$V(n) + \Lambda(n)\tilde{Q}(n) - \tilde{Q}(n)\Lambda(n) + V(n)\tilde{Q}(n) = [I + \tilde{Q}(n)]D(n),$$

i.e.,

$$[I + \tilde{Q}(n)]^{-1}[\Lambda(n) + V(n)][I + \tilde{Q}(n)] = \Lambda(n) + D(n),$$

then the transformation $y(n) = [I + \tilde{Q}(n)]\tilde{y}(n)$ takes (5.1) into

$$\tilde{y}(n+1) = [\Lambda(n) + D(n) + \tilde{V}(n)]\tilde{y}(n),$$

with

$$\tilde{V}(n) = \left\{[I + \tilde{Q}(n+1)]^{-1}[I + \tilde{Q}(n)] - I\right\}[\Lambda(n) + D(n)].$$

Depending upon the magnitude of $[\tilde{Q}(n+1) - \tilde{Q}(n)]\Lambda(n)$, this may or may not be an improvement to (5.1). This case and other similar situations will be discussed in Sect. 5.3. Another situation occurring in Sect. 5.6 concerns conditionally convergent perturbations $V(n)$, where choosing $Q(n) = -\sum_{k=n}^{\infty} V(k)$ and therefore $Q(n+1) - Q(n) = V(n)$, can sometimes lead to improvements.

5.3 Reduction to Eigenvalues

We begin with a discrete counterpart to Theorem 4.1 (see [9, Theorem 3.1]).

Theorem 5.1 *Let A be a constant $d \times d$ matrix with d distinct, nonzero eigenvalues λ_k $(1 \leq k \leq d)$. Let the $d \times d$ matrix $V(n)$ be defined for $n \geq n_0$ and satisfy*

$$V(n) \to 0 \qquad \text{as } n \to \infty,$$

and

$$\sum_{n=n_0}^{\infty} |V(n+1) - V(n)| < \infty.$$

Assume that the $d \times d$ matrix $R(n)$ is defined for $n \geq n_0$ and that

$$\sum_{n=n_0}^{\infty} |R(n)| < \infty.$$

Let the eigenvalues of $A + V(n)$ satisfy the discrete Levinson dichotomy conditions (3.8), (3.9). *Then*

$$y(n+1) = [A + V(n) + R(n)]\, y(n), \qquad n \geq n_0 \tag{5.6}$$

has, for all n sufficiently large, a fundamental matrix satisfying

$$Y(n) = [T + \mathrm{o}(1)] \prod^{n-1} \Lambda(k) \qquad \text{as } n \to \infty, \tag{5.7}$$

where $T^{-1}AT = \Lambda_0 = \operatorname{diag}\{\lambda_1, \ldots, \lambda_d\}$, $\Lambda(n) = \operatorname{diag}\{\lambda_1(n), \ldots, \lambda_d(n)\}$ is a diagonal matrix of eigenvalues of $A + V(n)$, ordered such that $\lambda_i(n) \to \lambda_i$ as $n \to \infty$ for all $1 \leq i \leq d$.

Proof In (5.6), put

$$y(n) = Tz(n), \tag{5.8}$$

where $T^{-1}AT = \Lambda_0 = \operatorname{diag}\{\lambda_1, \ldots, \lambda_d\}$. Then

$$z(n+1) = \left[\Lambda_0 + \hat{V}(n) + \hat{R}(n)\right] z(n),$$

where $\hat{V}(n) = T^{-1}V(n)T$ and $\hat{R}(n) = T^{-1}R(n)T$. $\hat{V}$ and $\hat{R}$ inherit the properties of V and R, respectively, i.e., $\hat{V}(n) \to 0$ as $n \to \infty$, and $\{\hat{V}(n) - \hat{V}(n+1)\}$ and $\hat{R}(n)$ are in $l^1[n_0, \infty)$.

As shown in Sect. 4.3 (cf. (4.10)), there exists a matrix $Q(n)$ and an integer $n_1 \geq n_0$ such that

$$[I + Q(n)]^{-1} [\Lambda_0 + \hat{V}(n)] [I + Q(n)] = \Lambda(n) \qquad \text{for } n \geq n_1,$$

where $\operatorname{diag} Q(n) = 0$ for all $n \geq n_1$ and $Q(n)$ satisfies

$$Q(n) = \mathrm{O}(|V(n)|) = \mathrm{o}(1) \qquad \text{as } n \to \infty.$$

It can also be shown using the same argument as in [69, Remark 2.1] that

$$|Q(n) - Q(n+1)| = \mathrm{O}\left(|V(n) - V(n+1)|\right) \qquad \text{as } n \to \infty.$$

Put

$$z(n) = [I + Q(n)]w(n) \tag{5.9}$$

for $n \geq n_1$. It follows that

$$\begin{aligned} w(n+1) &= [I + Q(n+1)]^{-1} [\Lambda_0 + \hat{V}(n) + \hat{R}(n)] [I + Q(n)] w(n) \\ &= \Big\{\Lambda(n) + \left([I + Q(n+1)]^{-1} - [I + Q(n)]^{-1}\right) [\Lambda_0 + \hat{V}(n)] [I + Q(n)] \\ &\quad + R_1(n)\Big\} w(n) \qquad n \geq n_1, \end{aligned}$$

where

$$R_1(n) = [I + Q(n+1)]^{-1} \hat{R}(n) [I + Q(n)] = \mathrm{O}(|R(n)|) \qquad \text{as } n \to \infty.$$

Since

$$\begin{aligned} &[I + Q(n+1)]^{-1} - [I + Q(n)]^{-1} \\ &\quad = [I + Q(n+1)]^{-1} \{[I + Q(n)] - [I + Q(n+1)]\} [I + Q(n)]^{-1} \\ &\quad = [I + Q(n+1)]^{-1} \{Q(n) - Q(n+1)\} [I + Q(n)]^{-1} \\ &\quad = \mathrm{O}(|Q(n) - Q(n+1)|) \qquad \text{as } n \to \infty, \end{aligned}$$

one can see that

$$w(n+1) = [\Lambda(n) + R_2(n)] w(n) \qquad n \geq n_1, \tag{5.10}$$

where (as $n \to \infty$)

$$R_2(n) = \mathrm{O}(|Q(n) - Q(n+1)|) + R_1(n),$$

hence $R_2(n) \in l^1[n_1, \infty)$.

Recall that $\Lambda(n) \to \Lambda_0$ as $n \to \infty$ whose entries are nonzero constants, it follows that $\Lambda(n)$ is invertible for n sufficiently large and that $R_2(n)$ satisfies the growth condition (3.10). Finally, we assumed that $\Lambda(n)$ satisfies the discrete Levinson dichotomy conditions (3.8), (3.9) and, therefore, by Theorem 3.4, (5.10) has, for all n sufficiently large, a fundamental matrix satisfying

$$W(n) = [I + \mathrm{o}(1)] \prod^{n-1} \Lambda(k) \qquad \text{as } n \to \infty.$$

By (5.8) and (5.9), (5.6) has a fundamental matrix of the form (5.7), and the proof is complete. □

We remark that this result under somewhat stronger dichotomy conditions can be found in Rapoport [131].

In the case when $\Lambda(n)$ does not necessarily have a constant limit, a discrete analogue of Theorem 4.8 was first considered by Ren, Shi, and Wang [133, Theorem 3.3]. Their result was considerably improved by R.J. Kooman [98, Theorem 3.1], who applied one of his earlier results (Theorem 3.8) in order to weaken the growth/decay assumption on the perturbation as well as also providing error bounds in the asymptotic representation. Here we provide a further modification of Kooman's result by slightly relaxing the growth/decay assumption, but not explicitly including error bounds.

Theorem 5.2 (Kooman) *Let $\Lambda(n) = \operatorname{diag}\{\lambda_1(n), \dots, \lambda_d(n)\}$ be a $d \times d$ diagonal matrix, in which $\lambda_i(n) - \lambda_j(n)$ is nowhere zero for $n \geq n_0$ for each $1 \leq i \neq j \leq d$. Let the $d \times d$ matrix $V(n)$ satisfy*

$$\frac{v_{ik}(n)}{\lambda_i(n) - \lambda_j(n)} \to 0 \qquad \text{as } n \to \infty \qquad \text{for } i \neq j \text{ and } k \neq i, \tag{5.11}$$

$$\Delta\left[\frac{v_{ij}(n)}{\lambda_i(n) - \lambda_j(n)}\right] \in l^1[n_0, \infty) \qquad \text{for } i \neq j. \tag{5.12}$$

Suppose that for $i \neq j$ and $k \neq i$ holds that

$$\Delta\left[\frac{v_{ik}(n)}{\lambda_i(n) - \lambda_j(n)}\right] \max_{i \neq j} \left|\frac{v_{ij}(n+1)}{\lambda_i(n+1) - \lambda_j(n+1)}\right| \in l^1[n_0, \infty), \tag{5.13}$$

where $\Delta F(n) = F(n+1) - F(n)$. Also, let the eigenvalues $\mu_k(n)$ of $\Lambda(n) + V(n)$ be nowhere zero for $n \geq n_0$ and satisfy the discrete Levinson dichotomy conditions (3.8), (3.9). *Finally, suppose that the $d \times d$ matrix $R(n)$ satisfies that*

$$\frac{|R(n)|}{|\mu_k(n)|} \in l^1[n_0, \infty), \qquad \forall\ 1 \leq k \leq d. \tag{5.14}$$

Then

$$y(n+1) = [\Lambda(n) + V(n) + R(n)]y(n) \qquad n \geq n_0, \tag{5.15}$$

has for n sufficiently large d solutions satisfying

$$y_i(n) = [e_i + o(1)] \prod^{n-1} \mu_i(k), \qquad \text{as } n \to \infty, \quad \forall\ 1 \leq i \leq d. \tag{5.16}$$

Proof Writing the diagonal matrix of eigenvalues of $\Lambda(n) + V(n)$ as

$$\tilde{\Lambda}(n) = \Lambda(n) + D(n) = \operatorname{diag}\{\mu_1(n), \dots, \mu_d(n)\},$$

the beginning of the proof follows exactly the proof of Theorem 4.8. In particular, one utilizes the implicit function theorem to show the existence of a matrix $Q(n)$ satisfying diag $Q(n) \equiv 0$ and

$$Q(n) \to 0 \qquad \text{as } n \to \infty, \tag{5.17}$$

such that

$$[I + Q(n)]^{-1}[\Lambda(n) + V(n)][I + Q(n)] = \tilde{\Lambda}(n). \tag{5.18}$$

In the following, we let $n \geq n_1$, where $n_1 \geq n_0$ is sufficiently large such that $I+Q(n)$ is invertible for $n \geq n_1$. The transformation

$$y(n) = [I + Q(n)]z(n), \tag{5.19}$$

and (5.18) imply that

$$\begin{aligned} z(n+1) &= [I + Q(n+1)]^{-1}[\Lambda(n) + V(n) + R(n)][I + Q(n)]z(n) \\ &= \left[[I + Q(n+1)]^{-1}\,[I + Q(n+1) - \Delta Q(n)]\,\tilde{\Lambda}(n) + R_2(n)\right] z(n) \\ &= \left[\tilde{\Lambda}(n) + R_1(n) + R_2(n)\right] z(n), \end{aligned} \tag{5.20}$$

where

$$\begin{aligned} R_1(n) &= -[I + Q(n+1)]^{-1}\,[\Delta Q(n)]\,\tilde{\Lambda}(n), \\ R_2(n) &= [I + Q(n+1)]^{-1}R(n)[I + Q(n)]. \end{aligned}$$

We seek to apply Corollary 3.9 to conclude that (5.20) has a fundamental solution matrix satisfying

$$Z(n) = [I + \mathrm{o}(1)]\prod^{n-1}\tilde{\Lambda}(k) = [I + \mathrm{o}(1)]\prod^{n-1}[\Lambda(k) + D(k)] \qquad \text{as } n \to \infty.$$

To do so, it is sufficient to show that $R_i(n)\tilde{\Lambda}^{-1}(n) \in l^1[n_1, \infty)$ for $i = 1, 2$. First considering R_1, it suffices to show that $\Delta Q(n) \in l^1[n_1, \infty)$. This claim follows with an argument analogous to the proof of Lemma 4.6 and for this we just indicate a brief outline. Making the same normalizations diag $V(n) \equiv 0$ (by replacing $\lambda_i(n)$ by $\lambda_i(n) + v_{ii}(n)$, which does not change the hypotheses), diag $Q(n) \equiv 0$, one sees that $D(n) = \text{diag }(V(n)Q(n))$ as in (4.25). Then writing (4.26) as (4.29) and continuing as in Lemma 4.6 one obtains

$$Q(n) = \mathrm{O}\left(\max_{i \neq j}\left|\frac{v_{ij}(n)}{\lambda_j(n) - \lambda_i(n)}\right|\right) \qquad \text{as } n \to +\infty. \tag{5.21}$$

Furthermore, arguments similar to the ones in the proof of Lemma 4.6 can be used to show (see below) that

$$\Delta Q(n) \in l^1[n_1, \infty),$$

and therefore that $R_1(n)\tilde{\Lambda}^{-1}(n) = -[I + Q(n+1)]^{-1}\,\Delta Q(n) \in l^1$. Secondly, (5.14) and (5.17) show that $R_2(n)\tilde{\Lambda}^{-1}(n) \in l^1[n_1, \infty)$.

We also note that

$$\tilde{\Lambda}(n) + R_1(n) + R_2(n) = \left[I + R_1(n)\tilde{\Lambda}^{-1}(n) + R_2(n)\tilde{\Lambda}^{-1}(n)\right]\tilde{\Lambda}(n),$$

thus $\tilde{\Lambda}(n) + R_1(n) + R_2(n)$ is invertible for large n since $\tilde{\Lambda}(n)$ was supposed to be invertible for all n. Corollary 3.9 now implies that (5.20) has for n sufficiently large a fundamental matrix of the form

$$Z(n) = [I + o(1)] \prod^{n-1} \tilde{\Lambda}(k) \qquad \text{as } n \to \infty.$$

Using the transformation (5.19) and recalling (5.17) implies (5.16).

To show that $\Delta Q(n) \in l^1$, we will use the notation and idea from Lemma 4.6 and just indicate the modifications in the argument for the discrete case. Beginning with

$$\mathbf{q}(n) = \Lambda^{-1}(n)\mathbf{v}(n) + \hat{\Lambda}^{-1}(n)\hat{V}(n)\mathbf{q}(n) - \hat{\Lambda}^{-1}(n)\hat{D}(V(n), Q(n))\mathbf{q}(n),$$

we apply the product rule for differences in the form

$$\Delta\,(a(n)b(n)) = (\Delta a(n))\,b(n+1) + a(n)\Delta b(n)$$

to obtain

$$\begin{aligned}\Delta\mathbf{q}(n) &= \Delta\left(\Lambda^{-1}(n)\mathbf{v}(n)\right) + \Delta\left(\hat{\Lambda}^{-1}(n)\hat{V}(n)\right)\mathbf{q}(n+1) + \hat{\Lambda}^{-1}(n)\hat{V}(n)\Delta\mathbf{q}(n)\\ &\quad -\Delta\left(\hat{\Lambda}^{-1}(n)\hat{D}(V(n), Q(n))\right)\mathbf{q}(n+1)\\ &\quad -\hat{\Lambda}^{-1}(n)\hat{D}(V(n), Q(n))\Delta\mathbf{q}(n). \end{aligned} \tag{5.22}$$

As we did in Lemma 4.6, we re-write this equation in the form

$$(I + E(n))\,\Delta\mathbf{q}(n) = L(n),$$

where we now show that $E(n) = o(1)$ as $n \to \infty$ and $L(n) \in l^1$. The assumption (5.12) implies that $\Delta\left(\Lambda^{-1}(n)\mathbf{v}(n)\right) \in l^1$ and (5.13) together with (5.21) imply that $\Delta\left(\hat{\Lambda}^{-1}(n)\hat{V}(n)\right)\mathbf{q}(n+1) \in l^1$, so the first two terms on the right hand side of (5.22) contribute to $L(n)$ (since $\mathbf{q}(n) = o(1)$ as $n \to \infty$). The third and fifth terms in (5.22) can be brought into $E(n)$ and can easily be seen using (5.11) to tend to zero as $n \to \infty$. It remains to discuss the fourth term, which is handled similarly to the differential case. Using the definition of $\hat{D}(V(n), Q(n))$ as in (4.27) and (4.28) one can again use the product rule above to show that the ith diagonal block $\Delta\left(\hat{\Lambda}^{-1}(n)\hat{D}(V(n), Q(n))\right)_i$ can be decomposed as

$$\Delta\left(\hat{\Lambda}^{-1}(n)\hat{D}(V(n), Q(n))\right)_i = G_i(n) + H_i(n),$$

where

$$G_i(n) = \sum_{k \neq i} \operatorname{diag}\left\{\Delta\left(\frac{v_{ik}(n)}{\lambda_i(n) - \lambda_1(n)}\right) q_{ki}(n+1), \cdots, \Delta\left(\frac{v_{ik}(n)}{\lambda_i(n) - \lambda_d(n)}\right) q_{ki}(n+1)\right\},$$

and

$$H_i(n) = \sum_{k \neq i} \operatorname{diag}\left\{\left(\frac{v_{ik}(n)}{\lambda_i(n) - \lambda_1(n)}\right) \Delta q_{ki}(n), \cdots, \left(\frac{v_{ik}(n)}{\lambda_i(n) - \lambda_d(n)}\right) \Delta q_{ki}(n),\right\}.$$

From the assumption (5.13) and (5.21) it follows that $G_i(n) \in l^1$, thus terms in $\Delta\left(\hat{\Lambda}^{-1}(n)\hat{D}(V(n), Q(n))\right)\mathbf{q}(n+1)$ arising from $G_i(n)$ can be brought into $L(n)$. Observe that $H_i(n)\mathbf{q}(n+1)$ can be expressed as $H_0(n)\Delta\mathbf{q}(n)$, where assumption (5.11) implies that $H_0(n) = o(1)$ as $n \to \infty$. Hence such terms can be brought into $E(n)$ and it follows that $\Delta\mathbf{q}(n) \in l^1$, hence $\Delta Q(n) \in l^1$, both provided n is sufficiently large so that $I + E(n)$ is invertible.

□

5.4 l^p-Perturbations for $p > 1$

The study of l^p-perturbations of diagonal systems with an appropriate dichotomy is usually associated with the names of Hartman and Wintner to honor their pioneering work concerning differential systems. For difference systems, an analogous result can be found in Benzaid and Lutz [9, Corollary 3.4]. In their proof, they used a conditioning transformation to reduce the perturbed system to one where Theorem 3.4 could be applied. It assumed point-wise dichotomy conditions and the requirement that the entries of the leading diagonal matrix are bounded away from zero. Bodine and Lutz [13, Theorem 5] introduced weaker "averaged" dichotomy conditions (5.23), (5.24), however, still requiring that the entries of $\Lambda(t)$ are bounded away from zero. Ren, Shi, and Wang [133, Theorem 3.1] published a result where this condition was relaxed, but they continued to require the stronger pointwise dichotomy conditions. We will combine all those improvements in Theorem 5.3 below.

First, however, we want to emphasize that Pituk [129, Theorem 2] studied the more general situation of l^2-perturbations of not necessarily diagonal, but blocked *constant* systems, and proved the existence of certain vector-valued solutions. Briefly afterwards, a similar result appeared in [56, Theorem 3.2]. This setting is easily generalized further to allow l^2-perturbations of certain *non-autonomous* difference systems that was already stated as Theorem 3.13. A major significance of this approach in Chap. 3 is that it leads to what could be called a "column-wise Hartman–Wintner result," i.e., the existence of a vector-valued solution (and not necessarily of a fundamental matrix) of certain l^2-perturbations of blocked matrices satisfying a certain dichotomy condition.

While the setting of the following theorem is less general by considering perturbations of *diagonal* matrices, it has two advantages. First, its proof is yet another example of the usefulness of conditioning transformations in investigating the asymptotic behavior of solutions of perturbed systems of difference equations. Secondly, this method lends itself to generalizations to l^p-perturbations with $p > 2$ (see Remark 5.4).

Theorem 5.3 (Discrete Hartman–Wintner) *Suppose that the diagonal matrix* $\Lambda(n) = \text{diag}\{\lambda_1(n), \ldots, \lambda_d(n)\}$ *is invertible for* $n \geq n_0$. *Assume that there exist constants* $K > 0$ *and* $q \in (0, 1)$ *such that for each* $1 \leq i \neq j \leq d$,

$$\text{either} \qquad \prod_{n_1}^{n_2-1} \left| \frac{\lambda_j(k)}{\lambda_i(k)} \right| \leq K q^{n_2 - n_1} \quad \forall \ n_0 \leq n_1 \leq n_2, \tag{5.23}$$

$$\text{or} \qquad \prod_{n_1}^{n_2-1} \left| \frac{\lambda_i(k)}{\lambda_j(k)} \right| \leq K q^{n_2 - n_1} \quad \forall \ n_0 \leq n_1 \leq n_2. \tag{5.24}$$

Assume that for all $1 \leq i \leq d$

$$\sum_{n=n_0}^{\infty} \left| \frac{V(n)}{\lambda_i(n)} \right|^p < \infty \qquad \text{for some } 1 < p \leq 2. \tag{5.25}$$

Then

$$y(n+1) = [\Lambda(n) + V(n)]y(n) \tag{5.26}$$

has for $n \geq n_1 \geq n_0$ *(*n_1 *sufficiently large) a fundamental solution matrix satisfying*

$$Y(n) = [I + o(1)] \prod_{k=n_1}^{n-1} [\Lambda(k) + \text{diag} V(k)] \qquad \text{as } n \to \infty. \tag{5.27}$$

Proof In (5.26), we set

$$y(k) = [I + Q(k)]z(k), \tag{5.28}$$

where $\operatorname{diag} Q(k) \equiv 0$ and the off-diagonal entries of Q will be chosen as appropriate solutions of

$$V(n) - \operatorname{diag}V(n) + \Lambda(n)Q(n) - Q(n+1)\Lambda(n) = 0. \tag{5.29}$$

If (5.23) holds for the ordered pair (i,j), $i \neq j$, choose a solutions of the form

$$q_{ij}(n) = -\sum_{k=n}^{\infty} \frac{v_{ij}(k)}{\lambda_j(k)} \prod_{l=n}^{k} \frac{\lambda_j(l)}{\lambda_i(l)}. \tag{5.30}$$

If (5.24) holds for the ordered pair (i,j), $i \neq j$, choose a solution of the form

$$q_{ij}(n) = \sum_{k=n_0}^{n-1} \frac{v_{ij}(k)}{\lambda_i(k)} \prod_{l=k}^{n-1} \frac{\lambda_i(l)}{\lambda_j(l)}. \tag{5.31}$$

In the following, we will show that the matrix $Q(n)$ with off-diagonal elements given by (5.30) and (5.31) is well defined and that it has the properties $Q(n) \to 0$ as $n \to \infty$ and $Q \in l^p[n_0, \infty)$. It is then straightforward to show that (5.30) and (5.31) are solutions of (5.29).

To show that (5.30) is well defined, note that Hölder's inequality implies for $1/p + 1/p' = 1$ that

$$\begin{aligned} |q_{ij}(n)| &\leq Kq^{1-n} \left(\sum_{k=n}^{\infty} q^{p'k} \right)^{1/p'} \left(\sum_{k=n}^{\infty} \left| \frac{V(k)}{\lambda_j(k)} \right|^p \right)^{1/p} \\ &\leq \hat{K} \left(\sum_{k=n}^{\infty} \left| \frac{V(k)}{\lambda_j(k)} \right|^p \right)^{1/p} \qquad \text{for some constant } \hat{K}, \end{aligned}$$

hence $q_{ij}(n)$ given in (5.30) is well defined and satisfies $q_{ij}(n) \to 0$ as $n \to \infty$.

Similarly, if $q_{ij}(n)$ is defined in (5.31), then it follows by (5.24) that

$$|q_{ij}(n)| \leq K \sum_{k=n_0}^{n-1} \left| \frac{v_{ij}(k)}{\lambda_i(k)} \right| q^{n-k}.$$

Given $\varepsilon > 0$, fix $n_1 \geq n_0$ sufficiently large such that

$$\left[\sum_{k=n_1}^{\infty}\left|\frac{v_{ij}(k)}{\lambda_i(k)}\right|^p\right]^{1/p} < \frac{\varepsilon}{2K}[1 - q^{p'}]^{1/p'}.$$

Hence applying Hölder's inequality once more for $n \geq n_1$ shows that

$$\begin{aligned}
|q_{ij}(n)| &\leq q^n K \sum_{k=n_0}^{n_1-1}\left|\frac{v_{ij}(k)}{\lambda_i(k)}\right| q^{-k} + K\left(\sum_{k=n_1}^{\infty}\left|\frac{v_{ij}(k)}{\lambda_i(k)}\right|^p\right)^{1/p}\left(\sum_{l=0}^{\infty} q^{p'l}\right)^{1/p'} \\
&\leq q^n K \sum_{k=n_0}^{n_1-1}\left|\frac{v_{ij}(k)}{\lambda_i(k)}\right| q^{-k} + \varepsilon/2,
\end{aligned}$$

hence $|q_{ij}(n)| < \varepsilon$ for all n sufficiently large. Therefore we have shown that $Q(n) \to 0$ as $n \to \infty$.

We next claim that $Q \in l^p[n_0, \infty)$. For $q_{ij}(n)$ given in (5.31), one can see using Hölder's inequality and Fubini's theorem that

$$\begin{aligned}
\sum_{n=n_0}^{\infty} |q_{ij}(n)|^p &\leq K^p \sum_{n=n_0}^{\infty}\left[\sum_{k=n_0}^{n-1}\left|\frac{v_{ij}(k)}{\lambda_i(k)}\right| q^{\frac{n-k}{p}} q^{\frac{n-k}{p'}}\right]^p \\
&\leq K^p \sum_{n=n_0}^{\infty}\left[\sum_{k=n_0}^{n-1}\left|\frac{v_{ij}(k)}{\lambda_i(k)}\right|^p q^{n-k}\right]\left[\sum_{k=n_0}^{n-1} q^{n-k}\right]^{p/p'} \\
&\leq \frac{K^p}{(1-q)^{p/p'}} \sum_{k=n_0}^{\infty}\left|\frac{v_{ij}(k)}{\lambda_i(k)}\right|^p q^{-k} \sum_{n=k+1}^{\infty} q^n \\
&\leq \left(\frac{K}{1-q}\right)^p \sum_{k=n_0}^{\infty}\left|\frac{v_{ij}(k)}{\lambda_i(k)}\right|^p < \infty.
\end{aligned}$$

For $q_{ij}(n)$ given in (5.30), very similar computations yield that

$$\begin{aligned}
\sum_{n=n_0}^{\infty} |q_{ij}(n)|^p &\leq K^p \sum_{n=n_0}^{\infty}\left(\sum_{k=n}^{\infty}\left|\frac{v_{ij}(k)}{\lambda_j(k)}\right| q^{\frac{n-k}{p}} q^{\frac{n-k}{p'}}\right)^p \\
&\leq \frac{K^p}{(1-q)^{p/p'}} \sum_{k=n_0}^{\infty}\left|\frac{v_{ij}(k)}{\lambda_j(k)}\right|^p q^{-k} \sum_{n=n_0}^{k} q^n \\
&\leq \left(\frac{K}{1-q}\right)^p \sum_{k=n_0}^{\infty}\left|\frac{v_{ij}(k)}{\lambda_j(k)}\right|^p < \infty.
\end{aligned}$$

Since $Q(n) \to 0$ as $n \to \infty$, $I + Q(n)$ is invertible for n sufficiently large, and then (5.26), (5.28) and (5.29) imply that

$$z(n+1) = [\Lambda(n) + \operatorname{diag} V(n) + R(n)]\, z(n), \tag{5.32}$$

where

$$R(n) = [I + Q(n+1)]^{-1} [V(n)Q(n) - Q(n+1) \operatorname{diag} V(n)] \tag{5.33}$$

is well defined for sufficiently large n. To show that (5.32) satisfies the growth condition (3.10) of Theorem 3.4, one needs to establish that

$$\sum_{n=n_0}^{\infty} \frac{|R(n)|}{|\lambda_i(n) + v_{ii}(n)|} < \infty \qquad \text{for } 1 \le i \le d. \tag{5.34}$$

For that purpose, note first that by (5.25)

$$\lim_{n\to\infty} \frac{v_{ii}(n)}{\lambda_i(n)} = 0 \qquad \text{for } 1 \le i \le d, \tag{5.35}$$

hence $|1 + v_{ii}(n)/\lambda_i(n)| \le 2$ for all $n \ge n_1$ for some $n_1 \ge n_0$. Then from Hölder's inequality it follows that

$$\begin{aligned}\sum_{n=n_1}^{\infty} \frac{|V(n)Q(n)|}{|\lambda_i(n) + v_{ii}(n)|} &\le 2 \sum_{n=n_1}^{\infty} \left|\frac{V(n)}{\lambda_i(n)}\right| |Q(n)| \\ &\le 2 \left(\sum_{n=n_1}^{\infty} \left|\frac{V(n)}{\lambda_i(n)}\right|^p\right)^{1/p} \left(\sum_{n=n_1}^{\infty} |Q(n)|^{p'}\right)^{1/p'} < \infty,\end{aligned}$$

since $Q \in l^{p'}[n_0, \infty)$ because it is in $l^p[n_0, \infty)$ and bounded. Recalling the boundedness of $[I + Q(n+1)]^{-1}$ yields that (5.34) holds.

Lastly, one needs to show $\Lambda + \operatorname{diag} V$ satisfies the discrete Levinson dichotomy conditions (3.8) and (3.9). However, since

$$\left|\frac{\lambda_i(n) + v_{ii}(n)}{\lambda_j(n) + v_{jj}(n)}\right| = \left|\frac{\lambda_i(n)}{\lambda_j(n)} \frac{1 + v_{ii}(n)/\lambda_i(n)}{1 + v_{jj}(n)/\lambda_j(n)}\right| \le \frac{2}{q+1} \left|\frac{\lambda_i(n)}{\lambda_j(n)}\right|,$$

for all n sufficiently large. Since $q < 2q/(q+1) < 1$, it is now easy to show that (5.23) and (5.24) imply (3.8) and (3.9), respectively.

Now Theorem 3.4 implies that (5.32) has for all $n \ge n_1$ sufficiently large ($n_1 \ge n_0$) a fundamental solution matrix satisfying

$$Z(n) = [I + o(1)] \prod_{k=n_1}^{n-1} [\Lambda(k) + \operatorname{diag} V(k)] \qquad \text{as } n \to \infty.$$

Then, by (5.28) and recalling that $Q(n) \to 0$ as $n \to \infty$, it follows that (5.26) has a fundamental matrix of the form (5.27), and the theorem is established. □

Originally [9] the dichotomy conditions for this theorem were stated in the more restrictive "point-wise" form that there exists a $\delta > 0$ such that for each $1 \leq i \neq j \leq d$ either

$$\left|\frac{\lambda_i(n)}{\lambda_j(n)}\right| \geq 1 + \delta \qquad \text{for all } n \geq n_0, \tag{5.36}$$

or

$$\left|\frac{\lambda_i(n)}{\lambda_j(n)}\right| \leq 1 - \delta \qquad \text{for all } n \geq n_0. \tag{5.37}$$

It is clear that (5.36) and (5.37) imply (5.23) and (5.24), respectively.

Remark 5.4 In the case that p given in (5.25) satisfies $p > 2$, the matrix Q satisfying (5.28) still exists, vanishes at infinity and satisfies $Q \in l^p[n_0, \infty)$. However, the perturbation $R(n)$ given in (5.33) in (5.32) satisfies now $R/(\lambda_i + v_{ii}) \in l^{p/2}$ instead of l^1, i.e., Theorem 3.4 does not necessarily apply. Note that by (5.35), the new diagonal matrix $\Lambda + \operatorname{diag} V$ satisfies the dichotomy conditions (5.23), (5.24) with slightly larger q, say $q_1 = (1 + q)/2$. The effect of the transformation (5.28) with (5.29) is therefore to replace system (5.26) in which V/λ_i is in l^p by system (5.32) in which $R/(\lambda_i + v_{ii})$ is in $l^{p/2}$. If $p > 2$ the procedure can be repeated with (5.32) as the new starting point.

5.5 Special Dichotomies and Triangular Perturbations

In Sect. 4.5, we discussed differential systems of the form

$$y' = [\Lambda(t) + V(t) + R(t)]\, y, \qquad t \geq t_0,$$

where Λ is a diagonal matrix satisfying some dichotomy condition, V is (up to a permutation) strictly triangular, and R is an L^1-perturbation. Hsieh and Xie [78, 79] discovered that for two classes of dichotomy conditions on $\Lambda(t)$, the triangular perturbation $V(t)$ was allowed to be significantly "larger" than a "full-size" perturbation. Their results were later generalized by Bodine and Lutz [15] to a general "balance" between the dichotomy condition on the diagonal and the growth condition of the triangular perturbation. In [15], results were also given for systems of difference equations and we will describe these results in the following.

We consider here strictly upper triangular perturbations V, but note that lower triangular perturbations would lead to the same results. The proof of the next theorem is related to the proof of Theorem 5.3 and begins by making the same conditioning transformation. Instead of employing techniques such as Hölders's inequality suitable for l^p functions, it is based on the algebraic structure of the perturbation V and an inductive argument.

Theorem 5.5 *Consider*

$$y(n+1) = [\Lambda(n) + V(n) + R(n)]\, y(n) \qquad n \geq n_0, \tag{5.38}$$

where $\Lambda(n) = \operatorname{diag}\{\lambda_1(n), \ldots, \lambda_d(n)\}$ *is invertible and satisfies Levinson's discrete dichotomy conditions* (3.8), (3.9). *Let* $V(n)$ *be a strictly upper triangular* $d \times d$ *matrix, and suppose that there exists a scalar sequence* $\varphi(n)$ *satisfying*

$$\varphi(n) \geq \left| \frac{v_{ij}(n)}{\lambda_k(n)} \right| \qquad \text{for all } 1 \leq i < j \leq d, \;\; 1 \leq k \leq d$$

such that the following hold: for each pair (i,j), $1 \leq i < j \leq d$, *either*

$$\lim_{n\to\infty} \sum_{m=n_0}^{n-1} \varphi(m) \prod_{l=m}^{n-1} \left| \frac{\lambda_i(l)}{\lambda_j(l)} \right| = 0, \tag{5.39}$$

or

$$\begin{cases} \displaystyle\sum_{m=n}^{\infty} \varphi(m) \prod_{l=n}^{m} \left| \frac{\lambda_j(l)}{\lambda_i(l)} \right| \text{ is well-defined for all } n \geq n_0 \\ \displaystyle\text{and } \lim_{n\to\infty} \sum_{m=n}^{\infty} \varphi(m) \prod_{l=n}^{m} \left| \frac{\lambda_j(l)}{\lambda_i(l)} \right| = 0. \end{cases} \tag{5.40}$$

Finally, suppose that R *satisfies the Levinson's growth condition* (3.10). *Then* (5.38) *has for sufficiently large* n *a fundamental matrix satisfying*

$$Y(n) = [I + o(1)] \prod^{n-1} \Lambda(k) \qquad \text{as } n \to \infty. \tag{5.41}$$

Proof In (5.38), we make the transformation

$$y(n) = [I + Q(n)]\, z(n), \tag{5.42}$$

where Q is a strictly upper triangular matrix satisfying (5.29), i.e.,

$$\lambda_i(n) q_{ij}(n) - q_{ij}(n+1)\lambda_j(n) + v_{ij}(n) = 0 \qquad \text{for } 1 \leq i < j \leq d. \tag{5.43}$$

If the ordered pair (i,j) satisfies (5.40), choose a solution $q_{ij}(n)$ of (5.43) of the form (5.30). If the ordered pair (i,j) satisfies (5.39), choose a solution $q_{ij}(n)$ of (5.43) of the form (5.31).

Using (5.39) and (5.40), one can show that Q is well-defined for all $n \geq n_0$ and satisfies $Q(n) = \mathrm{o}(1)$ as $n \to \infty$. Moreover, $I + Q(n)$ is invertible for all $n \geq n_0$ by the upper triangular nature of Q. Then it follows from (5.38) and (5.29) that

$$z(n+1) = [\Lambda(n) + V_1(n) + R(n)]\, z(n), \qquad n \geq n_0,$$

where

$$V_1(n) = [I + Q(n+1)]^{-1}\, V(n) Q(n)$$
$$R_1(n) = [I + Q(n+1)]^{-1}\, R(n)\, [I + Q(n)].$$

Since V and Q are strictly upper triangular matrices, VQ is strictly upper triangular with additional zeroes on the first super-diagonal and so is V_1.

As $Q(n)$ is bounded, R_1 satisfies again (3.10) (with $R(n)$ replaced by $R_1(n)$). Moreover, there exists a positive constant c_1 such that for all $n \geq n_0$

$$\left| \frac{(V_1)_{ij}(n)}{\lambda_k(n)} \right| \leq c_1 \varphi(n) =: \varphi_1(n) \qquad \forall 1 \leq i < j \leq d, \;\; 1 \leq k \leq d.$$

Hence, for every index pair $i \neq j$, (5.39) or (5.40) still holds with φ replaced by φ_1.

Arguing as in the proof of Theorem 4.14, an induction argument shows that (5.38) can be conditioned in $d-1$ steps to

$$z_{d-1}(n) = [\Lambda(n) + R_{d-1}(n)]\, z_{d-1}(n),$$

where R_L satisfies (3.10) (with $R(n)$ replaced by $R_{d-1}(n)$). Since the entries of Λ satisfy (3.8) and (3.9), Theorem 3.4 applies and yields the existence of a fundamental matrix satisfying for $n \geq n_1$ sufficiently large

$$Z_L(n) = [I + \mathrm{o}(1)] \prod^{n-1} \Lambda(k) \qquad \text{as } n \to \infty.$$

Recalling the asymptotic behavior of the conditioning transformations shows that (5.38) has a fundamental matrix of the form (5.41), which completes the proof. □

We refer to [15] for corollaries, examples, and further explanations.

5.6 Conditionally Convergent Perturbations

This section is concerned with conditionally convergent perturbations of systems of difference equations in several settings.

We begin with a general result which is a slight modification of the first such result which can be found in Benzaid and Lutz [9, Theorem 3.2]. It can be considered as a discrete counterpart of Theorem 4.18.

Theorem 5.6 (Benzaid/Lutz) *Consider*

$$y(n+1) = [\Lambda(n) + V(n) + R(n)]\, y(n), \qquad n \geq n_0, \tag{5.44}$$

where $\Lambda(n)$ and $\Lambda + V(n) + R(n)$ are invertible $d \times d$ matrices for all $n \geq n_0$. Suppose that $\Lambda(n)$ is of the form

$$\Lambda(n) = I + D(n), \quad \textit{where } D(n) = \operatorname{diag}\{d_1(n), \ldots, d_d(n)\},$$

and where $\Lambda(n)$ satisfies Levinson's discrete dichotomy conditions (3.8) *and* (3.9). *Suppose that $\sum_{n=n_0}^{\infty} V(n)$ is conditionally convergent and put*

$$Q(n) = -\sum_{k=n}^{\infty} V(k). \tag{5.45}$$

Assume that for $1 \leq i \leq d$

$$\begin{cases} \sum_{n_0}^{\infty} \left| \dfrac{A(n)}{1 + d_i(n)} \right| < \infty, \\ \textit{where } A(n) \in \{V(n)Q(n),\ D(n)Q(n) - Q(n+1)D(n),\ R(n)\}. \end{cases} \tag{5.46}$$

Then (5.44) *has for n sufficiently large a fundamental matrix satisfying*

$$Y(n) = [I + o(1)] \prod^{n-1} \Lambda(k). \tag{5.47}$$

Proof Note that (5.45) implies that $Q(n) \to 0$ as $n \to \infty$ and $V(n) = Q(n+1) - Q(n)$. In (5.44), make the conditioning transformation

$$y(n) = [I + Q(n)]z(n). \tag{5.48}$$

Since $I + Q(n)$ is invertible for large n, it follows for such n that

$$\begin{aligned} z(n+1) &= \left\{\Lambda(n) + (I + Q(n+1))^{-1} [\Lambda(n)Q(n) - Q(n+1)\Lambda(n) + V(n)\right. \\ &\qquad \left. + V(n)Q(n) + R(n)\{I + Q(n)\}]\right\} z(n) \\ &= \left\{\Lambda(n) + (I + Q(n+1))^{-1} [D(n)Q(n) - Q(n+1)D(n) + V(n)Q(n)\right. \\ &\qquad \left. + R(n)\{I + Q(n)\}]\right\} z(n). \end{aligned}$$

Now (5.46) and Theorem 3.4 imply that there exists for n sufficiently large a fundamental matrix satisfying

$$Z(n) = [I + o(1)] \prod^{n-1} \Lambda(k),$$

which together with (5.48) implies (5.47). □

Analyzing the three growth conditions assumed in (5.46), one can view the perturbation $R(n)$ as the smallest perturbation. In the case $A(n) = V(n)Q(n)$, this condition can be viewed as a natural condition since both $V(n)$ and $Q(n)$ are small for large n. However, the case where $A(n) = D(n)Q(n) - Q(n+1)D(n)$ is usually only realistic when $D(n)$ is "small" in some sense, e.g., vanishing at infinity. In this case, (5.44) is then a perturbation of the trivial system

$$x(n+1) = x(n). \tag{5.49}$$

We continue with a more general theorem on conditionally convergent perturbations of the trivial system (5.49). Such results in the context of differential equations were published by Wintner in [162].

Theorem 5.7 (Discrete Wintner) *Let $I+A(n)$ be invertible for $n \geq n_0$ and assume that the following inductively defined matrices $A_j(n)$ exist:*

$$A_j(n) = -\sum_{m=n}^{\infty} A(m)A_{j-1}(m) \qquad \text{for } j = 1, \ldots, k; \qquad A_0(n) \equiv I. \tag{5.50}$$

Furthermore, assume that

$$A(n)A_k(n) \in l^1[n_0, \infty). \tag{5.51}$$

Then

$$y(n+1) = [I + A(n)]y(n), \qquad n \geq n_0, \tag{5.52}$$

has for n sufficiently large a fundamental matrix satisfying

$$Y(n) = I + o(1) \qquad \text{as } n \to \infty. \tag{5.53}$$

Proof Let

$$y(n) = S_k(n)\tilde{y}(n), \tag{5.54}$$

where

$$S_k(n) = I + A_1(n) + \ldots + A_k(n).$$

We note that it follows from (5.50) that

$$A_j(n+1) = A_j(n) + A(n)A_{j-1}(n), \qquad 1 \le j \le k$$

which in turn yields that

$$S_k(n+1) = S_k(n) + A(n)S_{k-1}(n).$$

Since $A_j(n) = o(1)$ as $n \to \infty$, $\exists\, n_1 \ge n_0$ such that $S_k(n)$ is invertible for $n \ge n_1$. Then for each such n,

$$\begin{aligned}
\tilde{y}(n+1) &= [S_k(n+1)]^{-1}[I + A(n)]S_k(n)\tilde{y}(n) \\
&= [S_k(n+1)]^{-1}\left[S_k(n) + A(n)\{S_{k-1}(n) + A_k(n)\}\right]\tilde{y}(n) \\
&= [S_k(n+1)]^{-1}[S_k(n+1) + A(n)A_k(n)]\tilde{y}(n) \\
&= \left[I + [S_k(n+1)]^{-1}A(n)A_k(n)\right]\tilde{y}(n) =: [I + R(n)]\tilde{y}(n),
\end{aligned}$$

where by (5.51)

$$\frac{R(n)}{\lambda_i(n)} = R(n) \in l^1[t_0, \infty).$$

Interpreting this as a perturbation of the trivial system $x(n+1) = x(n)$, Theorem 3.4 implies there exists a solution $\tilde{Y}(n) = I + o(1)$ as $n \to \infty$, hence also

$$Y(n) = [I + A_1(n) + \ldots + A_k(n)][I + o(1)] = I + o(1).$$

□

Theorem 5.8 (The Other Discrete Wintner) *Let $I + A(n)$ be invertible for $n \ge n_0$ and assume that the following inductively defined matrices $\hat{A}_j(n)$ exist:*

$$\hat{A}_j(n) = \sum_{m=n}^{\infty} \hat{A}_{j-1}(m+1)A(m) \qquad \text{for } j = 1, \ldots, k; \qquad A_0(n) \equiv I. \tag{5.55}$$

Furthermore, assume that

$$\hat{A}_k(n+1)A(n) \in l^1[n_0, \infty). \tag{5.56}$$

Then (5.52) *has for n sufficiently large a fundamental matrix satisfying* (5.53).

Proof Let

$$y_1(n) = \hat{S}_k(n)y(n), \tag{5.57}$$

where

$$\hat{S}_k(n) = I + \hat{A}_1(n) + \ldots + \hat{A}_k(n). \tag{5.58}$$

We note that it follows from (5.55) that

$$\hat{A}_j(n+1) = \hat{A}_j(n) - \hat{A}_{j-1}(n+1)A(n), \qquad 1 \leq j \leq k$$

which in turn yields that

$$\hat{S}_k(n+1) = \hat{S}_k(n) - \hat{S}_{k-1}(n+1)A(n).$$

Since $\hat{A}_j(n) = o(1)$ as $n \to \infty$, $\exists\, n_1 \geq n_0$ such that $\hat{S}_k(n)$ is invertible for $n \geq n_1$. Then for each such n,

$$\begin{aligned} y_1(n+1) &= \hat{S}_k(n+1)][I + A(n)]\hat{S}_k^{-1}(n)y_1(n) \\ &= \left[\hat{S}_k(n) - \hat{S}_{k-1}(n+1)A(n) + \hat{S}_k(n+1)A(n)\right]\hat{S}_k^{-1}(n)y_1(n) \\ &= \left[\hat{S}_k(n) + \hat{A}_k(n+1)A(n)\right]\hat{S}_k^{-1}(n)\hat{y}(n) =: [I + \hat{R}(n)]y_1(n), \end{aligned} \tag{5.59}$$

where by (5.56)

$$\frac{\hat{R}(n)}{\lambda_i(n)} \equiv \hat{R}(n) \in l^1[n_0, \infty).$$

Interpreting this as a perturbation of the trivial system $x(n+1) = x(n)$, Theorem 3.4 implies there exists a solution $Y_1(n) = I + o(1)$ as $n \to \infty$, hence also

$$Y(n) = I + o(1).$$

□

Theorem 5.8 was considered by Trench. It was correctly stated without a proof in [157, Theorem 1], and stated incorrectly (but with a proof) in [156, Corollary 2].

We continue with a discrete counterpart of Theorem 4.24.

Theorem 5.9 (Bodine–Lutz) *Let $w(n)$ be positive for $n \geq n_0$ and satisfy*

$$\lim_{n\to\infty} w(n) = 0.$$

Suppose that $I + A(n)$ is invertible for $n \geq n_0$. Assume that the matrices $\hat{A}_j(n)$ defined in (5.55) *exist for $1 \leq j \leq k+1$ and $n \geq n_0$ and satisfy*

$$\left|\hat{A}_j(n)\right| = \mathrm{O}\left(w(n)\right) \qquad 1 \leq j \leq k+1, \quad n \geq n_0. \tag{5.60}$$

Furthermore, assume that

$$\sum_{m=n}^{\infty} \left|\hat{A}_k(m+1)A(m)\right| w(m) = \mathrm{O}\left(w(n)\right). \tag{5.61}$$

Then (5.52) *has for n sufficiently large a fundamental matrix satisfying* (5.53).

Proof As in the proof of Theorem 5.8, we first make the transformation (5.57), with $\hat{S}_k(n)$ defined in 5.58. Then (5.60) implies that $\hat{S}_k(n) = I + \mathrm{O}\left(w(n)\right)$, and as in the proof of Theorem 4.24 one can show that $\hat{S}_k^{-1}(n) = I + \mathrm{O}\left(w(n)\right)$. Then (5.59) can be re-written as

$$\begin{aligned} y_1(n+1) &= \left[I + \hat{A}_k(n+1)A(n) + \hat{A}_k(n+1)A(n)H(n)\right] y_1(n), \\ H(n) &= \mathrm{O}\left(w(n)\right). \end{aligned}$$

To remove the term

$$\hat{A}_k(n+1)A(n),$$

we make one more conditioning transformation of the form

$$y_1(n) = \left[I - \hat{A}_{k+1}(n)\right] y_2(n),$$

which together with $\hat{A}_{k+1}(n+1) = \hat{A}_{k+1}(n) - \hat{A}_k(n+1)A(n)$ leads to

$$\begin{aligned} y_2(n+1) &= \left\{I + \left[I - \hat{A}_{k+1}(n+1)\right]^{-1} \mathrm{O}\left(\left|\hat{A}_k(n+1)A(n)\right| w(n)\right)\right\} y_2(n) \\ &= \left\{I + \mathrm{O}\left(\left|\hat{A}_k(n+1)A(n)\right| w(n)\right)\right\} y_2(n). \end{aligned}$$

By Theorem 3.25, this has for n sufficiently large a fundamental solution matrix of the form

$$Y_2(n) = I + \mathrm{O}\left(\sum_{m=n}^{\infty} \left|\hat{A}_k(m+1)A(m)\right| w(m)\right) = I + \mathrm{O}\left(w(n)\right),$$

where (5.61) was used to establish the last identity. Observing that both conditioning transformations had the form $I + \mathrm{O}\left(w(n)\right)$ completes the proof. □

We conclude this section with a result on conditionally convergent perturbations of invertible diagonal matrices, which is a discrete counterpart to Theorem 4.26.

Theorem 5.10 (Bodine/Lutz) *Let $\Lambda = \text{diag}\{\lambda_1, \ldots, \lambda_d\}$ be invertible and let $R(n) = \{r_{ij}(n)\}$ be a $d \times d$ matrix such that $\Lambda + R(n)$ is invertible for all $n \geq n_0$. Suppose that there exists a nonincreasing sequence $\varphi(n)$ ($n \geq n_0$), $\lim_{n\to\infty} \varphi(n) = 0$ such that for all $n \geq n_0$*

$$\sum_{m=n}^{\infty} R(m) = \mathrm{O}\left(\varphi(n)\right), \tag{5.62}$$

and

$$\sum_{m=n}^{\infty} |R(m)|\varphi(m) = \mathrm{O}\left(\varphi(n)\right). \tag{5.63}$$

Let

$$q := \max_{1\leq i,j\leq d} \left\{ \left|\frac{\lambda_i}{\lambda_j}\right| : |\lambda_i| < |\lambda_j| \right\}. \tag{5.64}$$

(In case that $|\lambda_i| = |\lambda_j|$ for all i, j, put $q = 1$.) Whenever $|\lambda_i| = |\lambda_j|$, suppose that

$$\sum_{m=n}^{\infty} \left(\frac{\lambda_j}{\lambda_i}\right)^m r_{ij}(m) = \mathrm{O}\left(\varphi(n)\right). \tag{5.65}$$

If $q \in (0, 1)$, suppose there exists $b \in [1, 1/q)$ such that

$$\varphi(n_1)b^{n_1} \leq \varphi(n_2)b^{n_2}, \qquad \forall \; n_0 \leq n_1 \leq n_2. \tag{5.66}$$

Then

$$y(n+1) = [\Lambda + R(n)]\, y(n), \qquad n \geq n_0, \tag{5.67}$$

has for n sufficiently large a fundamental matrix satisfying

$$Y(n) = [I + \mathrm{O}\left(\varphi(n)\right)]\, \Lambda^n. \tag{5.68}$$

Proof Fix an index $r \in \{1, \ldots, d\}$. By making the change of variables $y(n) = (\lambda_r)^n\, \tilde{y}(n)$, if necessary, we can assume without loss of generality that $\lambda_r = 1$ (noting that changing $R(n)$ to $R(n)/\lambda_r$ does not change the hypotheses on $R(n)$). To reduce the magnitude of the perturbation from $R(n)$ to $|R(n)|\varphi(n)$, we make a conditioning transformation

$$y(n) = [I + Q(n)]z(n), \quad \text{where} \quad \Lambda Q(n) - Q(n+1)\Lambda + R(n) = 0. \tag{5.69}$$

It is interesting to note that this conditioning transformation is closely related to the one used in the proof of the discrete Hartman–Wintner theorem (Theorem 5.3), but here we do not make the normalization diag $Q(n) \equiv 0$. Define

$$p_{ij}(n) = \sum_{m=n}^{\infty} r_{ij}(m),$$

hence $p_{ij}(n) = \mathrm{O}(\varphi(n))$ by (5.62).

1. If $|\lambda_j| < |\lambda_i|$, we define

$$q_{ij}(n) = -\sum_{m=n}^{\infty} \frac{r_{ij}(m)}{\lambda_j} \left(\frac{\lambda_j}{\lambda_i}\right)^{m-n+1}. \tag{5.70}$$

   Then summation by parts implies that

$$q_{ij}(n) = -\frac{1}{\lambda_j}\left(\frac{\lambda_j}{\lambda_i}\right)^{1-n}\left(\sum_{m=n+1}^{\infty} p_{ij}(m)\left[\left(\frac{\lambda_j}{\lambda_i}\right)^m - \left(\frac{\lambda_j}{\lambda_i}\right)^{m-1}\right]\right.$$
$$\left. + p_{ij}(n)\left(\frac{\lambda_j}{\lambda_i}\right)^n\right).$$

   Hence, using the monotonicity of $\varphi(n)$, it follows that

$$\left|q_{ij}(n)\right| \leq \frac{1}{|\lambda_j|}\left|\frac{\lambda_j}{\lambda_i}\right|^{1-n}\left(\varphi(n)\left|\frac{\lambda_j}{\lambda_i} - 1\right|\sum_{m=n}^{\infty}\left|\frac{\lambda_j}{\lambda_i}\right|^m + \varphi(n)\left|\frac{\lambda_j}{\lambda_i}\right|^n\right)$$
$$= \mathrm{O}\left(\varphi(n)\right).$$

2. If $|\lambda_j| = |\lambda_i|$, we define $q_{ij}(n)$ again by (5.70) and note that from (5.65) follows that $|q_{ij}(n)| = \mathrm{O}\left(\varphi(n)\right)$.
3. If $|\lambda_j| > |\lambda_i|$, we define

$$q_{ij}(n) = \sum_{m=n_0}^{n-1} \frac{r_{ij}(m)}{\lambda_i}\left(\frac{\lambda_i}{\lambda_j}\right)^{n-m},$$

   and note that summation by parts implies that

$$q_{ij}(n) = \frac{1}{\lambda_i}\left(\sum_{m=n_0+1}^{n-1} p_{ij}(m)\left(\frac{\lambda_i}{\lambda_j}\right)^{n-m}\left[1 + \frac{\lambda_i}{\lambda_j}\right] - p_{ij}(n)\frac{\lambda_i}{\lambda_j}\right).$$

Therefore (5.64) and (5.66) yield that

$$\begin{aligned}
|q_{ij}(n)| &\leq \left|\frac{\lambda_j+\lambda_i}{\lambda_i\lambda_j}\right| \sum_{m=n_0+1}^{n-1} \varphi(m)q^{n-m} + \mathrm{O}\left(\varphi(n)\right) \\
&\leq \left|\frac{\lambda_j+\lambda_i}{\lambda_i\lambda_j}\right| \sum_{m=n_0+1}^{n-1} \varphi(m)b^m q^{n-m} b^{-m} + \mathrm{O}\left(\varphi(n)\right) \\
&\leq \left|\frac{\lambda_j+\lambda_i}{\lambda_i\lambda_j}\right| \varphi(n) \sum_{m=n_0+1}^{n-1} (bq)^{n-m} + \mathrm{O}\left(\varphi(n)\right) \\
&= \mathrm{O}\left(\varphi(n)\right).
\end{aligned}$$

Going back to (5.69), one can see that

$$Q(n) = \mathrm{O}\left(\varphi(n)\right),$$

and

$$z(n+1) = \left[\Lambda + \hat{R}(n)\right] z(n), \qquad n \geq n_0,$$

where

$$\hat{R}(n) = [I + Q(n+1)]^{-1} R(n) Q(n) = \mathrm{O}\left(|R(n)|\varphi(n)\right).$$

To be able to deduce the error estimate specified above, we make one more conditioning transformation

$$z(n) = [I + \hat{Q}(n)]\hat{z}(n), \qquad \text{where} \quad \Lambda\hat{Q}(n) - \hat{Q}(n+1)\Lambda + \hat{R}(n) = 0.$$

If $|\lambda_i| \neq |\lambda_j|$, one use summation by parts as done above to show that $\hat{q}_{ij}(n) = \mathrm{O}\left(\varphi(n)\right)$. If $|\lambda_i| = |\lambda_j|$, then one can use (5.63) to show that $\hat{q}_{ij}(n) = \mathrm{O}\left(\varphi(n)\right)$. Therefore $\hat{Q}(n) = \mathrm{O}\left(\varphi(n)\right)$, and

$$\hat{z}(n+1) = [\Lambda + R_1(n)]\,\hat{z}(n), \qquad n \geq n_0, \qquad \text{where } R_1(n) = \mathrm{O}\left(|R(n)|\varphi^2(n)\right).$$

By Theorem 3.21, for the fixed value of r and (w.l.o.g.) with $\lambda_r = 1$, there exists for n_0 sufficiently large a solution

$$\begin{aligned}
\hat{z}_r &= e_r + \mathrm{O}\left(\sum_{m=n_0}^{n-1} q^{n-m}|R_1(m)|\right) + \mathrm{O}\left(\sum_{m=n}^{\infty} |R_1(m)|\right) \\
&= e_r + \mathrm{O}\left(\sum_{m=n_0}^{n-1} q^{n-m}\varphi(m)\left(|R(m)|\varphi(m)\right)\right) + \mathrm{O}\left(\varphi(n)\sum_{m=n}^{\infty} |R(m)|\varphi(m)\right)
\end{aligned}$$

$$= e_r + \mathrm{O}\left(\varphi(n)\sum_{m=n_0}^{n-1}(qb)^{n-m}\left(|R(m)|\varphi(m)\right)\right) + \mathrm{O}\left(\varphi^2(n)\right)$$

$$= e_r + \mathrm{O}\left(\varphi(n)\sum_{m=n_0}^{n-1}(qb)^{n-m}\right) + \mathrm{O}\left(\varphi(n)\right) = e_r + \mathrm{O}\left(\varphi(n)\right),$$

where we used once more the slow growth condition (5.66) to simplify the finite sum.

Since $Q(n)$ and $\hat{Q}(n)$ were of order $\varphi(n)$, (5.67) has a solution $y_r(n) = e_r + \mathrm{O}\left(\varphi(n)\right)$. Recalling that we made a normalization to put $\lambda_r = 1$ and repeating this process for all $1 \leq r \leq d$ implies (5.68). □

Remark 5.11 As in the case of differential systems, Trench derived a related result. In [154, Theorem 2], he considered conditionally convergent perturbations of a diagonalizable invertible constant matrix A. Using a fixed point argument, he was just interested in the existence of a single vector-valued solution, and his conditions were specifically chosen to ensure the existence of just one solution (and not a fundamental matrix). However, comparing with the assumptions of Theorem 5.10, his approached required the stronger assumption

$$\sum_{m=n}^{\infty} |R(m)|\varphi(m) = \mathrm{o}\left(\varphi(n)\right)$$

in lieu of (5.63). On the other hand, he added generality by adding an exponential weight factor, which we didn't pursue here.

5.7 Preliminary Transformations for Difference Equations

Given a system of difference equations of the form

$$x(n+1) = A(n)x(n), \qquad n \geq n_0, \tag{5.71}$$

we now discuss the problem of constructing, if possible, a so-called preliminary transformation $x(n) = P(n)y(n)$ which could be used to take (5.71) into a suitably normalized form

$$y(n+1) = P^{-1}(n+1)A(n)P(n)y(n) = [\Lambda(n) + V(n)]y(n), \tag{5.72}$$

so that previous asymptotic representation results such as Theorem 3.4 might be applied. Here $\Lambda(n)$ is a diagonal matrix satisfying a dichotomy condition, and $V(n)$ is a sufficiently small perturbation. In general, without $A(n)$ having a special structure and/or its entries having appropriate asymptotic behavior, there does not appear to be any algorithm for doing this or any guarantee that it is even possible. However, in many interesting cases such as the ones discussed in detail in Chap. 9, such preliminary transformations can be constructed. Here we wish to indicate which transformation have been useful for constructing $P(n)$, the role they play in the normalization process, and a rough strategy that could be followed when applying them.

(a) Scalar transformations

$$x(n) = \alpha(n)y(n), \quad \alpha(n) \neq 0,$$

take (5.71) into

$$y(n+1) = \frac{\alpha(n)}{\alpha(n+1)}A(n)y(n).$$

By choosing $\alpha(n)$ appropriately, the rate of growth of the coefficient matrix can be uniformly affected. In cases say when $A(n) = m(n)[C + V(n)]$, where $m(n)$ is a scalar-valued function, C is a constant matrix, and $V(n) = o(1)$, this could be advantageous.

(b) Constant transformations $x(n) = Py(n)$, P invertible, take (5.71) into

$$y(n+1) = P^{-1}A(n)Py(n).$$

If the leading terms in $A(n)$, say as in the above case, have constant limiting behavior, and if this limiting matrix C would be diagonalizable, such a transformation is important as a first step toward constructing $\Lambda(n)$ and $V(n)$. If C would not be diagonalizable, then such a constant transformation could still be used to block-diagonalize the leading coefficient matrix and this is a potentially useful procedure.

(c) Diagonal transformations $x(n) = D(n)y(n)$, $D(n) = \text{diag}\{\alpha_1(n), \dots, \alpha_d(n)\}$, $\alpha_i(n) \neq 0$, take (5.71) into

$$y(n+1) = D^{-1}(n+1)A(n)D(n)y(n).$$

Such transformations can be useful in cases where simple scalar transformations are not effective. This could be the case because C has just one eigenvalue (which might be zero) or there are off-diagonal elements in $A(n)$ with different asymptotic behavior. By choosing the $\alpha_i(n)$ appropriately, one can try to "balance" this effect and create a different asymptotic behavior in the coefficient matrix which in a way averages different rates of growth. Observe that while the main role of such transformations concerns the off-diagonal elements of $A(n)$, such transformations also have an effect on the diagonal terms as well. In case such a transformation produces a system as in (a) or (b) above with C diagonalizable, this can be an effective first step.

Special cases of diagonal transformations which are especially useful in treating systems $A(n)$ which have *meromorphic behavior* at infinity, i.e., $A(n) = n^r \sum_{n=0}^{\infty} A_k n^{-k}$, $r \in \mathbb{Z}$, are $D(n) = \operatorname{diag}\{1, n^b, n^{2b}, \ldots, n^{(d-1)b}\}$. These are called *shearing transformations* and the parameter b is to be chosen optimally to balance the effect of terms involving different powers of n above and below the main diagonal. For such meromorphic difference equations, there are effective algorithms for producing preliminary transformations using the transformations described in (a)–(c) above as well as the next and final type. See [82] for details of such a procedure.

(d) Conditioning transformations $x(n) = [I + Q(n)]y(n)$, where either $Q(n) = o(1)$ as $n \to \infty$ or $Q(n)$ has a special structure (e.g., strictly upper or lower triangular) which insures invertibility. These transformations play a similar role to the ones discussed above in Sects. 5.2–5.6. They are mainly used after the above types already identify what the leading terms in the diagonal matrix $\Lambda(n)$ should be, but the perturbation terms $V(n)$ is too large to allow one of the general asymptotic results to be applied. Then depending upon special properties of the system, certain ad hoc $Q(n)$ might be constructed as solutions of approximate equations analogous to these described in Sect. 5.2. In Chap. 9 several examples of such conditioning transformations with special properties will be constructed.

Finally we mention that some of the above transformations are even useful for determining the asymptotic behavior of solutions of one-dimensional, i.e., scalar equations

$$y(n+1) = \lambda(n)y(n).$$

Here is a well-known, yet instructive example of what we mean by this: Consider the functional equation

$$y(n+1) = ny(n), \qquad n \geq n_0 \tag{5.73}$$

for the Gamma function. If $n_0 = 1$ and $y(1) = 1$, then the solution is, of course, $y(n) = \Gamma(n) = (n-1)!$. Without knowing Stirling's formula for the asymptotic behavior, the following sequence of transformations could be used. First reduce the coefficient n in (5.73) using $y(n) = n^n w(n)$, to obtain

$$w(n+1) = \left(\frac{n}{n+1}\right)^{n+1} w(n) = \left(1 + \frac{1}{n}\right)^{-(n+1)} w(n).$$

By the Taylor series of the logarithm,

$$\begin{aligned} \ln\left(1 + \frac{1}{n}\right)^{-(n+1)} &= -(n+1)\left[\frac{1}{n} - \frac{1}{2n^2} + \frac{1}{3n^3} + O\left(\frac{1}{n^4}\right)\right] \\ &= -1 - \frac{1}{2n} + \frac{1}{6n^2} + O\left(\frac{1}{n^3}\right), \end{aligned}$$

hence as $n \to \infty$

$$w(n+1) = \exp\left[-1 - \frac{1}{2n} + \frac{1}{6n^2} + \mathrm{O}\left(\frac{1}{n^3}\right)\right] w(n)$$

$$= \frac{1}{e}\left[1 - \frac{1}{2n} + \frac{7}{24n^2} + \mathrm{O}\left(\frac{1}{n^3}\right)\right] w(n),$$

which now is asymptotically constant. Then normalize the constant $1/e$ to 1 using $w(n) = e^{-n}u(n)$ to obtain

$$u(n+1) = \left[1 - \frac{1}{2n} + \frac{7}{24n^2} + \mathrm{O}\left(\frac{1}{n^3}\right)\right] u(n) \qquad \text{as } n \to \infty.$$

Next one removes the term $-1/(2n)$ using the transformation $u(n) = n^{-1/2}v(n)$ to obtain for $n \to \infty$

$$v(n+1) = \left[1 - \frac{1}{2n} + \frac{7}{24n^2} + \mathrm{O}\left(\frac{1}{n^3}\right)\right]\left(1 + \frac{1}{n}\right)^{1/2} v(n) \tag{5.74}$$

$$= \left[1 - \frac{1}{12n^2} + \mathrm{O}\left(\frac{1}{n^3}\right)\right] v(n). \tag{5.75}$$

One can also remove the $-1/(12n^2)$-term by putting $v(n) = [1 + 1/(12n)]\,z(n)$, noting that

$$\frac{1}{1 + \frac{1}{12(n+1)}} = \left[1 + \frac{1}{12n}\left(1 + \frac{1}{n}\right)^{-1}\right]^{-1}$$

$$= 1 - \frac{1}{12n} + \frac{1}{12n^2} + \frac{1}{144n^2} + \mathrm{O}\left(\frac{1}{n^3}\right),$$

to find

$$z(n+1) = \left[1 + \mathrm{O}\left(\frac{1}{n^3}\right)\right] z(n) \qquad \text{as } n \to \infty.$$

At this point one could use a well-known result on the asymptotic behavior of infinite products (see, e.g., [94]) or Theorem 3.25 below to obtain a solution of the form

$$z(n) = c + \mathrm{O}\left(\frac{1}{n^2}\right) \quad \text{as } n \to \infty.$$

Therefore (5.73) has a solution of the form

$$y(n) = c\left(\frac{n}{e}\right)^n n^{-1/2}\left[1 + \frac{1}{12n} + \mathrm{O}\left(\frac{1}{n^2}\right)\right] \qquad \text{as } n \to \infty. \tag{5.76}$$

Of course in this treatment one does not know the relation between the constant c (or in general a periodic function $c(n)$ of period 1) and the initial value $y(1) = 1$. But having a formula such as (5.76) and evaluating, for example, the sequence $\{y(n)(e/n)^n n^{1/2}\}$ for large n can yield a reasonable approximation to the constant $c = \sqrt{2\pi}$ appearing in Stirling's formula. To improve the approximation for c for a particular value of n, one can use more terms in the asymptotic expansion, e.g., $1 + \frac{1}{12n} + \frac{1}{288n^2} + \mathrm{O}\left(\frac{1}{n^3}\right)$ which also can be obtained from the above method, as well as any truncation of the Stirling series.

While the determination of the "exact value" of the connection constant (between a specific initial condition and the asymptotic representation) is very useful, it depends quite heavily upon the integral representation for Γ and such formulas are not routine or to be expected in general for one-dimensional equations. However, all *rational* first order equations $y(n+1) = [p(n)/q(n)]y(n)$ can be explicitly solved using products and quotients of Γ-functions. See [5] for an interesting discussion of Γ and $1/\Gamma$ and what additional conditions (in the complex plane) can be used to specify the asymptotic behavior, including potential periodic factors.

Chapter 6
Perturbations of Jordan Differential Systems

6.1 Chapter Overview

Whereas in Chaps. 2 and 4, we studied the asymptotic behavior of solutions of perturbations of diagonal systems of differential equations, we are now interested in the asymptotic behavior of solutions of systems of the form

$$y' = [J(t) + R(t)]\, y(t) \qquad t \geq t_0, \tag{6.1}$$

where $J(t)$ is now in Jordan form and $R(t)$ is again a perturbation. Early results on perturbations of constant Jordan blocks include works by Dunkel [50] and Hartman–Wintner [73]. The focus here is an approach, developed by Coppel and Eastham, to reduce perturbed Jordan systems to a situation where Levinson's fundamental theorem can be applied.

We will present this method first in the case of a single Jordan block in Sect. 6.2. The approach uses a preliminary transformation to reduce (6.1) to a system which can be considered as a perturbation of the trivial system $w' = 0$. This results in a system which has a perturbation that is unnecessarily large; a subsequent shearing transformation reduces its magnitude.

In Sect. 6.3 we extend this approach to the case where $J(t)$ consists of several blocks. To the authors' knowledge, this result has not been published before. As a corollary, we will derive an improvement of a result by Devinatz and Kaplan [49] on perturbed Jordan systems. Section 6.4 studies perturbations of Jordan systems under weak dichotomies, and Sect. 6.5 is concerned with certain triangular perturbations. Finally, Sect. 6.6 considers a somewhat special case investigated by Levinson [100, Theorem 3], who studied perturbations of a constant matrix whose simple eigenvalues are clearly "separated" from eigenvalues associated with Jordan blocks of bigger size.

S. Bodine, D.A. Lutz, *Asymptotic Integration of Differential and Difference Equations*, Lecture Notes in Mathematics 2129,
DOI 10.1007/978-3-319-18248-3_6

6.2 One Single Jordan Block

In this section, we consider systems with one Jordan block of the form

$$J(t) = \lambda(t)I + \mu(t)N,$$

where N is the $d \times d$ nilpotent matrix

$$N = \begin{pmatrix} 0 & 1 & & \\ & 0 & \ddots & \\ & & \ddots & 1 \\ & & & 0 \end{pmatrix}. \tag{6.2}$$

Note that the transformation

$$y = e^{\int^t \lambda(\tau)\,d\tau}\,\tilde{y},$$

in (6.1) leads to

$$\tilde{y}' = [\mu(t)N + R(t)]\,\tilde{y}, \qquad t \geq t_0.$$

Therefore, to simplify the exposition, we will assume without loss of generality that

$$\lambda(t) = 0 \text{ for all } t \geq t_0.$$

The following theorem was first published in Eastham [53, Theorem 1.10.1], and we give his proof here.

Theorem 6.1 (Eastham) *Consider*

$$y' = [\mu(t)N + R(t)]\,y, \qquad t \geq t_0, \tag{6.3}$$

where $\mu(t)$ is a continuous scalar-valued function and N is defined in (6.2). *Let $\sigma(t)$ be an antiderivative of $\mu(t)$ and suppose that $\sigma(t) \neq 0$ for all $t \geq t_0$ and that*

$$\left|\frac{\sigma(t)}{\sigma(s)}\right| \geq K > 0 \qquad t_0 \leq s \leq t < \infty, \tag{6.4}$$

for some positive constant K. Assume that the continuous $d \times d$ matrix $R(t)$ satisfies

$$D(t)R(t)D^{-1}(t) \in L_1[t_0, \infty), \tag{6.5}$$

where

$$D(t) = \text{diag}\left\{1, \sigma(t), \sigma^2(t), \dots, \sigma^{d-1}(t)\right\}.$$

Then (6.3) *has a fundamental matrix satisfying as* $t \to \infty$,

$$Y(t) = \begin{pmatrix} 1+\mathrm{o}(1) & \frac{\sigma}{1!}[1+\mathrm{o}(1)] & \frac{\sigma^2}{2!}[1+\mathrm{o}(1)] & \cdots & \frac{\sigma^{d-1}}{(d-1)!}[1+\mathrm{o}(1)] \\ \mathrm{o}\left(\frac{1}{\sigma}\right) & 1+\mathrm{o}(1) & \frac{\sigma}{1!}[1+\mathrm{o}(1)] & & \\ \mathrm{o}\left(\frac{1}{\sigma^2}\right) & \mathrm{o}\left(\frac{1}{\sigma}\right) & 1+\mathrm{o}(1) & \ddots & \\ \vdots & \vdots & \ddots & \ddots & \frac{\sigma}{1!}[1+\mathrm{o}(1)] \\ \mathrm{o}\left(\frac{1}{\sigma^{d-1}}\right) & \mathrm{o}\left(\frac{1}{\sigma^{d-2}}\right) & & \mathrm{o}\left(\frac{1}{\sigma}\right) & 1+\mathrm{o}(1) \end{pmatrix}. \tag{6.6}$$

Proof Define the $d \times d$-matrix

$$\Phi(t) = \exp\left[\int^t \mu(\tau) N\, d\tau\right] = \begin{pmatrix} 1 & \sigma & \frac{\sigma^2}{2!} & \cdots & \frac{\sigma^{d-1}}{(d-1)!} \\ 0 & 1 & \sigma & & \frac{\sigma^{d-2}}{(d-2)!} \\ \vdots & & \ddots & \ddots & \vdots \\ \vdots & & & \ddots & \sigma \\ 0 & & & 0 & 1 \end{pmatrix}, \qquad \sigma = \sigma(t). \tag{6.7}$$

Note that $\Phi'(t) = \mu(t)N\Phi(t)$. In (6.3), put

$$y(t) = \Phi(t)z(t), \tag{6.8}$$

which leads to

$$z' = \Phi^{-1}(t)R(t)\Phi(t)z(t). \tag{6.9}$$

Note that $\Phi^{-1}(t)R(t)\Phi(t)$ consists of sums of terms of the form $\sigma^k(t)r_{ij}(t)$ with $0 \leq k \leq (d-1)^2$, which are, in general, not in L^1 for perturbations $R(t)$ satisfying (6.5). To reduce the size of this perturbation, we make a shearing transformation which reduces the largest such power of σ from $2(d-1)$ to $d-1$. To this end, put

$$z(t) = D^{-1}(t)\tilde{z}(t), \tag{6.10}$$

where

$$D(t) = \text{diag}\left\{1, \sigma(t), \ldots, \sigma^{d-1}(t)\right\}, \tag{6.11}$$

which yields

$$\tilde{z}' = \left[\Lambda(t) + D(t)\Phi^{-1}(t)R(t)\Phi(t)D^{-1}(t)\right]\tilde{z},$$

where the diagonal matrix $\Lambda(t)$ is given by

$$\Lambda(t) = -D\left(D^{-1}\right)' = D^{-1}D' = \frac{\sigma'(t)}{\sigma(t)} \text{diag}\,\{0, 1, 2, \ldots, d-1\}. \tag{6.12}$$

To simplify, one can use the identity

$$\Phi(t)D^{-1}(t) = D^{-1}(t)e^N, \tag{6.13}$$

where e^N is the constant matrix

$$e^N = \begin{pmatrix} 1 & 1 & \frac{1}{2!} & \cdots & \frac{1}{(d-1)!} \\ 0 & 1 & 1 & & \frac{1}{(d-2)!} \\ \vdots & & \ddots & \ddots & \vdots \\ \vdots & & & \ddots & 1 \\ 0 & & & 0 & 1 \end{pmatrix}.$$

Therefore

$$\tilde{z}'(t) = \left[\Lambda(t) + e^{-N}D(t)R(t)D^{-1}(t)e^N\right]\tilde{z}(t), \tag{6.14}$$

and the perturbation $e^{-N}D(t)R(t)D^{-1}(t)e^N$ consists of linear combinations of terms in $D(t)R(t)D^{-1}(t)$ and is therefore absolutely integrable by (6.5). For further reference, we note that a straightforward computation shows that

$$DRD^{-1} = \begin{pmatrix} r_{11} & \frac{r_{12}}{\sigma} & \frac{r_{13}}{\sigma^2} & \cdots & \frac{r_{1d}}{\sigma^{d-1}} \\ \sigma r_{21} & r_{22} & \frac{r_{23}}{\sigma} & & \\ \sigma^2 r_{31} & & & & \\ \vdots & & & & \\ \sigma^{d-1} r_{d1} & \sigma^{d-2} r_{d2} & \sigma^{d-3} r_{d3} & \cdots & r_{dd} \end{pmatrix} = \left\{\sigma^{i-j}\, r_{ij}\right\}_{i,j=1}^{d}. \tag{6.15}$$

To apply Levinson's fundamental theorem, we note that

$$\operatorname{Re} \int_s^t \left(\lambda_j(\tau) - \lambda_i(\tau)\right) d\tau = (j-i) \int_s^t \operatorname{Re}\left(\frac{\sigma'(\tau)}{\sigma(\tau)}\right) d\tau = (j-i) \ln \left|\frac{\sigma(t)}{\sigma(s)}\right|,$$

and using (6.4) it follows that $\Lambda(t)$ satisfies (2.22) and (2.23), which were shown to be equivalent to Levinson's dichotomy conditions (2.13) and (2.14). Hence Theorem 2.7 implies that there exists a fundamental matrix of (6.14) of the form $[I + o(1)]\, e^{\int^t \Lambda(\tau)\, d\tau}$ or, recalling the definition of $\Lambda(t)$ in (6.12) and the assumption that $\sigma(t)$ does not change sign,

$$\tilde{Z}(t) = [I + o(1)]\, D(t), \qquad \text{as } t \to \infty.$$

By (6.8) and (6.10), (6.3) has a fundamental solution matrix satisfying, as $t \to \infty$,

$$Y(t) = \Phi(t) D^{-1}(t) \tilde{Z}(t). \tag{6.16}$$

Using (6.7), (6.11), and straightforward matrix multiplication, (6.16) can be shown to be of the form (6.6). □

In the important case that $\mu(t) = 1$ for all $t \geq t_0$, one can choose $\sigma(t) = t$. Then Theorem 6.1 reduces to the following corollary found already in Coppel [43, p. 91].

Corollary 6.2 *Consider*

$$y' = [N + R(t)]\, y \qquad t \geq t_0, \tag{6.17}$$

where the $d \times d$ matrix N is defined in (6.2), *and where the continuous $d \times d$ matrix $R(t)$ satisfies*

$$\left(t^{i-j}\, r_{ij}\right)_{i,j=1}^{d} \in L^1[t_0, \infty). \tag{6.18}$$

Then (6.3) *has a* $d \times d$ *fundamental matrix satisfying, as* $t \to \infty$,

$$Y(t) = \begin{pmatrix} 1+\mathrm{o}(1) & \frac{t}{1!}[1+\mathrm{o}(1)] & \frac{t^2}{2!}[1+\mathrm{o}(1)] & \cdots & \frac{t^{d-1}}{(d-1)!}[1+\mathrm{o}(1)] \\ \mathrm{o}\left(\frac{1}{t}\right) & 1+\mathrm{o}(1) & \frac{t}{1!}[1+\mathrm{o}(1)] & & \\ \mathrm{o}\left(\frac{1}{t^2}\right) & \mathrm{o}\left(\frac{1}{t}\right) & 1+\mathrm{o}(1) & \ddots & \\ \vdots & \vdots & \ddots & \ddots & \frac{t}{1!}[1+\mathrm{o}(1)] \\ \mathrm{o}\left(\frac{1}{t^{d-1}}\right) & \mathrm{o}\left(\frac{1}{t^{d-2}}\right) & & \mathrm{o}\left(\frac{1}{t}\right) & 1+\mathrm{o}(1) \end{pmatrix}. \tag{6.19}$$

This result can be applied to the following dth-order scalar differential equation

$$y^{(d)} + r_1(t)y^{(d-1)} + \ldots r_d(t)y = 0, \tag{6.20}$$

which can be considered as a perturbation of the trivial differential equation $x^{(d)} = 0$. (6.20) can be written in the standard way as a system in companion matrix form, i.e.,

$$y' = [N + R(t)]\,y,$$

with N given in (6.2) and

$$(R)_{ij}(t) = \begin{cases} -r_{d+1-j}(t) & \text{if } i = d, \\ 0 & \text{else.} \end{cases}$$

Then (6.18) is satisfied if

$$t^{k-1}r_k(t) \in L^1[t_0, \infty), \qquad 1 \le k \le d.$$

Therefore one obtains the following result:

Corollary 6.3 *If the coefficients* $r_k(t)$ *in* (6.20) *are continuous for* $t \ge t_0 > 0$ *and* $1 \le k \le d$ *and satisfy*

$$t^{k-1}\, r_k(t) \in L^1[t_0, \infty), \qquad 1 \le k \le d,$$

then the dth order scalar equation (6.20) *has a fundamental system of solutions* $\{y_k(t)\}_{k=1}^d$ *satisfying, as* $t \to \infty$,

$$y_k^{(i)}(t) = \begin{cases} [1+o(1)]\dfrac{t^{k-1-i}}{(k-1-i)!} & 0 \le i \le k-1, \\ \\ o\left(t^{k-i-1}\right) & k \le i \le d-1 \end{cases}, \tag{6.21}$$

for each $1 \le k \le d$.

To the authors' knowledge, Ghizzetti [65] was the first to study the asymptotic behavior of (6.20) in 1949. For $0 \le i \le k-1$, he found the asymptotic behavior described in the first part of (6.21), whereas for $k \le i \le d-1$, he derived the weaker statement $y_k^{(i)}(t) = o(1)$.

As noted by Eastham [53, p. 45], it is clear that one can also apply other asymptotic integration results to (6.14). For example, application of Theorem 4.11 requires that

$$\text{either} \qquad \left|\frac{\sigma(t)}{\sigma(s)}\right| \le Ke^{-\delta(s-t)}, \qquad t_0 \le s \le t,$$

$$\text{or} \qquad \left|\frac{\sigma(t)}{\sigma(s)}\right| \le Ke^{-\delta(t-s)}, \qquad t_0 \le t \le s,$$

and that

$$DRD^{-1} \in L^p, \qquad \text{for some } 1 < p \le 2.$$

Then (6.3) has a fundamental matrix satisfying as $t \to \infty$,

$$Y(t) = \Phi(t)D^{-1}(t)[I + o(1)]D(t)\exp\left[\int^t \operatorname{diag}\{e^{-N}D(\tau)R(\tau)D^{-1}(\tau)e^N\}\,d\tau\right].$$

Similarly, one could apply Theorem 4.8 to (6.14), and we leave it to the reader to explore details.

We finally consider the case where the perturbation R does not necessarily satisfy (6.5), but the weaker condition that DRD^{-1} is conditionally integrable. This was first studied by Medina and Pinto [107, Theorem 3.2] who considered (6.3) where J was a lower triangular Jordan matrices and $\mu(t) \equiv 1$. We make the necessary adjustments to bring it into a form corresponding to Theorem 6.1.

Theorem 6.4 *Consider the* $d \times d$ *system* (6.3), *where* $\mu(t)$ *is a continuous scalar-valued function and the* $d \times d$ *matrix* N *is defined in* (6.2). *Let* $\sigma(t)$ *be an antiderivative of* $\mu(t)$. *Suppose that* $\sigma(t) \neq 0$ *for all* $t \ge t_0$, *and assume that* (6.4) *holds for for some positive constant* K. *With* $D(t)$ *defined in* (6.11), *let* $R(t)$ *be a* *continuous $d \times d$ matrix such that*

$$V(t) := D(t)R(t)D^{-1}(t) = \left\{\sigma^{i-j}(t)\, r_{ij}(t)\right\}_{i,j=1}^{d} \tag{6.22}$$

is conditionally integrable for $t \geq t_0$. Let

$$Q(t) = -\int_t^\infty V(\tau)\, d\tau, \tag{6.23}$$

and assume that

$$\frac{\mu(t)}{\sigma(t)} q_{ij}(t) \text{ and } v_{ij}(t)\, q_{jk}(t) \in L_1[t_0, \infty). \tag{6.24}$$

Then (6.3) *has a fundamental matrix of the form* (6.6) *as $t \to \infty$.*

Proof As in the proof of Theorem 6.1, we make in (6.3) the transformation (6.8) and (6.10), arriving at (6.14), to which we want to apply Theorem 4.18. We note that $\Lambda(t)$ defined in (6.12) satisfies Levinson's dichotomy condition (2.13), (2.14). By (6.24), VQ, ΛQ, and $Q\Lambda$ are all in $L^1[t_0, \infty)$. Hence (6.14) has, by Theorem 4.18, a fundamental matrix of the form $\tilde{Z}(t) = [I + o(1)] \exp\left[\int^t \Lambda(s)\, ds\right]$ or, simplified,

$$\tilde{Z}(t) = [I + o(1)]D(t) \qquad \text{as } t \to \infty,$$

and the rest of the proof follows that of Theorem 6.1. □

We remark that there seems to be an error in one of the hypotheses in [107, Theorem 3.2] due to an incorrect assumption $Q(t)V(t) \in L^1$ instead of the correct assumption $V(t)Q(t) \in L^1$. In addition, the final result in this theorem does not seem to be correct due to a matrix multiplication error.

Medina and Pinto also applied their result to the dth-order equation (6.20). Note that their assumption $q_1(t)v_k(t) \in l^1$ should read $v_1(t)q_k(t) \in L^1$ for $1 \leq k \leq d$. Application of Theorem 6.4 gives the following corrected and slightly improved result:

Corollary 6.5 *Suppose that the coefficients $r_k(t)$ in* (6.20) *are continuous for $t \geq t_0 > 0$ and $1 \leq k \leq d$ and $v_k(t) := t^{k-1}\, r_k(t)$ are conditionally integrable for $1 \leq k \leq d$. Put $q_k(t) = \int_t^\infty v_k(s)\, ds$. Moreover, suppose that $q_k(t)/t \in L^1[t_0, \infty)$ and that $v_1(t)q_k(t) \in L^1[t_0, \infty)$ for $1 \leq k \leq d$. Then the dth order scalar equation* (6.20) *has a fundamental system of solutions $\{y_k(t)\}_{k=1}^d$ satisfying* (6.21).

The proof, analogous to the proof of Corollary 6.3, is a straightforward application of Theorem 6.4 in the special case $\sigma(t) = t$. A computation reveals that the entries of $V(t)Q(t)$ are of the form $v_1(t)q_k(t)$ for $1 \leq k \leq d$.

6.3 Multiple Jordan Blocks

The techniques and results from the previous section can be generalized the case where the leading matrix consists of several Jordan blocks. To simplify the exposition, we assume in each Jordan block $J_i(t)$ that the function $\mu_i(t) \equiv 1$. However, a result for time-dependent $\mu_i(t)$ can be derived similarly.

Theorem 6.6 *Consider,*

$$y' = [A(t) + R(t)]\, y, \qquad t \geq t_0 \tag{6.25}$$

where

$$A(t) = \sum_{i=1}^{m} \oplus J_i(t)\,, \qquad J_i(t) = \lambda_i(t) I_{n_i} + N_{n_i}. \tag{6.26}$$

Here I_{n_i} is the $n_i \times n_i$ identity matrix, $\lambda_i(t)$ is a continuous function for all $1 \leq i \leq m$, and N_{n_i} is the $n_i \times n_i$ matrix defined in (6.2)*. Let $d = \sum_{i=1}^{m} n_i$. Partition the continuous $d \times d$ matrix $R(t)$ into*

$$R(t) = \begin{pmatrix} R_{11}(t) & \dots & R_{1m}(t) \\ \vdots & & \vdots \\ R_{m1}(t) & \dots & R_{mm}(t) \end{pmatrix}, \tag{6.27}$$

where each block $R_{ij}(t)$ has size $n_i \times n_j$. For each positive integer k, put

$$D_k(t) = \operatorname{diag}\left\{1, t, t^2, \dots, t^{k-1}\right\}, \tag{6.28}$$

and assume that

$$D_{n_i} R_{ij} D_{n_j}^{-1} \in L^1[t_0, \infty) \qquad \text{for all } 1 \leq i, j \leq m. \tag{6.29}$$

Suppose that the diagonal matrix consisting of entries of the form

$$\left\{\lambda_i(t) + \frac{l_i - 1}{t}\right\} \qquad 1 \leq i \leq m\,, \quad 1 \leq l_i \leq n_i, \tag{6.30}$$

satisfies Levinson's dichotomy conditions (2.13), (2.14)*.*

Then (6.25) *has d linearly independent solutions satisfying, as $t \to \infty$,*

$$y_{i,l_i} = t^{l_i - 1} e^{\int^t \lambda_i(\tau)\, d\tau} \left[e_{[i]} + o(1)\right], \qquad 1 \leq i \leq m\,, \quad 1 \leq l_i \leq n_i, \tag{6.31}$$

where

$$[i] = 1 + \sum_{k=1}^{i-1} n_k.$$

Proof In (6.25), we make the change of variables

$$y = \left[\sum_{i=1}^{m} \oplus \Phi_{n_i}(t) \right] z, \tag{6.32}$$

where $\Phi_{n_i}(t)$ is the $n_i \times n_i$ matrix defined by

$$\Phi_{n_i}(t) = \begin{pmatrix} 1 & t & \frac{t^2}{2!} & \cdots & \frac{t^{n_i-1}}{(n_i-1)!} \\ 0 & 1 & t & & \frac{t^{n_i-2}}{(n_i-2)!} \\ \vdots & & \ddots & \ddots & \vdots \\ \vdots & & & \ddots & t \\ 0 & & & 0 & 1 \end{pmatrix} = e^{N_i t}. \tag{6.33}$$

Then $\Phi'_{n_i}(t) = N_i \Phi_{n_i}(t)$. It follows that

$$z' = \left[\sum_{i=1}^{m} \oplus \lambda_i(t) I_{n_i} + \tilde{R}(t) \right] z,$$

with

$$\tilde{R}(t) = \left[\sum_{i=1}^{m} \oplus \Phi_{n_i}(t) \right]^{-1} R(t) \left[\sum_{i=1}^{m} \oplus \Phi_{n_i}(t) \right] = \mathrm{O}\left(t^{2\rho} R \right),$$

where

$$\rho = \max_{1 \le i \le m} n_i - 1. \tag{6.34}$$

To reduce the magnitude of the perturbation, put

$$z = \left[\sum_{i=1}^{m} \oplus D_{n_i}^{-1}(t) \right] w, \tag{6.35}$$

where $D_{n_i}(t)$ was defined in (6.28). Since $\Phi_{n_i}(t) = D_{n_i}(t)e^{N_i}D_{n_i}(t)$ (see (6.13)), it follows that

$$w' = \left[\begin{pmatrix} \Lambda_1(t) & & \\ & \ddots & \\ & & \Lambda_m(t) \end{pmatrix} + \begin{pmatrix} e^{-N_1} & & \\ & \ddots & \\ & & e^{-N_m} \end{pmatrix} \hat{R}(t) \begin{pmatrix} e^{N_1} & & \\ & \ddots & \\ & & e^{N_m} \end{pmatrix}\right] w. \tag{6.36}$$

Here $\hat{R}(t)$ is given by

$$\hat{R}(t) = \left(\sum_{i=1}^{m} \oplus D_{n_i}(t)\right) R(t) \left(\sum_{i=1}^{m} \oplus D_{n_i}^{-1}(t)\right), \tag{6.37}$$

i.e., it consists of the form $D_{n_i}R_{ij}D_{n_j}^{-1}$, which are absolutely integrable by (6.29), and $\Lambda_i(t)$ are diagonal $n_i \times n_i$ matrices given by

$$\Lambda_i(t) = \lambda_i(t)I_{n_i} + \frac{1}{t}\operatorname{diag}\{0, 1, \ldots, n_i - 1\}.$$

The difference of any two elements in this leading diagonal matrix is given by

$$\left(\lambda_j(t) + \frac{l_j - 1}{t}\right) - \left(\lambda_i(t) + \frac{l_i - 1}{t}\right),$$

with $1 \le l_k \le n_k$ for all $1 \le k \le m$. By assumption (6.30), $\operatorname{diag}\{\Lambda_1(t), \ldots, \Lambda_m(t)\}$ satisfies Levinson's dichotomy conditions (2.13), (2.14). Hence Theorem 2.7 implies that (6.36) has a fundamental matrix satisfying

$$\begin{aligned} W(t) &= [I_{d\times d} + o(1)] \sum_{i=1}^{m} \oplus e^{\int^t \Lambda_i(\tau)\,d\tau} \\ &= [I_{d\times d} + o(1)] \sum_{i=1}^{m} \oplus e^{\int^t \lambda_i(\tau)\,d\tau} D_{n_i}(t). \end{aligned}$$

In particular, (6.36) has d linearly independent solutions of the form

$$w_{i,l_i} = t^{l_i - 1} e^{\int^t \lambda_i(\tau)\,d\tau} \left[e_{l_i + \sum_{k=1}^{i-1} n_k} + o(1)\right], \qquad \text{as } t \to \infty, \tag{6.38}$$

for $1 \leq i \leq m$, $1 \leq l_i \leq n_i$. By (6.32) and (6.35), (6.25) has d linearly independent solutions satisfying

$$y_{i,l_i} = \sum_{i=1}^{m} \oplus \left\{\Phi_{n_i}(t) D_{n_i}^{-1}(t)\right\} w_{i,l_i}, \tag{6.39}$$

which is identical to (6.31). □

Observe that a sufficient condition for (6.29) is that

$$t^{\rho} R \in L^1[t_0, \infty)$$

where ρ was defined in (6.34). One could (making additional assumptions) make a more quantitative statement about the error o(1) in (6.31) by analyzing the entries of the matrix $\sum_{i=1}^{m} \oplus \Phi_i(t) \sum_{i=1}^{m} \oplus D_i^{-1}(t)$ more carefully and using Theorem 2.25. We also refer to [73, Theorem (**)] for an early result on perturbations of several Jordan blocks including error estimates.

We conclude this section with a discussion of some results due to Devinatz and Kaplan [49], who were interested in extending Levinson's results (see Theorems 2.7 and 4.1) from the case of simple eigenvalues to Jordan blocks. As indicated in Sect. 2.1, Levinson's paper (as well as the treatment in Coddington/Levinson [42, pp. 91–97]) was concerned with a reduction of a certain asymptotically constant system (whose limiting matrix has simple eigenvalues) to L-diagonal form and, in addition, the asymptotic behavior of solutions of such a perturbed diagonal system satisfying Levinson's dichotomy conditions (see Theorems 4.1 and 2.7, respectively).

Devinatz and Kaplan generalized both of the two results by allowing the asymptotically constant matrix to have repeated eigenvalues. In the spirit of Levinson's work, they studied the reduction of perturbed systems to Jordan form (see [49, § 4]) and, moreover, the asymptotic behavior of solutions of perturbations of such Jordan systems (see [49, § 3]). We will refer to the original paper for the reduction to Jordan form, but we will show that their result on the asymptotic behavior of solutions of such perturbed Jordan systems follows as a corollary to Theorem 6.6. Their proof was based on directly applying the method of successive approximations to solutions of perturbed Jordan systems without utilizing a suitable diagonalization. The approach taken in this chapter allows for a weakening of two of their hypotheses.

Corollary 6.7 *Consider* (6.25)*, where R satisfies the assumptions of Theorem 6.6. Suppose that*

$$A(t) = \sum_{i=1}^{m} \oplus \hat{J}_i(t), \qquad \hat{J}_i(t) = \hat{\Lambda}_i(t) + N_{n_i}.$$

Here

$$\hat{\Lambda}_i(t) = \operatorname{diag}\{\hat{\lambda}_1^i(t), \hat{\lambda}_2^i(t), \ldots \hat{\lambda}_{n_i}^i(t)\},$$

and $\hat{\lambda}_{l_i}^i(t)$ *are continuous functions for all* $1 \leq i \leq m$, $1 \leq l_i \leq n_j$, *and* N_{n_i} *is the* $n_i \times n_i$ *matrix defined in* (6.2). *Assume that there exists constants* λ_i *for* $1 \leq i \leq m$ *such that*

$$\hat{\lambda}_{l_i}^i(t) - \lambda_i \in L^1[t_0, \infty) \qquad \text{for all } 1 \leq i \leq m, 1 \leq l_i \leq n_i. \tag{6.40}$$

Then (6.25) *has* d *linearly independent solutions satisfying* (6.31) *as* $t \to \infty$.

Proof We re-write (6.25) as

$$\begin{aligned} y' &= \left[\sum_{i=1}^{m} \oplus \left\{\hat{\Lambda}_i(t) + N_{n_i}\right\} + R(t)\right] y \\ &= \left[\sum_{i=1}^{m} \oplus \underbrace{\{\lambda_i I_{n_i} + N_{n_i}\}}_{=:J_i} + \underbrace{R(t) + \sum_{i=1}^{m} \oplus \left\{\hat{\Lambda}_i(t) - \lambda_i I_{n_i}\right\}}_{=:R_1(t)}\right] y. \end{aligned}$$

Then J_i is a special case of constant coefficients of $J_i(t)$ defined in (6.26), and $R_1(t)$ differs from $R(t)$ at most by L^1-terms on the diagonal. Since products of diagonal matrices commute,

$$\begin{aligned} &\left(\sum_{i=1}^{m} \oplus D_{n_i}(t)\right) R(t) \left(\sum_{i=1}^{m} \oplus D_{n_i}^{-1}(t)\right) \\ &= \left(\sum_{i=1}^{m} \oplus D_{n_i}(t)\right) R_1(t) \left(\sum_{i=1}^{m} \oplus D_{n_i}^{-1}(t)\right) + L^1, \end{aligned}$$

(cp. (6.37)) i.e., $D_{n_i} R_{ij} D_{nj}^{-1} = D_{n_i} (R_1)_{ij} D_{nj}^{-1} + L^1$ for $1 \leq i \neq j \leq m$. Therefore $R_1(t)$ satisfies the assumption (6.29) of Theorem 6.6. Clearly, the set of functions

$$\left\{\lambda_i + \frac{l_i - 1}{t}\right\} \qquad 1 \leq i \leq m\,, \quad 1 \leq l_i \leq n_i$$

satisfy Levinson's dichotomy condition (see Definition 2.6). By Theorem 6.6, (6.25) has d linearly independent solutions of the form (6.31). □

Several remarks are in order. First, the reason for considering functions $\hat{\lambda}_{l_i}^i(t)$ on the diagonal comes from transforming perturbations of asymptotically constant matrices into Jordan form, where terms of this nature occur naturally. For the same reason, Devinatz and Kaplan also required that $\hat{\lambda}^i_{l_i}(t) \to \lambda_i$ as $t \to \infty$. Secondly, they required the stronger assumption $t^\rho|\frac{d}{dt}\lambda^i_{l_i}(t)| \in L^1$ (see [49, Eq. (2.5)]), where ρ was defined in (6.34) (w.l.o.g. $\rho \geq 1$). They showed in Lemma 2.1 that this assumption implies (6.40), which is the (weaker) assumption used in Corollary 6.7. Furthermore, their hypothesis of certain terms satisfying Levinson's dichotomy conditions (see their Eqs. (2.6) and (2.7)) is superfluous since they are just L^1 perturbations of $\{\lambda_i + \frac{l_i-1}{t}\}$ which are immediately to be seen to satisfy these conditions. Finally, we can replace their assumption $t^\rho|R| \in L^1$ by the weaker assumption (6.29). They phrased their result slightly differently, using $e^{\int^t \hat{\lambda}^i_1(\tau)\,d\tau}$ instead of $e^{\lambda_i t}$ in (6.31), which has no impact on the asymptotic behavior of solutions because (6.40) implies $\exp\left[\int^t \left[\lambda_i + (L^1)\right]\right] = [c + o(1)]\exp[\lambda_i t]$, with $c \neq 0$.

We mention briefly that [49] improved an earlier result by Devinatz [47] who also considered systems consisting of several Jordan Blocks. He had already used the method of successive approximations for solutions of the system (6.25) without applying any preliminary diagonalizing transformations. This approach, however, seemed to require a significantly more restrictive hypothesis on the perturbation that $t^{2\rho}R \in L^1[t_0, \infty)$ with ρ defined in (6.34).

6.4 Weak Dichotomies

Perturbations of Jordan systems under so-called weak dichotomies were studied by Bodine and Lutz [18, Theorem 4]. To allow for an exposition following the methods of Sect. 6.3, we give here a slight modification of this result. We again use two preliminary transformations to diagonalize the system, but the terms of the leading diagonal matrix are only required to satisfy a "weak dichotomy" condition. An asymptotic result, made possible by a more restrictive appropriate growth condition, will then follow from Theorem 2.12.

Theorem 6.8 *Consider* (6.25), *where* $A(t)$ *was defined in* (6.26). *Here* I_{n_i} *is the* $n_i \times n_i$ *identity matrix,* $\lambda_i(t)$ *is a continuous function for all* $1 \leq i \leq m$, *and* N_{n_i} *is the* $n_i \times n_i$ *matrix defined in* (6.2). *Let* $d = \sum_{i=1}^m n_i$. *Partition the continuous* $d \times d$ *matrix* $R(t)$ *as in* (6.27), *where each block* $R_{ij}(t)$ *has size* $n_i \times n_j$. *For each* $k \in \mathbb{N}$, *let* $D_k(t)$ *be defined in* (6.28).

Let a pair of indices (i, l_i) *be given such that* $1 \leq i \leq m$ *and* $1 \leq l_i \leq n_i$. *Assume that there exists a continuous function* $\beta_i(t) \geq 1$ *for* $t \geq t_0$ *such that for each* $1 \leq j \leq m$ *either*

$$\left.\begin{array}{ll} e^{\int_{t_0}^t \left(\mathrm{Re}\{\lambda_j(\tau) - \lambda_i(\tau)\} + \frac{n_j - l_i}{\tau}\right) d\tau} \to 0 & \text{as } t \to \infty \\ \text{and} \quad e^{\int_s^t \left(\mathrm{Re}\{\lambda_j(\tau) - \lambda_i(\tau)\} + \frac{n_j - l_i}{\tau}\right) d\tau} \leq \beta_i(s) & \forall\ t_0 \leq s \leq t \end{array}\right\} \tag{6.41}$$

or

$$e^{\int_s^t \left(\operatorname{Re}\{\lambda_j(\tau)-\lambda_i(\tau)\}+\frac{1-l_i}{\tau}\right)d\tau} \leq \beta_i(s) \qquad \forall \quad t_0 \leq t \leq s. \tag{6.42}$$

Furthermore, assume that $R(t)$ is a continuous $d \times d$ matrix for $t \geq t_0$ and that

$$D_{n_j} R_{jk} D_{n_k}^{-1} \beta_i(t) \in L^1[t_0,\infty) \qquad \text{for all } 1 \leq j,k \leq m. \tag{6.43}$$

Then there exists a solution of (6.25) *satisfying* (6.31) *as $t \to +\infty$.*

Proof We follow the proof of Theorem 6.6 up to (6.36). We recall that $\hat{R}(t)$ consists of blocks of the form $D_{n_j} R_{jk} D_{n_k}^{-1}$ and that $\Lambda_j(t)$ are diagonal $n_j \times n_j$ matrices given by

$$\Lambda_j(t) = \lambda_j(t) I_{n_j} + \frac{1}{t}\operatorname{diag}\left\{0,1,\ldots,n_j-1\right\}.$$

By (6.43), $\hat{R}(t)\beta_i(t) \in L^1[t_0,\infty)$.

We want to apply Theorem 2.12 to (6.36), and we will show that (6.41) and (6.42) imply that the dichotomy conditions (2.32) and (2.33) of Theorem 2.12 are satisfied. To avoid possible confusion in notation due to different meanings of the variables $\lambda_j(t)$, we first wish to rewrite the dichotomy conditions (2.32) and (2.33) in the following equivalent manner, considering a diagonal matrix $\{\mu_1(t),\mu_2(t),\ldots,\mu_d(t)\}$:

For fixed $\nu \in \{1,\ldots,d\}$, assume that there exists a continuous function $\beta_\nu(t) \geq 1$ for $t \geq t_0$ such that for all $1 \leq k \leq d$ either

$$\left.\begin{aligned} & e^{\int_{t_0}^t \operatorname{Re}\{\mu_k(\tau)-\mu_\nu(\tau)\}\,d\tau} \to 0 && \text{as } t \to \infty \\ \text{and} \quad & e^{\int_s^t \operatorname{Re}\{\mu_k(\tau)-\mu_\nu(\tau)\}\,d\tau} \leq \beta_\nu(s) && \forall \ t_0 \leq s \leq t \end{aligned}\right\} \tag{6.44}$$

or

$$e^{\int_s^t \operatorname{Re}\{\mu_k(\tau)-\mu_\nu(\tau)\}\,d\tau} \leq \beta_\nu(s) \qquad \forall \quad t_0 \leq t \leq s. \tag{6.45}$$

Then $\nu = \sum_{\kappa=1}^{i-1} n_\kappa + l_i$, $\mu_\nu(t) = \lambda_i(t) + \frac{l_i-1}{t}$, and $\mu_k(t) = \lambda_j(t) + \frac{l_j-1}{t}$ for some $1 \leq j \leq m$, $1 \leq l_j \leq n_j$.

To show that (6.44) holds, note that from (6.41) follows that

$$
\begin{aligned}
e^{\int_{t_0}^{t} \mathrm{Re}\{\mu_k(\tau)-\mu_\nu(\tau)\}\, d\tau} &= e^{\int_{t_0}^{t} \left(\mathrm{Re}\{\lambda_j(\tau)-\lambda_i(\tau)\} + \frac{l_j - l_i}{\tau}\right) d\tau} \\
&\leq e^{\int_{t_0}^{t} \left(\mathrm{Re}\{\lambda_j(\tau)-\lambda_i(\tau)\} + \frac{n_j - l_i}{\tau}\right) d\tau} \to 0
\end{aligned}
$$

as $t \to \infty$. Similarly, for $t_0 \leq s \leq t$,

$$
e^{\int_{s}^{t} \mathrm{Re}\{\mu_k(\tau)-\mu_\nu(\tau)\}\, d\tau} \leq e^{\int_{t_0}^{t} \left(\mathrm{Re}\{\lambda_j(\tau)-\lambda_i(\tau)\} + \frac{n_j - l_i}{\tau}\right) d\tau} \leq \beta_i(s).
$$

Therefore (6.44) holds with $\beta_\nu(s)$ renamed $\beta_i(s)$. Also, for $t_0 \leq t \leq s$,

$$
\begin{aligned}
e^{\int_{s}^{t} \mathrm{Re}\{\mu_k(\tau)-\mu_\nu(\tau)\}\, d\tau} &= e^{\int_{s}^{t} \mathrm{Re}\{\lambda_j(\tau)-\lambda_i(\tau)\}} e^{\int_{t}^{s} \frac{l_i - l_j}{\tau}\, d\tau} \\
&\leq e^{\int_{s}^{t} \mathrm{Re}\{\lambda_j(\tau)-\lambda_i(\tau)\}} e^{\int_{t}^{s} \frac{l_i - 1}{\tau}\, d\tau} \\
&= e^{\int_{s}^{t} \mathrm{Re}\{\lambda_j(\tau)-\lambda_i(\tau)\} + \frac{1 - l_i}{\tau}} \leq \beta_i(s),
\end{aligned}
$$

by (6.42). Hence (6.45) holds. Now Theorem 2.12 implies the existence of a solution $w_{i,l_i}(t)$ of (6.36) of the form (6.38), and the rest of the proof coincides with the proof of Theorem 6.6. □

We conclude this section by mentioning that Chiba and Kimura [38, Theorem 5.1] first considered perturbations of Jordan systems in this setting of weak dichotomies. Their approach was based on a theorem by Hukuhara. The approach presented here in the framework of Levinson's theory allows us to simplify some of their hypotheses. In particular, various monotonicity requirements that they made are not needed in Theorem 6.8.

6.5 Block-Triangular Perturbations

Kimura [91] established the following extension of Theorem 2.7. He considered systems whose coefficient matrices are a sum of three components

$$
\Lambda(t) + A(t) + R(t).
$$

Here $\Lambda(t)$ is a block-diagonal matrix $\Lambda(t) = \operatorname{diag}\{\lambda_1(t)I_{n_1}, \ldots, \lambda_m(t)I_{n_m}\}$ with asymptotically constant $\lambda_i(t)$. $A(t)$ is a uniformly bounded perturbation in the form of certain strictly triangular blocks, and $R(t)$ is an L^1-perturbation. His result was only concerned with the existence of certain vector-valued solutions and not with the existence of a fundamental solution matrix. Vaguely speaking, he showed for such a vector-valued solution to exist, one can allow bounded strictly triangular perturbations in certain "other" blocks. To be precise, he considered

$$y' = \left[\sum_{i=1}^{m} \oplus \{\lambda_i(t)I_{n_i} + A_i(t)\} + R(t)\right] y, \qquad t \geq t_0 \tag{6.46}$$

where $A_i(t)$ are a strictly lower triangular $n_i \times n_i$ matrices. He established the following result:

Theorem 6.9 *For $1 \leq i \leq m$, let $\lambda_i(t)$ be continuous functions on $[t_0, \infty)$ such that*

$$\lim_{t\to\infty} \lambda_i(t) = \lambda_i^0 \qquad \text{as} \quad t \to \infty, \tag{6.47}$$

for some constants λ_i^0. For $1 \leq i \leq m$, let $A_i(t)$ be $n_i \times n_i$ strictly lower triangular continuous matrices satisfying

$$A_i(t) = O(1) \qquad \text{for all } 1 \leq i \leq m \quad \text{as} \quad t \to \infty. \tag{6.48}$$

Fix an index $h \in \{1, \ldots, m\}$ and suppose that $A_h(t) \equiv 0$. Suppose that for each $1 \leq i \leq m$, $i \neq h$,

$$\text{if} \quad \operatorname{Re}\lambda_i^0 = \operatorname{Re}\lambda_h^0 \qquad \text{then} \qquad A_i(t) \equiv 0. \tag{6.49}$$

Finally, for this fixed index h, suppose that for each ordered pair of indices (i, h) $(1 \leq i \leq m)$, the differences $\lambda_i(t) - \lambda_h(t)$ satisfy Levinson's dichotomy conditions, i.e., there exists a constant $K > 0$ such that either

$$\left.\begin{array}{ll} \int_{t_0}^t \operatorname{Re}\{\lambda_i(\tau) - \lambda_h(\tau)\}\, d\tau \to -\infty & \text{as } t \to \infty \\ \text{and} \quad \int_s^t \operatorname{Re}\{\lambda_i(\tau) - \lambda_h(\tau)\}\, d\tau < K & \forall\ t_0 \leq s \leq t \end{array}\right\} \tag{6.50}$$

or

$$\int_s^t \operatorname{Re}\{\lambda_i(\tau) - \lambda_h(\tau)\}\, d\tau > -K \qquad \forall \quad t_0 \leq s \leq t. \tag{6.51}$$

Assume that $R(t)$ is continuous for $t \geq t_0$ and $R(t) \in L^1[t_0, \infty)$. Then (6.46) *has n_h linearly independent solutions satisfying, as $t \to \infty$,*

$$y_k = \left[e_{\sum_{i=1}^{h-1} n_i + k} + o(1)\right] e^{\int^t \lambda_h(\tau)\, d\tau} \qquad \forall \quad 1 \leq k \leq n_h. \tag{6.52}$$

While originally Kimura used an existence theorem by Hukuhara for solutions of systems of initial value problems as the main tool of his proof, we will prove his result here by applying Theorem 2.2 to establish a correspondence between the bounded solutions of unperturbed and perturbed systems.

Proof For the fixed value of $h \in \{1, \ldots, m\}$, put, in (6.46),

$$y(t) = z(t) \exp\left[\int^{t} \lambda_h(\tau)\, d\tau\right]. \tag{6.53}$$

Then it follows from (6.46) that

$$z' = \left[\sum_{i=1}^{m} \oplus \{[\lambda_i(t) - \lambda_h(t)]\, I_{n_i} + A_i(t)\} + R(t)\right] z, \qquad t \geq t_0. \tag{6.54}$$

We also consider the unperturbed system

$$w' = \sum_{i=1}^{m} \oplus \{[\lambda_i(t) - \lambda_h(t)]\, I_{n_i} + A_i(t)\}\, w, \qquad t \geq t_0, \tag{6.55}$$

and we will apply Theorem 2.2 to (6.54) and (6.55). Since

$$A_h(t) \equiv 0,$$

(6.55) has n_h linearly independent bounded solutions in form of Euclidean vectors given by

$$w_k = e_{\sum_{i=1}^{h-1} n_i + k}, \qquad 1 \leq k \leq n_h.$$

To apply Theorem 2.2, we need to show that (6.55) possesses an ordinary dichotomy (see Definition 2.1). Towards that goal, we define index sets $\mathcal{I}_\nu$ for $\nu \in \{1, 2\}$ by

$$i \in \mathcal{I}_1 \iff \begin{cases} \text{either: } \operatorname{Re}\lambda_i^0 = \operatorname{Re}\lambda_h^0 \text{ and } \lambda_i(t) - \lambda_h(t) \text{ satisfies (6.50)} \\ \text{or: } \operatorname{Re}\lambda_i^0 < \operatorname{Re}\lambda_h^0 \end{cases}$$

and

$$i \in \mathcal{I}_2 \iff \begin{cases} \text{either: } \operatorname{Re}\lambda_i^0 = \operatorname{Re}\lambda_h^0 \text{ and } \lambda_i(t) - \lambda_h(t) \text{ satisfies (6.51)} \\ \text{or: } \operatorname{Re}\lambda_i^0 > \operatorname{Re}\lambda_h^0 \end{cases}$$

We also define two corresponding projection matrices P_1, P_2 satisfying $P_1 + P_2 = I$ by

$$P_\nu = \operatorname{diag}\{p_{1,\nu}\,,\ p_{2,\nu}\,,\ \ldots\,, p_{d,\nu}\} \qquad \text{for} \quad \nu \in \{1, 2\},$$

where

$$p_{i,\nu} = \begin{cases} 1 & \text{if } i \in \mathcal{I}_\nu \\ 0 & \text{otherwise.} \end{cases}$$

We will first show that

$$|W(t)P_1W^{-1}(s)| \leq c_1 \qquad \forall \quad t_0 \leq s \leq t, \tag{6.56}$$

for some positive constant c_1 and that

$$|W(t)P_1| \to 0 \qquad \text{as } t \to \infty. \tag{6.57}$$

To show (6.56), first assume that for some $i \in \{1, \dots, m\}$, $\operatorname{Re}\lambda_i^0 = \operatorname{Re}\lambda_h^0$ and that $\lambda_i(t) - \lambda_h(t)$ satisfies (6.50). Then $A_i(t) \equiv 0$ and the ith block of $W(t)P_1W^{-1}(s)$ contains only diagonal terms of the form $\exp\left[\int_s^t\{\lambda_i(\tau) - \lambda_h(\tau)\}\,d\tau\right]$ satisfying

$$\left|\exp\left[\int_s^t \{\lambda_i(\tau) - \lambda_h(\tau)\}\,d\tau\right]\right| < e^K \qquad \forall \quad t_0 \leq s \leq t.$$

Also note that from (6.50) it follows that $\exp\left[\int_{t_0}^t\{\lambda_i(\tau) - \lambda_h(\tau\}\,d\tau\right] \to 0$ as $t \to \infty$.

If for some i, $\lambda_i^0 < \lambda_h^0$, then the ith block of (6.55) can be written as

$$w_i' = \begin{pmatrix} \lambda_i(\tau) - \lambda_h(\tau) & & 0 \\ & & \\ A_i(t) & \ddots & \\ & & \lambda_i(\tau) - \lambda_h(\tau). \end{pmatrix} w_i.$$

By its triangular nature, this system is explicitly solvable. Furthermore, it is not hard to show that the components of $w_i(t)$ are of the form

$$(w_i(t))_k = f_k(t)\exp\left[\int_{t_0}^t \{\lambda_i(\tau) - \lambda_h(\tau)\}\ d\tau\right],$$

where $f_k(t)$ is a linear combination of (repeated) integrals over $a_{kj}^i(\tau)$, and hence $|f_k(t)|$ has at most polynomial growth. Since $\operatorname{Re}\lim_{t\to\infty}\{\lambda_i(t) - \lambda_h(t)\} = -\delta_i < 0$, it follows that the ith block of $W(t)P_1W^{-1}(s)$ is bounded in norm by $K_ie^{-\frac{\delta_i}{2}(t-s)}$ for some positive constant K_i and for all $t_0 \leq s \leq t$. Therefore (6.56) holds for some positive constant c_1 and, moreover, $W(t)P_1 \to 0$ as $t \to \infty$.

Similar arguments show that $|W(t)P_2W^{-1}(s)| \leq c_2$ for all $t_0 \leq t \leq s$, for some positive constant c_2. Hence (6.55) possesses an ordinary dichotomy and $W(t)P_1 \to 0$ as $t \to \infty$.

Since $R \in L^1[t_0, \infty)$, Theorem 2.2 implies that there exist n_h solutions of (6.54) of the form

$$z_k = e_{\sum_{i=1}^{h-1}+k} + o(1) \qquad \text{as } t \to \infty, \quad 1 \le k \le n_h.$$

It follows from (6.53) that (6.46) has n_h solutions of the form (6.52) ($1 \le k \le n_h$), which are easily seen to be linearly independent for t sufficiently large. □

Note that if there exists $h \in \{1, \ldots, m\}$ such that $\lambda_i^0 = \lambda_h^0$ for all $1 \le i \le m$, then $A_i(t) = 0$ for all i and all entries in the diagonal matrix $\sum_{i=1}^m \oplus \lambda_i(t) I_{n_i}$ satisfy Levinson's dichotomy conditions, hence Theorem 6.9 follows from Theorem 2.7 in this case.

6.6 An Early Result on L^1-Perturbations of Certain Jordan Matrices

The following early result on the asymptotic behavior of certain systems was given by Levinson [100, Theorem 3].

Theorem 6.10 *Consider*

$$y' = [A + R(t)]y, \qquad t \ge t_0 \tag{6.58}$$

where A is a constant $d \times d$ matrix similar to

$$J = \begin{pmatrix} \Lambda_0 & 0 & 0 & \cdot & \cdot & 0 \\ 0 & J_1 & 0 & \cdot & \cdot & \cdot \\ 0 & 0 & J_2 & \cdot & \cdot & \cdot \\ \cdot & \cdot & \cdot & \cdot & \cdot & \cdot \\ \cdot & \cdot & \cdot & \cdot & \cdot & 0 \\ 0 & \cdot & \cdot & \cdot & 0 & J_s \end{pmatrix}, \tag{6.59}$$

with $\Lambda_0, J_1, \ldots, J_s$ being matrices satisfying the following requirements:

$$\Lambda_0 = \operatorname{diag}\{\lambda_1, \ldots, \lambda_m\} \quad \textit{(not necessarily distinct)};$$

For $1 \le i \le s$, consider

$$J_i = \lambda_{m+i} I_{n_i} + N_{n_i},$$

where N_{n_i} is the $n_i \times n_i$ matrix defined in (6.2). *Assume that*

$$\operatorname{Re}\lambda_i \geq 0, \qquad 1 \leq i \leq m, \tag{6.60}$$

$$\operatorname{Re}\lambda_{m+i} < -\beta < 0, \qquad 1 \leq i \leq s. \tag{6.61}$$

Suppose that $R(t) \in L^1[t_0, \infty)$.

Then (6.58) *has for t sufficiently large d linearly independent solutions satisfying, as $t \to \infty$,*

$$y_k(t) = [\mathbf{v}_k + o(1)] \exp[\lambda_k t] \qquad 1 \leq k \leq m, \tag{6.62}$$

$$e^{\beta t} y_k(t) = o(1), \qquad m + 1 \leq k \leq d. \tag{6.63}$$

Here $\mathbf{v}_k$ is the kth column of the invertible matrix S such that $S^{-1}AS = J$.

Proof While Levinson's original proof used successive approximations, we give a simpler proof here based on Theorem 2.2. In a first step, we set $y = Sz$, where S is a constant matrix with linearly independent column vectors $\mathbf{v}_k$, $1 \leq k \leq d$, such that

$$z' = S^{-1}[A + R(t)]\,S(t)z = [J + R_1(t)]\,z, \qquad R_1 \in L^1[t_0, \infty). \tag{6.64}$$

First, to establish the existence of solutions of the form (6.62), fix $k \in \{1, \ldots, m\}$ and put $z = w \exp[\lambda_k t]$. Then

$$w' = [J - \lambda_k I + R_1(t)]\,w. \tag{6.65}$$

Also consider the unperturbed system

$$x' = [J - \lambda_k I]\,x, \tag{6.66}$$

which has the bounded solution $x_k = e_k$. Let $P_2 = \operatorname{diag}\{p_{21}, p_{22}, \ldots p_{2m}, 0, 0, \ldots 0\}$, where $p_{2i} = 1$ if $\operatorname{Re}[\lambda_i - \lambda_k] \geq 0$ and $p_{2i} = 0$ else. Set $P_1 = I - P_2$. Then (6.66) has an ordinary dichotomy of the form (2.1) (with P replaced by P_1) and, moreover, $X(t)P_1 \to 0$ as $t \to \infty$. As $R_1 \in L^1[t_0, \infty)$, Theorem 2.2 implies that (6.65) has a solution of the form $w_k = e_k + o(1)$ as $t \to \infty$. Repeating this process for all $k \in \{1, 2, \ldots, m\}$, one finds that (6.64) has m corresponding solutions

$$z_k(t) = [e_k + o(1)]e^{\lambda_k t} \qquad \text{as } t \to \infty, \qquad 1 \leq k \leq m. \tag{6.67}$$

Hence (6.58) has m solutions of the form (6.62).

Secondly, to establish the existence of solutions of the form (6.63), put $z = \hat{w} \exp[-\beta t]$ in (6.64). Then

$$\hat{w}' = [J + \beta I + R(t)]\,\hat{w}. \tag{6.68}$$

Also consider the unperturbed system

$$\hat{x}' = [J + \beta I]\,\hat{x}, \tag{6.69}$$

which has a fundamental matrix

$$\hat{X}(t) = \operatorname{diag}\{\hat{X}_0(t)\,,\, \hat{X}_1(t),\, \dots\,, \hat{X}_s(t)\}, \tag{6.70}$$

where

$$\hat{X}_0(t) = \operatorname{diag}\left\{e^{(\lambda_1+\beta)t}, \dots, e^{(\lambda_m+\beta)t}\right\},$$

and for $1 \le j \le s$,

$$\hat{X}_j(t) = e^{(\lambda_{m+j}+\beta)t}\begin{pmatrix} 1 & t & \frac{t^2}{2!} & \frac{t^3}{3!} & \cdots \\ 0 & 1 & t & \ddots & \\ \vdots & \ddots & \ddots & \ddots & \frac{t^2}{2!} \\ \vdots & & \ddots & \ddots & t \\ 0 & \cdots & & 0 & 1 \end{pmatrix}. \tag{6.71}$$

We note that by (6.61) $\hat{X}_j(t)$ is bounded for all $t \ge t_0$ and $1 \le j \le s$ and vanishes at infinity.

Fix any one of these $d - m$ bounded solutions, say $\hat{x}_k(t)$, for some $k \in \{m+1, \dots, d\}$. Put

$$\hat{P}_2 = \operatorname{diag}\{\underbrace{1, \dots, 1}_{m}, \underbrace{0, \dots, 0}_{d-m}\} \tag{6.72}$$

and $\hat{P}_1 = I - \hat{P}_2$. Then (6.69) possesses an ordinary dichotomy (see Definition 2.1, with P replaced by $\hat{P}_1$) and, moreover, $\hat{X}(t)\hat{P}_1 \to 0$ as $t \to \infty$. As R_1 is absolutely integrable, Theorem 2.2 implies that (6.68) has a solution of the

$$\hat{w}_k(t) = \hat{x}_k(t) + o(1) \qquad \text{as } t \to \infty. \tag{6.73}$$

Repeating this process for all $k \in \{m+1, \dots, d\}$, one finds that (6.64) has $d - m$ solutions of the form

$$z_k(t) = [\hat{x}_k(t) + o(1)]e^{-\beta t} \qquad \text{as } t \to \infty \qquad m+1 \le k \le d. \tag{6.74}$$

Hence (6.58) has $d - m$ solutions satisfying as $t \to \infty$

$$y_k(t) = S[\hat{x}_k(t) + o(1)]e^{-\beta t} = o(1)e^{-\beta t}, \qquad m+1 \le k \le d,$$

which yields (6.63) for all $m + 1 \le k \le d$.

We use Levinson's argument [100, pp. 125] to show the linear independence of the solutions $z_k(t)$ of (6.64) for all $1 \leq k \leq d$. Since S is nonsingular, this will immediately imply the linear independence of solutions of (6.58). For that purpose, suppose that there are constants c_k such that

$$\sum_{k=1}^{d} c_k z_k(t) = 0 \quad \text{for all t } \geq t_0. \tag{6.75}$$

Using components we have

$$\sum_{k=1}^{d} c_k (z_k)_i(t) = 0 \quad \text{for all t } \geq t_0, \qquad 1 \leq i \leq d.$$

Taking the first m equations and using the properties (6.67) and (6.74) we find, letting $t \to \infty$, that $c_k = 0$ for $1 \leq k \leq m$. We now consider only the equations with components $k > m$,

$$\sum_{k=m+1}^{d} c_k \left(z_k(t)\right)_i = 0 \quad \text{for all t } \geq t_0, \qquad m+1 \leq i \leq d.$$

Multiplying with $\exp[\beta t]$, this is equivalent to consider

$$\sum_{k=m+1}^{d} c_k \left(\hat{w}_k(t)\right)_i = 0 \qquad \text{for all } t \geq t_0, \qquad m+1 \leq i \leq d. \tag{6.76}$$

Going back to the proof of Theorem 2.2, (6.73) could be written in greater detail as

$$\hat{w}_k(t) = \hat{x}_k(t) - (T\hat{w}_k)(t), \tag{6.77}$$

where T is defined (see (2.11)) for all $t \geq t_1$ for some sufficiently large t_1 as

$$(T\hat{w}_k)(t) = \int_{t_1}^{t} \hat{X}(t)\hat{P}_1\hat{X}^{-1}(s)R_1(s)\hat{w}_k(s)\,ds - \int_{t}^{\infty} \hat{X}(t)\hat{P}_2\hat{X}^{-1}(s)R_1(s)\hat{w}_k(s)\,ds,$$

with $\hat{X}(t)$ and $\hat{P}_2$ defined in (6.70) and (6.72), respectively. It follows immediately that

$$(T\hat{w}_k)(t_1) = -\int_{t_1}^{\infty} \hat{X}(t_1)\hat{P}_2\hat{X}^{-1}(s)R_1(s)\hat{w}_k(s)\,ds.$$

By the definition of $\hat{P}_2$ in (6.72), the last $(d-m)$ components of the vector $\hat{P}_2\hat{X}^{-1}(s)R_1(s)\hat{w}_k(s)$ are equal to zero. This together with the structure of $\hat{X}(t)$

(see (6.70)) implies that the last $d - m$ components of $(T\hat{w}_k)(t_1)$ are equal to zero. Hence it follows from (6.76) and (6.77) with $t = t_1$ that

$$\sum_{k=m+1}^{d} c_k \left(\hat{x}_k(t_1)\right)_i = 0 \qquad m + 1 \leq i \leq d.$$

From the explicit structure of the vectors $\hat{x}_k(t_1)$ described in (6.71), it follows that $\{(\hat{x}_k(t_1))_i : m + 1 \leq i \leq d\}_{k=m+1}^{d}$ are linearly independent vectors in $\mathbb{R}^{d-m}$, hence $c_k = 0$ for all $m + 1 \leq k \leq d$, which completes the proof. □

Note that one can find statements similar to (6.62) and (6.63) in case that zero in (6.60) and (6.61) would be replaced by any real number κ. Moreover, the inequalities in (6.60) and (6.61) can be reversed (also with zero replaced by any real κ). However, the statement corresponding to (6.63) would be weaker.

Chapter 7
Perturbations of Jordan Difference Systems

In this brief chapter, we only consider perturbations of systems of difference equations with a single non-singular Jordan block. That is, we consider

$$y(n+1) = [\lambda I + N + R(n)]\, y(n), \qquad \lambda \neq 0, \qquad N = \begin{pmatrix} 0 & 1 & & \\ & 0 & \ddots & \\ & & \ddots & 1 \\ & & & 0 \end{pmatrix}, \qquad n \geq n_0. \tag{7.1}$$

Following the approach taken in Sect. 6.2, the next theorem can be considered as a discrete counterpart of Corollary 6.2, and its proof is parallel to the proof given in Theorem 6.1.

Theorem 7.1 *Consider the d-dimensional system* (7.1)*, where* $\lambda \neq 0$*. Let* $r_{ij}(n)$ *denote the entries of* $R(n)$ $(1 \leq i,j \leq d)$*, and suppose that*

$$r_{ij}(n)n^{i-j} \in l^1 \qquad \textit{for } 1 \leq i,j \leq d. \tag{7.2}$$

S. Bodine, D.A. Lutz, *Asymptotic Integration of Differential and Difference Equations*, Lecture Notes in Mathematics 2129,
DOI 10.1007/978-3-319-18248-3_7

Then (7.1) *has for n sufficiently large a fundamental matrix satisfying as n* $\to \infty$

$$Y(n) = \lambda^n \begin{pmatrix} 1+o(1) & \binom{n}{1}\frac{1}{\lambda}[1+o(1)] & \binom{n}{2}\frac{1}{\lambda^2}[1+o(1)] & \cdots & \binom{n}{d-1}\frac{1}{\lambda^{d-1}}[1+o(1)] \\ o\left(\frac{1}{n}\right) & 1+o(1) & \binom{n}{1}\frac{1}{\lambda}[1+o(1)] & & \binom{n}{d-2}\frac{1}{\lambda^{d-2}}[1+o(1)] \\ o\left(\frac{1}{n^2}\right) & o\left(\frac{1}{n}\right) & 1+o(1) & \ddots & \vdots \\ \vdots & & \ddots & \ddots & \binom{n}{1}\frac{1}{\lambda}[1+o(1)] \\ o\left(\frac{1}{n^{d-1}}\right) & & & o\left(\frac{1}{n}\right) & 1+o(1) \end{pmatrix}. \tag{7.3}$$

Proof We begin with the normalization $y(n) = \lambda^n \tilde{y}(n)$ and hence (7.1) leads to

$$\tilde{y}(n+1) = \left[I + \frac{N}{\lambda} + \frac{R(n)}{\lambda}\right]\tilde{y}(n).$$

Define the $d \times d$-matrix

$$\Phi(n) := \left[I + \frac{N}{\lambda}\right]^n = \begin{pmatrix} 1 & \frac{1}{\lambda}\binom{n}{1} & \frac{1}{\lambda^2}\binom{n}{2} & \cdots & \frac{1}{\lambda^{d-1}}\binom{n}{d-1} \\ & 1 & \frac{1}{\lambda}\binom{n}{1} & \cdots & \frac{1}{\lambda^{d-2}}\binom{n}{d-2} \\ & & 1 & \ddots & \\ & & & \ddots & \frac{1}{\lambda}\binom{n}{1} \\ & & & & 1 \end{pmatrix}. \tag{7.4}$$

Making the transformation

$$\tilde{y}(n) = \Phi(n)z(n) \tag{7.5}$$

implies that

$$z(n+1) = \left[I + \lambda^{-1}\Phi^{-1}(n+1)R(n)\Phi(n)\right], \tag{7.6}$$

diagonalizing the unperturbed part, but increasing the size of the perturbation significantly: since both $(\Phi(n))_{1d}$ and $(\Phi^{-1}(n+1))_{1d}$ are of order n^{d-1}, $|\Phi^{-1}(n+$

$1)R(n)\Phi(n)|$ is of order $n^{2(d-1)}|R(n)|$. To reduce the size of the perturbation, one continues with a shearing transformation

$$z(n) = D^{-1}(n)\tilde{z}(n), \tag{7.7}$$

where

$$D(n) = \operatorname{diag}\left\{1, \binom{n}{1}, \dots, \binom{n}{d-1}\right\}. \tag{7.8}$$

It follows that

$$\tilde{z}(n+1) = \left[D(n+1)D^{-1}(n) + \lambda^{-1}D(n+1)\Phi^{-1}(n+1)R(n)\Phi(n)D^{-1}(n)\right]\tilde{z}(n). \tag{7.9}$$

Using the component-wise representation of $\Phi(n)$ defined in (7.4)

$$\Phi_{ij}(n) = \begin{cases} 0 & i > j, \\ \dfrac{1}{\lambda^{j-i}}\dbinom{n}{j-i} & i \le j, \end{cases}$$

one can show that

$$\Phi(n)D^{-1}(n) = D^{-1}(n)\left[P + E(n)\right].$$

Here P is the upper triangular matrix given by

$$P_{ij}(n) = \begin{cases} 0 & i > j, \\ \dfrac{1}{\lambda^{j-i}}\dbinom{j-1}{i-1} & i \le j, \end{cases}$$

and $E(n)$ is strictly upper triangular matrix vanishing at infinity. In particular, $\det(P + E(n)) = 1$ and hence $P + E(n)$ is invertible for all n. A useful identity in this context is

$$\binom{n}{k} = \frac{n^k}{k!}[1 + \mathrm{o}(1)] \qquad 0 \le k \le n.$$

This allows to re-write (7.9) as

$$\tilde{z}(n+1) = \left[\Lambda(n) + \lambda^{-1}[P + E(n+1)]^{-1}D(n+1)R(n)D^{-1}(n)[P + E(n)]\right]\tilde{z}(n), \tag{7.10}$$

with

$$\Lambda(n) := D(n+1)D^{-1}(n) = \operatorname{diag}\left\{1, 1+\frac{1}{n}, 1+\frac{2}{n-1}, \ldots, 1+\frac{d-1}{n+2-d}\right\}. \tag{7.11}$$

Note that $\Lambda(n)$ satisfies the dichotomy conditions (3.8), (3.9). Observe that

$$\begin{aligned}\left(D(n+1)R(n)D^{-1}(n)\right)_{ij} &= \binom{n+1}{i-1} r_{ij}(n) \frac{1}{\binom{n}{j-1}} \\ &= \left[\frac{(j-1)!}{(i-1)!} + \mathrm{o}(1)\right] n^{i-j} r_{ij}(n).\end{aligned}$$

Moreover, since $\Lambda(n) \to I$ as $n \to \infty$, it follows from (7.2) that

$$\left|D(n+1)R(n)D^{-1}(n)\right| / |\lambda_i(n)| \in l^1 \text{ for each } 1 \leq i \leq d.$$

Thus Theorem 3.4 implies for n sufficiently large the existence of a fundamental matrix of (7.9) satisfying

$$\tilde{Z}(n) = [I + \mathrm{o}(1)] \prod^{n-1} \Lambda(k) = [I + \mathrm{o}(1)]\, D(n) \qquad \text{as } n \to \infty.$$

By (7.5) and (7.7), (7.1) has a fundamental matrix of the form

$$Y(n) = \lambda^n \Phi(n) D^{-1}(n) [I + \mathrm{o}(1)]\, D(n),$$

and straightforward, yet somewhat tedious matrix multiplication shows that this is of the form (7.3), which completes the proof. □

Theorem 7.1 is due to Elaydi. In [54, Theorem 5], he established a preliminary result which was based on applying Theorem 3.4 to (7.6). As noted above, the perturbation in (7.6) is unnecessarily large, hence this result was not optimal. In [56, Theorem 2.1] he improved his earlier work, and this new result is in essence given in Theorem 7.1.

Noting that the entries of $\Lambda(n)$ defined in (7.11) satisfy $\lambda_i(n) - \lambda_j(n) \neq 0$ for $1 \leq i \neq j \leq d$ and all $n \geq n_0$, one could derive another asymptotic result based on applying Theorem 5.2 to (7.10).

Theorem 7.1 can be generalized to several Jordan blocks. Such an extension is straightforward and follows closely the approach taken in the proof of Theorem 6.6. To avoid repetition, we refer the reader to [56, Theorem 2.2] for such a result. We also mention that some results concerning perturbations of a non-constant Jordan block can be found in [54, Sect. 5]

We conclude with remarking that the results in Sects. 6.4–6.6 also have discrete counterparts, and we leave it to the interested reader to work out the details.

Chapter 8
Applications to Classes of Scalar Linear Differential Equations

8.1 Chapter Overview

In this chapter we consider various classes of dth-order ($d \geq 2$) linear homogeneous equations

$$y^{(d)} + a_1(t)y^{(d-1)} + \ldots + a_d(t)y = 0, \tag{8.1}$$

which we treat as first-order, d-dimensional systems

$$\mathbf{y}' = \begin{pmatrix} 0 & 1 & & \\ & 0 & \ddots & \\ & & \ddots & 1 \\ -a_d(t) & -a_{d-1}(t) & \ldots & -a_1(t) \end{pmatrix} \mathbf{y}, \qquad \text{with } \mathbf{y} = \begin{pmatrix} y \\ y' \\ \vdots \\ y^{(d-1)} \end{pmatrix}. \tag{8.2}$$

The main goal of this chapter is to apply theorems from Chaps. 2 and 4 to obtain asymptotic representations for a fundamental solution matrix

$$Y(t) = \begin{pmatrix} y_1 & y_2 & \ldots & y_d \\ y_1' & y_2' & \ldots & y_d' \\ \vdots & & & \vdots \\ y_1^{(d-1)} & y_2^{(d-1)} & \ldots & y_d^{(d-1)} \end{pmatrix},$$

which gives asymptotic formulas for solutions y_i of (8.1) and their derivatives. We will compare these results with those obtained using scalar and often rather ad-hoc (especially when $d = 2$) techniques.

S. Bodine, D.A. Lutz, *Asymptotic Integration of Differential and Difference Equations*, Lecture Notes in Mathematics 2129,
DOI 10.1007/978-3-319-18248-3_8

In Sect. 8.2, we consider perturbations of dth order equations with constant coefficients, i.e., where $a_j(t)$ in (8.1) are of the form $a_j(t) = a_j + b_j(t)$, with $b_j(t)$ being "small" in a sense to be made precise. Under various combinations of assumptions concerning the zeros of the limiting characteristic polynomial

$$p(\lambda) = \lambda^d + a_1\lambda^{d-1} + \cdots + a_d \tag{8.3}$$

and behavior of the perturbations $b_j(t)$, one obtains several kinds of asymptotic representations for solutions of (8.1) and their derivatives.

For the special case of second-order equations

$$y'' + a_1(t)y' + a_2(t)y = 0, \tag{8.4}$$

there is a very extensive literature going back to the nineteenth century and originating with the well-known asymptotic formulas due to Liouville and Green. There are several reasons why the case $d = 2$ has been studied in great detail, among them chiefly that second-order equations have many important applications, especially to problems arising from mathematical physics, and that special, ad-hoc methods are often available that do not carry over to higher order equations.

For us, the main advantage when $d = 2$ comes from the fact that eigenvalues and eigenvectors can be explicitly calculated and effectively used.

We will treat several quite different classes of second-order equations that have appeared in the literature, each having quite different appearing assumptions. The main focus is then to

1. normalize the systems using a variety of matrix transformations,
2. identify conditions which correspond to appropriate dichotomy and growth (decay) assumptions in the general theorems, and
3. relate the resulting asymptotic formulas and possibly apply error estimates.

While the main focus of asymptotic integration concerns reduction to solutions of first order linear equations, the methods may also be applied to yield results of a *comparative nature*. By this we mean, for example, asymptotically representing solutions of a class of second-order linear differential equations in terms of solutions of a *special* or *standard* second-order equation. The question is then how "close" should the equations be for such asymptotic representation to hold. A particularly useful and well-studied situation occurs for so-called nonoscillatory equations. These are studied in Sect. 8.4.2 where we show how a wide variety of special cases can be systematically treated using results from Chaps. 2 and 4.

Another interesting way to view asymptotic integration of linear differential equations concerns what could be called *asymptotic factorization* of the corresponding operator. Motivated by a similar result for difference equations and operators (see Sect. 9.5) in Sect. 8.5 we see how a class of second-order linear differential operators can be factored modulo L^1-perturbations by means of asymptotic representations of solutions.

For second-order equation

$$y'' + \lambda^2 y = g(t)y,$$

corresponding to perturbations of the harmonic linear oscillator, A. Kiselev and M. Christ/A. Kiselev have obtained some deep and interesting results concerned with asymptotic representations for solutions that hold not necessarily for all $\lambda \in \mathbb{R}$, but for *almost all* such λ. Such results are related to the phenomenon of resonance (also recall results of Cassell, Harris–Lutz, etc.) and in Sect. 8.6.2 it will be shown how in some cases they can be obtained using some highly innovative conditioning transformations and Levinson's fundamental theorem. This chapter ends with some results for third-order equations and for second-order matrix equations in Sects. 8.7 and 8.8, respectively.

8.2 Perturbations of Constant Differential Equations

A special case of (8.1) are the asymptotically constant equations

$$y^{(d)} + [a_1 + b_1(t)]y^{(d-1)} \dots + [a_d + b_d(t)](t)y = 0, \qquad t \geq t_0 \tag{8.5}$$

where

$$b_i(t) = o(1) \qquad \text{as } t \to \infty, \ \ 1 \leq i \leq d.$$

Such equations were first considered by Poincaré [130] under the assumption that the zeros of the limiting characteristic polynomial (8.3) have all distinct real parts. He proved that if $y(t)$ is any solution of (8.5) either it is eventually identically zero or else $y(t) \neq 0$ for all t sufficiently large and

$$\lim_{t\to\infty} \frac{y^{(k)}(t)}{y(t)} = \lambda^k, \qquad 1 \leq k \leq d-1, \tag{8.6}$$

where λ is a root of (8.3). Hartman [72] has termed such a result "asymptotic integration on a logarithmic scale," and this comparatively weak statement corresponds to the very weak assumption $b_i(t) = o(1)$.

Perron [119, 120], motivated by Poincaré's result, showed that in fact (8.6) holds for a fundamental system $\{y_1(t), \dots y_d(t)\}$, i.e., for each i there exists a solution $y_i(t)$ of (8.5) such that $y_i(t) \neq 0$ for all large t and (8.6) holds with $\lambda = \lambda_i$.

Poincaré's result follows from Perron's, since any nontrivial solution $y(t)$ of (8.5) can be represented as

$$y(t) = \sum_{i=1}^{d} c_i y_i(t),$$

with at least one $c_i \neq 0$, and where we assume without loss of generality that $\operatorname{Re}\lambda_i < \operatorname{Re}\lambda_{i+1}$ for all $1 \leq i \leq d-1$. Let $j = \{\max i : c_i \neq 0\}$. Then it follows that $y(t) \neq 0$ for all t sufficiently large (using the fact that the real parts of λ_i are distinct) and

$$\frac{y'(t)}{y(t)} = \sum_{i=1}^{j} \frac{c_i y_i'(t)}{c_i y_i(t)} = \frac{y_j'(t)}{y_j(t)} + \mathrm{o}(1) = \lambda_j + \mathrm{o}(1),$$

and similarly for $y^{(k)}(t)/y(t)$.

Hartman and Wintner [73] improved upon Poincaré's result by weakening the assumption on the perturbation terms to just an averaged condition of the form

$$\sup_{s\geq t} \frac{1}{1+s-t} \int_t^s |b_i(\tau)|\, d\tau = \mathrm{o}(1) \text{ as } t \to \infty, \qquad i = 1, \ldots, d. \tag{8.7}$$

The results in Sect. 2.6 may be applied to (8.5), expressed in system form as

$$\mathbf{y}'(t) = [A + B(t)]\mathbf{y}, \qquad t \geq t_0, \tag{8.8}$$

where A is the companion matrix corresponding to the characteristic polynomial $p(\lambda)$ in (8.3) and $B(t)$ is "small."

(i) If all roots of (8.3) have distinct real parts, then Theorem 2.22 implies that for $b_i(t)$ satisfying (8.7), there exists a fundamental matrix $Y(t)$ of (8.8) satisfying, as $t \to \infty$,

$$Y(t) = [P + \mathrm{o}(1)] \exp\left[\int^t \left[\Lambda + \operatorname{diag}\{P^{-1}B(\tau)P\} + \mathrm{o}(|B(\tau)|)\right] d\tau\right], \tag{8.9}$$

where $A = P\Lambda P^{-1}$. This implies in particular for the ith column

$$\mathbf{y}_i(t) = [\mathbf{p}_i + \mathrm{o}(1)] \exp\left[\int^t \left[\lambda_i + (P^{-1}B(\tau)P)_{ii} + \mathrm{o}\left(|B(\tau)|\right)\right] d\tau\right]. \tag{8.10}$$

Since $\mathbf{p}_i$ can be normalized to the ith column of the Vandermonde matrix, $\mathbf{p}_i = [1, \lambda_i, \ldots, \lambda_i^{d-1}]^T$, and $\mathbf{y}_i(t) = [y_i, y_i', \ldots, y_i^{d-1}]^T$ one obtains immediately from the representation (8.10) that $y_i(t) \neq 0$ for all t sufficiently large and also that (8.6) holds for each $k = 1, \ldots, d-1$. This argument yields an independent proof of Perron's result and also under a weaker assumption (8.7) and a somewhat stronger conclusion (8.10).

In line with Theorem 2.22 and as Hartman also points out, the above conclusion for a single solution $y_i(t)$ holds under the weaker assumption that λ_i is a simple root of (8.3) such that $\text{Re}\,\lambda_i \neq \text{Re}\,\lambda_j$ for all other roots.

In order to obtain a more precise asymptotic integration of (8.5) than (8.10) generally requires stronger assumptions on the perturbations than (8.7) or $b_k(t) = \text{o}(1)$ for all k.

(ii) If, for fixed $i \in \{1, \ldots, d\}$, $\text{Re}\,\lambda_j \neq \text{Re}\,\lambda_i$ for all roots of (8.3) such that $i \neq j$, and if $b_j(t) \in L^p$ for all $1 \leq j \leq d$ and $1 < p \leq 2$, then applying Theorem 2.19 one obtains a solution $y_i(t)$ of (8.5)

$$y_i(t) = [1 + \text{o}(1)] \exp\left[\int^t \{\lambda_i + (P^{-1}B(\tau)P)_{ii}\}\, d\tau\right] \qquad \text{as } t \to \infty.$$

(iii) If the zeros λ_i of (8.3) would all be distinct, but not necessarily having distinct real parts, and $b_i(t) \in L^1$ for all $1 \leq i \leq d$, then Theorem 2.7 applies and yields a fundamental set of solution $y_i(t)$ of (8.5) of the form

$$y_i(t) = [1 + \text{o}(1)]e^{\lambda_i t}, \qquad 1 \leq i \leq d, \qquad \text{as } t \to \infty.$$

(iv) If (8.3) has all distinct zeros λ_i and in addition to $b_i(t) = \text{o}(1)$ for all k, one also assumes that $b_i'(t) \in L^1$ for all i, then applying Theorem 4.1 one obtains a solution $y_i(t)$ of (8.5) of the form

$$y_i(t) = [1 + \text{o}(1)] \exp\left[\int^t \lambda_i(\tau)\, d\tau\right], \qquad \text{as } t \to \infty,$$

provided that the roots $\lambda_1(t), \ldots \lambda_d(t)$ of

$$\lambda^d + [a_1 + b_1(t)]\lambda^{d-1} + \ldots + [a_d + b_d(t)] = 0$$

satisfy Levinson's dichotomy conditions (2.13), (2.14), and the roots are ordered such that $\lambda_i(t) \to \lambda_i$.

(v) If the zeros λ_i of (8.3) would all be distinct and if $\int_t^\infty b_i(\tau)\, d\tau$ converge (conditionally) for all $1 \leq i \leq d$, one could check if the other assumptions of Theorem 4.18 are satisfied. If this is the case, one obtains solutions of the form

$$y_i(t) = [1 + \text{o}(1)]e^{\lambda_i t} \qquad 1 \leq i \leq d.$$

8.3 Second-Order Linear Differential Equations

8.3.1 General Remarks

In this section we are concerned with linear second-order equation of the form (8.4). The goal is to achieve, under various assumptions on the coefficients, asymptotic representations for solutions. These come about from an asymptotic factorization of a fundamental solution matrix for the corresponding system (8.2) in the form

$$Y(t) = P(t)[I + Q(t)][I + E(t)] \exp\left[\int^t \Lambda(\tau)\, d\tau\right]$$

(see the discussion in Sect. 4.1). We will compare such asymptotic representations with ones in the literature that have been obtained using other methods which are especially well suited for second-order equations. Among these are the

1. Riccati transformation:

   This involves setting $u = y'/y$ to reduce (8.4) to a Riccati equation $u' + a_1(t)u + a_2(t) = -u^2$. Asymptotic formulas for u lead directly to ones for $y = \exp\left[\int u\right]$. (See, e.g., Hartman [72], Stepin [148].)
2. Prüfer transformation.

   This involves a kind of polar coordinate decomposition $(r(t), \theta(t))$ for y, y' to reduce (8.4) to a pair of first order nonlinear equations. Analysis of these equations leads to formulas for $y(t) = r(t)\cos\theta(t)$, etc., which are especially useful when discussing perturbations of a linear harmonic oscillator $y'' + [\lambda^2 + f(t)]y = 0$ (see [2] for a very extensive and readable treatment). While many results originally obtained using this method were later shown to be susceptible to the linear (Levinson-type) approach (see [69, 70]), other results appear to be more adequately handled using a Prüfer transformation.

As we have said, we pursue a Levinson approach here. Eastham [51, 53] has given a very thorough account of how this method has and can be applied to a wide variety of second order equations. Rather than duplicating the results discussed already by him, we refer to his treatment, especially for results prior to 1987. We will mainly focus on supplementing his results by discussing some others not included there.

It is well known that (8.4) can be transformed into either of the following normalized forms

$$y'' = [f(t) + g(t)]y, \tag{8.11}$$

or

$$(p(t)y')' = q(t)y, \tag{8.12}$$

which are often discussed in the literature. Eastham focuses on (8.12) and we will discuss some corresponding results for (8.11). Some very early and rather general asymptotic formulas for solutions of (8.11) were given by Liouville [101] and Green [67] in papers with quite different sounding titles. Now such results are referred to as Liouville–Green formulas. They are closely connected to what are called WKB or WKJB approximations, especially in the mathematical physics literature (see Olver [116, p. 228] for a discussion of how they are related).

In the classical approach to (8.11) it is assumed that $f(t)$ is real valued and $f(t) \neq 0$ for $t \geq t_0$. Then the so-called Liouville–Green transformations are used to further normalize (8.11) into the standard forms

$$\frac{d^2\tilde{y}}{d\tau^2} = [1 + r(\tau)]\tilde{y}, \qquad t \geq t_0, \tag{8.13}$$

or

$$\frac{d^2\tilde{y}}{d\tau^2} = [-1 + r(\tau)]\tilde{y}, \qquad t \geq t_0, \tag{8.14}$$

depending upon whether $f(t) > 0$ or $f(t) < 0$, and also assuming that $\int^t \sqrt{f(s)}\, ds \to +\infty$ as $t \to +\infty$. These are referred to as the nonoscillatory and oscillatory cases, respectively. The Liouville–Green transformations consist of a change of the independent variable as well as a linear transformation of the dependent variable. See [116, Chap. 6] for a modern and very careful treatment of this approach for (8.11), which involves assuming that $r(\tau)$ in (8.13) and (8.14) is in L^1 and then obtaining asymptotic representations for solutions. Olver also has given very good error bounds for the o(1) terms in the asymptotics for so-called recessive solutions, which he calls "realistic" and which are a substantial improvement to the more general error bounds described in Sect. 2.7. Also see Taylor [150] for an improvement of Olver's error bounds for (8.13) and (8.14). Taylor applied a variant of the Bellman–Gronwall inequality to a Volterra integral equation for the error terms to obtain his result, which amounts to placing the factor

$$1 - \exp(-2|t - a_j|)$$

inside the exponential in Olver's estimate (see Sect. 8.3.3).

We will present an alternative treatment to Olver's approach and show how using the T-operator introduced in Chap. 2, we can also obtain the classical Liouville–Green formulas together with Olver's error bounds. This will be the focus of Sects. 8.3.2 and 8.3.3. Before that, however, we mention that Eastham [53, Chap. 2] has an excellent discussion for (8.12) of how the various general asymptotic integration results in Chaps. 2 and 4 can be used to obtain several Liouville–Green type formulas which are variants of the classical ones, but has not included a discussion of Olver-type error bounds there.

8.3.2 Liouville–Green Approximations

Choosing a system's approach, we write (8.11) as the equivalent system

$$\mathbf{y}' = \begin{pmatrix} 0 & 1 \\ f+g & 0 \end{pmatrix} \mathbf{y}, \qquad \mathbf{y} = \begin{pmatrix} y \\ y' \end{pmatrix}. \tag{8.15}$$

We assume that $f(t)$is a twice continuously differentiable complex-valued function and $g(t)$ is a continuous complex function. We also assume that $f(t) \neq 0$ for $t \geq t_0$, and we choose the principal branch for the argument of $f(t)$. Next we choose the principal branch of the square root function so that

$$\operatorname{Re} f^{1/2}(t) \geq 0 \qquad \text{for } f(t) \neq 0. \tag{8.16}$$

We consider $g(t)$ as small compared to $f(t)$, and continue with a shearing transformation to achieve that the terms above and below the main diagonal have the same magnitude. Therefore, we put in (8.15)

$$\mathbf{y} = \operatorname{diag}\{1, f^{1/2}(t)\}\hat{y},$$

to find that

$$\hat{y}' = \left[\begin{pmatrix} 0 & f^{1/2} \\ f^{1/2} & 0 \end{pmatrix} + \begin{pmatrix} 0 & 0 \\ \frac{g}{f^{1/2}} & -\frac{f'}{2f} \end{pmatrix} \right] \hat{y}.$$

To diagonalize the leading matrix, we put $\hat{y} = \begin{pmatrix} 1 & 1 \\ -1 & 1 \end{pmatrix} \tilde{y}$, which leads to

$$\begin{aligned} \tilde{y}' &= \left[\begin{pmatrix} -f^{1/2} & 0 \\ 0 & f^{1/2} \end{pmatrix} + \frac{f'}{4f} \begin{pmatrix} -1 & 1 \\ 1 & -1 \end{pmatrix} + \frac{g}{2f^{1/2}} \begin{pmatrix} -1 & -1 \\ 1 & 1 \end{pmatrix} \right] \tilde{y} \\ &=: [\Lambda(t) + V_1(t) + V_2(t)]\tilde{y}. \end{aligned} \tag{8.17}$$

In order to eliminate the off-diagonal terms in V_1, we take into account the special symmetries in V_1 and V_2 to construct a conditioning transformation of the form

$$\tilde{y} = [I + Q(t)]z = \begin{pmatrix} 1 + h(t) & h(t) \\ -h(t) & 1 - h(t) \end{pmatrix} z. \tag{8.18}$$

Because $Q^2 = 0$, $[I + Q(t)]^{-1} = I - Q(t)$, and it is also easy to see that $QV_1 = QV_2 = V_2 Q = QQ' = 0$. In addition,

$$V_1 Q = \frac{hf'}{2f}\begin{pmatrix} -1 & -1 \\ 1 & 1 \end{pmatrix} \qquad \text{and} \qquad Q\Lambda Q = 2h^2 f^{1/2}\begin{pmatrix} -1 & -1 \\ 1 & 1 \end{pmatrix}.$$

Thus (8.18) leads to

$$z' = \left[(\Lambda + \operatorname{diag} V_1) + (\Lambda Q - Q\Lambda + V_1 - \operatorname{diag} V_1) + r(t)\begin{pmatrix} -1 & -1 \\ 1 & 1 \end{pmatrix}\right] z,$$

where

$$r(t) = \frac{g}{2f^{1/2}} + \frac{f'h}{2f} + h' - 2h^2 f^{1/2}.$$

Finally choosing $h(t)$ so that $\Lambda Q - Q\Lambda + V_1 - \operatorname{diag} V_1 = 0$, i.e., $h = f'/(8f^{3/2})$, we obtain

$$z' = \left[\begin{pmatrix} \lambda_1(t) & \\ & \lambda_2(t) \end{pmatrix} + r(t)\begin{pmatrix} -1 & -1 \\ 1 & 1 \end{pmatrix}\right] z, \tag{8.19}$$

where

$$\lambda_1(t) = -f^{1/2} - \frac{f'}{4f} \quad \text{and } \lambda_2(t) = f^{1/2} - \frac{f'}{4f}, \tag{8.20}$$

and

$$r(t) = \frac{g}{2f^{1/2}} + \frac{f''}{8f^{3/2}} - \frac{5}{32}\frac{(f')^2}{f^{5/2}}. \tag{8.21}$$

Considering (8.19) as a perturbation of the diagonal system

$$w' = \operatorname{diag}\{\lambda_1(t), \lambda_2(t)\} w, \tag{8.22}$$

one can obtain an asymptotic representation of a fundamental matrix of (8.19) by applying suitable theorems of Chaps. 2 and 4.

1. If the unperturbed system (8.22) satisfies Levinson's dichotomy conditions (2.13), (2.14), and if $r(t) \in L^1$, Theorem 2.7 yields for t sufficiently large the existence of a fundamental matrix of (8.19) of the form

$$Z(t) = [I + o(1)]\begin{pmatrix} e^{\int^t \lambda_1(s)\,ds} & \\ & e^{\int^t \lambda_2(s)\,ds} \end{pmatrix}.$$

Therefore (8.15) has a fundamental matrix with the asymptotic factorization

$$Y(t) = \begin{pmatrix} y_1 & y_2 \\ y_1' & y_2' \end{pmatrix} = \begin{pmatrix} 1 & \\ & f^{1/2} \end{pmatrix} \begin{pmatrix} 1 & 1 \\ -1 & 1 \end{pmatrix} \begin{pmatrix} 1+h & h \\ -h & 1-h \end{pmatrix} Z(t),$$

which shows that (8.11) has solutions of the form

$$y_{1,2}(t) = [1 + o(1)] f^{-1/4} \exp\left[\mp \int^t f^{1/2}(s)\, ds\right].$$

Since $\mathrm{Re}\{\lambda_1(t) - \lambda_2(t)\} = -2\mathrm{Re} f^{1/2}$, sufficient for (8.22) to satisfy Levinson's dichotomy condition is, for example, that $\mathrm{Re} f^{1/2}$ does not change sign.

2. If $f(t)$ is bounded away from $(-\infty, 0]$, then (8.22) satisfies an exponential dichotomy. Thus, if $r(t) \in L^p$ for some $1 < p \leq 2$, Theorem 4.11 would yield the asymptotic behavior of a fundamental matrix with $\lambda_1(t)$ and $\lambda_2(t)$ replaced by $\lambda_1(t) - r(t)$ and $\lambda_2(t) + r(t)$, respectively.
3. Other possibilities to determine the asymptotic behavior of solutions of (8.11) include Theorem 4.8.

We note that all these results provide the asymptotic behavior of solutions of (8.19) including an error term of the form o(1) or maybe, more quantitatively, $O(\varphi(t))$ for some positive decreasing function $\varphi(t)$ that vanishes at infinity. In what follows, we are interested in refining those general results to obtain more precise "realistic" error bounds in the spirit of Olver [116, pp. 193–196].

More specifically, in the next section, we will analyze the asymptotic behavior of solutions of (8.19) for $t \in [t_0, \infty)$. We will consider the general case $f(t) \neq 0$ for all $t \geq t_0$, and show how results can be improved in the so-called "nonoscillatory" case $f(t) > 0$. As we will see, this work will not only give an alternative approach to Olver's results, but offer some improvements in the nonoscillatory case. These improvements are comparable, although not identical, to results in Taylor [150]. We remark that these methods also carry over to $t \in [a_1, a_2]$, a finite or infinite interval, which is the setting Olver was interested in.

Also see [144] for an extension of such results (called then WKB-approximations) to second order equations of the form $y'' + [a + g(t)]y = 0$, where y is in a C^* algebra.

We also refer to [115] for a presentation of Liouville–Green formulas with error bounds on both bounded and unbounded intervals. In particular, the formulas are applied to equations of the form $y'' = \left[\pm u^2 + f(u,t)\right] y$ and $y'' = \pm u^2 p(u,t) y$, where u is a large positive parameter. In addition, this paper contains a comprehensive discussion of error bounds in Liouville–Green formulas, including early results of Blumenthal [10], uniformity of error bounds, and alternative forms for error bounds. Some interesting applications of the Liouville–Green formulas for parabolic cylinder and Bessel functions of large order are given. Finally, under additional conditions involving higher order derivatives of the coefficients, some asymptotic series representations for error bounds are presented.

Another procedure for obtaining Liouville–Green-type formulas for (8.11), but under different assumptions than above and with somewhat modified conclusions is due to D.R. Smith [138]. There, it is assumed that f is real-valued, nowhere zero on $[a, \infty)$, and piecewise $C^{(1)}$ (instead of $C^{(2)}$). For convenience and also to follow the notation in [138] let $f = \pm p^2$ and $g = q$ with $p > 0$. We will indicate the results in the nonoscillatory case corresponding to $f = +p^2$ and refer to [138] for the corresponding oscillatory case.

Also following a systems approach for (8.11), a (Riccati) transformation of the form

$$\tilde{y} = \begin{pmatrix} 1 & 0 \\ -\tau & 1 \end{pmatrix} \begin{pmatrix} 1 & -\sigma \\ 0 & 1 \end{pmatrix} z \tag{8.23}$$

is used to bring (8.17) into the diagonal system

$$z' = \mathrm{diag}\{\tilde{\lambda}_1(t), \tilde{\lambda}_2(t)\}z,$$

where

$$\tilde{\lambda}_1(t) = -\left[p + \frac{q + p'}{2p}\right] + \tilde{r},$$

$$\tilde{\lambda}_2(t) = \left[p + \frac{q - p'}{2p}\right] + \tilde{r},$$

$$\tilde{r}(t) = \left(\frac{p' - q}{2p}\right) \tau.$$

This complete diagonalization of (8.17) requires that σ and τ satisfy the equations

$$\tau' - \left[2p + \frac{q}{p}\right] \tau + \frac{p' + q}{2p} = \left[\frac{p' - q}{2p}\right] \tau^2, \tag{8.24}$$

$$\sigma' + \left[2p + \frac{q}{p} + \left\{\frac{p' - q}{2p}\right\} \tau\right] \sigma + \frac{p' - q}{2p} = 0. \tag{8.25}$$

Equation (8.24), being of Riccati type, motivates calling (8.23) a *Riccati transformation*. The idea is to first treat (8.24) as an equivalent integral equation

$$\tau(t) = \tau_0(t) - \int_t^\infty \exp\left[-\int_t^s (2p + q/p)\right] \left[\frac{p' - q}{2p}\right] \tau^2(s)\, ds,$$

where

$$\tau_0(t) = \int_t^\infty \exp\left[-\int_t^s (2p + q/p)\right] \left[\frac{p' + q}{2p}\right] ds.$$

Assuming that

$$\kappa(t) := \int_t^\infty \exp\left[-\int_t^s (2p + \operatorname{Re} q/p)\right] \frac{|p'| + |q|}{|p|}\, ds < \infty \tag{8.26}$$

for $t \in [a, \infty)$, it can be shown with the Banach fixed point theorem that for t sufficiently large so that $\kappa(t) < 1$, there exists a solution $\tau(t)$ satisfying $|\tau(t)| \leq \kappa(t)$. Then integrating (8.25) by quadrature leads to a solution $\tilde{y}(t)$ of the system (8.17) and solutions y_1, y_2 of (8.11).

Some further assumptions involving p, q and κ (see [138, Thm. 3.1]) then finally yield the formulas

$$y_1(t) = [1 + \varepsilon_1(t)]p^{-1/2} \exp\left[-\int \left(p + \frac{q}{2p}\right) dt\right]$$

$$y_2(t) = [1 + \varepsilon_2(t)]p^{-1/2} \exp\left[\int \left(p + \frac{q}{2p}\right) dt\right],$$

with corresponding expressions for y_1' and y_2'. Here $\epsilon_{1,2}(t) = o(1)$ as $t \to \infty$ and can be estimated in terms of the data. See [138] for details and the analogous oscillatory case as well as some application of the results to Weber's (parabolic cylinder) and Airy's equations. A main difference between these formulas and the more classical ones treated above is the appearance of the term $q/(2p)$ in the exponential instead of being absorbed into $\epsilon_{1,2}(t)$ by incorporating it into the perturbation $r(t)$ defined in (8.21). Also recall that this treatment only requires that f is piecewise $C^{(1)}$. Finally, we remark that the dichotomy and growth conditions are combined as indicated in (8.26), similarly to the coupled conditions discussed in Sect. 2.9.

8.3.3 Olver-Type Error Estimates

In this section, we will prove

Theorem 8.1 *Suppose that $f(t)$ is complex-valued, non-zero, and twice continuously differentiable, and $g(t)$ is complex-valued and continuous for $t \geq t_0$. Suppose that f is written in its principal branch form and let $f^{1/2}$ denote the principal branch of the square root function so that* (8.16) *holds. Suppose $r(t)$ defined in* (8.21) *satisfies*

$$r(t) \in L^1[t_0, \infty), \tag{8.27}$$

and define a function $R_1 : [t_0, \infty) \to \mathbb{R}^+$ *by*

$$R_1(t) := \int_t^\infty |r(s)|\, ds. \tag{8.28}$$

Then (8.11) *has a solution* $y_1(t)$ *on* $[t_0, \infty)$ *satisfying*

$$y_1(t) = f^{-1/4} e^{-\int^t f^{1/2}} [1 + \varepsilon_1(t)], \tag{8.29}$$

where

$$|\varepsilon_1(t)|\,, \left|\frac{\varepsilon_1'(t)}{f^{1/2}(t)}\right| \le e^{2R_1(t)} - 1, \qquad (i = 1, 2). \tag{8.30}$$

Moreover, in the special case that

$$f(t) > 0 \qquad \textit{for all } t \in [t_0, \infty),$$

the error estimate for this recessive solution $y_1(t)$ *can improved to*

$$|\varepsilon_1(t)| \le \exp\left[\int_t^\infty |r(s)| \left(1 - e^{-\int_{t_0}^s 2f^{1/2}\, du}\right) ds\right] - 1, \tag{8.31}$$

and

$$\left|\frac{\varepsilon_1'}{f^{1/2}}\right| \le 2\left[e^{R_1(t)} - 1\right]. \tag{8.32}$$

Remark 8.2 If $f(t) > 0$, then (8.31) is an improvement of Olver's results, whereas (8.32) coincides with his estimates for an infinite interval $[t_0, \infty)$ (cf. [116, Theorem 2.1, p. 193]).

The estimates in (8.30), which we will show to hold for general $f(t) \neq 0$, were already derived by Olver when he considered the case of $f(t) < 0$ in [116, Theorem 2.2.])

Proof Recall that Theorem 2.7 concerns the existence of asymptotic solutions for t sufficiently large, say $t \ge t_1$, where t_1 was selected so that the T-operator used in its proof was a contraction for $t \ge t_1$. To compare with Olver's results and error estimates on the entire interval $[t_0, \infty)$ (or even as in Olver's case to $(-\infty, \infty)$), we will modify the existence proof by introducing a so-called Bielecki norm (see [43, Chap. 1]). We will first show the existence of a solution corresponding to $\lambda_1(t)$ on the entire interval $[t_0, \infty)$ by working in such a suitable weighted Banach space. In a second step, we use a version of Gronwall's inequality to derive a quantitative error bound for the error on $[t_0, \infty)$.

We begin by making in (8.19) the normalization $z = \exp\left[\int^t \lambda_1(s)\,ds\right] u$, where $\lambda_1(t)$ was defined in (8.20). It then follows that

$$\begin{aligned} u' &= \left[\begin{pmatrix} 0 & \\ & \lambda_2(t) - \lambda_1(t) \end{pmatrix} + r(t)\begin{pmatrix} -1 & -1 \\ 1 & 1 \end{pmatrix}\right] u \\ &= \left[\begin{pmatrix} 0 & \\ & 2f^{1/2}(t) \end{pmatrix} + r(t)\begin{pmatrix} -1 & -1 \\ 1 & 1 \end{pmatrix}\right] u. \end{aligned} \tag{8.33}$$

We will show that (8.33) has, for $t \in [t_0, \infty)$, a solution in the form of the integral equation

$$u(t) = e_1 - \int_t^\infty \begin{pmatrix} 1 & \\ & e^{-\int_t^s 2f^{1/2}} \end{pmatrix} r(s) \begin{pmatrix} -1 & -1 \\ 1 & 1 \end{pmatrix} u(s)\,ds. \tag{8.34}$$

For that purpose, we make one final normalization

$$w(t) = u(t) - e_1. \tag{8.35}$$

Then (8.34) having a solution $u(t)$, $\lim_{t\to\infty} u(t) = e_1$ is equivalent to showing that

$$\begin{aligned} w(t) &= -\int_t^\infty \begin{pmatrix} 1 & \\ & e^{-\int_t^s 2f^{1/2}} \end{pmatrix} r(s) \begin{pmatrix} -1 & -1 \\ 1 & 1 \end{pmatrix} [e_1 + w(s)]\,ds \\ &= -\int_t^\infty \begin{pmatrix} -1 \\ e^{-\int_t^s 2f^{1/2}} \end{pmatrix} r(s)\,ds - \int_t^\infty \begin{pmatrix} -1 & -1 \\ e^{-\int_t^s 2f^{1/2}} & e^{-\int_t^s 2f^{1/2}} \end{pmatrix} r(s) w(s)\,ds \\ &=: (Tw)(t) \end{aligned} \tag{8.36}$$

has a solution $w(t)$ such that $\lim_{t\to\infty} w(t) = 0$.

[A]: Existence of a Solution of the Integral Equation (8.34) for $t \geq t_0$

In the following, let $|\cdot|$ denote the infinity vector norm in $\mathbb{C}^2$, that is $|w| = \max_{i=1,2} |w_i|$. Note that the induced matrix norm is the maximum of the "absolute row sum." We will assume w.l.o.g. that the perturbation $r(t)$ is not identical to zero for large t, i.e., we assume that the function $R_1(t)$ defined in (8.28) satisfies $R_1(t) > 0$ for all $t \geq t_0$.

Define the set

$$\mathscr{B}_1 = \left\{ w(t) \in \mathbb{C}^2 : \|w\|_1 = \sup_{t \geq t_0} \frac{|w(t)|}{e^{2R_1(t)} - 1} < \infty. \right\}$$

Then $\mathscr{B}$ is a Banach space. Note that $w \in \mathscr{B}_1$ implies that $w(t) \to 0$ as $t \to \infty$. We also define

$$\mathscr{W}_1 = \{w(t) \in \mathscr{B}_1 : \|w\|_1 \leq 1\},$$

hence $\mathscr{W}_1$ is a closed subset of $\mathscr{B}_i$. With the goal of applying the Banach Fixed Point principle, we claim that the operator T defined in (8.36) maps $\mathscr{W}_1$ into $\mathscr{W}_1$ and is a contraction. To show the former, let $w(t) \in \mathscr{W}_1$, and observe that

$$\begin{aligned} |(Tw)(t)| &\leq \int_t^\infty |r(s)|\, ds + \int_t^\infty 2|r(s)|\, |w(s)|\, ds \\ &\leq R_1(t) + 2\int_t^\infty |r(s)| \left[e^{2R_1(s)} - 1\right] ds \\ &\leq 2\int_t^\infty |r(s)| e^{2R_1(s)}\, ds = e^{2R_1(t)} - 1. \end{aligned}$$

Here we used that (8.16) and the choice of the infinity vector norm implies that

$$\left| e^{-\int_t^s 2f^{1/2}} \right| \leq 1 \quad \text{and} \quad \left\| \begin{pmatrix} -1 & -1 \\ e^{-\int_t^s 2f^{1/2}} & e^{-\int_t^s 2f^{1/2}} \end{pmatrix} \right\| = 2.$$

Therefore

$$\frac{|(Tw)(t)|}{e^{2R_1(t)} - 1} \leq 1,$$

and taking the supremum over all $t \geq t_0$ establishes that $\|(Tw)\|_1 \leq 1$. To establish that T is a contraction, let $w(t)$ and $\tilde{w}(t)$ be elements in $\mathscr{B}$. Then $w(t) - \tilde{w}(t) \in \mathscr{B}$ and it follows for $t \geq t_0$

$$\begin{aligned} |(Tw)(t) - (T\tilde{w})(t)| &= \left| \int_t^\infty \begin{pmatrix} -1 & -1 \\ e^{-\int_t^s 2f^{1/2}} & e^{-\int_t^s 2f^{1/2}} \end{pmatrix} r(s) \left[w(s) - \tilde{w}(s)\right] ds \right| \\ &\leq 2\|w - \tilde{w}\|_1 \int_t^\infty |r(s)| \left[e^{2R_1(s)} - 1\right] ds \\ &= \|w - \tilde{w}\|_1 \left(e^{2R_1(t)} - 1 - 2R_1(t)\right). \end{aligned}$$

Hence

$$\frac{|(Tw)(t) - (T\tilde{w})(t)|}{e^{2R_1(t)} - 1} \leq \|w - \tilde{w}\|_1 \left(1 - \frac{2R_1(t)}{e^{2R_1(t)} - 1}\right) \leq \|w - \tilde{w}\|_1 \left(1 - \frac{2R_1(t_0)}{e^{2R_1(t_0)} - 1}\right).$$

Here we used that $R_1(t)$ is a nonincreasing function on $[t_0,\infty)$ and the fact that the function $1-2x/(e^{2x}-1)$ is strictly increasing for positive values of x. Put

$$\theta_1 = 1-\frac{2R_1(t_0)}{e^{2R(t_0)}-1}.$$

Then $\theta_1 \in (0,1)$ and taking the supremum over all $t \geq t_0$ shows that

$$\|T(w-\tilde{w})\|_1 \leq \theta_1 \|w-\tilde{w}\|_1,$$

so T is a contraction on $\mathscr{W}_1$.

The Banach Fixed Point principle now implies that there exists a unique solution $w(t)$ of (8.36) such that $w(t) \in \mathscr{W}_1$. In particular, $w(t) \to 0$ as $t \to \infty$. By (8.35), (8.34) has a unique solution $u(t) = e_1 + w(t)$. We emphasize that $u(t)$ exists for all $t \geq t_0$. We also know that this solution $u(t)$ is bounded by $|u(t)| \leq |w(t)|+1 \leq e^{2R_1(t)}$, but we will use Gronwall's inequality to provide a sharper estimate.

[B]: Use of Gronwall's Inequality to Derive Bounds for the Error on $[t_0,\infty)$

Using (8.34), it follows that (8.15) has a solution, which we will call $(y_1,y_1')^T$, such that for $t \geq t_0$

$$\begin{pmatrix} y_1 \\ y_1' \end{pmatrix} = \begin{pmatrix} 1 & 1 \\ f^{1/2}[-1-2h] & f^{1/2}[1-2h] \end{pmatrix} f^{-1/4} e^{-\int^t f^{1/2}} \Bigg[e_1 - \int_t^\infty r(s)[u_1(s)+u_2(s)] \begin{pmatrix} -1 \\ e^{-\int_t^s 2f^{1/2}} \end{pmatrix} ds \Bigg]$$

Looking at the first component, (8.11) has a solution y_1 for $t \geq t_0$ having the form (8.29) where

$$\varepsilon_1(t) = \int_t^\infty r(s)[u_1(s)+u_2(s)] \left(1-e^{-\int_t^s 2f^{1/2}\,du}\right) ds. \tag{8.37}$$

To find an upper bound for $|\varepsilon_1(t)|$, we will use Gronwall's inequality to derive an estimate of the term u_1+u_2. To that end, adding the two components in (8.34) yields that

$$|u_1(t)+u_2(t)| \leq 1 + \int_t^\infty |r(s)| \left|1-e^{-\int_t^s 2f^{1/2}\,du}\right| |u_1(s)+u_2(s)|\, ds. \tag{8.38}$$

From (8.16) it follows for $f(t) \neq 0$ and $s \geq t$ that

$$\left|1 - e^{-\int_t^s 2f^{1/2}\,du}\right| \leq 1 + e^{-\int_t^s 2\mathrm{Re} f^{1/2}\,du} \leq 2. \tag{8.39}$$

(This estimate can and will be improved in the "nonoscillatory case" that $f(t) > 0$.) Therefore

$$|u_1(t) + u_2(t)| \leq 1 + \int_t^\infty 2|r(s)|\,|u_1(s) + u_2(s)|\,ds,$$

and from Lemma 8.3 it follows that

$$|u_1(t) + u_2(t)| \leq \exp\left[\int_t^\infty 2|r(s)|\,ds\right] = e^{2R_1(t)}, \tag{8.40}$$

where $R_1(t)$ is defined in (8.28). Now (8.37) and (8.39) yield

$$|\varepsilon_1(t)| \leq \int_t^\infty 2|r(s)|e^{2R_1(s)}\,ds = e^{2R_1(t)} - 1.$$

Moving to the estimate of $\varepsilon_1'/f^{1/2}$, observe that (8.37) implies that

$$\varepsilon_1'(t) = -2f^{1/2}(t) \int_t^\infty r(s)e^{-\int_t^s 2f^{1/2}}[u_1(s) + u_2(s)]\,ds,$$

and therefore, using (8.40),

$$\left|\frac{\varepsilon_1'(t)}{f^{1/2}(t)}\right| \leq \int_t^\infty 2|r(s)|\,|u_1(s) + u_2(s)|\,ds \leq \int_t^\infty 2|r(s)|\,e^{2R_1(s)}\,ds = e^{2R_1(t)} - 1,$$

which establishes (8.30).

In the special case that

$$f(t) > 0 \qquad \text{for all } t \geq t_0,$$

these estimates can be improved as follows: instead of (8.39), the sharper estimate

$$\left|1 - e^{-\int_t^s 2f^{1/2}\,du}\right| = 1 - e^{-\int_t^s 2f^{1/2}\,du} \leq 1 - e^{-\int_{t_0}^s 2f^{1/2}\,du}$$

holds for $s \geq t \geq t_0$. Hence by (8.38)

$$|u_1(t) + u_2(t)| \leq 1 + \int_t^\infty |r(s)|\left[1 - e^{-\int_{t_0}^s 2f^{1/2}\,du}\right]|u_1(s) + u_2(s)|\,ds,$$

and Lemma 8.3 implies that (8.40) can be sharpened to

$$|u_1(t) + u_2(t)| \leq \exp\left[\int_t^\infty |r(s)| \left(1 - e^{-\int_{t_0}^s 2f^{1/2}\,du}\right) ds\right].$$

Proceeding as before, one finds that

$$|\varepsilon_1(t)| \leq \exp\left[\int_t^\infty |r(s)| \left(1 - e^{-\int_{t_0}^s 2f^{1/2}\,du}\right) ds\right] - 1$$

and

$$\begin{aligned}\left|\frac{\varepsilon_1'}{f^{1/2}}\right| &\leq 2\int_t^\infty \left|r(s)e^{-\int_t^s 2f^{1/2}}[u_1(s) + u_2(s)]\right| ds \\ &\leq 2\int_t^\infty |r(s)|\; e^{R_1(s)}\, ds = 2\left[e^{R_1(t)} - 1\right].\end{aligned}$$

□

Noting that Theorem 8.1 is concerned with the solution $y_1(t)$ of (8.11) which is regressive at $+\infty$, a corresponding result for the solution that is regressive at $-\infty$ can be established by working on the interval $(-\infty, t_1]$.

We end this section with a review of the two Gronwall inequalities used above:

Lemma 8.3 *Let φ and χ: $[t_0, \infty) \to \mathbb{R}$ be continuous functions and $\chi(t) \geq 0$ for all $t \geq t_0$. Assume that*

$$\chi(t) \in L^1[t_0, \infty).$$

(a) If

$$\varphi(t) \leq 1 + \int_{t_0}^t \chi(s)\varphi(s)\, ds,$$

then

$$\varphi(t) \leq \exp\left[\int_{t_0}^t \chi(s)\, ds\right]. \tag{8.41}$$

(b) If

$$\varphi(t) \leq 1 + \int_t^\infty \chi(s)\varphi(s)\, ds,$$

then

$$\varphi(t) \leq \exp\left[\int_t^\infty \chi(s)\,ds\right]. \tag{8.42}$$

Proof

(a) Put $R(t) = \int_{t_0}^t \chi(s)\varphi(s)\,ds$. Then $R' - \chi R = \chi(\varphi - R) \leq \chi$. Thus

$$\left(Re^{-\int_{t_0}^t \chi(s)\,ds}\right)' \leq \chi(t)e^{-\int_{t_0}^t \chi(s)\,ds},$$

and integration over $[t_0, t]$ establishes (8.41).

(b) Similarly, let $\hat{R}(t) = \int_t^\infty \chi(s)\varphi(s)\,ds$, thus

$$-(\hat{R}' + \chi\hat{R}) = \chi(t)[\varphi(t) - \hat{R}(t)] \leq \chi(t).$$

Hence

$$-\left(\hat{R}(t)e^{\int_{t_0}^t \chi(u)\,du}\right)' \leq \chi(t)e^{\int_{t_0}^t \chi(u)\,du},$$

and integration over $[t, \infty)$ establishes (8.42).

□

8.4 Asymptotic Comparison Results for Second-Order Equations

While most of the asymptotic integration results discussed in this chapter involve representations using solutions of first-order equations, here we will modify that approach. We will consider instead a pair of second-order equations and ask when solutions of one (perturbed) equation can be asymptotically represented using solutions of another (normalized or unperturbed) equation. The main results concern quantifying the size of permissible permutations in terms of the behavior of solutions of the unperturbed equation. We begin with a rather general result of this type due to Trench in Sect. 8.4.1. Then in Sect. 8.4.2 we discuss some modifications and improvements in special cases when the unperturbed equation is nonoscillatory.

8.4.1 A General Comparison Result

In [151], Trench considered second-order equations

$$y'' = [f(t) + g(t)]y, \qquad t \geq t_0, \tag{8.43}$$

as perturbations of

$$x'' = f(t)x, \qquad t \geq t_0. \tag{8.44}$$

Theorem 8.4 (Trench) *Let x_1 and x_2 be two linearly independent solutions of (8.44). Assume that $g(t)$ is continuous on $[t_0, \infty)$ and*

$$g(t)M(t) \in L^1[t_0, \infty), \qquad \text{where } M(t) = \max\{|x_1(t)|^2, |x_2(t)|^2\}. \tag{8.45}$$

Then, for arbitrary complex-valued constants a and b, there exists a solution $y(t)$ of (8.43) of the form

$$y(t) = \alpha(t)x_1(t) + \beta(t)x_2(t), \tag{8.46}$$

with $\lim_{t\to\infty} \alpha(t) = a$ and $\lim_{t\to\infty} \beta(t) = b$.

While Trench used substitution and Gronwall's inequality in his proof, this result can also be derived from one preliminary transformation and Levinson's fundamental theorem as follows.

Proof Rewriting (8.43) in matrix form,

$$\mathbf{y}' = \begin{pmatrix} 0 & 1 \\ f(t) + g(t) & 0 \end{pmatrix} \mathbf{y}, \qquad t \geq t_0, \tag{8.47}$$

we make for $t \geq t_0$ the transformation

$$\mathbf{y}(t) = \begin{pmatrix} x_1(t) & x_2(t) \\ x_1'(t) & x_2'(t) \end{pmatrix} \hat{y}(t) = X(t)\hat{y}(t).$$

Letting

$$A(t) = \begin{pmatrix} 0 & 1 \\ f(t) & 0 \end{pmatrix},$$

it follows that $X^{-1}(t)A(t)X(t) - X^{-1}(t)X'(t) = 0$, so

$$X^{-1}(t) \begin{pmatrix} 0 & 1 \\ f(t) + g(t) & 0 \end{pmatrix} X(t) - X^{-1}(t)X'(t) = X^{-1}(t) \begin{pmatrix} 0 & 0 \\ g(t) & 0 \end{pmatrix} X(t).$$

Also observing that $\det X(t) = \det X(t_0)$ is constant, it follows that

$$\hat{y}' = \frac{g(t)}{\det X(t_0)} \begin{pmatrix} -x_1x_2 & -x_2^2 \\ x_1^2 & x_1x_2 \end{pmatrix} \hat{y}. \tag{8.48}$$

Under the assumption (8.45), (8.48) can be considered as an L^1-perturbation of the trivial system $w' = 0$, noting that $|x_1(t)x_2(t)| \leq M(t)$. Hence Theorem 2.7 implies that (8.48) has a fundamental system $\hat{Y}(t) = I + \mathrm{o}(1)$. Hence (8.47) has a fundamental matrix

$$Y(t) = \begin{pmatrix} y_1 & y_2 \\ y_1' & y_2' \end{pmatrix} = \begin{pmatrix} x_1 & x_2 \\ x_1' & x_2' \end{pmatrix} [I + \mathrm{o}(1)] , \qquad \text{as } t \to \infty.$$

Given arbitrary complex-valued constants a and b, the function $y(t)$ defined by $y(t) = ay_1(t) + by_2(t)$ is a solution of (8.43) satisfying (8.46). □

Remark 8.5 Observe that due to the very strong assumption (8.45) on the perturbation, no extra dichotomy condition is required and, moreover, $f(t)$ and $g(t)$ are allowed to be complex-valued. If one would also assume that $x_i(t) \neq 0$ for all t sufficiently large ($i = 1, 2$), then (8.45) can be somewhat weakened at the expense of adding an extra dichotomy condition. This follows by applying to (8.48) the shearing transformation $\hat{y} = \mathrm{diag}\,\{1, x_1(t)/x_2(t)\}\tilde{y}$, which leads to the system

$$\tilde{y}' = \left[\begin{pmatrix} 0 & \\ & \frac{x_2'}{x_2} - \frac{x_1'}{x_1} \end{pmatrix} + \frac{x_1 x_2 g}{\det X(t_0)} \begin{pmatrix} -1 & -1 \\ 1 & 1 \end{pmatrix} \right] \tilde{y} = [\Lambda(t) + R(t)]\tilde{y}.$$

If $z' = \Lambda(t)z$ would be known to satisfy Levinson's dichotomy conditions (2.13), (2.14), then the requirement $R \in L^1$, which is weaker than (8.45), would suffice for an asymptotic integration result. Observe that the dichotomy condition amounts to assuming that Re $(x_2(t)/x_1(t))$ does not tend to both $+\infty$ and $-\infty$ as $t \to +\infty$.

In the next subsection we will consider comparison results for nonoscillatory equations. For the purpose of conforming to the standard notation used in the literature, we will treat equations in a more general (self-adjoint) form and also switch the roles of x and y so that solutions y_1 and y_2 become the known or given ones. This should be kept in mind when comparing assumption (8.45) with the ones described in Sect. 8.4.2, also observing that g should be replaced by $g - f$.

8.4.2 Nonoscillatory Second-Order Differential Equations

Here we consider a pair of second-order linear differential equations in the (self-adjoint) form

$$[r(t)x']' + f(t)x = 0, \qquad t \geq t_0, \tag{8.49}$$

and

$$[r(t)y']' + g(t)y = 0, \qquad t \geq t_0. \tag{8.50}$$

Here r, f, and g are continuous for $t \geq t_0$ with $r(t) > 0$ and g is real-valued. Through the rest of this section we make the basic assumption that (8.50) is *nonoscillatory* on $[t_0, \infty)$, i.e., every nontrivial solution has at most a finite number of zeros on $[t_0, \infty)$ (see, e.g., [72, p. 351]). It is known [72, p. 355] that since (8.50) is nonoscillatory, it has solutions y_i which satisfy the following:

Assumption 8.6 *There exist solutions y_1, y_2 of* (8.50) *which are eventually positive and satisfy*

$$r(t)[y_1(t)y_2'(t) - y_1'(t)y_2(t)] \equiv 1, \tag{8.51}$$

and such that

$$\rho(t) = \frac{y_2(t)}{y_1(t)} \tag{8.52}$$

is strictly increasing and tending to infinity as $t \to +\infty$.

The goal is to find sufficient conditions on these solutions y_1, y_2, and on $f - g$ so that there exist solutions x_1, x_2 of (8.49) which exhibit the following asymptotic behavior: There exist scalar-valued functions $\gamma_i(t)$ and $\delta_i(t)$ such that for $i = 1, 2$, such that

$$\left.\begin{array}{c} x_i = [1 + \gamma_i(t)]y_i, \\[1em] \left(\dfrac{x_i}{y_i}\right)' = \delta_i(t)\left(\dfrac{\rho'}{\rho}\right), \\[1em] \gamma_i(t), \delta_i(t) \to 0 \quad \text{as} \quad t \to \infty, \end{array}\right\},$$

and, if possible, with error estimates for γ_i and δ_i.

Hartman–Wintner [72, p. 379], Trench [153, 158], Šimša [137], Chen [35], and Chernyavskaya–Shuster [36] considered this problem under various assumptions and using corresponding ad hoc methods. These methods involve for the most part a scalar approach in which the asymptotic behavior of a recessive or subdominate solution is obtained first and then utilizing the relation (8.51) or reduction of order for a dominant solution. In [19], we have presented a more unified, systems-based approach for such results which consists of making certain preliminary transformations, followed by special conditioning transformations (linked to the particular assumptions), and finally applying Levinson's fundamental theorem (with error estimates). This has led to improvements in several cases. In what follows we will summarize some of these results by identifying which conditioning transformations correspond to various assumptions and refer the reader to [19] for complete details and error estimates.

Recalling that r is only assumed to be continuous, but not necessarily continuously differentiable, we re-write (8.49) and (8.50) as first order systems in the form

$$\xi' = \begin{pmatrix} x \\ rx' \end{pmatrix}' = \begin{pmatrix} 0 & 1/r \\ -f & 0 \end{pmatrix} \xi \tag{8.53}$$

and

$$\eta' = \begin{pmatrix} y \\ ry' \end{pmatrix}' = \begin{pmatrix} 0 & 1/r \\ -g & 0 \end{pmatrix} \eta. \tag{8.54}$$

The preliminary transformation

$$\xi = \begin{pmatrix} y_1 & y_2 \\ ry_1' & ry_2' \end{pmatrix} \tilde{x} \tag{8.55}$$

takes (8.53) into [cf. (8.48)]

$$\tilde{x}' = (g-f) \begin{pmatrix} -y_1y_2 & -y_2^2 \\ y_1^2 & y_1y_2 \end{pmatrix} \tilde{x}.$$

As in Remark 8.5, the diagonal transformation

$$\tilde{x} = \begin{pmatrix} 1 & 0 \\ 0 & \frac{1}{\rho} \end{pmatrix} \hat{x} = \begin{pmatrix} 1 & 0 \\ 0 & \frac{y_1}{y_2} \end{pmatrix} \hat{x}, \tag{8.56}$$

leads to the system

$$\hat{x}' = \left\{ \frac{\rho'}{\rho} \begin{pmatrix} 0 & 0 \\ 0 & 1 \end{pmatrix} + (g-f)y_1y_2 \begin{pmatrix} -1 & -1 \\ 1 & 1 \end{pmatrix} \right\} \hat{x} =: \left[\Lambda(t) + \tilde{V}(t)\right] \hat{x}. \tag{8.57}$$

For the rest of the section, we will use the notation

$$\Lambda(t) = \begin{pmatrix} \lambda_1(t) & \\ & \lambda_2(t) \end{pmatrix} = \frac{\rho'}{\rho} \begin{pmatrix} 0 & 0 \\ 0 & 1 \end{pmatrix}. \tag{8.58}$$

Since $\Lambda(t)$ satisfies Levinson's dichotomy conditions (2.13), (2.14), Theorem 2.7 can applied immediately if $\tilde{V}$ is an L^1-perturbation. If, however, $(g-f)y_1y_2 \notin L^1$, then we will use various conditioning transformations of the form

$$\hat{x} = [I + Q(t)]u, \qquad Q(t) \to 0 \text{ as } t \to \infty, \tag{8.59}$$

with the goal of reducing the magnitude of the perturbation.

Because of the algebraic structure of $\tilde{V}$ in (8.57), we will employ conditioning transformations of the form

$$I + Q(t) = \begin{pmatrix} 1-q & -q \\ q & 1+q \end{pmatrix}. \tag{8.60}$$

Note that $(I+Q)^{-1} = I - Q$, which exists whether or not $q \to 0$. Then we obtain from (8.57)

$$\begin{aligned} u' &= \left\{\Lambda + [I+Q]^{-1}\left(\Lambda Q - Q\Lambda + \tilde{V}(I+Q) - Q'\right)\right\} u \\ &= \left\{\Lambda + q\frac{\rho'}{\rho}\begin{pmatrix} 0 & 1 \\ 1 & 0 \end{pmatrix} + \left[(g-f)y_1y_2 - q' - q^2\frac{\rho'}{\rho}\right]\begin{pmatrix} -1 & -1 \\ 1 & 1 \end{pmatrix}\right\} u. \end{aligned} \tag{8.61}$$

One approach is to determine q such that $(g-f)y_1y_2 - q' - q^2\frac{\rho'}{\rho}$ is "small" in a suitable sense, ideally absolutely integrable, and then reduce the size of the perturbation $q\frac{\rho'}{\rho}\begin{pmatrix} 0 & 1 \\ 1 & 0 \end{pmatrix}$ with the help of another conditioning transformation.

8.4.3 Special Cases

We now consider results from in [35, 72, 137, 153, 158] by choosing appropriate scalar functions q in (8.60). Frequently, a second conditioning transformation will be necessary to reduce the system.

1. Hartman and Wintner [72, XI, Theorem 9.1]) derived the following result with just o(1) estimates for $\gamma_i(t), \delta_i(t)$ in (8.4.2). Error estimates were provided in [19, Thm. 2].

   Theorem 8.7 *Suppose that Assumption 8.6 holds. Assume that*

   $$G(t) := \int_t^\infty (f-g)y_1^2\, ds \tag{8.62}$$

   is well-defined (i.e., conditionally integrable) and

   $$\int^\infty \rho'\Gamma\, dt < \infty \tag{8.63}$$

is satisfied, where

$$\Gamma(t) = \sup_{s \geq t} |G(s)| \,. \tag{8.64}$$

Then (8.49) *has solutions* x_i $(i = 1, 2)$ *satisfying* (8.4.2)*, and bounds of the error terms* $\gamma_i(t)$ *and* $\delta_i(t)$ *are given in [19, Thm. 2].*

The main idea of the proof is to make in (8.60) the choice

$$q = \rho G,$$

and it can be shown that (8.61) takes on the form

$$u' = \left\{ \frac{\rho'}{\rho} \begin{pmatrix} 0 & 0 \\ 0 & 1 \end{pmatrix} + \rho' G \begin{pmatrix} 1 & 2 \\ 0 & -1 \end{pmatrix} + \rho' \rho G^2 \begin{pmatrix} 1 & 1 \\ -1 & -1 \end{pmatrix} \right\} u \tag{8.65}$$

$$= \{ \quad \Lambda(t) \quad + \quad V(t) \quad + \quad R(t) \quad \} u. \tag{8.66}$$

Using (8.63) and (8.64), one can show that $V + R = O(\rho' |G|) \in L^1[t_0, \infty)$. Hence Levinson's fundamental theorem can be applied and, with more work, upper bounds for the error terms can be determined. For details, see [19, Thm. 2].

2. In the following theorem, due to Trench [152], the conditions on the size of the perturbation are relaxed. In particular, the existence of G_1 in (8.67) and hypothesis (8.69) below can be shown to follow from the assumptions of Theorem 8.7. This relaxation of the hypotheses leads to the situation that the matrix V in (8.66) is not necessarily absolutely integrable. Therefore, Levinson's fundamental theorem is not applicable at this point, and one reduces the size of the perturbations in (8.66) with the help of another conditioning transformation.

Theorem 8.8 (Trench 1986 [152]) *Suppose that Assumption 8.6 holds. Assume that*

$$G_1(t) := \int_t^\infty y_1 y_2 (f - g) \, ds \tag{8.67}$$

converges conditionally for $t \geq t_0$. *Define*

$$\Phi(t) = \sup_{s \geq t} |G_1(s)|. \tag{8.68}$$

Then it follows that $G(t)$ *as defined in* (8.62) *exists, and we further assume that*

$$\int^\infty \rho' |G| \Phi \, dt < \infty. \tag{8.69}$$

Then (8.49) *has solutions* x_i $(i = 1, 2)$ *satisfying* (8.4.2)*, and bounds of the error terms* $\gamma_i(t)$ *and* $\delta_i(t)$ *can be found in [153] or [19, Thm. 3].*

To outline the proof, integration by parts shows that the existence of G_1 in (8.67) implies the existence of G defined in (8.62) and, moreover, one can show that

$$|G| \leq 2\Phi/\rho. \tag{8.70}$$

In (8.60), we again put $q = \rho G$, thus (8.60) leads as before to (8.66). Note that the perturbation matrix R in (8.66) is absolutely integrable by (8.69) and (8.70) since

$$R = \mathrm{O}\left(\rho'\rho G^2\right) = \mathrm{O}(\rho'|G|\Phi).$$

However, the perturbation matrix V in (8.66) is not necessarily absolutely integrable, and we reduce the size of this perturbation by making a second conditioning transformation of the form

$$u = [I + \hat{Q}]\hat{u} = \begin{pmatrix} 1 + \hat{q}_{11} & \hat{q}_{12} \\ 0 & 1 + \hat{q}_{22} \end{pmatrix} \hat{u}, \tag{8.71}$$

with $\hat{q}_{ij}$ to be determined in such a way that $I + \hat{Q}$ is invertible for all t. In the proof of [19, Thm. 3], precise formulas for the entries of $\hat{Q}$ are derived such that conditioning transformation (8.71) takes (8.66) into

$$\hat{u}' = \left\{\Lambda(t) + \hat{R}(t)\right\} \tilde{u}, \qquad \hat{R}(t) = \mathrm{O}\left(\rho'|G|\Phi\right),$$

hence $\hat{R} \in L^1[t_0\infty)$ by (8.69). Levinson's fundamental theorem can now be applied and, in addition, one can determine upper bounds for the error terms by estimating the associated T-operator.

We note that Bohner and Stević generalized Trench's result to dynamic equations on time scales in [27].

3. The next contribution is due to Šimša [137, Theorem 1], but sharper error estimates can be found in [19, Thm. 5]. As in Šimša's exposition, only the asymptotic behavior of the recessive solution x_1 is considered (although the system's approach taken here also yields immediate results for a dominant solution x_2).

Theorem 8.9 (Šimša) *Suppose that Assumption 8.6 holds. Assume that* $G_1(t)$ *as defined in* (8.67) *converges conditionally and let* Φ *be defined as in* (8.68)*. If*

$$\int^\infty |G_1|\frac{\rho'}{\rho}\Phi\, dt < \infty, \tag{8.72}$$

then (8.49) *has a solution* x_1 *satisfying* (8.4.2), *and bounds of the error terms* $\gamma_1(t)$ *and* $\delta_1(t)$ *are given in [19, Thm. 5].*

To identify the major steps in the proof, we note that since Šimša's assumptions involve G_1, it is natural to use

$$q = G_1$$

in (8.60). With this choice (8.61) becomes

$$u' = \left[\frac{\rho'}{\rho}\begin{pmatrix} 0 & 0 \\ 0 & 1 \end{pmatrix} + \underbrace{G_1 \frac{\rho'}{\rho}\begin{pmatrix} 0 & 1 \\ 1 & 0 \end{pmatrix}}_{\hat{V}(t)} + \underbrace{G_1^2 \frac{\rho'}{\rho}\begin{pmatrix} 1 & 1 \\ -1 & -1 \end{pmatrix}}_{\hat{R}(t)} \right] u.$$

Note that $\hat{R} \in L^1[t_0, \infty)$ by (8.68) and (8.72), but one needs to modify the elements in $\hat{V}(t)$ with the help of another conditioning transformation. To that end, put

$$u = [I + \tilde{Q}]\tilde{u} = \begin{pmatrix} 1 + \tilde{q}_{11} & \tilde{q}_{12} \\ \tilde{q}_{21} & 1 + \tilde{q}_{22} \end{pmatrix} \tilde{u}, \tag{8.73}$$

with $\tilde{q}_{ij}$ to be determined $(i, j = 1, 2)$. If $I + \tilde{Q}(t)$ is invertible (which is eventually verified), then

$$\tilde{u}' = \left\{ \Lambda + (I + \tilde{Q})^{-1} \left[\Lambda\tilde{Q} - \tilde{Q}\Lambda + V(I + \tilde{Q}) - \tilde{Q}' + R(I + \tilde{Q}) \right] \right\} \tilde{u}, \tag{8.74}$$

and the goal is to find $\tilde{q}_{ij}(t)$ that make the term

$$\Lambda\tilde{Q} - \tilde{Q}\Lambda + V(I + \tilde{Q}) - \tilde{Q}' \tag{8.75}$$

"small." Starting with the off-diagonal elements, this first leads to linear differential equations in $\tilde{q}_{ij}$ for $i \neq j$, which can be solved precisely (and whose solutions can be shown to vanish at infinity). Focusing next on the diagonal elements, one finds that choosing $\tilde{q}_{11} \equiv 0$ and $\tilde{q}_{22} = \tilde{q}_{12}\tilde{q}_{21}$ allows to re-write (8.74) as

$$\tilde{u}' = \left\{ \Lambda(t) + \tilde{R}(t) \right\} \tilde{u}, \qquad \tilde{R}(t) = O\left(|G_1| \frac{\rho'}{\rho} \Phi \right),$$

hence $\tilde{R} \in L^1[t_0, \infty)$ by (8.72). Levinson's fundamental theorem can now be applied and, in addition, one can determine upper bounds for the error terms.

4. The next result is in essence due to Chen [35, Theorem 3.1], but the system's approach taken here (see [19, Theorem 6 and Corollary 2]) allows to drop Chen's

assumption of $f(t)$ in (8.49) and hence also $G(t)$ in (8.62) being real-valued. We now give such a generalized version.

Theorem 8.10 (Chen) *Suppose that Assumption 8.6 holds. Suppose that $G(t)$ exists as defined in* (8.62) *and that $\int_t^\infty \rho' G < \infty$. If*

$$\int_{t_0}^{\infty} \rho' \rho \, |G|^2 < \infty, \tag{8.76}$$

and $\rho G \to 0$ as $t \to \infty$, then (8.49) *has solutions x_i, $(i = 1, 2)$, satisfying* (8.4.2)*, and bounds of the error terms $\gamma_i(t)$ and $\delta_i(t)$ can be found in [35] or [19, Thm. 6].*

As an indication of the proof, we again choose $q = \rho G$ in (8.60), hence (8.60) leads as before to (8.66). While the perturbation R in (8.66) is absolutely integrable by (8.76), we need again to condition the matrix $V(t)$. Here it is possible to remove the perturbation V completely, i.e., to find matrix $\breve{Q}$ such that $I + \breve{Q}(t)$ is invertible for large t and satisfies [cf. (8.74)]

$$\Lambda \breve{Q} - \breve{Q} \Lambda + V[I + \breve{Q}] - \breve{Q}' = 0.$$

To this end, we make in (8.66) the substitution

$$u = [I + \breve{Q}(t)]z = \begin{pmatrix} 1 + \breve{q}_{11} & \breve{q}_{12} \\ 0 & 1 + \breve{q}_{22} \end{pmatrix} \breve{u}. \tag{8.77}$$

It can be shown that such $\breve{q}_{ij}$ exist as appropriately chosen solutions of differential equations and that $I + \breve{Q}$ is invertible for all $t \geq t_0$. Then (8.66) and (8.77) imply that

$$\breve{u}' = \left\{\Lambda + \breve{R}\right\} \breve{u}, \qquad \breve{R} = (I + \breve{Q})^{-1} R (I + \breve{Q}) = O\left(\rho' \rho |G|^2\right),$$

hence $\breve{R} \in L^1[t_0, \infty)$ by (8.76). Levinson's fundamental theorem can now be applied and, in addition, one can determine upper bounds for the error terms.

5. Finally, Trench considered another type of perturbation condition in [158] . A different proof and improved error estimates were given in [19, Thm. 7].

Theorem 8.11 (Trench 2003 [158]) *Suppose that Assumption 8.6 holds. Assume that*

$$I(t) := \int_t^\infty (f - g) y_2^2 \, ds \tag{8.78}$$

converges conditionally for $t \geq t_0$ and put

$$\sigma(t) = \sup_{s \geq t} |I(s)|.$$

Then (8.49) *has solutions x_i ($i = 1, 2$) satisfying* (8.4.2)*, and bounds of the error terms $\gamma_i(t)$ and $\delta_i(t)$ can be found in [158] or [19, Thm. 7].*

To indicate the main arguments, one can first show that the existence of I in (8.78) implies that G and G_1 as defined in (8.62) and (8.67), respectively, are well-defined and satisfy certain estimates involving σ and ρ.

In (8.60), we once more choose $q = \rho G$ in (8.60), hence (8.60) leads as before to (8.66), where we make a conditioning transformation of the form

$$u = [I + Q_0]\xi = \begin{pmatrix} e^{-H} & q_0 \\ 0 & e^H \end{pmatrix} \xi,$$

where $H = -\rho G + G_1$ and q_0 will satisfy a certain linear differential equation. This choice of H will make the diagonal terms in (8.75) vanish, and then determining q_0 as an appropriately chosen solution of a suitable differential equation will reduce the size of the off-diagonal term.

This is shown to lead to

$$\xi' = [\Lambda(t) + R_0(t)]\, \xi,$$

where $R_0 = O\left(\frac{\rho' \sigma^2}{\rho^3}\right)$, hence $R_0 = O\left(\frac{\rho'}{\rho^3}\right)$ and thus $R_0 \in L^1[t_0, \infty)$. Once again, Levinson's fundamental theorem can now be applied and, in addition, one can determine upper bounds for the error terms by estimating the T-operator involved.

We want to mention that Theorem 8.11 was generalized to dynamic equations on time scales by Bohner and Stević in [28].

We end this section by briefly mentioning two examples, referring to [19] for details. A classical example studied by Trench, Šimša, and Chen [35, 137, 153] is

$$x'' + K\left[\frac{\sin t}{t(\ln t)^\alpha}\right] x = 0, \tag{8.79}$$

where $K \neq 0$ and $\alpha > 0$ are constants, as a perturbation of $y'' = 0$. Here one can choose $y_1 = 1$, $y_2 = \rho(t) = t$. Focusing on solutions x_i and not their derivatives for the sake of brevity, one is interested in establishing the existence of solutions satisfying

$$x_1 = 1 + \gamma_1(t) \qquad \text{and } x_2 = [1 + \gamma_2(t)]\, t,$$

and deriving bounds on the error terms γ_i. It follows from (8.62), (8.67), and integration by parts that

$$G(t) = \int_t^\infty \frac{K \sin s}{s(\ln s)^\alpha}\, ds = \left[K \cos t + \mathrm{O}\left(\frac{1}{t}\right)\right] \frac{1}{t(\ln t)^\alpha}$$

$$G_1(t) = \left[K \cos t + \mathrm{O}\left(\frac{1}{t}\right)\right] \frac{1}{(\ln t)^\alpha}.$$

For $\alpha > 1$, Theorem 8.7 together with the error estimates derived in [19, Thm. 2] imply that (8.49) has solutions x_i satisfying

$$x_1 = 1 + \mathrm{O}\left(\frac{1}{(\ln t)^{\alpha-1}}\right) \qquad \text{and} \qquad x_2 = \left[1 + \mathrm{O}\left(\frac{1}{(\ln t)^{\alpha-1}}\right)\right] t.$$

Theorem 8.8 can be shown to hold for $\alpha > 1/2$ and implies that (8.49) has solutions x_i such that

$$x_1 = 1 + 2K \frac{\cos t}{(\ln t)^\alpha} + \mathrm{O}\left(\frac{1}{(\ln t)^{2\alpha-1}}\right),$$

$$x_2 = \left[1 + \mathrm{O}\left(\frac{1}{(\ln t)^\alpha}\right) + \mathrm{O}\left(\frac{1}{(\ln t)^{2\alpha-1}}\right)\right] t.$$

In the special case that $1/2 < \alpha < 1$, this reduces to

$$x_1 = 1 + \mathrm{O}\left(\frac{1}{(\ln t)^{2\alpha-1}}\right) \qquad \text{and} \qquad x_2 = \left[1 + \mathrm{O}\left(\frac{1}{(\ln t)^{2\alpha-1}}\right)\right] t.$$

Chen's result also applies if $0 < \alpha < 1/2$ and we refer to [35] for details.

8.5 Asymptotic Factorization of a Differential Operator

Motivated by results for second-order difference operators (see Sect. 9.5), we consider here linear scalar differential operators

$$L = D^2 + a(t)D + b(t), \qquad t \geq t_0, \tag{8.80}$$

where $D = d/dt$, acting on $C^{(2)}[t_0, \infty)$. It is easy to show that L can be factored as

$$L = [D - \alpha(t)]\,[D - \beta(t)], \tag{8.81}$$

(which means $Ly = [D+\alpha(t)]\,[D+\beta(t)]y$ for all $y \in C^2[t_0,\infty)$) if and only if $\alpha(t)$, $\beta(t)$ satisfy for $t \geq t_0$ the nonlinear system of equations

$$\begin{cases} a(t) = -\alpha(t) - \beta(t) \\ b(t) = \alpha(t)\beta(t) - \beta'(t). \end{cases} \tag{8.82}$$

On one hand, being able to factor L as (8.81) leads directly to solutions of $Ly = 0$. One can first solve, for example, $[D - \alpha(t)]w = 0$ and then solve $w = [D - \beta(t)]y$ to obtain a pair of linearly independent solutions. On the other hand, if we know a solution y_1 of $Ly = 0$ which is nonzero for all $t \geq t_0$, then letting $\beta = y_1'/y_1$ and $\alpha = -a - y_1'/y_1$ leads directly to the factorization (8.81). So in this sense finding an exact factorization of L is equivalent to solving $Ly = 0$.

Here we want to instead consider what could be called an "asymptotic factorization." As we shall see this means "finding" such functions $\alpha(t)$, $\beta(t)$ modulo L^1-perturbations. These have the effect of determining solutions only up to factors that are asymptotically constant, i.e., up to [c +o(1)], $c \neq 0$. We will consider here only a special case of (8.80), namely when the coefficients are asymptotically constant and of bounded variation on $[t_0,\infty)$, i.e.,

$$a(t) = a_0 + r_1(t), \qquad b(t) = b_0 + r_2(t), \tag{8.83}$$

where a_0 and b_0 are constant, $r_i'(t) \in L^1$ and $r_i(t) \to 0$ as $t \to \infty$ for $i = 1, 2$.

The limiting operator $L_0 = D^2 + a_0 D + b_0$ can be explicitly factored as

$$L_0 = (D - \alpha_0)(D - \beta_0),$$

where α_0, β_0 are zeros of the limiting characteristic polynomial

$$p_0(\lambda) = \lambda^2 + a_0\lambda + b_0 = \lambda^2 - (\alpha_0 + \beta_0)\lambda + \alpha_0\beta_0.$$

If $\lambda_1(t)$, $\lambda_2(t)$ are zeros of

$$p(\lambda) = \lambda^2 + [a_0 + r_1(t)]\lambda + [b_0 + r_2(t)], \tag{8.84}$$

then it is easy to construct examples showing that

$$L \neq [D - \lambda_1(t)]\,[D - \lambda_2(t)], \tag{8.85}$$

in general. However, we now want to show that the identity $L = [D - \zeta_1(t)][D - \zeta_2(t)]$ does hold for functions $\zeta_i(t)$ satisfying $\zeta_i(t) = \lambda_i(t) + (L^1)$ $(i = 1, 2)$, provided we also assume that $\operatorname{Re}\alpha_0 \neq \operatorname{Re}\beta_0$.

Theorem 8.12 *Assume* (8.83), $r_i'(t) \in L^1$, $r_i(t) \to 0$ *as* $t \to \infty$ *for* $i = 1, 2$, *and* $\operatorname{Re}\alpha_0 \neq \operatorname{Re}\beta_0$. *Then there exist functions* $\zeta_1(t)$, $\zeta_2(t) \in C^1$ *such that the operator* L *given in* (8.80) *can be factored as*

$$L = [D - \zeta_1(t)]\,[D - \zeta_2(t)], \tag{8.86}$$

with

$$\zeta_i(t) - \lambda_i(t) \in L^1 \qquad (i = 1, 2). \tag{8.87}$$

Here $\lambda_i(t)$ *are the zeros of* (8.84).

Proof Assume without loss of generality that $\operatorname{Re}\alpha_0 > \operatorname{Re}\beta_0$. Let $\lambda_1(t)$ and $\lambda_2(t)$ be the zeros of (8.84) such that $\lambda_1(t) \to \alpha_0$ and $\lambda_2(t) \to \beta_0$ as $t \to \infty$.

Observe that $Ly = 0$ is equivalent to

$$\mathbf{y}' = \begin{pmatrix} 0 & 1 \\ -[b_0 + r_2(t)] & -[a_0 + r_1(t)] \end{pmatrix} \mathbf{y}, \qquad \mathbf{y} = \begin{pmatrix} y \\ y' \end{pmatrix}. \tag{8.88}$$

Let

$$\mathbf{y} = \begin{pmatrix} 1 & 1 \\ \lambda_1(t) & \lambda_2(t) \end{pmatrix} \hat{y}.$$

Then

$$\begin{aligned} \hat{y}' &= \left[\begin{pmatrix} \lambda_1(t) & \\ & \lambda_2(t) \end{pmatrix} + \frac{1}{\lambda_2(t) - \lambda_1(t)} \begin{pmatrix} \lambda_1'(t) & \lambda_2'(t) \\ -\lambda_1'(t) & -\lambda_2'(t) \end{pmatrix} \right] \hat{y} \\ &=: [\Lambda(t) + R(t)]\hat{y}. \end{aligned} \tag{8.89}$$

Now $r_i' \in L^1$ and $\lambda_2(t) - \lambda_1(t) \to \beta_0 - \alpha_0 \neq 0$ imply that $\lambda_i'(t) \in L^1$ $(i = 1, 2)$ and

$$R(t) \in L^1. \tag{8.90}$$

Moreover, since $\operatorname{Re}[\lambda_1(t) - \lambda_2(t)] > 0$ for all t sufficiently large, $\Lambda(t)$ satisfies an exponential and hence ordinary dichotomy condition (see Definition 2.6), and Theorem 2.7 implies that there exists a fundamental solution matrix $Y(t)$ of (8.88) of the form

$$Y(t) = \begin{pmatrix} y_1 & y_2 \\ y_1' & y_2' \end{pmatrix} = \begin{pmatrix} 1 & 1 \\ \lambda_1(t) & \lambda_2(t) \end{pmatrix} \begin{pmatrix} 1 + \varepsilon_{11}(t) & \varepsilon_{12}(t) \\ \varepsilon_{21}(t) & 1 + \varepsilon_{22}(t) \end{pmatrix} e^{\int^t \Lambda(\tau)\, d\tau}, \tag{8.91}$$

with $\varepsilon_{ij}(t) \to 0$ as $t \to \infty$. Therefore $Ly = 0$ has a solution $y_1(t)$ satisfying

$$y_1(t) = [1 + \varepsilon_{11}(t) + \varepsilon_{21}(t)] \exp\left[\int^t \lambda_1(\tau)\, d\tau\right],$$

and

$$y_1'(t) = [\lambda_1(t)(1 + \varepsilon_{11}(t)) + \lambda_2(t)\varepsilon_{21}(t)] \exp\left[\int^t \lambda_1(\tau)\, d\tau\right].$$

Note that $y_1(t) \neq 0$ for all t sufficiently large, say $t \geq t_1$ for some $t_1 \geq t_0$. Defining now for $t \geq t_1$

$$\zeta_1(t) = \frac{y_1'(t)}{y_1(t)} = \frac{\lambda_1(t)[1 + \varepsilon_{11}(t)] + \lambda_2(t)\varepsilon_{21}(t)}{1 + \varepsilon_{11}(t) + \varepsilon_{21}(t)},$$

we see that

$$\begin{aligned} \zeta_1(t) - \lambda_1(t) &= \frac{\lambda_1(t)[1 + \varepsilon_{11}] + \lambda_2(t)\varepsilon_{21} - [1 + \varepsilon_{11} + \varepsilon_{21}]\lambda_1(t)}{1 + \varepsilon_{11} + \varepsilon_{21}} \\ &= \frac{[\lambda_2(t) - \lambda_1(t)]\varepsilon_{21}(t)}{1 + \varepsilon_{11}(t) + \varepsilon_{21}(t)}. \end{aligned}$$

Since $\varepsilon_{11}(t), \varepsilon_{21}(t) \to 0$ as $t \to \infty$, and $\lambda_2(t) - \lambda_1(t)$ is bounded, it follows that $\zeta_1(t) - \lambda_1(t) \in L^1$ if $\varepsilon_{21}(t) \in L^1$, which we will show in Lemma 8.14 below.

Define $\zeta_2(t) = -\zeta_1(t) - a(t)$. Since then $\zeta_1 + \zeta_2 = -[a_0 + r_1(t)] = \lambda_1(t) + \lambda_2(t)$, it follows that

$$\zeta_2(t) = \lambda_2(t) + [\lambda_1(t) - \zeta_1(t)] = \lambda_2(t) + (L^1).$$

Therefore (8.86) and (8.87) are established.

Moreover, if we had chosen instead $\tilde{\zeta}_2(t) = y_2'(t)/y_2(t)$, then a similar calculation and estimation shows that there also exists $\tilde{\zeta}_1(t)$ such that

$$L = [D - \tilde{\zeta}_2(t)]\,[D - \tilde{\zeta}_1(t)],$$

with $\tilde{\zeta}_i(t) = \lambda_i(t) + L^1$ $(i = 1, 2)$. But, in general, the factors cannot be commuted and so $\zeta_i(t) \neq \tilde{\zeta}_i(t)$ $(i = 1, 2)$. □

Remark 8.13 In case that $\alpha_0 \neq \beta_0$, but $\operatorname{Re}\alpha_0 = \operatorname{Re}\beta_0$, then there exist factorizations as above, with $\zeta_i(t) - \lambda_i(t) = \mathrm{o}(1)$ as $t \to \infty$ $(i = 1, 2)$, but the differences are not necessarily in L^1. This is because while $\varepsilon_{21}(t) = \mathrm{o}(1)$ and can be estimated in terms of r_1, r_2, these estimates are not sufficient for showing that the differences are in L^1.

We conclude this section with a technical lemma establishing the L^1-property of the function $\varepsilon_{21}(t)$ in Theorem 8.12.

Lemma 8.14 *Under the assumptions of Theorem 8.12, the function $\varepsilon_{21}(t)$ defined in* (8.91) *satisfies $\varepsilon_{21}(t) \in L^1$.*

Proof In (8.89), we make the normalization $\hat{y}(t) = \exp\left[\int^t \lambda_1(\tau)\,d\tau\right] z$. Then (8.89) is equivalent to

$$z' = \left[\begin{pmatrix} 0 & \\ & \lambda_2(t) - \lambda_1(t) \end{pmatrix} + R(t)\right] z.$$

Since $\mathrm{Re}\,[\lambda_2(t) - \lambda_1(t)] = \beta_0 - \alpha_0 =: -2\gamma_0 < 0$, pick t_2 sufficiently large such that $\mathrm{Re}\,[\lambda_2(t) - \lambda_1(t)] \leq -\gamma_0$ for all $t \geq t_2$. For $t \geq t_2$, we look (as in the proof of Levinson's fundamental theorem) for a solution of

$$z(t) = e_1 + \int_{t_2}^{t} W(t)PW^{-1}(s)R(s)z(s)\,ds - \int_t^{\infty} W(t)PW^{-1}(s)R(s)z(s)\,ds.$$

Here $W(t)$ is a fundamental matrix of $w' = \mathrm{diag}\,\{0, \lambda_2(t)-\lambda_1(t)\}w$, $P = \mathrm{diag}\,\{0, 1\}$, and $z(t)$ is bounded, say $|z(t)| \leq M$ for all $t \geq t_1$. The proof of Theorem 2.7 showed that this integral equation has a unique solution $z(t)$, and that the function $\varepsilon_{21}(t)$ is the second component of $z(t)$, i.e.,

$$\begin{aligned} |\varepsilon_{21}(t)| &\leq M \int_{t_2}^{t} \left|\exp\left[\int_s^t \{\lambda_2(\tau) - \lambda_1(\tau)\}\right]\right| |R(s)|\,ds \\ &\leq M \int_{t_2}^{t} e^{-\gamma_0(t-s)} |R(s)|\,ds. \end{aligned}$$

Then Fubini's theorem and (8.90) show that

$$\int_{t_2}^{\infty} |\varepsilon_{21}(t)|\,dt \leq M \int_{t_2}^{\infty} \int_s^{\infty} e^{-\gamma_0(t-s)} |R(s)|\,dt\,ds = \frac{M}{\gamma_0} \int_{t_2}^{\infty} |R(s)|\,ds < \infty.$$

□

8.6 Perturbations of the Harmonic Linear Oscillator

We consider now equations of the form

$$y'' + \lambda^2 y = g(t)y \qquad t \geq t_0, \tag{8.92}$$

where $\lambda > 0$ and g is real-valued and at least locally integrable on $[t_0, \infty)$. Recall from Sects. 8.2 and 8.3 that under various additional conditions on the perturbation g, results from Chaps. 2 and 4 may be applied to obtain asymptotic representations for solutions. For example, if $g \in L^1[t_0, \infty)$, then the Liouville-Green formulas apply (with $f = -\lambda^2$) to yield solutions $y_\lambda, \bar{y}_\lambda$ satisfying

$$y_\lambda(t) = [1 + o(1)] \exp[i\lambda t] \qquad \text{as } t \to \infty,$$

hence it follows that every solution $y(t)$ of (8.92) can be represented as

$$y(t) = A\cos(\lambda t + B) + o(1) \qquad \text{as } t \to \infty \tag{8.93}$$

for suitable constants A and B, which is sometimes alternatively expressed as $y(t) = A\cos(\lambda t + B + o(1))$.

For the remainder of this section, we focus our attention on classes of perturbations g not in L^1, which are of particular importance in applications and for which special transformations will be used to bring them into L-diagonal form.

In Sect. 8.6.1 perturbations g are discussed involving assumptions of conditional integrability and for which methods from Sect. 4.6 can be applied. In particular, special cases when $g(t) = \xi(t)p(t)$ with $\xi' \in L^1[t_0, \infty)$ and p periodic are studied, and some examples are discussed demonstrating particular instances of resonance and non-resonance for solutions.

In Sect. 8.6.2 we consider perturbations $g \in L^p[t_0, \infty)$ with $p < 2$ together with some additional growth restrictions. This class has special importance when discussing spectral properties of Schrödinger-type operators associated with (8.92). Observe that since λ is real, the unperturbed equation $x'' + \lambda^2 x = 0$ does not have an exponential dichotomy, and results of the Hartman-Wintner type (see Sect. 4.4) do not apply. However, it is surprising that the same asymptotic integration formulas can still be shown to hold, not for all $\lambda > 0$, but for all λ avoiding an exceptional set of values (depending upon a fixed g) that has Lebesgue measure zero. We refer to such a result as almost everywhere (a.e.) asymptotic integration.

In both sections we will treat (8.92) as an equivalent linear system

$$\mathbf{y}' = \begin{pmatrix} 0 & 1 \\ -\lambda^2 + g(t) & 0 \end{pmatrix} \mathbf{y}(t), \qquad t \geq t_0,$$

and apply the transformation

$$\mathbf{y} = \begin{pmatrix} 1 & 1 \\ -i\lambda & i\lambda \end{pmatrix} \hat{y}$$

to diagonalize the leading constant term and obtain

$$\hat{y}' = \left\{ \begin{pmatrix} -i\lambda & 0 \\ 0 & i\lambda \end{pmatrix} + \frac{g(t)}{2i\lambda} \begin{pmatrix} -1 & -1 \\ 1 & 1 \end{pmatrix} \right\} \hat{y}. \tag{8.94}$$

8.6.1 Perturbations of Harmonic Oscillators Involving Conditional Integrability

Asymptotic integration for Eq. (8.92) with conditionally integrable perturbations g has a long history and an extensive literature associated with it. Principally two different methods have been used for the asymptotic analysis. A Prüfer transformation can be used to replace (8.92) by a system of first-order (non-linear) equations for the amplitude $r(t)$ and phase $\theta(t)$ of solutions, i.e., by letting $y(t) = r(t)\cos(\theta(t))$ and $y' = -r(t)\sin(\theta(t))$. For a comprehensive treatment and results using this approach, see [2], which also contains many references to earlier results up to the 1950s, and [89] for some more recent ones.

Alternatively, certain preliminary and conditioning transformations have been used to bring the linear system (8.94) into an L-diagonal form and then apply Levinson's fundamental theorem. For an excellent treatment of this approach with more recent references and results, see [53, Chap. 4]. The treatment here will be based on this second approach and also follows some earlier results of Harris and Lutz.

First apply the diagonal transformation $\hat{y} = \operatorname{diag}\left\{e^{-i\lambda t}, e^{i\lambda t}\right\} z_1$ to remove the constant term in (8.94) and obtain

$$z_1' = \frac{g(t)}{2i\lambda} \begin{pmatrix} -1 & -e^{2i\lambda t} \\ e^{-2i\lambda t} & 1 \end{pmatrix} z_1 =: A(t) z_1. \tag{8.95}$$

To proceed further, it is helpful to distinguish the following situations:

(i) $g(t)$ and $g(t)e^{\pm 2i\lambda t}$ are conditionally integrable;
(ii) $g(t)e^{\pm 2i\lambda t}$ are conditionally integrable (but not necessarily g itself);
(iii) $g(t)e^{\pm 2i\lambda t}$ are not conditionally integrable.

(i) Consider applying Theorem 4.19. If $A\int_t^\infty A \in L^1$, i.e., $g(t)\int_t^\infty g(s)e^{\pm 2i\lambda s}\,ds \in L^1$, then Theorem 4.19 with $k = 1$ yields that

$$Z_1(t) = I + o(1) \qquad \text{as } t \to \infty,$$

which implies that (8.92) has solutions as in (8.93).

If $A\int_t^\infty A \notin L^1$, but at least conditionally integrable, consider applying Theorem 4.19 with $k > 1$. This involves checking conditionally integrability of products of integrals (see also (iii) below).

(ii) In this case, one can write $A(t) = \Lambda_1(t) + V_1(t)$ with $\Lambda_1(t) = \frac{g(t)}{2i\lambda}\text{diag}\,\{-1, 1\}$. Then V_1 is conditionally integrable and one can let $z_1 = [I + Q_1(t)]z_2$ with $Q_1(t) = -\int_t^\infty V_1(s)\,ds$ to obtain

$$z_2' = [\Lambda_2(t) + V_2(t) + R_2(t)]\,z_2,$$

where

$$\begin{aligned} \Lambda_2 &= \Lambda_1 + V_1 Q_1 \quad \text{is diagonal,} \\ V_2 &= \Lambda_1 Q_1 - Q_1 \Lambda_1, \quad \text{with diag}\, V_2 \equiv 0, \\ R_2 &= Q_1^2 (I + Q_1)^{-1} [\Lambda_1 + V_1 Q_1 + \Lambda_1 Q_1]. \end{aligned}$$

Letting $\hat{g}^{\pm}(t) = \int_t^\infty g(s) e^{\pm 2i\lambda s}\,ds$, if $g\hat{g}^{\pm} \in L^1$, it follows that $V_1 Q_1$, V_2, and R_2 are all in L^1. Since g is real-valued, $\Lambda_1(t)$ satisfies Levinson's dichotomy condition, and Theorem 2.7 implies that there exists a solution

$$Z_2(t) = [I + \mathrm{o}(1)] \exp\left[\int_{t_0}^t \Lambda_1(s)\,ds\right] \quad \text{as } t \to \infty.$$

Hence (8.92) has solutions which can be expressed as

$$y_1(t) = \cos[\lambda t + \sigma(t)] + \mathrm{o}(1), \tag{8.96}$$

$$y_2(t) = \sin[\lambda t + \sigma(t)] + \mathrm{o}(1), \tag{8.97}$$

as $t \to \infty$ with $\sigma(t) = \frac{1}{2\lambda} \int_{t_0}^t g(s)\,ds$ (cf. [70, Thm. 2.1]).

If, on the other hand, $g\hat{g}^{\pm}$ are only conditionally integrable, another transformation $z_2 = [I + Q_2(t)]z_3$ with $Q_2(t) = -\int_t^\infty V_2(s)\,ds$ can be made. This yields

$$z_3' = [\Lambda_2(t) + R_3(t)]\,z_3,$$

where

$$R_3 = (\Lambda_2 + V_2)Q_2 - Q_2[I + Q_2]^{-1}[\Lambda_2 + (\Lambda_2 + V_2)Q_2] + (I + Q_2)^{-1} R_2 (I + Q_2).$$

If $g\hat{g}^+\hat{g}^- \in L^1$, then $R_2 \in L^1$ and, furthermore, if $g|Q_2| \in L^1$, i.e., $g\int_t^\infty g\hat{g}^{\pm} \in L^1$, it follows that also $R_3 \in L^1$. Hence Theorem 2.7 can be applied and yields

$$Z_3(t) = [I + \mathrm{o}(1)] \exp\left[\int_{t_0}^t \Lambda_2(\tau)\,d\tau,\right] \quad \text{as } t \to \infty,$$

with

$$\Lambda_2 = \Lambda_1(t) - \frac{g(t)}{4\lambda^2}\operatorname{diag}\left\{\int_t^\infty g(s)e^{2i\lambda(t-s)}\,ds, \int_t^\infty g(s)e^{2i\lambda(s-t)}\,ds\right\},$$

(cf. [70, Thm. 3.2]).

(iii) In case $ge^{\pm 2i\lambda t}$ are not conditionally integrable, we say that *resonance* occurs between solutions $e^{\pm i\lambda t}$ of the unperturbed equation $x'' + \lambda^2 x = 0$ and the perturbation g. This also may occur in cases (i) and (ii) above, whenever non-conditionally convergent terms are introduced.

In order to treat situations in which non-conditionally integrable perturbations arise and other methods are not applicable, it appears to be necessary to restrict the class of admissible functions g in order to make progress toward asymptotic integration.

Recall that in connection with non-resonance criteria in Sect. 4.6.4, the class of functions g of the form

$$g(t) = \xi(t)p(t),$$

where $\xi(t) = o(1)$ as $t \to \infty$, ξ' and $\xi^q \in L^1[t_0,\infty)$, and p is periodic and continuous was discussed. For perturbations in this class, asymptotic integration results for (8.92) have been obtained by J.S. Cassell [32], M.S.P. Eastham et al. [53, Ch. 4] including cases where resonance occurs.

Another class of perturbations of the form

$$g(t) = \sum_{k=1}^{N} \frac{a_k e^{ib_k t}}{t^{p_k}},$$

with $b_k \in \mathbb{R}$ and $0 < p_k \le 1$ has been treated by V. Burd and P. Nesterov [31] using techniques mentioned in Sect. 4.6.5.

In order to asymptotically integrate (8.94) for perturbations in these classes (including cases where resonance arises, that is, when non-conditionally integrable terms are encountered), special conditioning $(I+Q)$ transformations are constructed and certain constant, invertible transformations are also used to reduce the system to L-diagonal form. For a complete discussions, see the above-mentioned references.

In order to just illustrate the ideas involved, we end this section with two examples when $g(t) = \sin(\theta t)/t^p$ to show how resonance is related to non-conditional integrability and how the transformations described above lead to asymptotic integration in such cases.

8.6.1.1 Example I

Applying criteria for non-resonance for the system (4.114) in Sect. 4.6.4 to (8.92) with

$$g(t) = \xi(t)p(t) = \sin(\theta t)/t^p,$$

the assumptions (4.115) take for positive λ and θ the form

$$\theta \neq \frac{2\lambda}{n}, \qquad n = 1, 2, 3, \ldots . \tag{8.98}$$

The first resonant value $\theta = 2\lambda$ corresponds to the case (iii) above when in (8.95)

$$g(t)e^{2i\lambda t} = \frac{e^{4i\lambda t} - 1}{2it^p} \quad \text{and} \quad g(t)e^{-2i\lambda t} = \frac{1 - e^{-4i\lambda t}}{2it^p}.$$

This allows the coefficient matrix $A(t)$ in (8.95) to be decomposed into the sum

$$A(t) = \frac{-1}{4\lambda t^p}\left\{\begin{pmatrix} 0 & 1 \\ 1 & 0 \end{pmatrix} + \begin{pmatrix} \bar{a} - a & a^2 \\ \bar{a}^2 & a - \bar{a} \end{pmatrix}\right\},$$

where $a = a(t) = \exp(2i\lambda t)$. The non-conditionally integrable part can be diagonalized using the constant matrix $P = \begin{pmatrix} 1 & 1 \\ -1 & 1 \end{pmatrix}$, so letting $z_1 = P\tilde{z}$ we obtain

$$\tilde{z}' = \left(\tilde{\Lambda}(t) + \tilde{V}(t)\right)\tilde{z},$$

with

$$\tilde{\Lambda}(t) = \frac{1}{4\lambda t^p}\text{diag}\{1, -1\},$$

and where

$$\tilde{V} = \frac{-1}{8\lambda t^p}\begin{pmatrix} -a^2 - \bar{a}^2 & -2(a - \bar{a}) + a^2 - \bar{a}^2 \\ -2(a - \bar{a}) - a^2 + \bar{a}^2 & a^2 + \bar{a}^2 \end{pmatrix}$$

is conditionally integrable. One could then use a conditioning transformation $\tilde{z} = [I + \tilde{Q}(t)]\hat{z}$ with

$$\tilde{Q}(t) = -\int_t^\infty \tilde{V}(s)\,ds$$

to obtain

$$\hat{z}' = \left[\tilde{\Lambda}(t) + \hat{V}(t)\right]\hat{z},$$

where

$$\hat{V} = [I + \tilde{Q}]^{-1}\left(\tilde{V}\tilde{Q} + \tilde{\Lambda}\tilde{Q} - \tilde{Q}\tilde{\Lambda}\right).$$

Since $\tilde{Q}(t) = \mathrm{O}\left(t^{-p}\right)$ as $t \to \infty$, if $1/2 < p \leq 1$ it follows that $\hat{V} \in L^1$, hence the system is in L-diagonal form.

If $p \leq 1/2$, it is advantageous to replace $\tilde{Q}$ by an approximation $\hat{Q}(t) = \frac{1}{t^p}\hat{P}(t)$, where $\hat{Q}(t) = \tilde{Q}(t) + (L^1)$ and $\hat{P}$ is an explicit sum of complex exponentials. This comes about integrating $\tilde{Q}$ by parts using

$$\int_t^\infty \frac{e^{ibs}}{s^p}\,ds = \frac{ie^{ibt}}{bt^p} + \frac{p}{ib}\int_t^\infty \frac{e^{ibs}}{s^{p+1}}\,ds \tag{8.99}$$

and integrating by parts again yields

$$\int_t^\infty \frac{e^{ibs}}{s^p}\,ds = \frac{ie^{ibt}}{bt^p} + \mathrm{O}\left(\frac{1}{t^{p+1}}\right).$$

Using the L^1-approximation $\hat{Q}$ instead of $\tilde{Q}$ replaces $\hat{V}$ by $\hat{V}_1 + \hat{R}_1$, where $\hat{R}_1 \in L^1$ and $\hat{V}_1$ is again $1/t^p$ times a matrix of complex exponentials. See [71, Ex. 3] where this procedure is applied to the normalized equation $x'' + \left(1 + \frac{\sin 2t}{t^{1/2}}\right)x = 0$ and yields solutions of the form

$$x_1(t) = \exp\left(\frac{1}{2}t^{1/2}\right)[\cos t + \mathrm{o}(1)],$$

$$x_2(t) = \exp\left(-\frac{1}{2}t^{1/2}\right)[\sin t + \mathrm{o}(1)],$$

as $t \to \infty$. This shows the significant effect of resonance on the amplitude of solutions.

8.6.1.2 Example II: $g(t) = \frac{\sin \lambda t}{t^p}$, $0 < p \leq 1$

The second resonant value in (8.98), $\theta = \lambda$ corresponds to case (i) above, since $A(t)$ in (8.95) is then conditionally integrable on $[1, \infty)$. Hence a conditioning transformation $z_1 = [I + Q(t)]z_2$ with $Q(t) = -\int_t^\infty A(s)\,ds$ reduces (8.95) to

$$z_2' = [I + Q]^{-1}AQ\,z_2. \tag{8.100}$$

Observing that $Q(t) = \mathrm{O}(t^{-p})$ as $t \to \infty$, it follows that if $1/2 < p \leq 1$, (8.100) is L-diagonal form with $\Lambda \equiv 0$ and $R = [I + Q]^{-1}AQ$, so Theorem 2.7 applies and yields the existence of a fundamental matrix $Z_2 = I + \mathrm{o}(1)$ as $t \to \infty$.

If, however, $0 < p \leq 1/2$, resonance occurs in the transformed system since AQ is not conditionally integrable. To see this, we express $A(t)$ in (8.95) as

$$A(t) = \frac{1}{4\lambda t^p}\begin{pmatrix} e^{i\lambda t} - e^{-i\lambda t} & e^{3i\lambda t} - e^{i\lambda t} \\ e^{-3i\lambda t} - e^{-i\lambda t} & e^{-i\lambda t} - e^{i\lambda t} \end{pmatrix},$$

and use integration by parts, i.e., (8.99), to obtain the following L^1-approximation of $-\int_t^\infty A(s)\,ds$,

$$\hat{Q}(t) = \frac{-i}{4\lambda^2 t^p}\begin{pmatrix} e^{i\lambda t} + e^{-i\lambda t} & \frac{e^{3i\lambda t}}{3} - e^{i\lambda t} \\ -\frac{e^{-3i\lambda t}}{3} + e^{-i\lambda t} & -e^{-i\lambda t} - e^{i\lambda t} \end{pmatrix}.$$

Note that $\hat{Q}' = A + \hat{R}$ with $\hat{R} \in L^1$. Therefore $z_1 = [I + \hat{Q}(t)]z_2$ takes (8.95) into

$$z_2' = \left(A\hat{Q} + \tilde{R}\right) z_2 \quad \text{with } \tilde{R} \in L^1.$$

A short calculation leads to

$$A(t)\hat{Q}(t) = \frac{i}{16\lambda^3 t^{2p}}\hat{A} + \hat{V}(t), \quad \text{where } \hat{A} = \begin{pmatrix} 4/3 & -2 \\ 2 & -4/3 \end{pmatrix},$$

and $\hat{V}(t) = \frac{1}{t^{2p}}P(t)$, with

$$P(t) = P_1 e^{2i\lambda t} + P_2 e^{-2i\lambda t} + P_3 e^{4i\lambda t} + P_4 e^{-4i\lambda t}$$

and P_j are real and constant 2×2 matrices. Observe that the constant matrix $\hat{A}$ has eigenvalues $\pm 2i\sqrt{5}/3$, hence letting $P_0 = \begin{pmatrix} 1 & 1 \\ -2i\sqrt{5}/3 & 2i\sqrt{5}/3 \end{pmatrix}$ and $z_2 = P_0 z_3$, we obtain

$$z_3' = \left[\Lambda(t) + \tilde{V}(t)\right] z_3,$$

where

$$\Lambda(t) = \frac{1}{\lambda^3 t^{2p}}\begin{pmatrix} \frac{\sqrt{5}}{24} & \\ & -\frac{\sqrt{5}}{24} \end{pmatrix}$$

and $\tilde{V}(t) = P_0^{-1}P(t)P_0 = \mathrm{O}\left(1/t^{2p}\right)$ as $t \to \infty$. Note that $\tilde{V}(t)$ is conditionally integrable. Hence a further transformation $z_3 = [I + \tilde{Q}(t)]z_4$ can be made with

$\tilde{Q}' = \tilde{V}$, which leads to

$$z_4' = \left[\Lambda + (I+Q)^{-1}(\Lambda Q - Q\Lambda + VQ)\right] z_4 = \left[\Lambda + \mathrm{O}\left(\frac{1}{t^{3p}}\right)\right] z_4.$$

If $p > 1/3$, this results in a system in L_1-diagonal form and the asymptotic integration is complete, leading to a fundamental solution matrix for (8.94) of the form

$$\hat{Y}(t) = \begin{pmatrix} e^{-i\lambda t} & \\ & e^{i\lambda t} \end{pmatrix} \left[I + \hat{Q}(t)\right] P_0 \left[I + \mathrm{o}(1)\right] \exp\left[\int^t \Lambda(s)\, ds\right],$$

as $t \to \infty$. Thus the effect of resonance in this case can be seen to lead to solutions of (8.92) of the form

$$y_1(t) = \rho(t)\left[\cos(\lambda t - \beta_1) + \mathrm{o}(1)\right],$$
$$y_2(t) = \rho^{-1}(t)\left[\sin(\lambda t - \beta_2) + \mathrm{o}(1)\right],$$

as $t \to \infty$, where $\beta_i \in \mathbb{R}$ and

$$\rho(t) = \begin{cases} \exp\left(\frac{\sqrt{5}}{24\lambda^3}\log t\right) & \text{if } p = 1/2, \\ \exp\left(\frac{\sqrt{5}\, t^{1-2p}}{24\lambda^3(1-2p)}\right) & \text{if } 1/3 < p < 1/2. \end{cases}$$

See [71, Ex. 2] for the result for the normalized equation $x'' + \left(1 + \frac{\sin t}{t^{1/2}}\right) x = 0$ and [53, Sect. 4.10] for another comparable treatment, which is also applicable to the more general case $y'' + (1 + \xi(t)\sin t)\, y = 0$, where $\xi' \in L^1$ and $\xi^3 \in L^1$.

Other possible resonant values $\theta = \frac{2\lambda}{3}, \frac{\lambda}{2}$, etc. arise for p in the range $0 < p \leq 1/3$. For such cases, the procedure of V. Burd and P. Nesterov (see [31] and remarks in Sect. 4.6.5) is especially well-suited for finding the asymptotic behavior of solutions. See also [31] for an interesting application to equations of the form

$$x'' + \left[1 + a\frac{\sin\varphi(t)}{t^p}\right] x = 0,$$

where $a \in \mathbb{R} \setminus \{0\}$, $p > 0$, and either $\varphi(t) = t + \alpha \log t$ or $\varphi(t) = t + \alpha t^\beta$, where $\alpha \in \mathbb{R} \setminus \{0\}$, and $0 < \beta < 1$.

8.6.2 *Almost Everywhere Asymptotic Integration*

In Sect. 8.6.1, case (iii), it was shown that if $g(t) \int_t^\infty g(s)e^{\pm 2i\lambda s}\, ds$ exist and are in $L^1[t_0, \infty)$, then (8.92) has solutions $y_\lambda, \bar{y}_\lambda$ which can be expressed in the form

$$y_\lambda(t) = [1 + o(1)] \exp\left[i\lambda t - \frac{i}{2\lambda} \int_{t_0}^t g(s)\, ds \right]. \tag{8.101}$$

Another set of conditions which imply the same conclusions are

$$g(t) = o(1) \text{ as } t \to \infty, \quad g' \in L^1, \text{ and } g \in L^2.$$

In this case one can apply Theorem 4.1 to the system (8.94), noting that the eigenvalues

$$\mu^\pm(t) = \sqrt{-\lambda^2 + g(t)}$$

can be approximated up to an L^1-perturbation by $\pm\left(i\lambda - \frac{i}{2\lambda} g(t)\right)$ since $g \in L^2$. Since g is real-valued, the dichotomy conditions automatically hold.

This section concerns other types of assumptions on g which imply (8.101) *almost everywhere* in the following sense: Given g (called in this context a "potential") satisfying certain conditions, there exists a set $S(g) \subset \mathbb{R}^+$ of Lebesgue measure zero such that (8.101) holds for all $\lambda \notin S(g)$. These interesting and remarkable results are due to A. Kiselev [92] and M. Christ and A. Kiselev [40], who applied them for the spectral analysis of Schrödinger operators associated with (8.92). Here, we give a brief overview of these results as they pertain to rather novel constructions of conditioning transformations, followed by applying Levinson's theorem. For complete proofs of these results and extensions not included here, for applications, and for other references, we refer the reader to [92] and [40]. See also [53, pp. 176–178] for a discussion of applications of asymptotic integration formulas to the spectral analysis of associated operators on L^2.

Considering again the system (8.95), let

$$q(t, \lambda) = \frac{1}{2i\lambda} \int_t^\infty g(\tau) e^{2i\lambda\tau}\, d\tau, \tag{8.102}$$

for those $\lambda \in \mathbb{R}^+$ such that the integral converges for $t \geq t_0$. In what follows g will always be fixed satisfying certain conditions, and we will be interested in the λ-dependence as indicated by the notation.

If $q(\lambda, t)$ exists and if $q(t, \lambda) g \in L^1$, then the conditioning transformation

$$z_1 = \begin{pmatrix} 1 & q(t, \lambda) \\ \bar{q}(t, \lambda) & 1 \end{pmatrix} \tilde{z} \tag{8.103}$$

is invertible for $t \geq t_1$ sufficiently large and takes (8.95) into a system

$$\tilde{z}' = \left\{ \begin{pmatrix} \frac{ig(t)}{2\lambda} & \\ & \frac{-ig(t)}{2\lambda} \end{pmatrix} + R(t) \right\} \tilde{z}, \qquad \text{where } R(t) \in L^1[t_1, \infty). \tag{8.104}$$

Hence Theorem 2.7 would apply and yield (8.101).

It is easy to construct elementary examples of functions $g(t) \in L^2[t_0, \infty)$ where $q(t, \lambda)$ fails to exist for a certain discrete set of values λ, e.g., where resonance occurs. Moreover, even if $q(t, \lambda)$ would exist for such a $g(t)$, the function $q(t, \lambda)$ might fail to be in L^2 so that qg might not be in L^1. A. Kiselev, in a private communication [93], has described examples of potentials $g \in L^2$ such that for λ in a set of positive measure, $q(t, \lambda)$ decays too slowly to be in L^2.

Therefore for this approach involving (8.102) to succeed, g is required to satisfy stronger decay conditions. The first class of potentials g studied by Kiselev satisfy

$$|g(t)| \leq ct^{-\frac{3}{4}-\epsilon} \tag{8.105}$$

for $t \geq t_0$ and some $c, \epsilon > 0$.

Observe that (8.105) implies that $f(t) := g(t)t^{1/4} \in L^2[t_0, \infty)$. Letting $\Phi(f)(k)$ denote the L^2-limit of the Fourier transform $\lim_{N\to\infty} \int_{-N}^{N} \exp(ik\tau)f(\tau)\, d\tau$, and letting

$$M^+ (F(t)) = \sup_{h>0} \frac{1}{h} \int_0^h |F(t+\tau) - F(t-\tau)|\, d\tau,$$

denote a kind of "maximal function" for F, a set S is defined as

$$S = \left\{ t : M^+(\Phi(f(t)) < \infty \right\}.$$

It is known that the complement of S has Lebesgue measure zero. It can be shown that for $2\lambda \in S$, the integral (8.102) exists and satisfies $q(t, \lambda) = O\left(t^{-1/4} \log t\right)$ as $t \to \infty$. See Lemmas 1.3 and 1.5 in [92] for details. It follows that for $2\lambda \in S$, the transformation (8.103) takes (8.95) into (8.104) with $R \in L^1[t_2, \infty)$, where t_2 is so large such that $|q(t, \lambda)| \leq 1/2$ for $t \geq t_2$. Hence Theorem 2.7 applies and leads to solutions y_λ, $\bar{y}_\lambda$ satisfying (8.101) for almost all $\lambda \in \mathbb{R}^+$. Moreover, using the above estimate on q, it can be further shown (using arguments such as we have used involving the T-operator) that o(1) in (8.101) can be more precisely estimated by $O\left(t^{-\varepsilon} \log t\right)$ as $t \to \infty$.

This result was then extended by Kiselev for more general classes of potentials that are not power decreasing. This is because they contain gaps, i.e., intervals, where the function is identically zero. But when the gaps are removed and the function is compressed, it satisfies (8.105). In this case (8.101) also holds, but the set S is not explicitly described as in the previous case. See [92, Sect. 2] for details.

Next, in a joint paper with M. Christ [40], they consider a much weaker decay condition of the form

$$g(t)\,(1+|t|)^{\varepsilon} \in L^p(\mathbb{R}^+), \qquad \text{for some } 0 < p \le 2 \text{ and } \epsilon > 0.$$

For this class the approach involving the Fourier transform is replaced by a completely different method which still involves using a conditioning transformation (8.103), but without an explicit representation such as (8.102) for $q(t,\lambda)$. In this situation, the diagonal transformation

$$\tilde{y} = \begin{pmatrix} e^{-p(t,\lambda)} & \\ & e^{p(t,\lambda)} \end{pmatrix} z,$$

is first applied to (8.95) with $p(t,\lambda) = \frac{1}{2\lambda i}\int_{t_0}^{t} g(\tau)\,d\tau$. This results in the system

$$z' = \frac{ig(t)}{2\lambda}\begin{pmatrix} 0 & -\exp[2i\lambda t + 2p(t,\lambda)] \\ \exp[-2i\lambda t - 2p(t,\lambda)] & 0 \end{pmatrix} z. \tag{8.106}$$

Then for almost all $\lambda \in \mathbb{R}^+$, a function $q(t,\lambda)$ is constructed which satisfies $q(t,\lambda) = o(1)$ and also the nonlinear differential equation

$$\begin{aligned} q'(t,\lambda) &- \frac{i}{2\lambda}g(t)\exp[2i\lambda t + 2p(t,\lambda)] \\ &- \frac{i}{2\lambda}g(t)\exp[-2i\lambda t - 2p(t,\lambda)]q^2(t,\lambda) = f(t,\lambda), \end{aligned} \tag{8.107}$$

where $f \in L^1$. This is accomplished by means of a finite number of iterations (like successive approximations) involving almost everywhere convergence for a certain class of integral operators on L^p spaces. This technique goes far beyond the rather elementary constructions used in Chap. 4 to solve simpler types of approximate equations. We refer to [40] for details of their proof.

Using $q(t,\lambda)$ as in (8.107), the transformation (8.103) takes (8.106) into an equivalent system with off-diagonal elements in L^1. A separate argument shows that the integrals of elements appearing on the diagonal of this new system can be estimated as $c(\lambda)[1 + o(1)]$ and therefore Theorem 2.7 applies and yields (8.101) for almost all $\lambda \in \mathbb{R}^+$ but with no explicit description of how the exceptional values λ depend upon g. Actually in their main result [40, Theorem 1.2], a somewhat more general statement concerning solutions of

$$y'' - [U(t) + g(t)]y = -\lambda^2 y$$

is made, under the assumption that all solutions of $y'' - U(t)y = -\lambda^2 y$ are bounded for all t and for almost all λ in a set of positive Lebesgue measure. Their proof concerns this more general situation but with $U(t) \equiv 0$ reduces to the treatment above.

In a later paper, [41] Christ and Kiselev extend the above result to an even wider class of perturbations, namely $g \in l^p(L^1)(\mathbb{R})$ for $p < 2$. This means there exists $M > 0$ such that for all $N \in \mathbb{N}$

$$\sum_{-N}^{N} \left(\int_n^{n+1} |f(t)|\, dt \right)^p \leq M.$$

It follows from Hölder's inequality that $L^p(\mathbb{R}) \subset l^p(L^1)(\mathbb{R})$, hence their result implies an a.e. asymptotic integration formula for $g \in L^p$, $p < 2$. To achieve this, they construct a series solution of an associated T-operator equation which avoids the intermediate step of constructing $q(t, \lambda)$ and applying Levinson's fundamental theorem. See also [46] for some more recent contributions to this subject.

8.7 Some Liouville–Green Formulas for Third-Order Differential Equations

The literature concerning third-order linear differential equations is considerably sparser than for second-order equations for several reasons including tractability and applications. See, however, [68] for a general discussion of third-order equations and several interesting applications. Here we are concerned with some results concerning asymptotic integration formulas for solutions which are natural analogues of the Liouville–Green formulas in the second-order case. These were studied in [123] for two classes of normalized equations of the form

$$y''' + q(t)y' + r(t)y = 0, \qquad t \geq t_0 \tag{8.108}$$

say for q and r locally integrable on $[t_0, \infty)$. Pfeiffer treated (8.108) using a systems approach and applying some results within Levinson's theory to obtain his formulas. The focus here is to give a brief, slightly different treatment and indicate some of the main ideas behind this approach, in particular how certain preliminary transformations may be used, followed by applying results from Chap. 4.

In system form, (8.108) is equivalent to

$$\mathbf{y}' = \begin{pmatrix} 0 & 1 & 0 \\ 0 & 0 & 1 \\ -r & -q & 0 \end{pmatrix} \mathbf{y}, \qquad \mathbf{y} = \begin{pmatrix} y \\ y' \\ y'' \end{pmatrix}. \tag{8.109}$$

When r and q are both asymptotically constant, results from Sect. 8.2 can be applied. If that is not the case, then it is important to identify which coefficient contributes the dominant effect to the asymptotic behavior of solutions. We will discuss the following two cases: In case (a), we assume that $r(t) \neq 0$ for all $t \geq t_0$ and $q/r^{2/3} = o(1)$ as $t \to \infty$ and consider r as dominant, while in case (b) when $q(t) \neq 0$ and $r/q^{2/3} = o(1)$ as $t \to \infty$, we think of q as the dominant coefficient. The critical cases such as q and $r^{2/3}$ having roughly the same magnitude are more difficult to analyze except say when both are non-zero and $q/r^{2/3}$ is asymptotically constant as $t \to \infty$. In Pfeiffer's treatment the dominant term is assumed to be real-valued while the other is allowed to take on complex values, but we do not make that assumption here.

In case (a), when $r(t) \neq 0$, let $\mathbf{y} = \operatorname{diag}\{1, r^{1/3}, r^{2/3}\}\hat{y}$ to obtain

$$\hat{y}' = \begin{pmatrix} 0 & r^{1/3} & 0 \\ 0 & -\frac{1}{3}\frac{r'}{r} & r^{1/3} \\ -r^{1/3} & -\frac{q}{r^{1/3}} & -\frac{2}{3}\frac{r'}{r} \end{pmatrix} \hat{y}. \tag{8.110}$$

The "leading coefficient" matrix

$$\begin{pmatrix} 0 & 1 & 0 \\ 0 & 0 & 1 \\ -1 & 0 & 0 \end{pmatrix}$$

has distinct eigenvalues $-1, 1/2 \pm i\frac{\sqrt{3}}{2}$, the cubed roots of -1, which can be expressed as $w_k = \exp\left(\frac{i}{3}(\pi + 2k\pi)\right)$, $k = 1, 2, 3$. Using the constant transformation

$$\hat{y} = \begin{pmatrix} 1 & 1 & 1 \\ w_1 & w_2 & w_3 \\ w_1^2 & w_2^2 & w_3^2 \end{pmatrix} \tilde{y} =: P\tilde{y}, \tag{8.111}$$

one obtains

$$\tilde{y}' = [\Lambda(t) + V_1(t) + V_2(t)]\,\tilde{y}, \tag{8.112}$$

where

$$\Lambda(t) = r^{1/3}\operatorname{diag}\{w_1, w_2, w_3\},$$

$$V_1(t) = \frac{r'}{r}P^{-1}\begin{pmatrix} 0 & 0 & 0 \\ 0 & -1/3 & 0 \\ 0 & 0 & -2/3 \end{pmatrix} P, \text{ and } V_2(t) = \frac{q}{r^{1/3}}P^{-1}\begin{pmatrix} 0 & 0 & 0 \\ 0 & 0 & 0 \\ 0 & -1 & 0 \end{pmatrix} P.$$

If $r'/r^{4/3} = o(1)$ and $q/r^{2/3} = o(1)$ as $t \to \infty$ and if $\left(r'/r^{4/3}\right)'$ and $\left(q/r^{2/3}\right)' \in L^1$, and if the eigenvalues $\hat{\Lambda}(t)$ of $\Lambda(t) + V_1(t) + V_2(t)$ satisfy Levinson's dichotomy conditions, it follows from Theorem 4.8 that there exists a fundamental solution matrix satisfying

$$Y(t) = \operatorname{diag}\{1, r^{1/3}, r^{2/3}\}[P + o(1)] \exp\left[\int^t \hat{\Lambda}(\tau)\, d\tau\right]. \tag{8.113}$$

Some assumptions which imply this are discussed by Pfeiffer and these also allow the eigenvalues $\hat{\Lambda}(t)$ to be explicitly computed up to L^1-perturbations in terms of $r^{1/3}$ and $q/r^{1/3}$. See [123, Theorem 4] for details.

In case (b), when $q(t) \neq 0$ and $r/q^{3/2} = o(1)$ as $t \to \infty$, the shearing transformation

$$\mathbf{y} = \operatorname{diag}\{1, q^{1/2}, q\}\hat{y},$$

leads to

$$\begin{aligned}\hat{y}' &= \left(q^{1/2}\begin{pmatrix} 0 & 1 & 0 \\ 0 & 0 & 1 \\ 0 & -1 & 0 \end{pmatrix} + \frac{q'}{q}\begin{pmatrix} 0 & 0 & 0 \\ 0 & -1/2 & 0 \\ 0 & 0 & -1 \end{pmatrix} + \frac{r}{q}\begin{pmatrix} 0 & 0 & 0 \\ 0 & 0 & 0 \\ -1 & 0 & 0 \end{pmatrix}\right)\hat{y} \\ &= \left(q^{1/2}A_0 + \frac{q'}{q}A_1 + \frac{r}{q}A_2\right)\hat{y}.\end{aligned}$$

The leading matrix A_0 has eigenvalues $0, \pm i$ and the transformation

$$\hat{y} = \tilde{P}\tilde{y} = \begin{pmatrix} 1 & 1 & 1 \\ 0 & i & -i \\ 0 & -1 & -1 \end{pmatrix}\tilde{y},$$

leads to the system

$$\tilde{y}' = \left[\tilde{\Lambda}(t) + \tilde{V}_1(t) + \tilde{V}_2(t)\right]\tilde{y},$$

where $\tilde{\Lambda}(t) = \operatorname{diag}\{0, iq^{1/2}, -iq^{1/2}\}$, $\tilde{V}_1(t) = \frac{q'}{q}\tilde{P}^{-1}A_1\tilde{P}$, and $\tilde{V}_2(t) = \frac{r}{q}\tilde{P}^{-1}A_2\tilde{P}$. In this case it is natural to distinguish between two subcases (b_1), when q is real and positive, and (b_2), when $|\mathrm{Im}\,(q^{1/2})| \geq a > 0$. In the first case, if $q'/q^{3/2} = o(1)$ and $r/q^{3/2} = o(1)$, $(q'/q^{3/2})' \in L^1$, and $\left(r/q^{3/2}\right)' \in L^1$, and if also the eigenvalues $\hat{\Lambda}(t)$ of $\tilde{\Lambda} + \tilde{V}_1(t) + \tilde{V}_2(t)$ satisfy Levinson's dichotomy conditions, then Theorem 4.8 implies there is a fundamental matrix satisfying

$$Y(t) = \operatorname{diag}\{1, q^{1/2}, q\}\left[\tilde{P} + o(1)\right]\exp\left[\int^t \hat{\Lambda}(\tau)\, d\tau\right].$$

Again, sufficient conditions are given for which the eigenvalues are explicit in terms of $q^{1/2}$ and r/q, up to L^1-perturbations (see [123, Theorem 8]). In case (b_2) , $\tilde{\Lambda}(t)$

satisfies an exponential dichotomy condition so the conditions on the perturbations can be weakened to L^2.

We conclude this section by mentioning [134] for an analogous approach to discussing the asymptotic behavior of solutions of a fourth-order equation

$$y^{(4)} + q(t)y' + r(t)y = 0,$$

under various conditions on q and r.

8.8 Second-Order Matrix Equations

In [145], Spigler and Vianello studied the second-order matrix equation

$$Y'' = [D(t) + G(t)]Y, \qquad t \geq t_0, \tag{8.114}$$

where $D(t)$ and $G(t)$ are continuous $d \times d$ matrices. They showed that under various assumptions (8.114) has on the entire interval $[t_0, \infty)$ solutions

$$Y_k(t) = D^{-1/4} \exp\left[(-1)^{k+1} \int_a^t D^{1/2}(s)\, ds\right] [I + E_k(t)], \qquad k = 1, 2, \tag{8.115}$$

where $E_1(t) \to 0$ as $t \to (t_0)^+$ and $E_2(t) \to 0$ as $t \to \infty$. Moreover, they derive precise computable bounds for these error terms. Their approach consists of making an Ansatz of the form (8.115) in (8.114), leading to a second-order matrix differential equation for $E_k(t)$. They establish a solution of this differential equations by solving a related integral equation using the method of successive approximations.

We take a different approach and derive the asymptotic behavior of solutions and their error bounds by re-writing (8.114) as a $2d \times 2d$ first order system and follow closely the approach taken in Sects. 8.3.2 and 8.3.3. We note that one can alternatively consider a finite or infinite interval of the form $[a_1, a_2]$, and refer to Sect. 8.3.3 for the details. More specifically, we will show the following

Theorem 8.15 *In (8.114), assume that $D(t) = \operatorname{diag}\{\lambda_1(t), \ldots, \lambda_d(t)\}$ is an invertible and twice continuously differentiable $d \times d$ matrix, and that $G(t)$ is a continuous $d \times d$ matrix. Define a branch of the square root function by making a cut in the complex plane from 0 to $-\infty$ along the negative real axis and choosing a branch of the square root function so that*

$$\operatorname{Re} \sqrt{\lambda_j(t)} \geq 0 \qquad \text{for all } 1 \leq j \leq d. \tag{8.116}$$

Suppose that

$$R_i(t) \in L^1[t_0, \infty),$$

where

$$R_1(t) := \frac{5}{32} D^{-\frac{5}{2}} (D')^2 + \frac{D^{-\frac{3}{2}} D''}{8} + \frac{D^{-\frac{1}{4}} e^{-\int^t D^{1/2}} G e^{\int^t D^{1/2}} D^{-\frac{1}{4}}}{2}, \tag{8.117}$$

$$R_2(t) := -\frac{5}{32} D^{-\frac{5}{2}} (D')^2 + \frac{D^{-\frac{3}{2}} D''}{8} + \frac{D^{-\frac{1}{4}} e^{\int^t D^{1/2}} G e^{-\int^t D^{1/2}} D^{-\frac{1}{4}}}{2}. \tag{8.118}$$

Then there exists solutions $Y_k(t)$ $(k = 1, 2)$ of the form (8.115) *on* $[t_0, \infty)$ *such that*

$$|E_1(t)|\,,\ |D^{-1/2}(t) E_1'(t)| \leq \exp\left[2 \int_{t_0}^t |R_1(\tau)|\, d\tau\right] - 1,$$

$$|E_2(t)|\,,\ |D^{-1/2}(t) E_2'(t)| \leq \exp\left[2 \int_t^\infty |R_2(\tau)|\, d\tau\right] - 1.$$

Moreover, in the case that

$$\lambda_i(t) > 0 \qquad \text{for all } 1 \leq i \leq d, \quad t \geq t_0,$$

these estimates can be improved to

$$|E_1(t)| \leq \exp\left[\int_{t_0}^t M_1(\tau) |R_1(\tau)|\, d\tau\right] ds - 1, \tag{8.119}$$

$$|E_2(t)| \leq \exp\left[\int_t^\infty M_2(\tau) |R_2(\tau)|\, d\tau\right] ds - 1, \tag{8.120}$$

where

$$M_1(t) = \max_{1 \leq i \leq d} \left\{1 - e^{-\int_t^\infty 2\sqrt{\lambda_i(\tau)}\, d\tau}\right\}$$

$$M_2(t) = \max_{1 \leq i \leq d} \left\{1 - e^{-\int_{t_0}^t 2\sqrt{\lambda_i(\tau)}\, d\tau}\right\}. \tag{8.121}$$

Furthermore, in this case,

$$\left|D^{-1/2} E_1'\right| \leq 2\, e^{\int_{t_0}^t |R_1(\tau)|\, d\tau} - 1, \qquad \left|D^{-1/2} E_2'\right| \leq 2\, e^{\int_t^\infty |R_2(\tau)|\, d\tau} - 1. \tag{8.122}$$

Remark 8.16 We note that (8.119) and (8.120) are improvements of results in [145, Theorem 2.1], where the weaker estimate $M(\tau) \equiv 1$ was used. Moreover, the approach taken here allows us to correct a misstatement in [145], where the multiplicative factor "2" was neglected. (This was caused by an incorrect estimate of the norm of $\exp\left[-2\int_t^s D^{1/2}(\tau)\,d\tau\right]$ on p. 272.) Also see [146] for a discussion of the case $Y'' = [f(t)A + G(t)]Y$, where A is constant matrix, $G(t)$ is a matrix-valued and f is a scalar-valued function. This also includes some interesting examples and applications to semi-discretized wave and convection-diffusion equations.

Proof As mentioned in the introduction, the proof follows closely the outline of the approach taken in Sects. 8.3.2 and 8.3.3, and the notation chosen will reflect this. We start by writing (8.114) as the equivalent $2d \times 2d$ system

$$\mathbf{Y}' = \begin{pmatrix} 0 & I \\ D(t) + G(t) & 0 \end{pmatrix} \mathbf{Y}, \qquad \mathbf{Y} = \begin{pmatrix} Y \\ Y' \end{pmatrix}. \tag{8.123}$$

Since $G(t)$ is considered small compared to $D(t)$, one continues with a shearing transformation to achieve that the terms in the leading matrix have the same magnitude. Therefore, we put in (8.123)

$$\mathbf{Y} = \operatorname{diag}\{I, D^{1/2}(t)\}\hat{\mathbf{Y}},$$

to find that

$$\hat{\mathbf{Y}}' = \left[\begin{pmatrix} 0 & D^{1/2} \\ D^{1/2} & 0 \end{pmatrix} + \begin{pmatrix} 0 & 0 \\ D^{-1/2}G & -\frac{1}{2}D^{-1}D' \end{pmatrix}\right]\hat{\mathbf{Y}}.$$

In this proof we use capital letters to denote $d \times d$ matrices and bold-faced capital letters for $2d \times 2d$ matrices.

To diagonalize the leading matrix, let $\hat{\mathbf{Y}} = \begin{pmatrix} I & I \\ -I & I \end{pmatrix} \tilde{\mathbf{Y}}$ to find

$$\begin{aligned}\tilde{\mathbf{Y}}' &= \left[\begin{pmatrix} -D^{1/2} & 0 \\ 0 & D^{1/2} \end{pmatrix} + \frac{1}{4}\begin{pmatrix} -D^{-1}D' & D^{-1}D' \\ D^{-1}D' & -D^{-1}D' \end{pmatrix} + \frac{1}{2}\begin{pmatrix} -D^{-1/2}G & -D^{-1/2}G \\ D^{-1/2}G & D^{-1/2}G \end{pmatrix}\right]\tilde{\mathbf{Y}} \\ &= [\quad \tilde{\mathbf{D}}(t) \quad + \quad \mathbf{V}_1(t) \quad + \quad \mathbf{V}_2(t) \quad]\tilde{\mathbf{Y}}.\end{aligned} \tag{8.124}$$

Using the symmetry in both $\tilde{\mathbf{D}}(t)$ and $\mathbf{V}_1(t)$ in (8.124), one diagonalizes $\tilde{\mathbf{D}}(t) + \mathbf{V}_1(t)$ in the next step with a conditioning transformation of the form

$$\tilde{\mathbf{Y}} = [\mathbf{I} + \tilde{\mathbf{Q}}(t)]z = \begin{pmatrix} I + Q(t) & Q(t) \\ -Q(t) & I - Q(t) \end{pmatrix} \mathbf{Z}. \tag{8.125}$$

Here $Q(t)$ is determined by putting

$$\tilde{\mathbf{D}}\tilde{\mathbf{Q}} - \tilde{\mathbf{Q}}\tilde{\mathbf{D}} + \mathbf{V}_1 - \operatorname{diag}\mathbf{V}_1 = 0,$$

leading to $Q = \frac{1}{8}D^{-3/2}D'$, in particular Q is a diagonal $d \times d$-matrix.

It follows from (8.124) and this choice of $\tilde{\mathbf{Q}}(t)$ that

$$\mathbf{Z}' = \left[\tilde{\mathbf{D}}(t) + \operatorname{diag} \mathbf{V}_1(t) + \begin{pmatrix} -R(t) & -R(t) \\ R(t) & R(t) \end{pmatrix}\right] \mathbf{Z}$$

$$= \left[\begin{pmatrix} \Lambda_1(t) & \\ & \Lambda_2(t) \end{pmatrix} + \quad \mathbf{R}(t) \quad \right] \mathbf{Z}, \tag{8.126}$$

where

$$\Lambda_1(t) = -D^{1/2} - \frac{1}{4}D^{-1}D' \quad \text{ and } \Lambda_2(t) = D^{1/2} - \frac{1}{4}D^{-1}D', \tag{8.127}$$

and

$$R(t) = \frac{1}{2}D^{-1/2}G - \frac{5}{32}D^{-5/2}(D')^2 + \frac{1}{8}D^{-3/2}D''. \tag{8.128}$$

We note that the unperturbed $2d \times 2d$-system $\mathbf{X}' = \tilde{\mathbf{D}}(t)\mathbf{X}$ does not necessarily satisfy Levinson's dichotomy conditions. The reason is, that while $\Lambda_2(t) - \Lambda_1(t) = 2D^{1/2}$ is "well-behaved" because d "corresponding" diagonal entries are subtracted from each other, this is in general not true when subtracting any possible difference of the diagonal entries of $2d \times 2d$ matrix $\tilde{\mathbf{D}}(t)$. For example, $\Lambda_1(t) = \operatorname{diag}\{\mu_1(t), \mu_2(t)\}$ could be such that $\operatorname{Re} \int_{2n}^{2n+1} [\mu_1(t) - \mu_2(t)]\, dt \to +\infty$ as $n \to \infty$, whereas $\operatorname{Re} \int_{2n-1}^{2n} [\mu_1(t) - \mu_2(t)]\, dt \to -\infty$ as $n \to \infty$.

To establish the existence of a solution corresponding to $\Lambda_1(t)$, we begin by making in (8.126) the normalization

$$\mathbf{Z}(t) = \begin{pmatrix} \exp\left[\int^t \Lambda_1(s)\, ds\right] & \\ & \exp\left[\int^t \Lambda_1(s)\, ds\right] \end{pmatrix} \mathbf{U}(t).$$

It then follows that

$$\mathbf{U}' = \left[\begin{pmatrix} 0 & \\ & \Lambda_2(t) - \Lambda_1(t) \end{pmatrix} + \begin{pmatrix} -R_2(t) & -R_2(t) \\ R_2(t) & R_2(t) \end{pmatrix}\right] \mathbf{U}$$

$$= \left[\begin{pmatrix} 0 & \\ & 2D^{1/2}(t) \end{pmatrix} + \begin{pmatrix} -R_2(t) & -R_2(t) \\ R_2(t) & R_2(t) \end{pmatrix}\right] \mathbf{U}, \tag{8.129}$$

where

$$R_2(t) = e^{-\int^t \Lambda_1(\tau)\, d\tau} R(t) e^{\int^t \Lambda_1(\tau)\, d\tau}.$$

Using (8.128) and (8.127), $R_2(t)$ can be re-written as in (8.118).

We will show that (8.129) has, for all $t \geq t_0$, a $2d \times d$ matrix solution $\mathbf{u}(t)$ of the integral equation

$$\mathbf{u}(t) = \begin{pmatrix} I_{d\times d} \\ 0_{d\times d} \end{pmatrix} - \int_t^\infty \begin{pmatrix} I_{d\times d} & \\ & e^{-\int_t^s 2D^{1/2}} \end{pmatrix} \begin{pmatrix} -R_2(s) & -R_2(s) \\ R_2(s) & R_2(s) \end{pmatrix} \mathbf{u}(s)\, ds. \tag{8.130}$$

For that purpose, we make one final normalization

$$\mathbf{w}(t) = \mathbf{u}(t) - \begin{pmatrix} I_{d\times d} \\ 0_{d\times d} \end{pmatrix}. \tag{8.131}$$

Then (8.130) having a solution $\mathbf{u}(t)$, $\lim_{t\to\infty} \mathbf{u}(t) = \begin{pmatrix} I_{d\times d} \\ 0_{d\times d} \end{pmatrix}$ is equivalent to showing that

$$\begin{aligned} \mathbf{w}(t) &= -\int_t^\infty \begin{pmatrix} I_{d\times d} & \\ & e^{-\int_t^s 2D^{1/2}} \end{pmatrix} \begin{pmatrix} -R_2(s) & -R_2(s) \\ R_2(s) & R_2(s) \end{pmatrix} \left[\begin{pmatrix} I_{d\times d} \\ 0_{d\times d} \end{pmatrix} + \mathbf{w}(s) \right] ds \\ &= \int_t^\infty \begin{pmatrix} I_{d\times d} & \\ & e^{-\int_t^s 2D^{1/2}} \end{pmatrix} \begin{pmatrix} R_2(s) \\ -R_2(s) \end{pmatrix} ds \\ &\quad - \int_t^\infty \begin{pmatrix} I_{d\times d} & \\ & e^{-\int_t^s 2D^{1/2}} \end{pmatrix} \begin{pmatrix} -R_2(s) & -R_2(s) \\ R_2(s) & R_2(s) \end{pmatrix} \mathbf{w}(s)\, ds \\ &=: (\mathbf{Tw})(t) \end{aligned} \tag{8.132}$$

has a solution $\mathbf{w}(t)$ such that $\lim_{t\to\infty} \mathbf{w}(t) = 0_{2d\times d}$.

[A]: Existence of a Solution of the Integral Equation (8.130) *on* $[t_0, \infty)$

In what follows, we choose for an $m \times n$ matrix A the sub-multiplicative matrix norm

$$|A| = \max_{1\leq i\leq m} \sum_{j=1}^n |a_{ij}|. \tag{8.133}$$

Define the function $\rho_2 : [t_0, \infty) \to \mathbb{R}^+$ by

$$\rho_2(t) := \int_t^\infty |R_2(s)|\, ds, \tag{8.134}$$

where $R_2(t)$ was defined in (8.118).

We will assume w.l.o.g. that the perturbation $R_2(t)$ does not vanish on an interval of the form $[t_1, \infty)$ for some $t_1 \geq t_0$, i.e., we assume that $\rho_2(t) > 0$ for all $t \geq t_0$.

Define the set

$$\mathbf{B} = \left\{ \mathbf{w}(t) \in \mathbb{C}^{2d \times d} : \|\mathbf{w}\| = \sup_{t \geq t_0} \frac{|\mathbf{w}(t)|}{e^{2\rho_2(t)} - 1} < \infty. \right\}$$

Then $\mathbf{B}$ is a Banach space. We also define

$$\mathbf{W} = \{\mathbf{w}(t) \in \mathbf{B} : \|\mathbf{w}\| \leq 1\},$$

hence $\mathbf{W}$ is a closed subset of $\mathbf{B}$. We will show that the operator $\mathbf{T}$ defined in (8.132) maps $\mathbf{W}$ into $\mathbf{W}$ and is a contraction. Given $\mathbf{w}(t) \in \mathbf{W}$, note that

$$\begin{aligned}
|(\mathbf{Tw})(t)| &\leq \int_t^\infty \left| \begin{pmatrix} I_{d\times d} & \\ & e^{-\int_t^s 2D^{1/2}} \end{pmatrix} \right| \left| \begin{pmatrix} R_2(s) \\ -R_2(s) \end{pmatrix} \right| ds \\
&\quad + \int_t^\infty \left| \begin{pmatrix} I_{d\times d} & \\ & e^{-\int_t^s 2D^{1/2}} \end{pmatrix} \begin{pmatrix} -R_2(s) & -R_2(s) \\ R_2(s) & R_2(s) \end{pmatrix} \mathbf{w}(s) \right| ds \\
&\leq \int_t^\infty |R_2(s)|\, ds + 2\int_t^\infty |R_2(s)| \left[e^{2\rho_2(s)} - 1 \right] ds \\
&= e^{2\rho_2(t)} - \rho_2(t) - 1 < e^{2\rho_2(t)} - 1.
\end{aligned}$$

Here we used that from the matrix norm defined in (8.133) follows that

$$\left| \begin{pmatrix} R_2(s) \\ -R_2(s) \end{pmatrix} \right| = |R_2(s)|, \quad \left| \begin{pmatrix} -R_2(s) & -R_2(s) \\ R_2(s) & R_2(s) \end{pmatrix} \right| = 2|R_2(s)|, \quad \left| \begin{pmatrix} I_{d\times d} & \\ & e^{-\int_t^s 2D^{1/2}} \end{pmatrix} \right| = 1.$$

Therefore

$$\frac{|(\mathbf{Tw})(t)|}{e^{2\rho_2(t)} - 1} < 1,$$

and taking the supremum over all $t \geq t_0$ establishes that $\|\mathbf{Tw}\| \leq 1$. To establish that $\mathbf{T}$ is a contraction, let $\mathbf{w}(t)$ and $\tilde{\mathbf{w}}(t)$ be elements in $\mathbf{W}$. Then it follows for $t \geq t_0$ that

$$\begin{aligned}
|(\mathbf{Tw})(t) - (\mathbf{T}\tilde{\mathbf{w}})(t)| &= \int_t^\infty \left| \begin{pmatrix} I_{d\times d} & \\ & e^{-\int_t^s 2D^{1/2}} \end{pmatrix} R_2(s) \begin{pmatrix} -1 & -1 \\ 1 & 1 \end{pmatrix} \right| |\mathbf{w}(s) - \tilde{\mathbf{w}}(s)|\, ds \\
&\leq \|\mathbf{w} - \tilde{\mathbf{w}}\| \int_t^\infty 2|R_2(s)| \left[e^{2\rho_2(s)} - 1 \right] ds \\
&= \|\mathbf{w} - \tilde{\mathbf{w}}\| \left(e^{2\rho_2(t)} - 1 - 2\rho_2(t) \right).
\end{aligned}$$

Hence

$$\frac{|(\mathbf{T}\mathbf{w})(t) - (\mathbf{T}\tilde{\mathbf{w}})(t)|}{e^{2\rho_2(t)} - 1} \leq \|\mathbf{w} - \tilde{\mathbf{w}}\| \left(1 - \frac{2\rho_2(t)}{e^{2\rho_2(t)} - 1}\right) \leq \|\mathbf{w} - \tilde{\mathbf{w}}\| \left(1 - \frac{2\rho_2(t_0)}{e^{2\rho_2(t_0)} - 1}\right).$$

Here we used that $\rho_2(t)$ is a non-increasing function on $[t_0, \infty)$ and the fact that the function $1 - 2x/(e^{2x} - 1)$ is strictly increasing for positive values of x. Put

$$\hat{\theta} = 1 - \frac{2\rho_2(t_0)}{e^{2\rho_2(t_0)} - 1}.$$

Then $\hat{\theta} \in (0, 1)$ and taking the supremum over all $t \geq t_0$ shows that

$$\|\mathbf{T}(\mathbf{w} - \tilde{\mathbf{w}})\| \leq \hat{\theta} \|\mathbf{w} - \tilde{\mathbf{w}}\|,$$

so $\mathbf{T}$ is a contraction.

The Banach Fixed Point principle now implies that there exists for $t \geq t_0$ a unique function $\mathbf{w}(t) \in \mathbf{W}$ satisfying (8.132). By (8.131), (8.129) then has for $t \geq t_0$ a unique solution $\mathbf{u}(t)$ satisfying (8.130) for all $t \geq t_0$. Next, we use Gronwall's inequality to provide a sharper error estimate.

[B]: Use of Gronwall's Inequality to Derive Bounds for the Error on $[t_0, \infty)$

It follows that (8.123) has a solution $\mathbf{Y}_2 = \left(Y_2, Y_2'\right)$ satisfying for $t \geq t_0$

$$\mathbf{Y}_2 = \begin{pmatrix} Y_2 \\ Y_2' \end{pmatrix} = \begin{pmatrix} I & \\ & D^{1/2} \end{pmatrix} \begin{pmatrix} I & I \\ -I & I \end{pmatrix} \begin{pmatrix} I + Q & Q \\ -Q & I - Q \end{pmatrix} \begin{pmatrix} e^{\int^t \Lambda_1(s)\, ds} & \\ & e^{\int^t \Lambda_1(s)\, ds} \end{pmatrix} \mathbf{u}(t),$$

with $\mathbf{u}(t)$ given in (8.130). Looking at the first $d \times d$ matrix component, (8.114) has a solution Y_2 for $t \geq t_0$ satisfying

$$Y_2 = D^{-1/4}(t) e^{-\int^t D^{1/2}(\tau)\, d\tau} \left[I + E_2(t)\right], \tag{8.135}$$

where

$$E_2(t) = \int_t^\infty \left[I - e^{-\int_t^s 2D^{1/2}}\right] R_2(s)[\mathbf{u}_1(s) + \mathbf{u}_2(s)]\, ds, \tag{8.136}$$

where $\mathbf{u} = (\mathbf{u}_1, \mathbf{u}_2)^T$, $\mathbf{u}_i$ being $d \times d$ matrices. Adding up the components of u in (8.130), it follows that

$$|\mathbf{u}_1(t) + \mathbf{u}_2(t)| \leq 1 + \int_t^\infty \left| \left[I - e^{-\int_t^s 2D^{1/2}} \right] R_2(s) \left[\mathbf{u}_1(s) + \mathbf{u}_2(s) \right] \right| ds.$$

From (8.116) it follows for $s \geq t$ that

$$\left| I - e^{-\int_t^s 2D^{1/2}} \right| = \max_{1 \leq i \leq d} \left| 1 - e^{-\int_t^s 2\sqrt{\lambda_i(\tau)}\, d\tau} \right| \leq 2, \tag{8.137}$$

hence $|\mathbf{u}_1(t) + \mathbf{u}_2(t)| \leq 1 + \int_t^\infty 2|R_2(s)|\, |\mathbf{u}_1(s) + \mathbf{u}_2(s)|\, ds$, and Lemma 8.3 yields that

$$|\mathbf{u}_1(t) + \mathbf{u}_2(t)| \leq \exp\left[\int_t^\infty 2|R_2(s)|\, ds \right] = e^{2\rho_2(t)}, \tag{8.138}$$

where $\rho_2(t)$ was defined in (8.134). Applying this estimate to (8.136), one finds that

$$|E_2(t)| \leq \int_t^\infty 2|R_2(s)| \exp\left[\int_s^\infty 2|R_2(u)|\, du \right] ds = e^{2\rho_2(t)} - 1.$$

Moving to the estimate of $D^{-1/2}(t)E_2'(t)$, observe that (8.136) implies that for $t \geq t_0$

$$E_2' = -2D^{1/2} \int_t^\infty e^{-\int_t^s 2D^{1/2}} R_2(s)[\mathbf{u}_1(s) + \mathbf{u}_2(s)]\, ds,$$

and therefore (8.116) and (8.138) imply that

$$\left| D^{-1/2} E_2' \right| \leq 2 \int_t^\infty |R_2(s)| e^{2\rho_2(s)}\, ds = e^{2\rho_2(t)} - 1.$$

In case $\lambda_i(t) \geq 0$ for all $1 \leq i \leq d$, (8.137) can be sharpened to

$$\begin{aligned} \left| I - e^{-\int_t^s 2D^{1/2}} \right| &= \max_{1 \leq i \leq d} \left\{ 1 - e^{-\int_t^s 2\sqrt{\lambda_i(\tau)}\, d\tau} \right\} \\ &\leq \max_{1 \leq i \leq d} \left\{ 1 - e^{-\int_{t_0}^s 2\sqrt{\lambda_i(\tau)}\, d\tau} \right\} = M_2(s), \end{aligned}$$

where $M_2(t)$ was defined in (8.121), and it follows that $M_2(s) \in [0, 1)$. Proceeding as above, one finds that

$$|\mathbf{u}_1(t) + \mathbf{u}_2(t)| \leq \exp\left[\int_t^\infty M_2(s)|R_2(s)|\, ds \right],$$

$$|E_2(t)| \leq \exp\left[\int_t^\infty M_2(\tau)|R_2(\tau)|\, d\tau \right] ds - 1,$$

and

$$\left|D^{-1/2}E_2'\right| \le 2\int_t^\infty |R_2(s)| \exp\left[\int_s^\infty |R_2(\tau)|\,d\tau\right] ds = 2e^{\rho_2(t)} - 1,$$

hence all statements concerning the solution $Y_2(t)$ in Theorem 8.15 are established.

As to a solution corresponding to $\Lambda_2(t)$, one makes in (8.126) the normalization

$$\mathbf{Z}(t) = \begin{pmatrix} \exp\left[\int^t \Lambda_2(s)\,ds\right] & \\ & \exp\left[\int^t \Lambda_2(s)\,ds\right] \end{pmatrix} \mathbf{V}(t)$$

which implies that

$$\mathbf{V}' = \left[\begin{pmatrix} -2D^{1/2}(t) & \\ & 0 \end{pmatrix} + \begin{pmatrix} -R_1(t) & -R_1(t) \\ R_1(t) & R_1(t) \end{pmatrix}\right]\mathbf{V},$$

where

$$R_1(t) = e^{-\int^t \Lambda_2(\tau)\,d\tau} R(t) e^{\int^t \Lambda_2(\tau)\,d\tau},$$

which can be shown to coincide with (8.117). It also can be shown that this has for $t \ge t_0$ a solution of the integral equation

$$\mathbf{v}(t) = \begin{pmatrix} 0_{d\times d} \\ I_{d\times d} \end{pmatrix} + \int_{t_0}^t \begin{pmatrix} e^{-\int_s^t -2D^{1/2}} & \\ & I_{d\times d} \end{pmatrix} \begin{pmatrix} -R_2(s) & -R_2(s) \\ R_2(s) & R_2(s) \end{pmatrix} \mathbf{v}(s)\,ds,$$

and the rest of the proof follows the outline above. □

For a system of second-order equations of the form

$$y'' = F(t)y, \qquad t \ge a, \;\; y \in \mathbb{C}^d, \tag{8.139}$$

D.R. Smith [139] has given a procedure for obtaining some Liouville-Green formulas for solutions analogous to his treatment of the scalar case (see Sect. 8.3.2). It is assumed that $F(t)$ is a $d \times d$ matrix of piecewise $C^{(1)}$ functions satisfying the following properties:

(a) $F(t) = G(t)\Lambda^2(t)G^{-1}(t)$, where $G(t)$ is invertible for $t \ge a$;
(b) $\Lambda(t) = \operatorname{diag}\{\lambda_1(t), \ldots, \lambda_d(t)\}$ with $\operatorname{Re}\lambda_i(t) > 0$ for $t \ge a$;
(c) $\sup_{t\ge a} \int_t^\infty |\lambda_i'(s)/\lambda_i(s)| \exp\left[-2\int_t^s \operatorname{Re}\lambda_i\right] ds < \infty$;
(d) $G^{-1}(t)G'(t)$ is "sufficiently small."

The goal is to represent solution matrices of (8.139) in the form

$$Y_1(t) = G(t)\,[I + E_1(t)]\,\Lambda^{-1/2} \exp\left[-\int^t \Lambda(s)\,ds\right][I + \tilde{E}_1(t)]$$

$$Y_1(t) = G(t)\,[I + E_2(t)]\,\Lambda^{-1/2} \exp\left[+\int^t \Lambda(s)\,ds\right][I + \tilde{E}_2(t)],$$

(with corresponding expressions for Y_1' and Y_2'), where E_i, $\tilde{E}_i = o(1)$ as $t \to \infty$ and can be estimated in terms of G and Λ. The assumption (d) above consists of a number of conditions which are sufficient for the method described briefly below to work.

Treating (8.139) as the equivalent $2d$-dimensional system

$$\begin{pmatrix} y \\ y' \end{pmatrix}' = \begin{pmatrix} 0 & I_d \\ F & 0 \end{pmatrix} \begin{pmatrix} y \\ y' \end{pmatrix},$$

a preliminary transformation

$$\begin{pmatrix} y \\ y' \end{pmatrix} = R \begin{pmatrix} z_1 \\ z_2 \end{pmatrix} \qquad \text{with } R = \begin{pmatrix} G & G \\ -G\Lambda & G\Lambda \end{pmatrix},$$

yields

$$\begin{pmatrix} z_1 \\ z_2 \end{pmatrix}' = \left[\begin{pmatrix} -\Lambda & 0 \\ 0 & \Lambda \end{pmatrix} - R^{-1}R'\right] \begin{pmatrix} z_1 \\ z_2 \end{pmatrix}, \tag{8.140}$$

where

$$R^{-1}R' = \frac{1}{2}\begin{pmatrix} G^{-1}G' + \Lambda^{-1}G^{-1}(G\Lambda)' & G^{-1}G' - \Lambda^{-1}G^{-1}(G\Lambda)' \\ G^{-1}G' - \Lambda^{-1}G^{-1}(G\Lambda)' & G^{-1}G' + \Lambda^{-1}G^{-1}(G\Lambda)' \end{pmatrix}.$$

Then (as in the scalar case, see Sect. 8.3.2), a Riccati transformation

$$\begin{pmatrix} z_1 \\ z_2 \end{pmatrix} = \begin{pmatrix} I_d & 0 \\ T & I_d \end{pmatrix} \begin{pmatrix} I_d & S \\ 0 & I_d \end{pmatrix} \begin{pmatrix} w_1 \\ w_2 \end{pmatrix}$$

is used to completely diagonalize the system. For this purpose, the $d \times d$ matrices T and S are required to satisfy first-order systems of differential equations. The first system for T is independent of S and is a matrix Riccati differential equation corresponding to (8.24). The second system for S is linear with coefficients also depending linearly on T. For the existence of solutions $S, T = o(1)$ as $t \to \infty$, assumptions (d) are made [similar to (8.26)] which allow for employing the contraction mapping principle and also making error estimates. See [139] for details and precise statements of these assumptions and conclusions. The results are applied to a Weber–Airy equation and several references to concrete applications are given, including one to hydrodynamic stability.

Chapter 9
Applications to Classes of Scalar Linear Difference Equations

9.1 Chapter Overview

In this chapter we are interested in scalar dth-order linear difference equations (also called linear recurrence relations) of the form

$$y(n+d) = c_1(n)y(n) + \cdots + c_d(n)y(n+d-1), \qquad n \in \mathbb{N}, \tag{9.1}$$

and we wish to apply the results of Chaps. 3 and 5 to obtain asymptotic representations for solutions. We will always treat (9.1) as the equivalent system

$$\mathbf{y}(n+1) = C(n)\mathbf{y}(n), \tag{9.2}$$

where $C(n)$ is the companion matrix

$$C(n) = \begin{pmatrix} 0 & 1 & & 0 \\ 0 & 0 & \ddots & \\ \vdots & & & 1 \\ c_1(n) & c_2(n) & \cdots & c_d(n) \end{pmatrix}, \qquad \mathbf{y}(n) = \begin{pmatrix} y(n) \\ y(n+1) \\ \vdots \\ y(n+d-1) \end{pmatrix}.$$

Under various assumptions on $c_i(n)$ ($1 \leq i \leq d$) we obtain asymptotic representations for either one vector-valued solution or a fundamental matrix (which exists provided $c_1(n) \neq 0$). Then we specialize this information to obtain asymptotic formulas for solutions of (9.1).

We first consider in Sect. 9.2 the important and from a historical perspective earliest general results on asymptotically constant scalar equations, i.e., $c_i(n) = a_i + b_i(n)$, $b_i(n) = o(1)$ as $n \to \infty$. Such equations were first considered by Poincaré [130] and later by O. Perron in a series of papers [117, 118, 121]. We

S. Bodine, D.A. Lutz, *Asymptotic Integration of Differential and Difference Equations*, Lecture Notes in Mathematics 2129,
DOI 10.1007/978-3-319-18248-3_9

apply results from Chap. 3 and discuss extensions of Perron's results, leading to not only a more general setting, but including somewhat more precise estimates and also an alternative operator-type proof.

Next, in Sect. 9.3, we discuss the important case of second-order linear recurrence relations. We present an analogue of a continuous Liouville–Green-type result (see Sect. 8.3.2) for equations of the form

$$\Delta^2 y(n) = [f(n) + g(n)]y(n).$$

We recall from Sect. 8.3.3 that Olver has obtained much more precise error estimates for certain solutions of second-order differential equations, but we will not discuss the analogues of such estimates here. Only in the special case of Sect. 9.4.1 we will provide precise error estimates.

A difficulty in applying a general Liouville–Green result to a given equation usually rests on appropriately normalizing the equation and then finding a suitable decomposition into $f(n) + g(n)$ which satisfy the assumptions. We observe that any one Liouville–Green type statement does not appear to fit all the kinds of special classes of second order equations we wish to consider.

In Sect. 9.4, we will particularly investigate certain classes of second-order equations corresponding to special kinds of perturbations. For second-order equations, finding suitable preliminary and normalizing transformations are more tractable than for higher-order equations. In addition, 2×2 matrices have the advantage that eigenvalues can be easily calculated explicitly. Second-order equations are also very important for applications from other areas, notably in the asymptotic theory of orthogonal polynomials. Due to Favard's theorem (see, e.g., [59]), all orthogonal polynomials corresponding to a positive measure $d\alpha$ satisfy equations

$$xp_n(x) = a_{n+1}p_{n+1}(x) + b_n p_n(x) + a_n p_{n-1}(x),$$

with coefficients a_n, b_n generated by their moments. Certain important classes of such polynomials correspond to sequences $\{a_n\}$, $\{b_n\}$ satisfying various asymptotic properties (see, e.g., [113]). We will examine several kinds of results from the literature in Sect. 9.4 to determine how the application of general principles from Chaps. 3 and 5 compares with using either Riccati-type or certain kinds of ad hoc techniques.

Riccati-based approaches involve reduction to an equivalent nonlinear first order equation, similar to the well-known Riccati differential equation. We will discuss in Sect. 9.4.4 an example originally treated in this manner. Another technique somewhat related to the Riccati approach uses an iteration to obtain "formal" solutions as continued fraction expansions. This leads to some interesting asymptotic representations for solutions of equations with what could be called "averaged growth conditions." We discuss this in Sect. 9.4.3.

Other applications to second-order equations arise from the analysis of Jacobi transformations. There, asymptotic representation results are used to determine behavior of eigenfunctions of the operators and applied to determine spectral information. One particularly interesting and deep result we discuss in Sect. 9.6 concerns a kind of "almost everywhere" asymptotic representation for solutions analogous to the Christ–Kiselev results described in Sect. 8.6.2.

Finally we mention that there are many other important applications of asymptotic behavior of higher-order scalar recurrence relations arising from combinatorial sequences (see, e.g., [161]). These interesting examples could also be treated using the results of Chaps. 3 and 5, but in fact have polynomial or rational coefficients. Hence the meromorphic methods (see, e.g., [82]) can and have been used there and we will not discuss any specific examples here. We mention, however, that in cases where just the main asymptotic terms and not a complete asymptotic expansion is required, results from Chaps. 3 and 5 could be simpler and more direct.

9.2 Asymptotically Constant Scalar Equations

In what could be considered the first general asymptotic treatment for solutions of linear difference equations, H. Poincaré [130] discussed for $n \geq n_0$ equations of the form

$$\begin{cases} y(n+d) + [a_d + b_d(n)]\,y(n+d-1) + \cdots + [a_1 + b_1(n)]\,y(n) = 0, \\ \qquad \text{where } b_i(n) = o(1) \text{ as } n \to \infty, \ 1 \leq i \leq d, \end{cases} \tag{9.3}$$

and proved the following

Theorem 9.1 (Poincaré) *Consider* (9.3) *and assume that the roots of the limiting characteristic equation*

$$p(\lambda) := \lambda^d + a_d\lambda^{d-1} + \ldots + a_1 = 0 \tag{9.4}$$

satisfy the condition

$$0 \leq |\lambda_1| < |\lambda_2| < \cdots < |\lambda_d|. \tag{9.5}$$

If $y(n)$ is a solution of (9.3)*, then either*

(a) $y(n) \equiv 0$ for all n sufficiently large, or
(b) $y(n) \neq 0$ for all n sufficiently large and

$$\lim_{n\to\infty} \frac{y(n+1)}{y(n)} = \lambda, \tag{9.6}$$

where λ is a root of (9.4).

The main elements of difficulty in Poincaré's proof are first showing the alternatives (a) and (b) and then proving that the limit (9.6) exists. Then it follows immediately that $\lim_{n\to\infty} y(n+k)/y(n) = \lambda^k$ for $1 \leq k \leq d$, and dividing (9.3) by $y(n)$ one sees that λ must be a root of (9.4).

In [130], Poincaré actually discussed a more special case of (9.3) than in the above statement. Because of his interest both in linear difference as well as differential equations, and the interaction between solutions (especially in the classical cases of special functions), he considered equations of the form

$$p_d(n)y(n+d) + p_{d-1}(n)y(n+d-1) + \cdots + p_0(n)y(n) = 0, \tag{9.7}$$

where $p_i(n)$ were assumed to be polynomials, all of the *same degree*. Normalizing by dividing by $p_d(n)$ lead him to (9.3), where the $b_i(n)$ would be rational functions in n, all satisfying $O(1/n)$ as $n \to \infty$. But because Poincaré did not use the rational behavior of the coefficients to obtain his result (9.6), he is given credit for that more general statement including the case that a_1 could be zero.

Inspired by Poincaré's result (9.6), O. Perron [117] went on to prove a somewhat stronger result concerning a fundamental system of linearly independent solutions of (9.3) in the following case. Namely, he also considered the possibility of one the roots of (9.4) being zero, but made the additional assumption that

$$\text{if } \; a_1 = 0, \quad \text{then } b_1(n) \neq 0 \text{ for all } n \text{ sufficiently large.} \tag{9.8}$$

He then showed that there exists for each root λ_i of (9.4) a non-trivial solution $y_i(n)$ of (9.3) satisfying (9.6) with $y(n) = y_i(n)$ and $\lambda = \lambda_i$ $(1 \leq i \leq d)$. Assumption (9.8) implies that the coefficient matrix of the corresponding companion system is invertible for all n sufficiently large and so a fundamental system of solutions exists.

In case (9.3), (9.5), and (9.8) are all satisfied, Poincaré's result immediately follows from Perron's work. This is because a fundamental system $\{y_1(n), \ldots, y_d(n)\}$ of solutions satisfying $\frac{y_i(n+1)}{y_i(n)} = \lambda_i + o(1)$ as $n \to +\infty$ exists and any nontrivial solution $y(n)$ can be expressed as $y(n) = \sum_{i=1}^{d} c_i y_i(n)$, where not all the c_i are zero. Letting

$$j = \max_i \{ i \,|\, c_i \neq 0 \},$$

it follows immediately from (9.5) that

$$\frac{y(n+1)}{y(n)} = \lambda_j + o(1) \qquad \text{as } n \to \infty.$$

However, it is important to observe that Poincaré did not assume (9.8) and so his more general statement is not a consequence of Perron's.

We will now delay the proof of Poincaré's theorem until after a discussion of Perron's result. We first consider a more general setting than assumed by Perron and show that the assumption (9.5) can be weakened to just assuming that a single non-zero root of (9.4) has modulus different from all the others. Then we obtain a single solution of Perron's type, but as it turns out by applying Theorem 3.17, a somewhat more explicit asymptotic statement also follows. We state this more general form of a Perron-type of result as

Theorem 9.2 *Suppose that there exists a root $\lambda_j \neq 0$ of* (9.4) *such that*

$$0 \leq |\lambda_1| \leq |\lambda_2| \leq \ldots \leq |\lambda_{j-1}| < |\lambda_j| < |\lambda_{j+1}| \leq \ldots \leq |\lambda_d|.$$

Then (9.3) *has for n sufficiently large a solution satisfying*

$$y_j(n) = [1 + o(1)] \prod^{n-1} \left[\lambda_j + l_j(k)\right], \tag{9.9}$$

where

$$l_j(k) = r_{jj}(k) + o\left(\sum_{\nu \neq j} |r_{j\nu}(k)|\right).$$

Here $r_{j\nu}$ are elements of the matrix $R(n) = o(1)$ defined in (9.12) *given below and, in particular, $r_{jj}(n)$ is given in* (3.85) *and* (3.84).

Proof We express (9.3) as an equivalent system of the form

$$\mathbf{y}(n+1) = [A + B(n)]\mathbf{y}(n), \tag{9.10}$$

where

$$A = \begin{pmatrix} 0 & 1 & & 0 \\ 0 & 0 & \ddots & \\ \vdots & & & 1 \\ -a_1 & -a_2 & \cdots & -a_d \end{pmatrix}, \quad B(n) = \begin{pmatrix} 0 & 0 & \cdots & 0 \\ \vdots & & & \vdots \\ 0 & 0 & \cdots & 0 \\ -b_1(n) & -b_2(n) & \cdots & -b_d(n), \end{pmatrix}. \tag{9.11}$$

Note that the characteristic equation of A is (9.4). We bring A into the form (3.73) by a similarity transformation

$$U^{-1}AU = J = \begin{pmatrix} A_{-1} & & \\ & \lambda_j & \\ & & A_1 \end{pmatrix}.$$

Here A_{-1} is a $(j-1) \times (j-1)$ matrix whose eigenvalues satisfy $|\lambda| < |\lambda_j|$ and A_1 is a $(d-j) \times (d-j)$ matrix whose eigenvalues satisfy $|\lambda| > |\lambda_j|$. Observe that the jth column of U, $\mathbf{u}_j$, is an eigenvector of A corresponding to λ_j and because A is a companion matrix, one can take

$$\mathbf{u}_j = (1, \lambda_j, \lambda_j^2, \ldots, \lambda_j^{d-1})^T.$$

Then $\mathbf{y}(n) = U\tilde{y}(n)$ yields

$$\tilde{y}(n+1) = [J + R(n)]\tilde{y}(n)$$

with

$$R(n) = U^{-1}B(n)U = \mathrm{o}(1) \text{ as } n \to \infty. \tag{9.12}$$

Then Theorem 3.17 implies for n sufficiently large the existence of a solution vector $\tilde{y}_j(n)$ satisfying

$$\tilde{y}_j(n) = [e_j + \mathrm{o}(1)] \prod^{n-1} [\lambda_j + l_j(k)],$$

where $l_j(n) = r_{jj}(n) + \mathrm{o}\left(\sum_{\nu \neq j} |r_{j\nu}(n)|\right)$, and $r_{jj}(n)$ can be expressed in terms of the $b_j(n)$ using (3.85) and (3.84).

Therefore (9.10) has for n sufficiently large a solution

$$\mathbf{y}_j(n) = [\mathbf{u}_j + \mathrm{o}(1)] \prod^{n-1} [\lambda_j + l_j(k)],$$

and looking at the first component of $\mathbf{y}_j(n)$ yields (9.9). □

One therefore obtains (9.6) for this one solution $\mathbf{y}_j(n)$ under considerably weaker conditions than (9.5) and (9.8).

In case (9.4) satisfies (9.5) and, in addition, that each $\lambda_j \neq 0$, one obtains as a corollary that (9.3) has a fundamental system of solutions $y_j(n)$ satisfying (9.6) for each λ_j and with a somewhat more explicit asymptotic representation (9.9). The proof of this, which follows from Theorems 3.17 and 3.12, offers an alternative to Perron's approach. See also [117] and [109], the other sources we know of in the literature.

In case $a_1 = 0$ and one of the λ_i, say λ_1, is equal to zero, as remarked above it is necessary to assume Perron's condition (9.8) to ensure that the coefficient matrix in (9.10) is invertible. Letting in (9.10)

$$\mathbf{y}(n) = V_0 \hat{y}(n),$$

where

$$V_0 = \begin{pmatrix} 1 & 1 & \cdots & 1 \\ 0 & \lambda_2 & \cdots & \lambda_d \\ \vdots & & \vdots & \\ 0 & \lambda_2^{d-1} & \cdots & \lambda_d^{d-1} \end{pmatrix},$$

and where we also assume (9.5), we obtain

$$\hat{y}(n+1) = [\Lambda_0 + R(n)]\,\hat{y}(n),$$

with $\Lambda_0 = \operatorname{diag}\{0, \lambda_2, \ldots, \lambda_d\}$ and $R(n) = V_0^{-1}B(n)V_0 = o(1)$. It can be seen that $r_{11}(n)$ is a non-zero scalar multiple of $b_1(n)$ (the lower left entry in $B(n)$ in (9.11)) and therefore it is nonzero by (9.8). In order to apply Theorem 3.18 to this situation, we furthermore require that

$$b_j(n) = O(|b_1(n)|) \qquad \text{for each } j > 1. \tag{9.13}$$

This yields the asymptotic representation

$$y_1(n) = [1 + o(1)] \prod^{n-1} \left[r_{11}(k) + o\left(\sum_{\nu \neq 1} |r_{1\nu}(k)| \right) \right],$$

which is somewhat more explicit than Perron's result $y_1(n+1)/y_1(n) = o(1)$ as $n \to \infty$, but at the expense of making the stronger additional hypothesis (9.13).

In spite of the fact that in the general case (that is, just assuming (9.5), but not (9.8)), Perron's result does not directly imply Poincaré's Theorem 9.1, we will now show that Theorem 9.2 together with the principle of mathematical induction is sufficient to give a new proof of Poincaré's theorem.

(Proof of Theorem 9.1). First consider $d = 1$. The sequence $\{a_1 + b_1(n)\}$ is either nonzero for all n sufficiently large or else there exists an integer subsequence $\{n_k\}$ for which $a_1 + b_1(n_k) = 0$. (Observe that since $b_1(n) = o(1)$, the latter case only occurs when $a_1 = 0$.) In the former case $\hat{y}(n) = (-1)^n \prod_{n_1}^{n-1} [a_1 + b_1(k)]$ is nonzero for n_1 sufficiently large, hence any other solution $y(n)$ is either also nonzero or identically zero. If $y(n)$ is not identically zero, then $y(n+1)/y(n) = -[a_1 + b_1(n)] \to -a_1$, the eigenvalue of $\lambda + a_1 = 0$. In the latter case, every solution will be eventually identically zero.

Now let $d \geq 2$ and suppose that Theorem 9.1 holds for all Eq. (9.3) of order $d-1$. We want to make a reduction of order using a solution $y_d(n)$ of (9.3) corresponding to the root λ_d of largest modulus. Since $\lambda_d \neq 0$ and there are no other roots of (9.4) with modulus $|\lambda_d|$, one can rewrite (9.3) as a corresponding companion matrix system and apply Theorem 3.12 which yields for n sufficiently large (say $n \geq n_1$) the existence of a solution $y_d(n)$ satisfying

$$y_d(n) = [1 + o(1)] \prod_{n_1}^{n-1} [\lambda_d + \varepsilon(k)],$$

where $\varepsilon(k) \to 0$ as $k \to \infty$. Since $\lambda_d \neq 0$, $y_d(n) \neq 0$ for n_1 sufficiently large and thus

$$\lim_{n\to\infty} \frac{y_d(n+1)}{y_d(n)} = \lim_{n\to\infty} [1 + o(1)][\lambda_d + \varepsilon(n)] = \lambda_d. \tag{9.14}$$

Let $y(n)$ be any other solution of (9.3) and define for $n \geq n_1$

$$u(n) = \frac{y(n)}{y_d(n)}. \tag{9.15}$$

We want to show that either $u(n) \equiv 0$ for all n sufficiently large or else $u(n) \neq 0$ for all large n and in the latter case

$$\frac{u(n+1)}{u(n)} \to \frac{\lambda}{\lambda_d}, \qquad \text{as } n \to \infty, \tag{9.16}$$

where λ is a root of (9.4). Therefore either $y(n) \equiv 0$ for n sufficiently large or

$$\lim_{n\to\infty} \frac{y(n+1)}{y(n)} = \lim_{n\to\infty} \frac{u(n+1)}{u(n)} \frac{y_d(n+1)}{y_d(n)} = \lambda.$$

To substantiate this claim, we make the substitution $y(n) = u(n)y_d(n)$ into (9.3). Using

$$y_d(n+d) = -[a_1 + b_1(n)]y_d(n+d-1) - \ldots - [a_d + b_d(n)]y_d(n)$$

one finds for $n \geq n_1$ that

$$\sum_{k=1}^{d} [a_k + b_k(n)] \{u(n+d-k) - u(n+d)\} y_d(n+d-k) = 0. \tag{9.17}$$

Defining for $n \geq n_1$,

$$v(n) := u(n+1) - u(n) \tag{9.18}$$

and thus $u(n+d) - u(n+d-k) = \sum_{l=1}^{k} v(n+d-l)$ for $k \geq 1$, (9.17) implies that

$$\sum_{k=1}^{d} [a_k + b_k(n)] \left\{ \sum_{l=1}^{k} v(n+d-l) \right\} y_d(n+d-k) = 0.$$

Changing the order of the summation leads to

$$\sum_{l=1}^{d}\left(\sum_{k=l}^{d}[a_k + b_k(n)]y_d(n+d-k)\right)v(n+d-l) = 0,$$

which is a scalar linear difference equation of order $d-1$ in v. Dividing by $y_d(n)$ and noting that (9.14) implies that $y_d(n+d-k)/y_d(n) = \lambda_d^{d-k} + o(1)$ as $n \to \infty$, one sees that

$$\sum_{l=1}^{d}\left(\sum_{k=l}^{d}a_k\lambda_d^{d-k} + o(1)\right)v(n+d-l) = 0.$$

The leading coefficient $\sum_{k=1}^{d}a_k\lambda_d^{d-k} + o(1) = -\lambda_d^d + o(1)$ (since λ_d is a root of (9.4)) is non-zero for large n and division by it implies that

$$v(n+d-1) + \sum_{l=2}^{d}\left[1 + \frac{a_1}{\lambda_d} + \frac{a_2}{\lambda_d^2} + \ldots + \frac{a_{l-1}}{\lambda_d^{l-1}} + o(1)\right]v(n+d-l) = 0. \tag{9.19}$$

Observe that the limiting characteristic equation of (9.19) is

$$q(\lambda) := \lambda^{d-1} + \left[1+\frac{a_1}{\lambda_d}\right]\lambda^{d-2} + \left[1+\frac{a_1}{\lambda_d}+\frac{a_2}{\lambda_d^2}\right]\lambda^{d-3} + \ldots + \left[1+\frac{a_1}{\lambda_d}+\cdots+\frac{a_{d-1}}{\lambda_d^{d-1}}\right] = 0,$$

or, putting $a_0 := 1$,

$$q(\lambda) = \sum_{k=1}^{d}\lambda^{d-k}\sum_{m=0}^{k-1}\frac{a_m}{\lambda_d^m} = 0.$$

It can be shown that the roots of $q(\lambda)$ are given by

$$\mu_i = \frac{\lambda_i}{\lambda_d}, \qquad 1 \le i \le d-1.$$

This follows from a straightforward calculation establishing for the original limiting equation given in (9.4) that

$$\frac{p(\lambda)}{\lambda - \lambda_d} = \sum_{k=1}^{d}\lambda^{d-k}\sum_{\mu=1}^{k}a_{k-\mu}\lambda_d^{\mu-1},$$

and that

$$q\left(\frac{\lambda_i}{\lambda_d}\right) = \frac{1}{\lambda_d^{d-1}} \frac{p(\lambda_i)}{\lambda - \lambda_d},$$

hence $q\left(\frac{\lambda_i}{\lambda_d}\right) = 0$ for $1 \le i \le d-1$.

Since the moduli of these roots are distinct, we can apply the induction hypothesis to conclude that any solution of (9.19) either satisfies

(α) $v(n) \equiv 0$ for all n sufficiently large, or
(β) $v(n) \neq 0$ for all n sufficiently large and

$$\lim_{n\to\infty} \frac{v(n+1)}{v(n)} = \frac{\lambda_i}{\lambda_d},$$

for some $i \in \{1, \ldots, d-1\}$.

First we consider the case (α). By (9.18), $u(n)$ is eventually constant, that is $\exists\ c \in \mathbb{C}$ such that $u(n) \equiv c$ for all n sufficiently large. If $c = 0$, then $u(n) \equiv 0$ and also $y(n) \equiv 0$ for all large n, which implies (a). If $c \neq 0$, then $y(n) = cy_d(n)$ and the conclusion (b) follows with $\lambda = \lambda_d$.

Now consider the case (β). That is, $v(n) \neq 0$ for all large n and $v(n+1)/v(n) = \frac{\lambda_i}{\lambda_d}$, for some $i \in \{1, \ldots, d-1\}$. Put $\mu_i = \lambda_i/\lambda_d$ for $1 \le i \le d-1$. Any solution $u(n)$ of (9.18) can be expressed as $u(n) = \sum_{k=n_1}^{n-1} v(k)$. Since $|\mu_i| < 1$ for $1 \le i \le d-1$ by (9.5) and $v(n) = \prod_{k=n_1}^{n-1} [\mu_i + \varepsilon(k)]$ with $\varepsilon(k) \to 0$ as $k \to \infty$, the terms $v(n)$ tend to zero exponentially, so $S = \sum_{k=n_1}^{\infty} v(k)$ converges. Therefore for $n \ge n_1$

$$u(n) = S - \sum_{k=n}^{\infty} v(k).$$

If $S \neq 0$, then $u(n) = S + o(1)$ and the conclusion (b) follows with $\lambda = \lambda_d$. If $S = 0$, an application of Lemma 9.3 implies that $u(n) \neq 0$ for all large n and

$$\lim_{n\to\infty} \frac{u(n+1)}{u(n)} = \lim_{n\to\infty} \frac{v(n+1)}{v(n)} = \frac{\lambda_i}{\lambda_d},$$

which completes the proof. □

Lemma 9.3 *Let $\{v(n)\}$ be a complex-valued sequence satisfying $v(n) \neq 0$ for $n \ge n_1$ and assume that*

$$\lim_{n\to\infty} v(n+1)/v(n) = \mu, \qquad \text{where } 0 \le |\mu| < 1. \tag{9.20}$$

If $\{u(n)\}$ is defined for $n \geq n_1$ by

$$u(n) = -\sum_{k=n}^{\infty} v(k),$$

then for n sufficiently large, $u(n) \neq 0$ and

$$\lim_{n\to\infty} \frac{u(n+1)}{u(n)} = \mu.$$

Proof We begin by considering

$$\sum_{k=n+1}^{\infty} \frac{v(k)}{v(n)} = \sum_{k=1}^{\infty} \frac{v(n+k)}{v(n)} =: \sum_{k=1}^{\infty} f_k(n).$$

We will use the Weierstraß M-test to show that $\sum_{k=1}^{\infty} f_k(n)$ is uniformly convergent for all $n \geq N$ for some fixed $N \geq n_1$ and hence

$$\lim_{n\to\infty} \sum_{k=1}^{\infty} \frac{v(n+k)}{v(n)} = \lim_{n\to\infty} \sum_{k=1}^{\infty} f_k(n) = \sum_{k=1}^{\infty} \lim_{n\to\infty} f_k(n) = \sum_{k=1}^{\infty} \lim_{n\to\infty} \frac{v(n+k)}{v(n)}$$

$$= \sum_{k=1}^{\infty} \mu^k = \frac{\mu}{1-\mu}, \tag{9.21}$$

where (9.20) was used. To establish the uniform convergence, note that (9.20) implies that

$$\frac{v(n+1)}{v(n)} = \mu + \varepsilon(n),$$

where $\varepsilon(n) \to 0$ as $n \to \infty$. Fix $N \geq n_1$ such that

$$|\varepsilon(n)| \leq \frac{1-|\mu|}{2} \qquad \text{for all } n \geq N.$$

Since $v(n)$ can be written as

$$v(n) = \prod_{l=n_1}^{n-1} [\mu + \varepsilon(k)],$$

it follows for $k \geq 1$ and $n \geq N$ that

$$\left|\frac{v(n+k)}{v(n)}\right| = \prod_{l=n}^{n+k-1} |\mu + \varepsilon(k)| \leq \prod_{l=n}^{n+k-1} \left(|\mu| + \frac{1-|\mu|}{2}\right)$$
$$= \prod_{l=n}^{n+k-1} \frac{1+|\mu|}{2} = \left(\frac{1+|\mu|}{2}\right)^k.$$

Since $\frac{1+|\mu|}{2} < 1$, the uniform convergence of $\sum_{k=1}^{\infty} \frac{v(n+k)}{v(n)} = \sum_{k=1}^{\infty} f_k(n)$ follows for all $n \geq N$ and (9.21) holds.

To show that $u(n) \neq 0$, observe that from (9.21) it also follows that

$$\frac{u(n)}{v(n)} = -\frac{\sum_{k=n}^{\infty} v(k)}{v(n)} = -1 - \sum_{k=1}^{\infty} \frac{v(n+k)}{v(n)} \to \frac{-1}{1-\mu} \neq 0,$$

hence $u(n) \neq 0$ for all n sufficiently large.

Finally, observe that

$$\frac{u(n+1)}{u(n)} = \frac{\sum_{k=n+1}^{\infty} v(k)}{\sum_{k=n}^{\infty} v(k)} = 1 - \frac{v(n)}{\sum_{k=n}^{\infty} v(k)}$$
$$= 1 - \frac{1}{1 + \sum_{k=1}^{\infty} \frac{v(n+k)}{v(n)}} \to 1 - \frac{1}{1 + \frac{\mu}{1-\mu}} = \mu$$

as $n \to \infty$, and the proof is complete. □

The proof given above is not as elementary in nature as Poincaré's original argument or others in the literature based on similar ideas (see, e.g., [110, Chap. XVII] for the case $d = 3$). We chose to include it here since it shows how solutions of Perron's type together with reduction of order can be utilized to yield an alternative argument even it case that Perron's extra assumption (9.8) would not be satisfied.

While we observed in Sect. 9.1 that the natural flow of asymptotic information in this chapter goes from systems to scalar equations, here we make an exception and show how Poincaré's result for systems can be obtained from that for scalar equations. Consider

$$\mathbf{y}(n+1) = [A + R(n)]\mathbf{y}(n), \qquad R(n) = \mathrm{o}(1) \quad (n \to \infty), \tag{9.22}$$

under the conditions that A has eigenvalues λ_i all having distinct moduli (but one of them could be equal to zero) and its characteristic polynomial is of the form (9.4). Consider the transformation

$$\mathbf{y}(n) = PT^{-1}(n)\mathbf{z}(n),$$

where $P^{-1}AP = \Lambda_0$ and $T(n)$ is the matrix

$$T(n) = \begin{pmatrix} \mathbf{t}_1(n) \\ \vdots \\ \mathbf{t}_d(n) \end{pmatrix}, \tag{9.23}$$

with the rows inductively defined by $\mathbf{t}_1(n) = (1, 1, \ \ldots \ , 1)$ and $\mathbf{t}_k(n) = \mathbf{t}_{k-1}(n+1)\left[\Lambda_0 + P^{-1}R(n)P\right]$ for $k = 2, 3, \ldots, d$. It can easily be shown that since Λ_0 is diagonal with distinct eigenvalues, then as $n \to \infty$

$$T(n) \to \begin{pmatrix} 1 & 1 & \cdots & 1 \\ \lambda_1 & \lambda_2 & \cdots & \lambda_d \\ \vdots & & & \vdots \\ \lambda_1^{d-1} & \lambda_2^{d-1} & \cdots & \lambda_d^{d-1} \end{pmatrix} = V,$$

the Vandermonde matrix. Hence $T(n)$ is invertible for all n sufficiently large. Because of the structure of $T(n)$, it follows that

$$T(n+1)[\Lambda_0 + P^{-1}R(n)P]T^{-1}(n) = C(n),$$

where $C(n)$ is a companion matrix. Moreover, $C(n) = C + B(n)$, where C is the companion matrix corresponding to (9.4) and $B(n) = o(1)$ as $n \to \infty$.

From (9.22) now follows that

$$\mathbf{z}(n+1) = C(n)\mathbf{z}(n),$$

and since $C(n)$ is a companion matrix, solution vectors have the special form

$$\mathbf{z}(n) = (z(n), z(n+1), \ \ldots \ , z(n+d-1))^T,$$

where $z(n)$ is a solution of the corresponding scalar equation (9.3). If $z(n)$ is a solution that is not eventually trivial, then according to Theorem 9.1, it follows that $z(n) \neq 0$ for all n sufficiently large and that there exists a root λ_i of (9.4) (i.e., an eigenvalue of A) such that (9.6) holds. Therefore $z(n+j) = \left(\lambda_i^j + o(1)\right) z(n)$ for each $j \geq 1$ and $z(n) = \prod^{n-1}[\lambda_i + o(1)]$. Hence any non-trivial solution of (9.22) satisfies

$$\mathbf{y}(n) = PT^{-1}(n)\mathbf{z}(n) = PT^{-1}(n) \begin{pmatrix} 1 \\ \lambda_i + o(1) \\ \lambda_i^2 + o(1) \\ \vdots \\ \lambda_i^{d-1} + o(1) \end{pmatrix} \prod^{n-1}[\lambda_i + o(1)].$$

Since $T^{-1}(n) \to V^{-1}$ and since $[1, \lambda_i, \ldots, \lambda_i^{d-1}]^T$ is the ith column of V, it follows that

$$T^{-1}(n)[1+\mathrm{o}(1), \lambda_i + \mathrm{o}(1), \ldots, \lambda_i^{d-1} + \mathrm{o}(1)]^T = e_i + \mathrm{o}(1),$$

where e_i is the ith Euclidean vector. And thus it follows that

$$\mathbf{y}(n) = [\mathbf{p}_i + \mathrm{o}(1)]\mathbf{z}(n) = [\mathbf{p}_i + \mathrm{o}(1)] \prod^{n-1} [\lambda_i + \mathrm{o}(1)], \tag{9.24}$$

where $\mathbf{p}_i$ is the ith column of P.

Thus Poincaré's result for (9.3) implies (9.24) for a nontrivial solution $\mathbf{y}(n)$ of a "Poincaré system" (9.22). Compare this with other independent proofs not appealing to Poincaré's result for scalar equations (9.3) (see, e.g., [105, 159]) and other equivalent statements to (9.24). We also refer to [55, Sect. 8.8] for further results of Poincaré/Perron type.

Several natural questions arise in light of the statements above.

The first is how important it is that the roots of (9.4) have all distinct moduli, or in the case of fixed i that $|\lambda_i| \neq |\lambda_j|$ for all other roots of (9.4). Perron [118] already gave an example with $d = 2$ showing that in case $|\lambda_1| = |\lambda_2|$, (9.3) could have a nontrivial solution $y(n)$ for which $y(n_k) = 0$ for an infinite sequence $\{n_k\} \in \mathbb{N}$. Thus (9.6) is meaningless for this solution. One can also construct similar examples showing that $y(n) = \prod^{n-1}[\lambda_i + \mathrm{o}(1)]$ is also false. Perron [121] also considered such cases without any assumptions on the roots of (9.4) and obtained weaker formulas than (9.6) such as $\limsup_{n\to\infty} \sqrt[n]{|y_i(n)|} = |\lambda_i|$. But those results are not directly related to asymptotic representations and we will not consider them here.

The next question concerns what extra conditions on $b_i(n)$ are sufficient for (9.6) to hold, say under the weaker condition that the roots of (9.4) are distinct, but might have equal moduli. This can be answered by applying Theorem 3.4 or 5.1. Sufficient assumptions together with an appropriate dichotomy condition are that either $b_i(n) \in l^1$ for all $1 \leq i \leq d$ (or, more generally, that $b_i(n) = \mathrm{o}(1)$ and $\Delta b_i(n) \in l^1$ for all $1 \leq i \leq d$). In these cases, one can do much better than (9.6), namely in the case $b_i(n) \in l^1$, one finds that (9.3) has solutions $y_i(n)$ satisfying $y_i(n) = [1+\mathrm{o}(1)]\lambda_i^{n-1}$. In case when $b_i(n) = \mathrm{o}(1)$, $\Delta b_i(n) \in l^1$, and the eigenvalues $\lambda_i(n)$ of $A + B(n)$ satisfy Levinson's dichotomy conditions (3.8), (3.9) one sees that (9.3) has solutions $y_i(n)$ satisfying

$$y_i(n) = [1+\mathrm{o}(1)] \prod^{n-1} \lambda_i(k). \tag{9.25}$$

In Poincaré's polynomial-coefficient case (9.7) mentioned above, we observe that one can easily check that if the roots of (9.4) are just assumed to be all distinct, then the conditions $\Delta b_i(n) \in l^1$ and the eigenvalues $\lambda_i(n)$ possessing a Levinson dichotomy are always satisfied, hence for any nontrivial solution of (9.7), (9.6) always holds even without assuming that $|\lambda_i| \neq |\lambda_j|$.

A third question related to the ones above also concerns cases when more explicit results such as (9.25) are possible, but under other assumptions on the $b_i(n)$. For example, again if the roots of (9.4) have all distinct moduli and the coefficients $b_i(n) \in l^2$ for $1 \leq i \leq d$, then making in (9.10) a transformation $\mathbf{y}(n) = P\hat{\mathbf{y}}(n)$, where $P^{-1}AP = \Lambda$ and $P^{-1}B(n)P = \hat{B}(n)$ and applying Theorem 5.3 to $\mathbf{z}(n+1) = [\Lambda + \hat{B}(n)]\hat{\mathbf{y}}(n)$, one one obtains in case $\lambda_i \neq 0$ a solution $y_i(n)$ of (9.3) of the form

$$y_i(n) = [1 + o(1)] \prod^{n-1} \left[\lambda_i + \hat{b}_{ii}(k)\right], \qquad \text{as } n \to \infty, \tag{9.26}$$

for a fundamental system. Under weaker l^p-conditions on the $b_i(n)$, statements such as (9.26) can still be obtained (see Theorem 5.3) with further explicit modifications of terms in the product.

Note that general results on error estimates (see Sect. 3.7) might also apply, and we will discuss some examples involving such error estimates in Sect. 9.4.

Finally, one could ask what happens in (9.7) if the coefficients $p_i(n)$ do not all have the same degree and therefore (9.7) does not reduce to an equation such as (9.3). In this polynomial coefficient case of Poincaré (or the more general case of coefficients meromorphic at ∞), the theory involving calculations of formal solutions and asymptotic expansions (see, e.g., [82]) can be applied and yields rather complete asymptotic information about a fundamental system of solutions.

9.3 Liouville–Green Results of Second-Order Linear Scalar Difference Equations

A reasonable difference equation analogue corresponding to $y'' = [f(t)+g(t)]y = 0$ $(t \geq t_0)$ treated in Chap. 8 is an equation of the form

$$\Delta^2 y(n) = [f(n) + g(n)]y(n), \qquad n \geq n_0. \tag{9.27}$$

Letting $\mathbf{y}(n) = [y(n), \Delta y(n)]^T$, (9.27) can be written for $n \geq n_0$ as

$$\Delta \mathbf{y}(n) = \begin{pmatrix} 0 & 1 \\ f(n) + g(n) & 0 \end{pmatrix} \mathbf{y}(n),$$

or

$$\mathbf{y}(n+1) = \begin{pmatrix} 1 & 1 \\ f(n) + g(n) & 1 \end{pmatrix} \mathbf{y}(n).$$

We now assume that $f(n) \neq 0$ for all $n \geq n_0$ and, loosely speaking, think of $f(n)$ being "large" with respect to $g(n)$ as $n \to \infty$. Letting $\mathbf{y}(n) = \text{diag}\,\{1, \sqrt{f(n)}\}\hat{\mathbf{y}}(n)$

yields

$$\hat{y}(n+1) = \left[\begin{pmatrix} 1 & \sqrt{f(n)} \\ \sqrt{f(n)} & 1 \end{pmatrix} + \begin{pmatrix} 0 & 0 \\ \sqrt{f(n)}\alpha(n) + \frac{g(n)}{\sqrt{f(n+1)}} & \alpha(n) \end{pmatrix} \right] \hat{y}(n),$$

where

$$\alpha(n) = \frac{\sqrt{f(n)}}{\sqrt{f(n+1)}} - 1 = -\frac{\Delta f(n)}{f(n+1) + \sqrt{f(n)f(n+1)}}.$$

(We make a fixed choice for a branch of $\sqrt{\ }$ such that if $f(n) > 0$, $\operatorname{Re}\sqrt{f(n)} > 0$.) We will assume in what follows that

$$\alpha(n) = \mathrm{o}(1) \qquad \text{as } n \to \infty.$$

This is implied by a certain type of regularity condition on $f(n)$, e.g., if $f(n) = n^p$ for $p \geq 0$, then $\alpha(n) = \mathrm{O}(1/n)$ as $n \to \infty$.

Now the constant transformation

$$\hat{y}(n) = \begin{pmatrix} 1 & 1 \\ -1 & 1 \end{pmatrix} \tilde{y}(n)$$

diagonalizes the leading terms and leads to

$$\tilde{y}(n+1) = [\Lambda(n) + V(n)]\,\tilde{y}(n), \tag{9.28}$$

where

$$\Lambda(n) = \operatorname{diag}\{1 - \sqrt{f(n)},\ 1 + \sqrt{f(n)}\}$$

$$V(n) = \frac{\alpha(n)}{2}\begin{pmatrix} 1 & -1 \\ -1 & 1 \end{pmatrix} + \frac{\beta(n)}{2}\begin{pmatrix} -1 & -1 \\ 1 & 1 \end{pmatrix},$$

with $\beta(n) = \alpha(n)\sqrt{f(n)} + g(n)/\sqrt{f(n+1)}$.

Similar to what Eastham has done in the continuous case [53, Chap.2], there are several choices now for applying results from Chaps. 3 and 5 to (9.28) such as Theorems 3.4, 5.2, and 5.3.

For example, to apply Theorem 5.2, we could assume that $|f(n)|$ stays bounded away from 0, and observe that the condition $V(n)/\sqrt{f((n)} = \mathrm{o}(1)$ is satisfied provided, as we said above, that $\alpha(n) = \mathrm{o}(1)$ and also $g(n)/\sqrt{f(n)f(n+1)} = \mathrm{o}(1)$.

Next, the condition $\Delta\left(V(n)/\sqrt{f(n)}\right) \in l^1$ is implied by the assumptions

$$\Delta\left(\frac{\alpha(n)}{\sqrt{f(n)}}\right) \in l^1 \qquad \text{and} \tag{9.29}$$

$$\Delta\left(\alpha(n) + \frac{g(n)}{\sqrt{f(n)f(n+1)}}\right) \in l^1, \tag{9.30}$$

or if we assume $\Delta(\alpha(n)) \in l^1$, then $\Delta\left(\frac{g(n)}{\sqrt{f(n)f(n+1)}}\right) \in l^1$ suffices. Finally, in order to obtain (for n sufficiently large) the existence of a fundamental matrix of (9.28)

$$\tilde{Y}(n) = [I + o(1)] \prod^{n-1} \tilde{\Lambda}(k),$$

requires assuming that $\tilde{\Lambda}(k)$, a diagonal matrix of eigenvalues of $\Lambda(k) + V(k)$, is invertible and satisfies the Levinson-type dichotomy conditions (3.8), (3.9). We observe that conditions (9.29) and (9.30) involve second differences of $f(n)$ and are analogous to the Liouville–Green conditions in Sect. 8.3.2 in the case of differential equations, which involve second derivatives of $f(x)$ and similar powers.

If $V(n)$ does not satisfy the respective growth condition, one could seek to apply conditioning transformations to modify $V(n)$. Depending upon whether $\alpha(n)$ or $\beta(n)$ dominates, there are several choices for conditioning transformations, but the difference analogue of (8.18) does not seem to be useful due to the different symmetry in the two constant matrices in $V(n)$.

Moreover, for a particular equation $\Delta^2 y(n) = a(n)y(n)$, it is not necessarily clear, a priori, how to decompose $a(n)$ as $f(n) + g(n)$ satisfying such assumptions as above, or other ones required to facilitate the application of Theorem 3.4 or 5.3.

So in what follows, instead of pursuing other general discrete Liouville–Green type results which might be difficult to apply, we prefer to look at several special cases of (9.27) and show how certain special preliminary and conditioning transformations together with various general results from Chaps. 3 and 5 may be utilized.

9.4 Some Special Second-Order Linear Difference Equations

9.4.1 l^1-Perturbations and Liouville–Green–Olver Error Estimates

In this section, we consider the second-order linear difference equation

$$\Delta^2 y(n) + [a + g(n)]\, y(n) = 0\,, \qquad a \in \mathbb{C} \setminus \{-1\}, \quad n \geq n_0, \tag{9.31}$$

where

$$g(n) \in l^1[n_0, \infty). \tag{9.32}$$

The unperturbed equation can be re-written as $y(n+2)-2y(n+1)+[1+a]y(n) = 0$, which explains why the case $a = -1$ is excluded. This equation was studied by Spigler and Vianello [141, 142] and by Spigler, Vianello, and Locatelli [147]. Each of the three papers is concerned with a certain range for the parameter a.

One of the main emphases of their papers is to find precise Liouville–Green–Olver type estimates of the error o(1) in terms of the data a and $\{g(n)\}$. In what follows, we will first study (9.31) in the usual context of matrix systems and later on compare these results with the ones in [141, 142, 147]. As we will see, this approach leads to some sharpened error estimates.

Also see [140] for a generalization of such results to second-order linear difference equations in Banach algebras.

We write (9.31) as the equivalent matrix system

$$\mathbf{y}(n+1) = \left[\begin{pmatrix} 0 & 1 \\ -a-1 & 2 \end{pmatrix} + \begin{pmatrix} 0 & 0 \\ -g(n) & 0 \end{pmatrix}\right]\mathbf{y}(n) = [A + R(n)]\,\mathbf{y}(n), \tag{9.33}$$

where $\mathbf{y}(n) = [y(n), y(n+1)]^T$. Note that the eigenvalues of A are

$$\lambda_{\pm} = 1 \pm \sqrt{-a}\,, \tag{9.34}$$

and that A is non-singular if $a \neq -1$. If $a \neq 0$, we diagonalize this matrix with help of the Vandermonde matrix by setting

$$\mathbf{y}(n) = \begin{pmatrix} 1 & 1 \\ \lambda_+ & \lambda_- \end{pmatrix} z(n),$$

which yields

$$z(n+1) = \left[\begin{pmatrix} \lambda_+ & 0 \\ 0 & \lambda_- \end{pmatrix} + \frac{g(n)}{2\sqrt{-a}}\begin{pmatrix} -1 & -1 \\ 1 & 1 \end{pmatrix}\right] z(n). \tag{9.35}$$

Since $g(n) \in l^1$, Theorem 3.4 applies to (9.35) and it follows that (9.31) has for n sufficiently large solutions $y_{\pm} = [1 + \mathrm{o}(1)]\lambda_{\pm}^n$, but without providing upper bounds for the error o(1). To obtain error bounds, a study of the T-operator is used with a type of Gronwall's inequality similar to the Olver-type estimates in Sect. 8.3.3. We will consider the cases $a > 0$, $a \in \mathbb{C} \setminus [0, +\infty)$ (and $a \neq -1$), and $a = 0$, respectively.

9.4.1.1 The Oscillatory Case: $a > 0$

In this case, the eigenvalues in (9.34) are $\lambda_{\pm} = 1 \pm i\sqrt{a}$ and it will be important to note that

$$|\lambda_+| = |\lambda_-| = \sqrt{1+a}. \tag{9.36}$$

First, to find the asymptotic behavior of the solution corresponding to λ_-, we set $z(n) = (\lambda_-)^n u(n)$. Then (9.35) yields

$$u(n+1) = \left[\begin{pmatrix} \frac{\lambda_+}{\lambda_-} & 0 \\ 0 & 1 \end{pmatrix} + \frac{\chi_-(n)}{2}\begin{pmatrix} 1 & 1 \\ -1 & -1 \end{pmatrix}\right] u(n), \tag{9.37}$$

where

$$\chi_-(n) := -\frac{g(n)}{\sqrt{-a}\,\lambda_-}. \tag{9.38}$$

Hence $\chi_-(n) \in l^1$. Recalling $\left|\frac{\lambda_+}{\lambda_-}\right| = 1$, we look for a solution of (9.37) satisfying

$$u(n) = e_2 - \sum_{k=n}^{\infty} \begin{pmatrix} \left(\frac{\lambda_+}{\lambda_-}\right)^{n-k-1} & 0 \\ 0 & 1 \end{pmatrix} \frac{\chi_-(k)}{2}\begin{pmatrix} 1 & 1 \\ -1 & -1 \end{pmatrix} u(k) := e_2 + (Tu)(n). \tag{9.39}$$

With $\tilde{n}_1$ fixed (see below), it is straightforward to show that T maps the Banach space $\mathscr{B}$ of bounded (for $n \geq \tilde{n}_1$) sequences $\{u(n)\}$ with the supremum-norm $\|u\| = \sup_{n \geq \tilde{n}_1} |u(n)|$ into itself. As a vector norm, we choose $|u| = |u_1| + |u_2|$, where u_i denote the components of the vector u. The associated matrix norm is then the absolute column norm. Then it follows for $u, \tilde{u} \in \mathscr{B}$ that

$$|T(u - \tilde{u})(n)| \leq \|u - \tilde{u}\| \sum_{k=n}^{\infty} |\chi_-(k)| = \|u - \tilde{u}\| V_-(n),$$

where

$$V_-(n) = \sum_{k=n}^{\infty} |\chi_-(k)|. \tag{9.40}$$

Fix $\tilde{n}_1 \geq n_0$ such that $V_-(n) < 1$ for all $n \geq \tilde{n}_1$. Therefore $(Tu)(n)$ is a contraction for all $n \geq \tilde{n}_1$, and the fixed point of (9.39) is a solution of (9.37). Introducing the notation $u(n) = (u_1(n), u_2(n))^T$, we see that (9.33) has for $n \geq \tilde{n}_1$

a vector-valued solution

$$\mathbf{y}(n) = \begin{pmatrix} 1 & 1 \\ \lambda_+ & \lambda_- \end{pmatrix} (\lambda_-)^n \left[e_2 - \sum_{k=n}^{\infty} \frac{\chi_-(k)}{2} \begin{pmatrix} \left(\frac{\lambda_+}{\lambda_-}\right)^{n-k-1} \\ -1 \end{pmatrix} [u_1(k) + u_2(k)] \right],$$

whose first component $y(n) = y_-(n)$ is a solution of (9.31):

$$\begin{aligned} y_-(n) &= (\lambda_-)^n \left[1 - \sum_{k=n}^{\infty} \frac{\chi_-(k)}{2} [u_1(k) + u_2(k)] \left\{ \left(\frac{\lambda_+}{\lambda_-}\right)^{n-k-1} - 1 \right\} \right] \\ &=: (\lambda_-)^n \left[1 + \varepsilon_-(n) \right]. \end{aligned}$$

Since

$$\left| \left(\frac{\lambda_+}{\lambda_-}\right)^{n-k-1} - 1 \right| \leq 2, \qquad k \geq n, \tag{9.41}$$

by (9.36), it follows that

$$|\varepsilon_-(n)| \leq \sum_{k=n}^{\infty} |\chi_-(k)||u_1(k) + u_2(k)|. \qquad n \geq \tilde{n}_1. \tag{9.42}$$

Next, we will use Gronwall's inequality to find an estimate of $|u_1(k) + u_2(k)|$. For that purpose, adding the two components of (9.39) yields

$$u_1(n) + u_2(n) = 1 - \sum_{k=n}^{\infty} \frac{\chi_-(k)}{2} \left\{ \left(\frac{\lambda_+}{\lambda_-}\right)^{n-k-1} - 1 \right\} [u_1(k) + u_2(k)], \tag{9.43}$$

and taking absolute values shows that

$$|u_1(n) + u_2(n)| \leq 1 + \sum_{k=n}^{\infty} |\chi_-(k)|\, |u_1(k) + u_2(k)|, \qquad n \geq \tilde{n}_1. \tag{9.44}$$

Using the notation $R_n = \sum_{k=n}^{\infty} |\chi_-(k)|\, |u_1(k) + u_2(k)|$, Gronwall's inequality established in Lemma 9.8 implies now that

$$R_n \leq \frac{1}{\prod_{k=n}^{\infty} (1 - |\chi_-(k)|)} - 1.$$

By (9.42),

$$|\varepsilon_-(n)| \le R_n \le \frac{1}{\prod_{k=n}^{\infty}(1-|\chi_-(k)|)} - 1. \tag{9.45}$$

A solution corresponding to λ_+ is derived similarly and we will be briefer here: setting $z(n) = (\lambda_+)^n w(n)$, (9.35) yields

$$w(n+1) = \left[\begin{pmatrix} 1 & 0 \\ 0 & \frac{\lambda_-}{\lambda_+} \end{pmatrix} + \frac{\chi_+(n)}{2}\begin{pmatrix} 1 & 1 \\ -1 & -1 \end{pmatrix}\right] w(n), \tag{9.46}$$

where

$$\chi_+(n) := -\frac{g(n)}{\lambda_+\sqrt{-a}}. \tag{9.47}$$

One can show that (9.46) has a solution satisfying

$$w(n) = e_1 - \sum_{k=n}^{\infty}\begin{pmatrix} 1 & \\ & \left(\frac{\lambda_-}{\lambda_+}\right)^{n-k-1} \end{pmatrix} \frac{\chi_+(k)}{2}\begin{pmatrix} 1 & 1 \\ -1 & -1 \end{pmatrix} w(k). \tag{9.48}$$

This can be used to show that (9.31) has for $n \ge \hat{n}_1$ a solution $y_-(n)$ of the form

$$\begin{aligned} y_+(n) &= \lambda_+^n \left[1 - \sum_{k=n}^{\infty}\frac{\chi_+(k)}{2}[u_1(k) + u_2(k)]\left\{1 - \left(\frac{\lambda_-}{\lambda_+}\right)^{n-k-1}\right\}\right] \\ &=: \lambda_+^n \left[1 + \varepsilon_+(n)\right]. \end{aligned} \tag{9.49}$$

Here $\hat{n}_1$ is fixed sufficiently large such that $V_+(n) = \sum_{k=n}^{\infty}|\chi_+(k)| < 1$ for $n \ge \hat{n}_1$. Gronwall's inequality can be used to show that

$$|\varepsilon_+(n)| \le \frac{1}{\prod_{k=n}^{\infty}(1-|\chi_+(k)|)} - 1.$$

We note that it was the fact $|\lambda_+| = |\lambda_-|$ that made it possible to look for two linearly independent solutions using operators involving only sums $\sum_n^{\infty}$. As we will see in the following nonoscillatory case, only the recessive solution will involve such a representation, and such a precise error estimate as above will only be possible for this solution. We also note that (9.36) allows us to introduce the abbreviation

$$|\chi(n)| := |\chi_+(n)| = |\chi_-(n)| = \frac{|g(n)|}{\sqrt{a(1+a)}}.$$

This also shows that $\tilde{n}_1$ and $\hat{n}_1$ can be chosen to be the same, and we put $n_1 := \tilde{n}_1 = \hat{n}_1$. We state a summary of these results as

Proposition 9.4 *Consider* (9.31), *where $g(n)$ satisfies* (9.32). *If $a > 0$, then* (9.31) *has, for $n \geq n_1$, two linearly independent solutions satisfying*

$$y_\pm(n) = (\lambda_\pm)^n \left[1 + \varepsilon_\pm(n)\right],$$

where $\lambda_\pm$ are given in (9.34) *and*

$$|\varepsilon_\pm(n)| \leq \frac{1}{\prod_{k=n}^{\infty}(1 - |\chi(k)|)} - 1 = \frac{1}{\prod_{k=n}^{\infty}\left(1 - \frac{|g(k)|}{\sqrt{a(1+a)}}\right)} - 1.$$

Here n_1 is fixed sufficiently large such that for $n \geq n_1$

$$V(n) = \sum_{k=n}^{\infty} |\chi(k)| = \frac{1}{\sqrt{a(a+1)}} \sum_{k=n}^{\infty} |g(k)| < 1.$$

9.4.1.2 The Nonoscillatory Case: $a \in \mathbb{C} \setminus [0, +\infty), a \neq -1$

As in [147, Theorem 2.1], we next consider (9.31) under the assumption $a \in \mathbb{C} \setminus [0, +\infty)$, $a \neq -1$, and $g(n) \in l^1[n_0, \infty)$.

Recalling that $\lambda_\pm = 1 \pm \sqrt{-a}$ by (9.34), it follows (using the principal branch of the square root function) that

$$|\lambda_-| < |\lambda_+|. \tag{9.50}$$

Now $|\lambda_-| / |\lambda_+| < 1$ implies that (9.41) holds, and the arguments given above still apply to the (recessive) solution associated to λ_-. That is, (9.31) has a solution

$$y_-(n) = (\lambda_-)^n \left[1 + \varepsilon_-(n)\right], \qquad |\varepsilon_-(n)| \leq \frac{1}{\prod_{k=n}^{\infty}(1 - |\chi_-(k)|)} - 1, \qquad n \geq \tilde{n}_1.$$

Here $\chi_-(n)$ is defined in (9.38) and $\tilde{n}_1$ is such that $V_-(n) < 1$ for all $n \geq \tilde{n}_1$, where $V_-(n)$ is defined in (9.40).

As for a solution corresponding to λ_+, instead of (9.49), we consider

$$\begin{aligned} w(n) &= e_1 + \sum_{k=n_1}^{n-1} \begin{pmatrix} 0 & \\ & \left(\frac{\lambda_-}{\lambda_+}\right)^{n-k-1} \end{pmatrix} \frac{\chi_+(k)}{2} \begin{pmatrix} 1 & 1 \\ -1 & -1 \end{pmatrix} w(k) \\ &\quad - \sum_{k=n}^{\infty} \begin{pmatrix} 1 & \\ & 0 \end{pmatrix} \frac{\chi_+(k)}{2} \begin{pmatrix} 1 & 1 \\ -1 & -1 \end{pmatrix} w(k) \\ &= e_1 + \hat{T}(w)(n). \end{aligned}$$

While $\hat{T} : \mathscr{B} \to \mathscr{B}$ can be seen to be a contraction for $n \geq \hat{n}_1$ sufficiently large by (9.32) and (9.47), the argument involving Gronwall's inequality does not seem to hold in this setting. From Theorem 3.4 one obtains the weaker result that (9.46) has a solution $w(n) = e_i + o(1)$ as $n \to \infty$, which implies that that (9.31) has a solution

$$y_+(n) = \lambda_+^n \left[1 + \varepsilon_+(n)\right], \qquad \varepsilon_+(n) = o(1).$$

We remark that to estimate $\varepsilon_+(n)$ using Theorem 3.25 would require to additionally assume a slow-decay condition, i.e., that there exists a constant $b \in [1, |\lambda_+|/|\lambda_-|)$ such that

$$|g(k_1)|b^{k_1} \leq |g(k_2)|b^{k_2} \qquad \text{for all } n_1 \leq k_1 \leq k_2.$$

Such a requirement would imply the estimate

$$|\varepsilon_+(n)| = O\left(\sum_{k=n}^{\infty} |\chi_+(k)|\right) = O\left(\sum_{k=n}^{\infty} |g(k)|\right),$$

but without an explicit estimate for the O-function. We refer the reader to Theorem 3.25 for details for this argument.

We state a summary of the results in this case as

Proposition 9.5 *Consider* (9.31), *where* $g(n)$ *satisfies* (9.32). *If* $a \in \mathbb{C} \setminus [0, +\infty)$, $a \neq -1$, *then* (9.31) *has for* $n \geq n_1$ *two linearly independent solutions satisfying*

$$\begin{aligned} y_+(n) &= \lambda_+^n \left[1 + \varepsilon_+(n)\right], &\quad \varepsilon_+(n) &= o(1), \qquad \textit{as } n \to \infty, \\ y_-(n) &= \lambda_-^n \left[1 + \varepsilon_-(n)\right], &\quad |\varepsilon_-(n)| &\leq \frac{1}{\prod_{k=n}^{\infty} (1 - |\chi_-(k)|)} - 1, \end{aligned}$$

where $\lambda_{\pm}$ are given in (9.34)*, and $\chi_{-}(n)$ is defined in* (9.38)*. Here n_1 is fixed sufficiently large such that for $n \geq n_1$*

$$V_{-}(n) = \sum_{k=n}^{\infty} |\chi_{-}(k)| < 1.$$

In the nonoscillatory case, one can see that the unperturbed system associated with (9.35), $v(n+1) = \begin{pmatrix} \lambda_{+} & 0 \\ 0 & \lambda_{-} \end{pmatrix} v(n)$, satisfies an exponential dichotomy, so the assumption $g(n) \in l^1$ could be considerably weakened to $g(n) \in l^p$ for $p < \infty$, by applying other results from Chaps. 3 and 5, leading to a modified asymptotic representation. Explicit error bounds for the recessive solution can also be obtained by modifying the approach above.

9.4.1.3 The Special Case a = 0

Finally, we consider the case where the parameter a in (9.31) takes on the value $a = 0$. This implies that the eigenvalues of the constant matrix A in (9.33) are given by $\lambda_{+} = \lambda_{-} = 1$, and the constant matrix A in (9.33) is not diagonalizable. To bring A in Jordan form, we put in (9.33)

$$\mathbf{y}(n) = \begin{pmatrix} 1 & -1 \\ 1 & 0 \end{pmatrix} \zeta(n). \tag{9.51}$$

Then (9.33) is equivalent to

$$\zeta(n+1) = \left[\begin{pmatrix} 1 & 1 \\ 0 & 1 \end{pmatrix} + g(n) \begin{pmatrix} -1 & 1 \\ -1 & 1 \end{pmatrix} \right] \zeta(n). \tag{9.52}$$

We consider (9.52) as a perturbation of the system

$$x(n+1) = \begin{pmatrix} 1 & 1 \\ 0 & 1 \end{pmatrix} x(n), \tag{9.53}$$

which has a fundamental matrix of the form

$$X(n) = \begin{pmatrix} 1 & n \\ 0 & 1 \end{pmatrix}.$$

We are first looking for a solution of (9.52) corresponding to the unperturbed solution $x_1(n) = e_1$ in T-operator form

$$\begin{aligned}\zeta(n) &= e_1 - \sum_{k=n}^{\infty} \begin{pmatrix} 1 & n \\ 0 & 1 \end{pmatrix} \begin{pmatrix} 1 & -(k+1) \\ 0 & 1 \end{pmatrix} g(k) \begin{pmatrix} -1 & 1 \\ -1 & 1 \end{pmatrix} \zeta(k) \\ &= e_1 - \sum_{k=n}^{\infty} \begin{pmatrix} k-n & n-k \\ -1 & 1 \end{pmatrix} g(k)\zeta(k) := e_1 + (\tilde{T}\zeta)(n). \end{aligned} \tag{9.54}$$

From (9.54) it follows that

$$\left|(\tilde{T}\zeta)(n)\right| \leq \|\zeta\| \sum_{k=n}^{\infty} [k-n+1]|g(k)|,$$

and

$$\left|(\tilde{T}\zeta)(n) - (\tilde{T}\tilde{\zeta})(n)\right| \leq \|\zeta - \tilde{\zeta}\| \sum_{k=n}^{\infty} [k-n+1]\,|g(k)|.$$

Thus one can see that $\tilde{T}$ is a contraction for $n \geq \tilde{n}_2$ if $\sum_{k=n_2}^{\infty} [k-n+1]|g(k)| < 1$, so we define

$$V^{(1)}(n) := \sum_{k=n}^{\infty} [k-n+1]|g(k)| < 1, \qquad \text{for all } n \geq \tilde{n}_2. \tag{9.55}$$

Therefore, for $n \geq \tilde{n}_2$, (9.54) has a fixed point solution, which is also a solution of (9.52). Recalling the preliminary transformation (9.51), (9.33) has a solution satisfying

$$\mathbf{y}(n) = \begin{pmatrix} 1 \\ 1 \end{pmatrix} - \sum_{k=n}^{\infty} \begin{pmatrix} k-n+1 & n-k-1 \\ k-n & n-k \end{pmatrix} g(k)\zeta(k),$$

and considering the first component, one finds for $n \geq \tilde{n}_2$ a scalar solution of (9.31) of the form

$$y_1(n) = 1 - \sum_{k=n}^{\infty} (k-n+1) g(k) [\zeta_1(k) - \zeta_2(k)] =: 1 + \tilde{\varepsilon}_n^{(1)}.$$

Here we used the notation $\zeta(n) = (\zeta_1(n), \zeta_2(n))^T$. Hence

$$\left|\tilde{\varepsilon}_n^{(1)}\right| \leq \sum_{k=n}^{\infty} |g(k)|k|\zeta_1(k) - \zeta_2(k)|.$$

To find an upper bound of the error $\tilde{\varepsilon}_n^{(1)}$, we slightly modify the argument used above in the case $a > 0$. Looking now at the difference (not the sum) of the components of $\zeta(n)$ in (9.54), one can see that

$$|\zeta_1(n) - \zeta_2(n)| \leq 1 + \sum_{k=n}^{\infty} k|g(k)||\zeta_1(k) - \zeta_2(k)|.$$

Then from Gronwall's inequality (see Lemma 9.8) it follows that

$$\left|\tilde{\varepsilon}_n^{(1)}\right| \leq \frac{1}{\prod_{k=n}^{\infty}\{1 - k|g(k)|\}} - 1, \qquad n \geq \tilde{n}_2.$$

Secondly, to find a solution of (9.52) corresponding to the unperturbed solution $x_2(n) = (n, 1)^T$ of (9.53), it is necessary to make a normalization since this unperturbed solution is not bounded. For that purpose, we put in (9.52) and (9.53)

$$\zeta(n) = (n-1)\tilde{\zeta}(n) \qquad x(n) = (n-1)\tilde{x}(n). \tag{9.56}$$

The particular choice of the factor "$n - 1$" stems from being able to work in the Banach space $\mathscr{B}$ of bounded sequences and, moreover, leading to the constant term with value 1, when applying Lemma 9.8. Then (9.52) and (9.53) imply that

$$\tilde{\zeta}(n+1) = \frac{n-1}{n}\left[\begin{pmatrix} 1 & 1 \\ 0 & 1 \end{pmatrix} + g(n)\begin{pmatrix} -1 & 1 \\ -1 & 1 \end{pmatrix}\right]\tilde{\zeta}(n), \tag{9.57}$$

and

$$\tilde{x}(n+1) = \frac{n-1}{n}\begin{pmatrix} 1 & 1 \\ 0 & 1 \end{pmatrix}\tilde{x}(n),$$

which has a fundamental matrix

$$\tilde{X}(n) = \frac{1}{n-1}\begin{pmatrix} 1 & n \\ 0 & 1 \end{pmatrix}.$$

To obtain a solution of (9.57) corresponding to the unperturbed solution $\tilde{x}_2(n) = \frac{1}{n-1}(n, 1)^T$ is equivalent to solving

$$\begin{aligned}\tilde{\xi}(n) &= \frac{1}{n-1}\begin{pmatrix} n \\ 1 \end{pmatrix} - \sum_{k=n}^{\infty} \tilde{X}(n)\tilde{X}^{-1}(k+1)\frac{k-1}{k} g(k) \begin{pmatrix} -1 & 1 \\ -1 & 1 \end{pmatrix} \tilde{\xi}(k) \\ &= \frac{1}{n-1}\begin{pmatrix} n \\ 1 \end{pmatrix} - \sum_{k=n}^{\infty} \frac{k-1}{n-1}\, g(k) \begin{pmatrix} k-n & n-k \\ -1 & 1 \end{pmatrix} \tilde{\xi}(k) \\ &:= \frac{1}{n-1}\begin{pmatrix} n \\ 1 \end{pmatrix} + (\hat{T}\tilde{\xi})(n). \end{aligned} \tag{9.58}$$

From (9.58) follows that

$$\left|(\hat{T}\tilde{\xi})(n)\right| \le \|\tilde{\xi}\| \sum_{k=n}^{\infty} \frac{k-1}{n-1}[k-n+1]|g(k)|,$$

and

$$\left|(\hat{T}\tilde{\xi})(n) - (\hat{T}\hat{\xi})(n)\right| \le \|\tilde{\xi} - \hat{\xi}\| \sum_{k=n}^{\infty} \frac{k-1}{n-1}\, [k-n+1]\, |g(k)|.$$

Therefore $\tilde{T}$ is a contraction on $\mathscr{B}$ for $n \ge \hat{n}_2$ if $\sum_{k=n}^{\infty} \frac{k-1}{n-1}\, [k-n+1]|g(k)| < \infty$ and, moreover, if $\hat{n}_2$ is fixed sufficiently large such that

$$V^{(2)}(n) := \sum_{k=n}^{\infty} \frac{k-1}{n-1}\, [k-n+1]|g(k)| < 1 \qquad \text{for all } n \ge \hat{n}_2.$$

It follows that (9.33) has for $n \ge \hat{n}_2$ a solution satisfying

$$\mathbf{y}(n) = \begin{pmatrix} n-1 \\ n \end{pmatrix} - \sum_{k=n}^{\infty} (k-1)g(k) \begin{pmatrix} k-n+1 \\ k-n \end{pmatrix} [\tilde{\xi}_1(k) - \tilde{\xi}_2(k)],$$

where $\tilde{\xi}(k) = (\tilde{\xi}_1(k), \tilde{\xi}_2(k))^T$. From the first component it follows that (9.31) has for $n \ge \hat{n}_2$ a scalar solution of the form

$$y_2(n) = n - 1 - \sum_{k=n}^{\infty} (k-1)(k-n+1)g(k)[\zeta_1(k) - \zeta_2(k)] =: n - 1 + \tilde{\varepsilon}_n^{(2)},$$

and

$$\left|\tilde{\varepsilon}_n^{(2)}\right| \leq \sum_{k=n}^{\infty} k^2 |g(k)||\tilde{\xi}_1(k) - \tilde{\xi}_2(k)|.$$

To find an upper bound of the error $\tilde{\varepsilon}_n^{(2)}$, (9.58) implies that

$$\left|\tilde{\xi}_1(n) - \tilde{\xi}_2(n)\right| \leq 1 + \sum_{k=n}^{\infty} k^2 |g(k)|\, |\tilde{\xi}_1(k) - \tilde{\xi}_2(k)|,$$

hence Lemma 9.8 yields that

$$\tilde{\varepsilon}_n^{(2)} \leq \frac{1}{\prod_{k=n}^{\infty} \{1 - k^2 |g(k)|\}} - 1.$$

We note that $0 \leq V^{(1)}(n) \leq V^{(2)}(n)$, hence one can put $n_2 = \max\{\tilde{n}_2, \hat{n}_2\} = \hat{n}_2$. Then a summary of the results in this case is

Proposition 9.6 *Consider* (9.31) *with* $a = 0$*, where* $g(n)$ *satisfies*

$$\sum_{k=n_0}^{\infty} k^2 |g(k)| < \infty. \tag{9.59}$$

Then (9.31) *has for* $n \geq n_2$ *two linearly independent solutions satisfying*

$$y_1(n) = 1 + \tilde{\varepsilon}_n^{(1)}, \qquad |\tilde{\varepsilon}_n^{(1)}| \leq \frac{1}{\prod_{k=n}^{\infty}\{1 - k|g(k)|\}} - 1, \tag{9.60}$$

$$y_2(n) = n - 1 + \tilde{\varepsilon}_n^{(2)}, \qquad |\tilde{\varepsilon}_n^{(2)}| \leq \frac{1}{\prod_{k=n}^{\infty}\{1 - k^2|g(k)|\}} - 1.$$

Here n_2 *is fixed sufficiently large such that for* $n \geq n_2$

$$V^{(2)}(n) = \sum_{k=n}^{\infty} \frac{k-1}{n-1} [k - n + 1]|g(k)| < 1 \qquad \textit{for all } n \geq n_2.$$

Remark 9.7 If (9.59) does not hold, but $k|g(k)| \in l^1[n_0, \infty)$, then the calculations above only yield the asymptotic behavior of the recessive solution $y_1(n)$ given in (9.60) and not of $y_2(n)$. Note that then (9.60) holds for $n \geq \tilde{n}_2$, where $\tilde{n}_2$ is defined in (9.55).

To compare the error estimates with the ones obtained in [141, 147], we will only consider the oscillatory case $a > 0$ since the estimates in the other case can be compared similarly. Recall from Proposition 9.4 that (9.31) has, for $n \geq n_1$, two linearly independent solutions satisfying

$$y_{\pm}(n) = (\lambda_{\pm})^n \left[1 + \varepsilon_{\pm}(n)\right],$$

where

$$|\varepsilon_{\pm}(n)| \leq \frac{1}{\prod_{k=n}^{\infty}(1 - |\chi(k)|)} - 1 = \frac{1}{\prod_{k=n}^{\infty}\left(1 - \frac{|g(k)|}{\sqrt{a(1+a)}}\right)} - 1.$$

Here n_1 is fixed sufficiently large such that

$$V(n) = \sum_{k=n}^{\infty} |\chi(k)| = \frac{1}{\sqrt{a(a+1)}} \sum_{k=n}^{\infty} |g(k)| < 1, \qquad \text{for } n \geq n_1. \tag{9.61}$$

In this oscillatory case, Spigler and Vianello [141] derived solutions

$$\hat{y}(n) = (\lambda_{\pm})^n \left[1 + \hat{\varepsilon}_{\pm}(n)\right], \qquad \text{where } |\hat{\varepsilon}_{\pm}(n))| \leq \frac{V(n)}{1 - V(n)},$$

with $V(n)$ and n_1 defined in (9.61). We emphasize that $0 \leq |\chi(n)| \leq V(n) < 1$ for all $n \geq n_1$.

Now

$$\begin{aligned}
& |\varepsilon_{\pm}(n)| < |\hat{\varepsilon}_{\pm}(n)| \\
\Longleftrightarrow \quad & \frac{1}{\prod_{k=n}^{\infty}(1 - |\chi(k)|)} < \frac{1}{1 - \sum_{k=n}^{\infty}|\chi(k)|} \\
\Longleftrightarrow \quad & 1 - \sum_{k=n}^{\infty}|\chi(k)| < \prod_{k=n}^{\infty}(1 - |\chi(k)|) \\
\Longleftrightarrow \quad & \log\left(1 - \sum_{k=n}^{\infty}|\chi(k)|\right) < \sum_{k=n}^{\infty}\log(1 - |\chi(k)|) \\
\Longleftrightarrow \quad & -\sum_{m=2}^{\infty}\frac{1}{m}\left(\sum_{k=n}^{\infty}|\chi(k)|\right)^m < -\sum_{k=n}^{\infty}\sum_{m=2}^{\infty}\frac{|\chi(k)|^m}{m},
\end{aligned}$$

which holds true since $\left(\sum_{k=n}^{\infty}|\chi(k)|\right)^m > \sum_{k=n}^{\infty}|\chi(k)|^m$ for $m \geq 2$. Thus the approach taken here improves the error estimates given in [141], and similar arguments can be made for the other cases we considered.

We conclude this section by stating and proving the version of Gronwall's inequality we needed.

Lemma 9.8 *Suppose that for all $n \geq n_0$, $0 \leq |\chi(n)| < 1$, and $\sum_{n_1}^{\infty}|\chi(n)| < \infty$, and $0 \leq \varphi(n) \leq M$, for some $M > 0$. If*

$$\varphi(n) \leq 1 + \sum_{k=n}^{\infty}|\chi(k)|\varphi(k) =: 1 + R(n), \qquad n \geq n_1 \tag{9.62}$$

then

$$R(n) \leq \frac{1}{\prod_{k=n}^{\infty}(1 - |\chi(k)|)} - 1, \tag{9.63}$$

and

$$\varphi(n) \leq \frac{1}{\prod_{k=n}^{\infty}(1 - |\chi(k)|)}, \qquad n \geq n_1. \tag{9.64}$$

Proof The hypotheses imply that $\sum_{k=n}^{\infty}|\chi(k)|\varphi(k)$ converges, thus $R(n) \to 0$ as $n \to \infty$. Since $\Delta R(m) = R(m+1) - R(m) = -|\chi(m)|\varphi(m)$, it follows from (9.62) that

$$-\Delta R(m) - |\chi(m)|R(m) \leq |\chi(m)|,$$

and therefore also

$$-[\Delta R(m) + |\chi(m)|R(m)] \prod_{k=m+1}^{\infty}(1 - |\chi(k)|) \leq |\chi(m)| \prod_{k=m+1}^{\infty}(1 - |\chi(k)|).$$

This can be re-written in an equivalent way as

$$-\Delta\left(R(m)\prod_{k=m}^{\infty}(1 - |\chi(k)|)\right) \leq \Delta\left(\prod_{k=m}^{\infty}(1 - |\chi(k)|)\right).$$

Summing this for $n \le m < \infty$ implies that

$$R(n)\prod_{k=n}^{\infty}(1-|\chi(k)|) \le 1-\prod_{k=n}^{\infty}(1-|\chi(k)|),$$

which establishes (9.63) and (9.64). □

9.4.2 Coefficients with Regular Growth

This section concerns (9.31) with $a = 0$. We recall that this case was also considered in Sect. 9.4.1, but under the assumption that $n^2 g(n) \in l^1$ or at least $ng(n) \in l^1$ (see Proposition 9.6 and Remark 9.7, respectively). Kooman [97] complemented these results by supposing that $ng(n) \notin l^1$. In order to obtain results in this case, he assumed that the coefficients $g(n)$ possess a certain type of *regular growth* (see (9.65) below) and studied three interesting special cases. We change the notation in (9.31) somewhat (i.e., $g(n)$ is replaced by $-c_n$) to conform to Kooman's. These results in a slightly modified form we now state as Theorem 9.9 (cf. Kooman [97, Theorem 1]). Note that c_n given in (9.65) satisfy $nc_n \notin l^1$ if and only if $p \ge -2$.

Theorem 9.9 *Let $\{c_n\}_{n=1}^{\infty}$ be a sequence of complex numbers such the $c_n \neq 1$ for all n and*

$$c_n = [c + \text{o}(1)]n^p \tag{9.65}$$

for some $p \ge -2$ and $c \neq 0$. Then the linear difference equation

$$\Delta^2 y(n) = c_n y(n), \qquad n \ge n_0 \tag{9.66}$$

has for n sufficiently large two linearly independent solutions $\{y_i(n)\}_{n=1}^{\infty}$, $(i = 1, 2)$, such that

1. If $p > -2$ and

$$\frac{c_{n+1}}{c_n} = 1 + \frac{p}{n} + \varepsilon_n, \qquad \text{where } \varepsilon_n \in l^1, \tag{9.67}$$

then as $n \to \infty$

$$y_1(n) = [1 + \text{o}(1)]n^{-\frac{p}{4}} \prod^{n-1}[1 + \sqrt{c_k}],$$

$$y_2(n) = [1 + \text{o}(1)]n^{-\frac{p}{4}} \prod^{n-1}[1 - \sqrt{c_k}],$$

and

$$y_1(n+1) - y_1(n) = [1 + o(1)]\sqrt{c_n}n^{-\frac{p}{4}} \prod^{n-1}[1 + \sqrt{c_k}],$$

$$y_2(n+1) - y_2(n) = [1 + o(1)]\sqrt{c_n}n^{-\frac{p}{4}} \prod^{n-1}[1 - \sqrt{c_k}],$$

where $\sqrt{c_n}$ *is defined such that* $\mathrm{Re}\sqrt{c_n} > 0$ *if* c *is not a negative real number, and* $\sqrt{c_n} = i\sqrt{-c_n}$ *if* $c < 0$. *Further, in the case that* $c < 0$ *the additional condition that*

$$\prod_{k=n_1}^{n_2} \left| \frac{1 + i\sqrt{-c_k}}{1 - i\sqrt{-c_k}} \right| \tag{9.68}$$

is bounded from above or bounded away from zero for all $n_0 \leq n_1 \leq n_2$ *is imposed.*

2. *If* $p = -2$, $c_n = \frac{c}{n^2} + \frac{\varepsilon_n}{n}$ *for some* $c \in \mathbb{C}$, $c \neq -\frac{1}{4}$ *and* $\varepsilon_n \in l^1$, *then there exist solutions satisfying*

$$y_1(n) = [1 + o(1)]n^{b_1}, \qquad y_2(n) = [1 + o(1)]n^{b_2}, \qquad \text{as } n \to \infty,$$

where b_1 *and* b_2 *are the (distinct) roots of* $z^2 - z - c = 0$, *and*

$$y_1(n+1) - y_1(n) = \frac{b_1 + o(1)}{n} y_1(n), \qquad y_2(n+1) - y_2(n) = \frac{b_2 + o(1)}{n} y_2(n).$$

3. *If* $p = -2$, $c_n = -\frac{1}{4n^2} + \frac{\varepsilon_n}{n \log n}$ *and* $\varepsilon_n \in l^1$, *then there exist solutions satisfying*

$$y_1(n) = [1 + o(1)]\sqrt{n}, \qquad y_2(n) = [1 + o(1)]\sqrt{n} \log n, \qquad \text{as } n \to \infty,$$

and

$$y_1(n+1) - y_1(n) = \frac{1}{n}\left(\frac{1}{2} + \frac{o(1)}{\log n}\right) y_1(n),$$

$$y_2(n+1) - y_2(n) = \frac{1}{n}\left(\frac{1}{2} + \frac{1 + o(1)}{\log n}\right) y_2(n).$$

Remark 9.10 In addition to considering (9.66) as a special case of (9.31) when $a = 0$, the first case of Theorem 9.9 it can also be considered as a special case of (9.27) where one could choose $f(n) = c_n$ and $g(n) \equiv 0$. When treating this case ($p > -2$) below, we will adapt the procedure indicated in Sect. 9.3 to this special situation.

In what follows, we give a somewhat independent proof of Kooman's results showing that they follow from an application of general system results from Chaps. 3 and 5.

Proof (9.66) can be re-written as

$$\mathbf{y}(n+1) = \begin{pmatrix} 0 & 1 \\ c_n - 1 & 2 \end{pmatrix} \mathbf{y}(n), \qquad \mathbf{y}(n) = \begin{pmatrix} y(n) \\ y(n+1) \end{pmatrix}. \tag{9.69}$$

To bring the constant matrix into its Jordan form, we set

$$\mathbf{y}(n) = \begin{pmatrix} 1 & 0 \\ 1 & 1 \end{pmatrix} \tilde{y}(n),$$

hence

$$\tilde{y}(n+1) = \begin{pmatrix} 1 & 1 \\ c_n & 1 \end{pmatrix} \tilde{y}(n). \tag{9.70}$$

Since the off-diagonal elements can have significantly different magnitude, we continue with the shearing transformation

$$\tilde{y}(n) = \begin{pmatrix} 1 & \\ & \sqrt{c_n} \end{pmatrix} \hat{y}(n),$$

for large enough n such that $c_n \neq 0$ (recall that $c \neq 0$). Then

$$\hat{y}(n+1) = \begin{pmatrix} 1 & \sqrt{c_n} \\ \sqrt{c_n}\sqrt{\frac{c_n}{c_{n+1}}} & \sqrt{\frac{c_n}{c_{n+1}}} \end{pmatrix} \hat{y}(n). \tag{9.71}$$

1. $p > -2$.
   Then (9.67) implies that

$$\sqrt{\frac{c_n}{c_{n+1}}} = 1 - \frac{p}{2n} + \varepsilon_n. \tag{9.72}$$

Here and throughout this exposition we use the term ε_n to denote generic l^1-sequences. Then (9.71) can be written as

$$\hat{y}(n+1) = \left[\begin{pmatrix} 1 & \sqrt{c_n} \\ \sqrt{c_n} & 1 \end{pmatrix} - \frac{p}{2n} \begin{pmatrix} 0 & 0 \\ \sqrt{c_n} & 1 \end{pmatrix} + \begin{pmatrix} 0 & 0 \\ \sqrt{c_n}\varepsilon_n & \varepsilon_n \end{pmatrix} \right] \hat{y}(n).$$

To diagonalize the leading matrix, we put

$$\hat{y}(n) = \begin{pmatrix} 1 & 1 \\ 1 & -1 \end{pmatrix} z(n),$$

leading to

$$z(n+1) = \Bigg[\underbrace{\begin{pmatrix} 1+\sqrt{c_n} & \\ & 1-\sqrt{c_n} \end{pmatrix}}_{\Lambda(n)} + \underbrace{\varepsilon_n \begin{pmatrix} 1 & -1 \\ -1 & 1 \end{pmatrix}}_{R(n)}$$

$$+ \underbrace{\sqrt{c_n}\left(\frac{p}{4n} - \varepsilon_n\right)\begin{pmatrix} -1 & -1 \\ 1 & 1 \end{pmatrix} + \frac{p}{4n}\begin{pmatrix} 1 & -1 \\ -1 & 1 \end{pmatrix}}_{V(n)} \Bigg] z(n). \quad (9.73)$$

Since c_n does not need to vanish at infinity, results such as Theorem 3.4 or 5.3 do not apply. However, the ratio $V(n)/(\lambda_1(n) - \lambda_2(n))$ contains more information, and therefore it is appropriate to apply Theorem 5.2 to (9.73). We note that the difference of the entries of $\Lambda(n)$ satisfies $\lambda_1(n) - \lambda_2(n) = 2\sqrt{c_n}$. Moreover, $V(n)/[\lambda_1(n) - \lambda_2(n)]$ has terms of magnitude $\{1/n, \varepsilon_n, 1/(n\sqrt{c_n})\}$. Thus it follows immediately that

$$\frac{V(n)}{\lambda_1(n) - \lambda_2(n)} \to 0 \qquad \text{as } n \to \infty. \quad (9.74)$$

We next claim that

$$\Delta\left(\frac{V(n)}{\lambda_1(n) - \lambda_2(n)}\right) \in l^1[n_0, \infty). \quad (9.75)$$

It is easy to show that $\Delta(1/n), \Delta(\varepsilon_n) \in l^1[n_0, \infty)$. Moreover,

$$\frac{1}{(n+1)\sqrt{c_{n+1}}} - \frac{1}{n\sqrt{c_n}} = \frac{1 - (1+1/n)\sqrt{c_{n+1}/c_n}}{(n+1)\sqrt{c_{n+1}}} = \frac{O(1/n) + \varepsilon_n}{(n+1)\sqrt{c_{n+1}}} \in l^1,$$

which follows from a straightforward calculation using (9.72) and the fact $p > -2$, and (9.75) is established. Note that (9.74) and (9.75) imply that all three conditions (5.11), (5.12) , and (5.13) of Theorem 5.2 are satisfied.

Next we claim that the eigenvalues $\mu_1(n)$ and $\mu_2(n)$ of $\Lambda(n) + V(n)$ satisfy Levinson's dichotomy conditions (3.8), (3.9). We begin by noting that

$$\mu_{1,2}(n) = 1 - \frac{p}{4n} \pm \sqrt{c_n}\sqrt{1 - \frac{p}{2n} + \varepsilon_n + \frac{p^2}{16n^2 c_n}},$$

and we will show that for all values $p > -2$ it holds that

$$\prod_{k=n_1}^{n_2} \left|\frac{\mu_1(k)}{\mu_2(k)}\right| = [\hat{c} + \mathrm{o}(1)] \prod_{k=n_1}^{n_2} \left|\frac{1 + \sqrt{c_k}}{1 - \sqrt{c_k}}\right|, \tag{9.76}$$

for some $\hat{c} \neq 0$. For $-2 < p < 0$, $\mu_{1,2}(n)$ simplifies to $\mu_{1,2}(n) = 1 - \frac{p}{4n} \pm \sqrt{c_n} + \varepsilon_n$ and it follows for $n_1 \leq n_2$ that

$$\begin{aligned}
\prod_{k=n_1}^{n_2} \left|\frac{\mu_1(k)}{\mu_2(k)}\right| &= \prod_{k=n_1}^{n_2} \left|\frac{1 + \sqrt{c_k}/[1 - p/(4k)] + \varepsilon_k}{1 - \sqrt{c_k}/[1 - p/(4k)] + \varepsilon_k}\right| \\
&= \prod_{k=n_1}^{n_2} \left|\frac{1 + \sqrt{c_k}}{1 - \sqrt{c_k}}[1 + \varepsilon_k]\right| \\
&= [\hat{c} + \mathrm{o}(1)] \prod_{k=n_1}^{n_2} \left|\frac{1 + \sqrt{c_k}}{1 - \sqrt{c_k}}\right|,
\end{aligned}$$

for some $\hat{c} \neq 0$, hence (9.76) holds in this case. It is easy to show that (9.76) hold for $p = 0$. Finally, if $p > 0$, then a Taylor expansion shows that

$$\begin{aligned}
\left|\frac{\mu_1(n)}{\mu_2(n)}\right| &= \left|\frac{\sqrt{1 - \frac{p}{2n} + \varepsilon_n + \frac{p^2}{16n^2 c_n}} + \frac{1}{\sqrt{c_n}} + \varepsilon_n}{-\sqrt{1 - \frac{p}{2n} + \varepsilon_n + \frac{p^2}{16n^2 c_n}} + \frac{1}{\sqrt{c_n}} + \varepsilon_n}\right| \\
&= \left|\frac{1 - \frac{p}{4n} + \frac{1}{\sqrt{c_n}} + \varepsilon_n}{-1 + \frac{p}{4n} + \frac{1}{\sqrt{c_n}} + \varepsilon_n}\right| = [1 + \varepsilon_n]\left|\frac{1 + \sqrt{c_n}}{1 - \sqrt{c_n}}\right|,
\end{aligned}$$

and (9.76) follows also in this case.

If c is not a negative real number, and taking the principal branch of the square root function, it follows that $\mathrm{Re}\sqrt{c_k} > 0$ for all k sufficiently large, which can be used to show that $\left|\frac{1+\sqrt{c_k}}{1-\sqrt{c_k}}\right| \geq 1 + \delta$ for some $\delta > 0$ for all k sufficiently large, and therefore Levinson's dichotomy conditions (3.8), (3.9) are satisfied. If $c < 0$, then hypothesis (9.68) ensures that (3.8), (3.9) hold. Finally, it holds that $|R(n)|/\mu_k(n)| \in l^1$ for $k = 1, 2$.

Now Theorem 5.2 implies that (9.73) has for n sufficiently large a fundamental matrix of the form

$$Y_3(n) = [I + o(1)] \prod^{n-1} \begin{pmatrix} \mu_1(k) & \\ & \mu_2(k) \end{pmatrix},$$

and therefore (9.69) has a fundamental matrix of the form

$$Y(n) = \begin{pmatrix} 1 & 0 \\ 1 & 1 \end{pmatrix} \begin{pmatrix} 1 & \\ & \sqrt{c_n} \end{pmatrix} \begin{pmatrix} 1 & 1 \\ 1 & -1 \end{pmatrix} [I + o(1)] \prod^{n-1} \begin{pmatrix} \mu_1(k) & \\ & \mu_2(k) \end{pmatrix},$$

which we prefer to write as

$$\begin{pmatrix} 1 & 0 \\ -1 & 1 \end{pmatrix} \begin{pmatrix} y_1(n) & y_2(n) \\ y_1(n+1) & y_2(n+1) \end{pmatrix}$$
$$= \begin{pmatrix} 1 & \\ & \sqrt{c_n} \end{pmatrix} \begin{pmatrix} 1 & 1 \\ 1 & -1 \end{pmatrix} [I + o(1)] \prod^{n-1} \begin{pmatrix} \mu_1(k) & \\ & \mu_2(k) \end{pmatrix},$$

or

$$\begin{pmatrix} y_1(n) & y_2(n) \\ y_1(n+1) - y_1(n) & y_2(n+1) - y_2(n) \end{pmatrix}$$
$$= \begin{pmatrix} 1 + o(1) & 1 + o(1) \\ \sqrt{c_n}[1 + o(1)] & -\sqrt{c_n}[1 + o(1)] \end{pmatrix} \prod^{n-1} \begin{pmatrix} \mu_1(k) & \\ & \mu_2(k) \end{pmatrix}.$$

Similar to the computations above, it is straightforward to show that for $-2 < p < 0$,

$$\mu_1(k) = 1 - \frac{p}{4k} + \sqrt{c_k} + \varepsilon_k = [1 - \frac{p}{4k}][1 + \sqrt{c_k}][1 + \varepsilon_k],$$

and hence

$$y_1(n) = [\hat{c} + o(1)] \prod^{n-1} [1 - \frac{p}{4k}][1 + \sqrt{c_k}] = [\hat{c} + o(1)] n^{-p/4} \prod^{n-1} [1 + \sqrt{c_k}],$$

and, moreover,

$$y_1(n+1) - y_1(n) = \sqrt{c_n} y_1(n) = \sqrt{c_n}[\hat{c} + o(1)] n^{-p/4} \prod^{n-1} [1 + \sqrt{c_k}],$$

for some $\hat{c} \neq 0$. Similar calculations establish the statement of Theorem 9.9 in the case $p = 0$ and $p > 0$ for the two linearly independent solutions, which completes the proof of part 1.

2. $p = -2$ and $c_n = \frac{c}{n^2} + \frac{\varepsilon_n}{n}$ for some $c \in \mathbb{C}$, $c \neq -1/4$ and $\varepsilon_n \in l^1$:
   In (9.70) we make the shearing transformation

$$\tilde{y}(n) = \begin{pmatrix} 1 & \\ & 1/n \end{pmatrix} w(n), \tag{9.77}$$

leading to

$$w(n+1) = \left[I + \frac{1}{n} \underbrace{\begin{pmatrix} 0 & 1 \\ c & 1 \end{pmatrix}}_{=:A} + R(n) \right] w(n), \tag{9.78}$$

where $R(n) \in l^1$. Note that the matrix A has distinct eigenvalues

$$b_{1,2} = \frac{1}{2} \pm \sqrt{c + \frac{1}{4}},$$

since $c \neq -1/4$. We diagonalize A by putting

$$w(n) = \begin{pmatrix} 1 & 1 \\ b_1 & b_2 \end{pmatrix} \tilde{w}(n),$$

and it follows that

$$\tilde{w}(n+1) = \left[I + \frac{1}{n} \begin{pmatrix} b_1 & \\ & b_2 \end{pmatrix} + \tilde{R}(n) \right] \tilde{w}(n).$$

Note that $\tilde{\Lambda}(n) = I + \frac{1}{n} \mathrm{diag}\{b_1, b_2\}$ satisfies Levinson's dichotomy conditions (3.8), (3.9) and that $\tilde{R}(n)/\tilde{\lambda}_i(n) \in l^1$, hence applying Theorem 3.4 we obtain for n sufficiently large a fundamental solutions matrix

$$\tilde{Y}(n) = [I + o(1)] \prod^{n-1} \tilde{\Lambda}(k) = [I + o(1)] \mathrm{diag}\{n^{b_1}, n^{b_2}\}.$$

Arguing as in case 1., we see that (9.69) has a fundamental matrix satisfying

$$\begin{pmatrix} 1 & 0 \\ -1 & 1 \end{pmatrix} \begin{pmatrix} y_1(n) & y_2(n) \\ y_1(n+1) & y_2(n+1) \end{pmatrix} = \begin{pmatrix} 1 & \\ & 1/n \end{pmatrix} \begin{pmatrix} 1 & 1 \\ b_1 & b_2 \end{pmatrix} [I + o(1)] \begin{pmatrix} n^{b_1} & \\ & n^{b_2} \end{pmatrix},$$

or

$$\begin{pmatrix} y_1(n) & y_2(n) \\ y_1(n+1) - y_1(n) & y_2(n+1) - y_2(n) \end{pmatrix}$$
$$= \begin{pmatrix} 1 + o(1) & 1 + o(1) \\ \frac{b_1}{n}[1 + o(1)] & \frac{b_2}{n}[1 + o(1)] \end{pmatrix} \begin{pmatrix} n^{b_1} & \\ & n^{b_1} \end{pmatrix}.$$

Therefore $y_1(n) = [1 + o(1)]n^{b_1}$ and

$$y_1(n+1) - y_1(n) = \frac{b_1 + o(1)}{n} n^{b_1} = \frac{b_1 + o(1)}{n} y_1(n).$$

Analogous calculations lead to Kooman's results for $y_2(n)$, which establishes part 2.

3. $p = -2$ and $c_n = -\frac{1}{4n^2} + \frac{\varepsilon_n}{n \log n}$, where $\varepsilon_n \in l^1$: In (9.70) we make again the shearing transformation (9.77) leading to

$$w(n+1) = \left[I + \frac{1}{n} \underbrace{\begin{pmatrix} 0 & 1 \\ -\frac{1}{4} & 1 \end{pmatrix}}_{=:\hat{A}} + \begin{pmatrix} 0 & 0 \\ b(n) & 0 \end{pmatrix} \right] w(n), \tag{9.79}$$

where $b(n) = \frac{\varepsilon_n}{\log n} + \frac{\varepsilon_n}{n \log n} - \frac{1}{4n^2}$. Noting that $\hat{A}$ is not diagonalizable, but one can bring $\hat{A}$ into Jordan form by putting

$$w(n) = \begin{pmatrix} 1 & 0 \\ \frac{1}{2} & 1 \end{pmatrix} \hat{w}(n),$$

and it follows that

$$\hat{w}(n+1) = \left[I + \frac{1}{2n} \begin{pmatrix} 1 & 2 \\ 0 & 1 \end{pmatrix} + \begin{pmatrix} 0 & 0 \\ b(n) & 0 \end{pmatrix} \right] \hat{w}(n).$$

To diagonalize the second term,we set

$$\hat{w}(n) = \begin{pmatrix} 1 & \log n \\ 0 & 1 \end{pmatrix} u(n)$$

to see that

$$u(n+1) = \left[I + \frac{1}{2n} I + \begin{pmatrix} -b(n)\log(n+1) & -b(n)\log n \log(n+1) + O(n^{-2}) \\ b(n) & b(n)\log n \end{pmatrix} \right] u(n).$$

Finally we apply

$$u(n) = \operatorname{diag}\left\{1, \frac{1}{\log n}\right\} \tilde{u}(n)$$

to obtain after some simplifications

$$\tilde{u}(n+1) = \left[\begin{pmatrix} 1 + \frac{1}{2n} & 0 \\ 0 & 1 + \frac{1}{2n} + \frac{1}{n\log n} \end{pmatrix} + \hat{R}(n) \right] \tilde{u}(n) = \left[\hat{\Lambda}(n) + \hat{R}(n) \right] \tilde{u}(n).$$

Since the diagonal matrix $\hat{\Lambda}$ satisfies Levinson's dichotomy conditions (3.8), (3.9) and $\hat{R}(n)/\hat{\lambda}_i \in l^1$, applying Theorem 3.4 implies that there exists for n sufficiently large a fundamental matrix of the form

$$\tilde{U}(n) = [I + o(1)] \prod^{n-1} \operatorname{diag}\left\{1 + \frac{1}{2k}, 1 + \frac{1}{2k} + \frac{1}{k\log k}\right\}.$$

Observing that $\prod^{n-1}\left[1 + \frac{1}{2k} + \frac{1}{k\log k}\right] = [\hat{c} + o(1)] n^{1/2} \log n$ as $n \to \infty$ for some $\hat{c} \neq 0$ (see, e.g., [97, Lemma 4]), we obtain that (9.69) has a fundamental matrix satisfying

$$\begin{pmatrix} 1 & 0 \\ -1 & 1 \end{pmatrix} \begin{pmatrix} y_1(n) & y_2(n) \\ y_1(n+1) & y_2(n+1) \end{pmatrix}$$

$$= \begin{pmatrix} 1 & \\ & 1/n \end{pmatrix} \begin{pmatrix} 1 & 0 \\ \frac{1}{2} & 1 \end{pmatrix} \begin{pmatrix} 1 & \log n \\ 0 & 1 \end{pmatrix} \begin{pmatrix} 1 & 0 \\ 0 & \frac{1}{\log n} \end{pmatrix} [I + o(1)] \begin{pmatrix} n^{1/2} & \\ & n^{1/2}\log n \end{pmatrix},$$

or

$$\begin{pmatrix} y_1(n) & y_2(n) \\ y_1(n+1) - y_1(n) & y_2(n+1) - y_2(n) \end{pmatrix}$$

$$= \begin{pmatrix} 1 & 1 \\ \frac{1}{2n} & \frac{1}{2n} + \frac{1}{n\log n} \end{pmatrix} n^{1/2} \begin{pmatrix} 1 + o(1) & o(\log n) \\ o(1) & [1 + o(1)]\log n \end{pmatrix}.$$

Hence $y_1(n) = [1 + o(1)]n^{1/2}$ and

$$y_1(n+1) - y_1(n) = \frac{1}{n}\left(\frac{\sqrt{n}[1+o(1)]}{2} + \frac{\sqrt{n}}{\log n}\right) = \frac{1}{n}\left(\frac{1}{2} + o\left(\frac{1}{\log n}\right)\right) y_1(n).$$

Analogous calculations lead to Kooman's results for $y_2(n)$, which establishes part 3. □

We conclude with two remarks:

1. We note that by one of the results of Chap. 7 (see Theorem 7.1) the asymptotics of a fundamental matrix of (9.70) are immediately known if $nc_n \in l^1$, i.e., if $p < -2$, which explains Kooman's choice $p \geq -2$.
2. Lemma 3 in [97] corresponds to making various choices for the preliminary and conditioning transformations we have used above.

9.4.3 Perturbations with Averaged Growth Conditions

In the previous section, the assumption of regular growth of the coefficients in (9.65) (or more generally that $\lim_{n\to\infty} c_{n+1}/c_n$ exists) was crucial. In this section we consider some second order equations treated by Stepin and Titov [149] in which the coefficients could possess a much more erratic behavior than regular growth. However, in an averaged sense described below, certain products of successive terms display a behavior which allows the corresponding matrix system to be brought into a form where results in Chaps. 3 and 5 are applicable.

More specifically, they considered equations of the form

$$w_{n+1} + a_n w_{n-1} - b_n w_n = 0, \qquad n \geq 0, \tag{9.80}$$

where $a_n > 0$ and $b_n > 0$. After a preliminary normalizing transformation, (9.80) becomes

$$u_{n+1} + u_{n-1} - q_n u_n = 0, \qquad n \geq 0 \tag{9.81}$$

where $q_n > 0$ for all n (i.e., $a_n = q_n$ and $b_n = 1$). While they state their results [149, Theorems 1,2] in terms of solutions of (9.80), we will for simplicity's sake work directly with the normalized equation (9.81).

They first (see [149, p. 136]) considered (9.81) under the assumption that $q_n > 0$ for all $n \geq 0$ and satisfies

$$\sum_{n=1}^{\infty} \frac{1}{q_{n-1}q_n} < \infty.$$

Then they showed that (9.81) has for n sufficiently large solutions u_n^+ and u_n^- such that

$$u_n^+ = [1+o(1)]\prod_{k=1}^{n-1} q_k \quad \text{and} \quad u_n^- = [1+o(1)]\prod_{k=1}^{n} \frac{1}{q_k} \quad \text{as } n \to \infty. \tag{9.82}$$

Relaxing their assumption of q_n being positive, we will show the slightly improved

Theorem 9.11 (cf. [149, Theorem 1]) *Suppose that $q_n \neq 0$ for all $n \geq 0$ and suppose that*

$$\sum_{n=1}^{\infty} \frac{1}{|q_{n-1}q_n|} < \infty. \tag{9.83}$$

Then (9.81) *has for n sufficiently large solutions u_n^+ and u_n^- satisfying* (9.82).

As Stepin/Titov remark, Theorem 3.4 or the results in Chap. 5 do not immediately apply to (a systems version of) (9.81) because while (9.83) means that $|q(n)|$ is large *on the average*, it could also be sometimes quite small (for example, $q_{2n} = 2^n$, $q_{2n+1} = 1/n$). While in [149] Stepin and Titov refer to such assumptions on q_n as "regularly varying in a certain sense," we prefer the terminology "average growth." Their approach is based on explicit matrix formulas which they acknowledge are somewhat related to the Levinson approach, but they do not explicitly use preliminary and conditioning transformations to achieve a reduction to L-diagonal form. We will use such transformations now to obtain an independent proof of Theorem 9.11.

Proof As usual, we re-write (9.81) as the equivalent matrix equation

$$\mathbf{u}(n+1) = \begin{pmatrix} 0 & 1 \\ -1 & q_n \end{pmatrix} \mathbf{u}(n), \qquad \mathbf{u}(n) = \begin{pmatrix} u_{n-1} \\ u_n \end{pmatrix}. \tag{9.84}$$

We first employ a shearing transformation of the form

$$\mathbf{u}(n) = \begin{pmatrix} 1 & 0 \\ 0 & q_{n-1} \end{pmatrix} \mathbf{u}_1(n),$$

leading to

$$\mathbf{u}_1(n+1) = \begin{pmatrix} 0 & q_{n-1} \\ -\frac{1}{q_n} & q_{n-1} \end{pmatrix} \mathbf{u}_1(n).$$

To bring a nonzero entry into the upper left corner, we continue with the preliminary transformation

$$\mathbf{u}_1(n) = \begin{pmatrix} 1 & 1 \\ 0 & 1 \end{pmatrix} \mathbf{u}_2(n),$$

which yields

$$\mathbf{u}_2(n+1) = \begin{pmatrix} \frac{1}{q_n} & \frac{1}{q_n} \\ -\frac{1}{q_n} & q_{n-1} - \frac{1}{q_n} \end{pmatrix} \mathbf{u}_2(n).$$

Next, the normalization

$$\mathbf{u}_2(n) = \left(\prod^{n-1} \frac{1}{q_k} \right) \mathbf{u}_3(n),$$

leads to

$$\mathbf{u}_3(n+1) = \begin{pmatrix} 1 & 1 \\ -1 & q_{n-1}q_n - 1 \end{pmatrix} \mathbf{u}_3(n). \tag{9.85}$$

It will be convenient to introduce the notation

$$\gamma_n = \frac{1}{q_{n-1}q_n - 1}, \tag{9.86}$$

where we note that γ_n is well-defined for n sufficiently large (say $n \geq n_1$) due to (9.83), and the consequence that $1/q_n q_{n+1} = o(1)$ as $n \to \infty$. Therefore, $\gamma_n = O\left(\frac{1}{|q_{n-1}q_n|}\right)$ and from (9.83) it follows that

$$\gamma_n \in l^1. \tag{9.87}$$

Therefore (9.85) can be re-written as

$$\mathbf{u}_3(n+1) = \begin{pmatrix} 1 & 1 \\ -1 & \frac{1}{\gamma_n} \end{pmatrix} \mathbf{u}_3(n).$$

Next, we make use of the difference in magnitude between the diagonal entries to reduce the size of the off-diagonal terms. For that purpose, we put

$$\mathbf{u}_3(n) = \begin{pmatrix} 1 & c_n \\ d_n & 1 \end{pmatrix} \mathbf{u}_4(n), \tag{9.88}$$

with c_n, d_n to be determined. Assuming for the moment invertibility, one can see that

$$\mathbf{u}_4(n+1) \\ = \frac{1}{1-c_{n+1}d_{n+1}} \begin{pmatrix} 1+d_n+c_{n+1}-\frac{c_{n+1}d_n}{\gamma_n} & 1-\frac{c_{n+1}}{\gamma_n}+c_n+c_nc_{n+1} \\ \frac{d_n}{\gamma_n}-d_{n+1}-1-d_nd_{n+1} & -d_{n+1}-d_{n+1}c_n-c_n+\frac{1}{\gamma_n} \end{pmatrix} \mathbf{u}_4(n). \tag{9.89}$$

Focusing on the off-diagonal entries and disregarding the quadratic terms, one can show that $1-\dfrac{\hat{c}_{n+1}}{\gamma_n}+\hat{c}_n = 0$ has a solution

$$\hat{c}_n = \sum_{k=1}^{n-1}\prod_{l=k}^{n-1}\gamma_l = \gamma_{n-1}+\gamma_{n-1}\gamma_{n-2}+\cdots, \tag{9.90}$$

and that $\dfrac{\hat{d}_n}{\gamma_n}-\hat{d}_{n+1}-1=0$ has a formal solution

$$\hat{d}_n = \sum_{k=n}^{\infty}\prod_{l=n}^{k}\gamma_l = \gamma_n+\gamma_n\gamma_{n+1}+\cdots. \tag{9.91}$$

For a "first order" approximation, we now choose

$$c_n = \gamma_{n-1} \quad \text{and} \quad d_n = \gamma_n. \tag{9.92}$$

Then $c_n \to 0$ and $d_n \to 0$ as $n \to \infty$ by (9.87), and the invertibility of the transformation matrix in (9.88) is justified for n sufficiently large. Moreover,

$$\begin{aligned} \mathbf{u}_4(n+1) &= \frac{1}{1-\gamma_n\gamma_{n+1}}\left[\begin{pmatrix} 1 & 0 \\ 0 & \frac{1}{\gamma_n}\end{pmatrix}\right. \\ &\quad \left. + \begin{pmatrix} \gamma_n & \gamma_{n-1}+\gamma_{n-1}\gamma_n \\ -\gamma_{n+1}-\gamma_n\gamma_{n+1} & -\gamma_{n+1}-\gamma_{n+1}\gamma_{n-1}-\gamma_{n-1}\end{pmatrix}\right]\mathbf{u}_4(n) \\ &=: [\Lambda(n)+R(n)]\,\mathbf{u}_4(n). \end{aligned} \tag{9.93}$$

Since by (9.87)

$$\left|\frac{\lambda_2(n)}{\lambda_1(n)}\right| = \left|\frac{1}{\gamma_n}\right| \geq 2,$$

for n sufficiently large, $\Lambda(n)$ satisfies an exponential dichotomy (and hence ordinary dichotomy condition). Moreover, $R(n)/\lambda_i(n) \in l^1[n_1, \infty)$ for $i = 1, 2$. Hence Theorem 3.4 implies that (9.93) has for large n a fundamental matrix of the form

$$U_4(n) = [I + o(1)] \begin{pmatrix} 1 & 0 \\ 0 & \prod^{n-1} \frac{1}{\gamma_k} \end{pmatrix}.$$

Therefore (9.84) has for large n a fundamental matrix

$$\begin{aligned} \begin{pmatrix} u_{n-1}^- & u_{n-1}^+ \\ u_n^- & u_n^+ \end{pmatrix} &= \begin{pmatrix} 1 & 0 \\ 0 & q_{n-1} \end{pmatrix} \begin{pmatrix} 1 & 1 \\ 0 & 1 \end{pmatrix} \left(\prod^{n-1} \frac{1}{q_k} \right) \begin{pmatrix} 1 & \gamma_{n-1} \\ \gamma_n & 1 \end{pmatrix} [I + o(1)] \begin{pmatrix} 1 & 0 \\ 0 & \prod^{n-1} \frac{1}{\gamma_k} \end{pmatrix} \\ &= \begin{pmatrix} 1 + \gamma_n & 1 + \gamma_{n-1} \\ q_{n-1}\gamma_n & q_{n-1} \end{pmatrix} \begin{pmatrix} 1 + o(1) & o(1) \\ o(1) & 1 + o(1) \end{pmatrix} \begin{pmatrix} \prod^{n-1} \frac{1}{q_k} & 0 \\ 0 & \prod^{n-1} \frac{1}{q_k \gamma_k} \end{pmatrix} \end{aligned}$$

It follows that

$$u_{n-1}^- = [1 + o(1)] \prod^{n-1} \frac{1}{q_k},$$

and

$$\begin{aligned} u_n^+ &= [1 + o(1)] q_{n-1} \prod^{n-1} \frac{1}{q_k \gamma_k} = [1 + o(1)] q_{n-1} \left(\prod^{n-1} q_{k-1} \right) \underbrace{\prod^{n-1} \left(1 - \frac{1}{q_{k-1} q_k} \right)}_{c + o(1)} \\ &= [c + o(1)] \prod^{n-1} q_k, \end{aligned}$$

which establishes (9.82) and finishes the proof. We remark that working with u_n^- instead of u_{n-1}^- would require a quantitative estimate of some of the error terms in the matrix $I + o(1)$ to derive (9.82). □

Remark 9.12 Since $\Lambda(n)$ satisfies an exponential dichotomy condition, (9.83) could be weakened to

$$\frac{1}{|q_n q_{n-1}|} \in l^2,$$

(hence $\gamma_n \in l^2$). Then Theorem 5.3 would imply a similar asymptotic result with modified terms in the product from diag $R(n)$.

Next, Stepin and Titov considered again (9.81), but now under the assumptions that $q_n > 2$ (instead of $q_n > 0$) and a somewhat weaker averaged growth condition of the form

$$\sum_{n=1}^{\infty} \frac{1}{q_{n-1} q_n^2 q_{n+1}} < \infty.$$

In this case they show ([149, p. 136]) that there exist solutions u_n^+ and u_n^- of (9.82) satisfying, as $n \to \infty$,

$$u_n^+ = [1 + o(1)] \prod_{k=1}^{n-1} \left(q_k - \frac{1}{q_{k-1}} \right) \quad \text{and} \quad u_n^- = [1 + o(1)] \prod_{k=1}^{n} \left(q_k - \frac{1}{q_{k+1}} \right)^{-1}, \tag{9.94}$$

i.e., the weaker growth condition requires modifications in the main terms in the products.

For (9.94) (and an extension of (9.96) involving similar products of more consecutive terms), they employ continued fraction expansions for solutions and this method appears to be substantially different in nature to the one they use in the case above and not related to an application of Theorem 3.4.

We will use again the approach taken in the proof of Theorem 9.11, but with suitable modifications to the choice of the coefficients c_n and d_n in the conditioning transformation (9.88). This also allows us to relax the assumptions on q_n.

Theorem 9.13 (cf. [149, Theorem 2]) *Assume that*

$$|q_n| \geq 1 + \delta \qquad \textit{for some } \delta > 0, \tag{9.95}$$

and suppose that

$$\sum_{n=1}^{\infty} \frac{1}{|q_{n-1} q_n^2 q_{n+1}|} < \infty. \tag{9.96}$$

Then (9.81) *has for large n solutions u_n^+ and u_n^- satisfying* (9.94).

Proof With γ_n defined in (9.86), note that (9.95) implies that

$$0 < |\gamma_n| \leq \frac{1}{\delta^2 + 2\delta} := \hat{\delta}, \tag{9.97}$$

and $\gamma_n = O\left(\frac{1}{q_{n-1} q_n}\right)$, hence by (9.96),

$$\rho_n := \gamma_n \gamma_{n+1} \in l^1. \tag{9.98}$$

We follow the proof of Theorem 9.11 up to (9.89), where we choose "2nd order approximations" of (9.90) and (9.91), namely

$$c_n = \gamma_{n-1}(1+\gamma_{n-2}), \qquad d_n = \gamma_n(1+\gamma_{n+1}). \tag{9.99}$$

A straightforward calculation shows that from (9.89), (9.98), and (9.99) it follows that

$$\hat{u}_4(n+1) = \left[\hat{\Lambda}(n) + \hat{R}(n)\right]\hat{u}_4(n), \tag{9.100}$$

where

$$\hat{\Lambda}(n) = [1+\eta_n]\begin{pmatrix} 1+\gamma_n & 0 \\ 0 & \frac{1}{\gamma_n}[1-\rho_{n-1}-\rho_n-\gamma_{n-1}\rho_n] \end{pmatrix},$$

and

$$\frac{\hat{R}(n)}{1+\eta_n} = -\begin{pmatrix} \gamma_{n-1}\rho_n & -\rho_{n-2}-\rho_{n-1}[1+\gamma_{n-2}][1+\gamma_{n-1}] \\ \rho_{n+1}+\rho_n[1+\gamma_{n+2}][1+\gamma_{n+1}] & \rho_{n-2}[1+\gamma_{n+1}+\rho_{n+1}]+\rho_{n+1}[1+\gamma_{n-1}] \end{pmatrix},$$

where ρ_n was defined in (9.98), $\rho_n \in l^1$. Here

$$1+\eta_n = \frac{1}{1-c_{n+1}d_{n+1}} = \frac{1}{1-\rho_n(1+\gamma_{n-1})(1+\gamma_{n+2})},$$

thus $\eta_n \in l^1$ by (9.98). Hence $\lambda_1(n) = 1+\gamma_n+o(1)$ and $\lambda_2(n) = [1+\delta(n)]\frac{1}{\gamma_n}$ with $\delta(n) \in l^1$. It follows that $\hat{R}(n)/\hat{\lambda}_i(n) \in l^1$ for $i = 1,2$, therefore the growth condition (3.10) of Theorem 3.4 is satisfied.

In checking the dichotomy conditions (3.8), (3.9), notice first that (9.97) implies that

$$\left|\frac{\lambda_2(n)}{\lambda_1(n)}\right| = [1+o(1)]\left|\frac{1}{\gamma_n(1+\gamma_n)}\right| \geq \frac{1}{2}\frac{1}{\hat{\delta}(1+\hat{\delta})} := \tilde{\delta},$$

for n sufficiently large. However, since it is not known if $\tilde{\delta} \geq 1$, we need one further step to establish that the dichotomy condition holds. For that purpose, note that (9.97) together with (9.98) imply that

$$\left|\frac{\lambda_2(n)\lambda_2(n+1)}{\lambda_1(n)\lambda_1(n+1)}\right| = [1+o(1)]\left|\frac{1}{\gamma_n\gamma_{n+1}}\frac{1}{(1+\gamma_n)(1+\gamma_{n+1})}\right| \geq \frac{1}{2}\frac{1}{|\rho_n|}\frac{1}{(1+\hat{\delta})^2} \geq 2,$$

for all n sufficiently large. Hence, for $n_2 > n_1$ and with n_1 sufficiently large, it follows that

$$\prod_{k=n_1}^{n_2-1} \left| \frac{\lambda_2(k)}{\lambda_1(k)} \right| \geq \tilde{\delta} \left(2^{\frac{n_2-n_1}{2}} \right).$$

Now Theorem 3.4 implies that for n sufficiently large (9.100) has a fundamental matrix satisfying

$$U_4(n) = [I + o(1)] \begin{pmatrix} \prod^{n-1}(1+\gamma_k) & 0 \\ 0 & \prod^{n-1} \frac{1}{\gamma_k} \end{pmatrix}, \qquad \text{as } n \to \infty.$$

Therefore (9.84) has for large n a fundamental matrix

$$\begin{pmatrix} u_{n-1}^- & u_{n-1}^+ \\ u_n^- & u_n^+ \end{pmatrix} = \begin{pmatrix} 1 & 0 \\ 0 & q_{n-1} \end{pmatrix} \begin{pmatrix} 1 & 1 \\ 0 & 1 \end{pmatrix} \left(\prod^{n-1} \frac{1}{q_k} \right) \begin{pmatrix} 1 & \gamma_{n-1}(1+\gamma_{n-2}) \\ \gamma_n(1+\gamma_{n+1}) & 1 \end{pmatrix} U_4(n)$$

$$= \left(\prod^{n-1} \frac{1}{q_k} \right) \begin{pmatrix} 1+\gamma_n + o(1) & 1 + \gamma_{n-1} + o(1) \\ q_{n-1}\gamma_n(1+\gamma_{n+1}) & q_{n-1} \end{pmatrix} U_4(n).$$

Hence (9.81) has a solution u_{n-1}^- given by

$$\begin{aligned} u_{n-1}^- &= \left(\prod^{n-1} \frac{1}{q_k} \right) [1 + \gamma_n + o(1)] \prod^{n-1} (1+\gamma_k) \\ &= \left(\prod^{n-1} \frac{1}{q_k} \right) [1 + o(1)] \prod^{n} (1+\gamma_k) = [1+o(1)] \prod^{n-1} \frac{1+\gamma_{k+1}}{q_k} \\ &= [1+o(1)] \prod^{n-1} \frac{q_{k+1}}{q_k q_{k+1} - 1} = [1+o(1)] \prod^{n-1} \frac{1}{q_k - \frac{1}{q_{k+1}}}, \end{aligned}$$

which establishes the asymptotics for u_n^- given in (9.94).

Focusing on the second row, one sees that (9.81) has a solution u_n^+ satisfying

$$\begin{aligned} u_n^+ &= [1+o(1)] \left(\prod^{n-2} \frac{1}{q_k} \right) \prod^{n-1} \frac{1}{\gamma_k} = [1+o(1)] \prod^{n-1} \left(\frac{1}{q_{k-1}\gamma_k} \right) \\ &= [1+o(1)] \prod^{n-1} \left(q_k - \frac{1}{q_{k-1}} \right), \end{aligned}$$

which establishes (9.94) and completes the proof. □

Remark 9.14 In [149, Theorem 2], Stepin and Titov considered even weaker averaged growth conditions involving products of more than three terms. They deduce asymptotic formulas like (9.94) involving more "correction terms" which are rational functions in the translates of q_n. These are most easily expressed using continued fractions and arise naturally from the Riccati approach and the corresponding equation

$$z_n z_{n-1} + 1 - q_n z_{n-1} = 0 \qquad (z_n = u_{n+1}/u_n).$$

While we believe that the approach of using conditioning transformations could also lead to such formulas, the algebraic technicalities appear formidable and we will not pursue that here. We observe however, that the more general formulas in [149, Theorem 2] must be interpreted as continued fractions.

Remark 9.15 We reiterate that the approach taken in Theorem 9.13, however, only requires the assumption $|q_n| \geq 1 + \delta$ for some $\delta > 0$ instead of $q_n > 2$. The (Prigsheim) condition $q_n > 2$ seems to be necessary for the convergence of a continued fraction expansion of solutions of the Riccati equation.

Remark 9.16 Recalling the results in Sect. 3.7, we remark that more quantitative estimates for the error term in (9.94) could be achieved under suitable slow-decay conditions on the perturbation.

Remark 9.17 Since $\hat{\Lambda}(n)$ satisfies an exponential dichotomy condition, one could apply Theorem 5.3 to (9.100), just assuming that $\hat{R}(n)/\hat{\lambda}_i(n) \in l^2$ and, of course, necessarily modifying the terms in the product by diag $\hat{R}(n)$. This gives another result similar to the ones considered in [149], but with a modified and not necessarily comparable asymptotic representation of solutions.

9.4.4 Perturbations Having Regular Variation

Second-order scalar difference equations of the form

$$d(n+1)y(n+1) - q(n)y(n) + y(n-1) = 0, \qquad n \geq n_0, \tag{9.101}$$

have been studied by Geronimo and Smith [61] under various assumptions on the complex-valued sequences $\{q(n)\}$ and $\{d(n)\}$ involving what is called either *regular* or *bounded variation* (see hypotheses (H_2) and (H_3) below). The resulting asymptotic representation for solutions were derived from an interesting analysis of solutions of related Riccati-type difference equations. Since their asymptotic formulae appear to be similar in nature to ones derived using a Levinson type of approach, the focus here is to investigate relations between results in their Theorems 2.3 and 2.5 and an approach using general results from Chaps. 3 and 5.

In [61, Theorem 2.3], the following assumptions were made:

(H_0) Let $\{q(n)\}$ and $\{d(n)\}$ be complex-valued sequences such that $d(n) \neq 0$ for $n \geq n_0$;

(H_1) There exists a compact set $K \subset \mathbb{C}$ such that K contains an open set U and $[-2,2] \subset U$, and suppose that $q(n) \notin K$ for all $n \geq n_0$;

(H_2) Letting $\delta(n) := 1 - d(n)$ and $r(n) := q(n+1) - q(n)$, assume that $\delta(n)$ and $r(n)$ are in l^2;

(H_3) $\delta(n+1) - \delta(n)$ and $r(n+1) - r(n)$ are in l^1;

(H_4) Finally suppose that

$$\frac{u_0(n) - d(n+1)u_0(n+1)}{u_0(n) - 1/u_0(n)} \neq 1, \qquad n \geq n_0,$$

and

$$\frac{v_0(n) - d(n+1)v_0(n+1)}{v_0(n) - 1/v_0(n)} \neq 1, \qquad n \geq n_0,$$

where

$$u_0(n) = \frac{q(n) + \sqrt{q^2(n) - 4}}{2} \qquad \text{and } v_0(n) = \frac{q(n) - \sqrt{q^2(n) - 4}}{2}. \tag{9.102}$$

Here the branch of the square root function is chosen so that

$$\left| z + \sqrt{z^2 - 4} \right| > 2, \qquad \forall\, z \in \mathbb{C} \backslash [-2, 2]. \tag{9.103}$$

Under these assumptions, they showed in [61, Theorem 2.3] that there exist two solutions $y_{\pm}$ of (9.101) such that

$$y_+(n) = [1 + \varepsilon_+(n)] \prod_{k=n_0}^{n} u_2(k), \qquad n \geq n_0 \tag{9.104}$$

where

$$|\varepsilon_+(n)| \leq \exp\left\{ c \sum_{j=n_0}^{n} |\xi(j)| \right\} - 1, \tag{9.105}$$

and

$$y_-(n) = [1 + \varepsilon_-(n)] \prod_{k=n_0}^{n} v_2(k), \qquad n \geq n_0, \tag{9.106}$$

where

$$|\varepsilon_-(n)| \leq \exp\left\{c \sum_{j=n+1}^{\infty} |\zeta(j)|\right\} - 1. \tag{9.107}$$

Here

$$\sum_{n_0}^{\infty} |\xi(n)| < \infty \qquad \text{and} \sum_{n_0}^{\infty} |\zeta(n)| < \infty, \tag{9.108}$$

with $\xi(n)$, $\zeta(n)$ computable in terms of the data $d(n)$ and $q(n)$, c is a non-zero constant, and

$$u_2(n) = u_0(n)\left\{1 - \frac{u_0(n) - d(n+1)u_0(n+1)}{u_0(n) - 1/u_0(n)}\right\}^{-1}, \tag{9.109}$$

$$v_2(n) = v_0(n)\left\{1 - \frac{v_0(n) - d(n+1)v_0(n+1)}{v_0(n) - 1/v_0(n)}\right\}^{-1}. \tag{9.110}$$

The functions $u_2(n)$ and $v_2(n)$ are a type of "approximate solution" of the related Riccati difference equation

$$d(n+1)u(n+1)u(n) - q(n)u(n) + 1 = 0, \tag{9.111}$$

where $u(n) = y(n)/y(n-1)$. A rationale for this procedure is that if $u(n)$ would be an exact solution of (9.111), then $y(n) = \prod_{n_0}^{n} u(k)$ would be an exact solution of (9.101). So by suitably approximating a solution of (9.111) (to within an l^1-perturbation), this should lead to an asymptotic representation of the form (9.104). The dominant solution $u_0(n)$ of the limiting equation

$$u^2(n) - q(n)u(n) + 1 = 0,$$

is a first approximation. Then an iteration and some clever heuristic reasoning lead to a better approximation $u_1(n)$. Then they re-write $u_1(n)$ in a form more suitable for their applications and to obtain a further approximation $u_2(n)$ (see (9.109)). Similar reasoning leads to (9.110).

We are here interested in comparing the results (9.104) and (9.106) with an approach involving applications of general results from Chaps. 3 and 5.

Setting $\mathbf{y}(n) = \begin{pmatrix} y(n-1) \\ y(n) \end{pmatrix}$, we write (9.101) as

$$\begin{aligned}
\mathbf{y}(n+1) &= \begin{pmatrix} 0 & 1 \\ -\frac{1}{1-\delta(n+1)} & \frac{q(n)}{1-\delta(n+1)} \end{pmatrix} \mathbf{y}(n) \\
&= \left[\begin{pmatrix} 0 & 1 \\ -1 & q(n) \end{pmatrix} + \frac{\delta(n+1)}{1-\delta(n+1)} \begin{pmatrix} 0 & 0 \\ -1 & q(n) \end{pmatrix} \right] \mathbf{y}(n) \\
&= [A(n) \qquad + \qquad V(n)]\, \mathbf{y}(n).
\end{aligned} \tag{9.112}$$

Note that the eigenvalues of $A(n)$ are given by

$$\lambda_{1,2}(n) = \frac{q(n) \pm \sqrt{q^2(n) - 4}}{2}, \tag{9.113}$$

which coincide with the quantities $u_0(n)$ and $v_0(n)$ given in (9.102). In what follows, the equation $\lambda_1(n)\lambda_2(n) = 1$ is often useful.

Since $q(n)$ is bounded away from $[-2, 2]$ by assumption (H_1), it follows from (9.103) that

$$|\lambda_1(n)| \geq 1+\varepsilon \qquad \text{and } |\lambda_2(n)| = \frac{1}{|\lambda_1(n)|} \leq 1-\varepsilon, \qquad \text{for some } \varepsilon > 0. \tag{9.114}$$

Thus

$$\left| \frac{\lambda_1(n)}{\lambda_2(n)} \right| \geq 1 + 2\varepsilon, \qquad \text{for some } \varepsilon > 0, \tag{9.115}$$

and $\Lambda(n) = \mathrm{diag}\{\lambda_1(n), \lambda_2(n)\}$ satisfies an exponential dichotomy condition.

To diagonalize $A(n)$, we put

$$\mathbf{y}(n) = \begin{pmatrix} 1 & 1 \\ \lambda_1(n-1) & \lambda_2(n-1) \end{pmatrix} \mathbf{y}_1(n) := T(n)\mathbf{y}_1(n).$$

Then it follows from (9.112) that

$$\begin{aligned}
\mathbf{y}_1(n+1) &= T^{-1}(n+1)\,[A(n) + V(n)]\, T(n)\mathbf{y}_1(n) \\
&= \left\{ \underbrace{T^{-1}(n+1)A(n)T(n+1)}_{=:\Lambda(n)} \right. \\
&\qquad \left. + \underbrace{T^{-1}(n+1)\Big[V(n)T(n) + A(n)\,\{T(n) - T(n+1)\}\Big]}_{=:R(n)} \right\} \mathbf{y}_1(n).
\end{aligned} \tag{9.116}$$

One can show that

$$T^{-1}(n+1)V(n)T(n) = \frac{-1}{\sqrt{q^2(n)-4}}\frac{\delta(n+1)}{1-\delta(n+1)}\begin{pmatrix}1-q(n)\lambda_1(n-1) & -1-q(n)\lambda_2(n-1)\\ 1+q(n)\lambda_1(n-1) & 1+q(n)\lambda_2(n-1)\end{pmatrix},$$

and

$$T^{-1}(n+1)A(n)\left[T(n)-T(n+1)\right] = \frac{1}{\sqrt{q^2(n)-4}}\begin{pmatrix}[\lambda_1(n-1)-\lambda_1(n)]\lambda_1(n) & [\lambda_2(n-1)-\lambda_2(n)]\lambda_1(n)\\ -[\lambda_1(n-1)-\lambda_1(n)]\lambda_2(n) & -[\lambda_2(n-1)-\lambda_2(n)]\lambda_2(n)\end{pmatrix}.$$

We make the following preliminary observations:

1. For $q(n)\notin K$, assumption (H_1) implies that there exists a constant $M>0$ such that

$$\left|\frac{1}{\sqrt{q^2(n)-4}}\right|\leq M,$$

and

$$\left|\frac{q(n)}{\sqrt{q^2(n)-4}}\right|\leq M. \tag{9.117}$$

Note that this even holds for unbounded $q(n)$.

2. Recalling assumption (H_2), we observe that

$$\begin{aligned}\frac{q(n)+q(n-1)}{\sqrt{q^2(n)-4}+\sqrt{q^2(n-1)-4}} &= \frac{2q(n)}{\sqrt{q^2(n)-4}+\sqrt{q^2(n-1)-4}}+\underbrace{\frac{q(n-1)-q(n)}{\sqrt{q^2(n)-4}+\sqrt{q^2(n-1)-4}}}_{o(1)}\\ &= \frac{2q(n)}{\sqrt{q^2(n)-4}+\sqrt{\{q(n)-r(n-1)\}^2-4}}+o(1)\\ &= \frac{2q(n)}{\sqrt{q^2(n)-4}}\frac{1}{1+\sqrt{1+\frac{r^2(n-1)-2q(n)r(n-1)}{q^2(n)-4}}}+o(1)\\ &= \frac{2q(n)}{\sqrt{q^2(n)-4}}\frac{1}{1+o(1)}+o(1)=O(1),\end{aligned}$$

where the boundedness follows from (9.117).

Therefore it follows that

$$
\begin{aligned}
&2\left[\lambda_1(n) - \lambda_1(n-1)\right] \\
&\quad = q(n) - q(n-1) + \sqrt{q^2(n) - 4} - \sqrt{q^2(n-1) - 4} \\
&\quad = q(n) - q(n-1) + [q(n) - q(n-1)]\frac{[q(n) + q(n-1)]}{\sqrt{q^2(n) - 4} + \sqrt{q^2(n-1) - 4}} \\
&\quad = r(n-1)\left[1 + \frac{q(n) + q(n-1)}{\sqrt{q^2(n) - 4} + \sqrt{q^2(n-1) - 4}}\right],
\end{aligned}
$$

so from hypothesis (H_2) one concludes that

$$\lambda_1(n) - \lambda_1(n-1) \in l^2.$$

It can be shown similarly that $\lambda_2(n) - \lambda_2(n-1) \in l^2$.

3. Since $\delta(n) \in l^2$ by assumption (H_2), it follows that $\delta(n) \to 0$ as $n \to \infty$ and that

$$\frac{\delta(n+1)}{1 - \delta(n+1)} = \delta(n+1) + l^1. \tag{9.118}$$

4. From (9.117) it also follows that

$$\left|\frac{\lambda_k(n)}{\sqrt{q^2(n) - 4}}\right| \leq \hat{M}, \qquad \text{for some } \hat{M} > 0$$

for $k = 1, 2$, hence

$$T^{-1}(n+1)A(n)\left[T(n) - T(n+1)\right] \in l^2.$$

While it is not necessarily true that $T^{-1}(n+1)V(n)T(n) \in l^2$, a short calculation shows that $T^{-1}(n+1)V(n)T(n)/\lambda_1(n) \in l^2$, and hence by (9.114)

$$\frac{R(n)}{\lambda_1(n)} \in l^2.$$

Asymptotic Representation for a Dominant Solution

Since $\Lambda(n)$ satisfies an exponential dichotomy condition and $R(n)/\lambda_1(n) \in l^2$, an application of Theorem 3.13 with $j = 1$ implies that (9.116) has a *dominant* solution

$\hat{y}_1(n)$ corresponding to $\lambda_1(n)$ of the form

$$\mathbf{y}_1(n) = [e_1 + \varepsilon_1(n)] \prod^{n-1} [\lambda_1(k) + r_{11}(k)],$$

where $\varepsilon_1(n) \to 0$ as $n \to \infty$ and

$$r_{11}(k) = -\frac{\delta(k+1)\,[1 - q(k)\lambda_1(k-1)]}{[1 - \delta(k+1)]\,\sqrt{q^2(k) - 4}} + \frac{\lambda_1(k)[\lambda_1(k) - \lambda_1(k-1)]}{\sqrt{q^2(k) - 4}}.$$

This leads to a corresponding solution $\mathbf{y}(n) = T(n)\mathbf{y}_1(n)$, and looking at the first component of $\mathbf{y}(n)$ shows that (9.101) has a solution

$$y(n-1) = [1 + \varepsilon_1(n)] \prod^{n-1} [\lambda_1(k) + r_{11}(k)].$$

Hence (9.101) has a solution $y_1(n)$ having the asymptotic representation

$$y_1(n) = [1 + \varepsilon_1(n))] \prod_{n_0}^{n} [\lambda_1(k) + r_{11}(k)], \qquad \text{as } n \to \infty. \tag{9.119}$$

We note that this statement requires only assumptions (H_1) and (H_2), not (H_3) and (H_4).

We next want to compare a dominant solution $y_1(n)$ in (9.119) with a dominant solution $y_+(n)$ obtained in (9.104). While both $y_1(n)$ and $y_+(n)$ are so-called "dominant solutions" of (9.101), such solutions are by no means unique. Nevertheless, it is interesting to compare the two solutions since their ratio should tend asymptotically to a non-zero constant as $n \to \infty$.

First observe that from (9.108) it only follows that the error term $\varepsilon_+(n)$ in (9.105) is bounded, but does not necessarily vanish at infinity, which yields a weaker asymptotic statement than (9.119).

We first want to compare $\lambda_1(n) + r_{11}(n)$ with $u_2(n)$ and find conditions such that the ratio has the form $1 + (l^1)$. We rewrite

$$\frac{\lambda_1(n) + r_{11}(n)}{\lambda_1(n)} = 1 + \frac{\lambda_1(n) - \lambda_1(n-1)}{\sqrt{q^2(n) - 4}} - \frac{\delta(n+1)\,[1 - q(n)\lambda_1(n-1)]}{[1 - \delta(n+1)]\,\sqrt{q^2(n) - 4}\,\lambda_1(n)}.$$

Using (9.118), we further simplify this to

$$\frac{\lambda_1(n) + r_{11}(n)}{\lambda_1(n)} = 1 + \frac{\lambda_1(n-1) - \lambda_1(n)}{\sqrt{q^2(n) - 4}} - \frac{\delta(n+1)\,[1 - q(n)\lambda_1(n-1)]}{\sqrt{q^2(n) - 4}\,\lambda_1(n)} + (l^1).$$

Then it follows from (9.109) that

$$
\begin{aligned}
&\frac{\lambda_1(n) + r_{11}(n)}{u_2(n)} \\
&\quad = \left\{ 1 + \frac{\lambda_1(n-1) - \lambda_1(n)}{\sqrt{q^2(n) - 4}} - \frac{\delta(n+1)\,[1 - q(n)\lambda_1(n-1)]}{\sqrt{q^2(n) - 4}\,\lambda_1(n)} + (l^1) \right\} \\
&\qquad \times \left\{ 1 - \frac{\lambda_1(n) - d(n+1)\lambda_1(n+1)}{\sqrt{q^2(n) - 4}} \right\} \\
&\quad = 1 + \frac{\lambda_1(n+1) + \lambda_1(n-1) - 2\lambda_1(n)}{\sqrt{q^2(n) - 4}} \\
&\qquad - \frac{\delta(n+1)}{\lambda_1(n)\sqrt{q^2(n) - 4}} \left[\lambda_1(n+1)\lambda_1(n) + 1 - \left(\lambda_1(n) + \frac{1}{\lambda_1(n)} \right) \lambda_1(n-1) \right],
\end{aligned}
$$

and therefore

$$
\begin{aligned}
\frac{\lambda_1(n) + r_{11}(n)}{u_2(n)} &= 1 + \frac{\Delta^2 \lambda_1(n)}{\sqrt{q^2(n) - 4}} \\
&\quad - \frac{\delta(n+1)}{\sqrt{q^2(n) - 4}} \left[\lambda_1(n+1) - \lambda_1(n-1) + \frac{\lambda_1(n) - \lambda_1(n-1)}{\lambda_1(n)} \right] + (l^1) \\
&= 1 + \frac{\Delta^2 \lambda_1(n)}{\sqrt{q^2(n) - 4}} + (l^1).
\end{aligned}
$$

Hence one can see that the additional assumption $\Delta^2 \lambda_1(n) \in l^1$ implies that

$$
\frac{\lambda_1(n) + r_{11}(n)}{u_2(n)} = 1 + (l^1),
$$

i.e., the two asymptotic representations are equivalent under this additional hypothesis. If, on the other hand $\Delta^2 \lambda_1(n) \notin l^1$, the asymptotic representation (9.119) shows which additional terms are required beyond those in [61, Theorem 2.3]. One can show that $\Delta^2 q(n) \in l^1$ suffices for $\Delta^2 \lambda_i(n) \in l^1$ for $i = 1, 2$. This also shows that under the additional assumption $\Delta^2 q(n) \in l^1$, there is a solution of the form (9.104) with an improved error estimate which, moreover, can be quantitatively estimated under appropriate extra slow decay conditions using the results in Sect. 3.7.

Asymptotic Representation for the Recessive Solution

For a recessive solution $\hat{y}_2(n)$ corresponding to $\lambda_2(n)$, an application of Theorem 3.13 with $j = 2$ requires that $R(n)/\lambda_2(n) \in l^2$. As we remarked above, it is not necessarily true that $T^{-1}(n+1)V(n)T(n)$ is in l^2, hence it is not necessarily true that $R(n) \in l^2$. Since $\lambda_2(n)$ could even tend to zero, it is also not necessarily true that $R(n)/\lambda_2(n) \in l^2$. This is the situation in the general case, where in particular $q(n)$ could be unbounded.

However, in the case $|q(n)|$ is bounded above, then so is $|\lambda_1(n)|$ and, moreover, $|\lambda_2(n)|$ stays bounded away from zero. In this case, using the boundedness of $q(n)$, $\lambda_1(n)$ and $\lambda_2(n)$, it is straightforward to show that $R(n)$ (and hence also $R(n)/\lambda_2(n)$) are in l^2, and now Theorem 3.13 with $j = 2$ implies that there exists for n sufficiently large a solution

$$\hat{y}_2(n) = [1 + \varepsilon_2(n)] \prod^{n-1} [\lambda_2(k) + r_{22}(k)], \qquad \varepsilon_2(n) \to 0. \tag{9.120}$$

Furthermore, a calculation similar to the one above would show that

$$\frac{\lambda_2(n)}{v_2(n)} = 1 + (l^1),$$

provided that $\Delta^2\lambda_2(n) \in l^1$, where $v_2(n)$ was defined in (9.110). This shows that under this additional assumption $\Delta^2\lambda_2(n) \in l^1$ the two asymptotic representations (9.106) and (9.120) are equivalent. Note that we have not used the second assumption in (H_3) concerning $\Delta^2\delta(n) \in l^1$ nor assumption (H_4).

In a private communication from J. Geronimo, he informed us that the proof of Theorem 2.3 in fact requires that $q(n)$ should be bounded in addition to the assumptions (H_i) for $1 \leq i \leq 4$ (also see [62]). Note that the condition

$$r(n) = q(n+1) - q(n) \in l^2$$

implies that $r(n) = o(1)$ but not that $q(n)$ is bounded. However, re-writing

$$q(n) = q(n_0) + \sum_{k=1}^{n-1} [q(k+1) - q(k)],$$

the Cauchy–Schwarz inequality implies that

$$\begin{aligned} |q(n)| &\leq |q(n_0)| + \left(\sum_{n_0}^{n-1} |q(k+1) - q(k)|^2 \right)^{1/2} \left(\sum_{n_0}^{n-1} 1 \right)^{1/2} \\ &\leq |q(n_0)| + \left(\sum_{n_0}^{\infty} |q(k+1) - q(k)|^2 \right)^{1/2} \sqrt{n-1-n_0}, \end{aligned}$$

so $|q(n)| \leq M\sqrt{n}$ at worst.

Observe, for example, that $q(n) = n^{1/3} + n^{-(1/3)}$ satisfies all the assumptions (H_1) through (H_3), and leads to $\lambda_1(n) = n^{1/3}$ and $\lambda_2(n) = n^{-(1/3)}$. It also satisfies (H_4) for n sufficiently large.

As an additional remark to Theorem 2.3, one could relax the l^2-assumption to l^p with $p > 2$ to obtain similar conclusions using iterations of conditioning transformations (see Remark 5.4).

As a second application of the Riccati approach for (9.101), in [61, Theorem 2.5] the assumption (H_1) is weakened to

(H_1') $\{q(n)\}$ in $\mathbb{C}$ is bounded away from the critical points ± 2,

and assumptions (H_2) and (H_3) are strengthened to

(H_{23}) $\delta(n)$ and $r(n)$ are in l^1.

Under these new assumptions, Theorem 2.5 asserts that (9.101) has two solutions $y_{\pm}(n)$ satisfying assertions (9.104)–(9.110).

We will briefly compare this to what can be concluded about solutions of (9.116) under these new assumptions. Observe first that (H_1') allows $q(n)$ to assume values in the critical interval $(-2, 2)$, where $\lambda_1(n)$ and $\lambda_2(n)$ have modulus equal to 1 as well as approaching $(-2, 2)$ from above and below in $\mathbb{C}$. Thus the exponential dichotomy condition (9.115) does not necessarily hold. Next, it follows from (H_{23}) above that $q(n) = c + \mathrm{o}(1)$ as $n \to +\infty$ and (H_1') implies $c \neq \pm 2$. This leads to $\lambda_1(n) = \hat{c} + \mathrm{o}(1)$ and $\lambda_2(n) = 1/\hat{c} + \mathrm{o}(1)$ with $\hat{c} = \left(c + \sqrt{c^4 - 4}\right)/2 \neq 0$. It then can be shown as above that $\left[\lambda_j(n) - \lambda_j(n-1)\right] \in l^1$ for $j = 1, 2$, which implies that $V(n)$ in (9.116) is in l^1. This suggests that Theorem 3.4 might be applicable, however, that requires the extra assumption:

(H_1'') $\Lambda(n) = \operatorname{diag}\{\lambda_1(n), \lambda_2(n)\}$ satisfies Levinson's dichotomy conditions (3.8), (3.9).

Hence under the assumptions (H_1'), (H_1''), (H_{23}), it follows that (9.116) has a fundamental solution matrix satisfying

$$Y_1(n) = [I + \mathrm{o}(1)] \prod^{n-1} \Lambda(k) \qquad \text{as } n \to \infty.$$

This leads to a result which is comparable, but not equivalent to [61, Theorem 2.5] which does not require (H_1''), but in turn shows that $u_2(n)$, $v_2(n)$ in (9.104) and (9.106) could be replaced by $\lambda_1(n)$ and $\lambda_2(n)$, respectively, under this additional assumption (H_1'').

As far as the error terms are concerned, they can be estimated from the T-operator as in Theorem 3.2, however, it does not make sense to compare such estimates with (9.105) and (9.107) since those implicitly involve $u_2(n)$ and $v_2(n)$ instead of the simpler $\lambda_1(n) = u_0(n)$ and $\lambda_2(n) = v_0(n)$.

We end this section by mentioning that specific examples were covered in [61, Section 3], and that more specialized second order difference equations were investigated in [63] and [64].

9.4.5 Another Class of Perturbations with Bounded Variation

Asymptotic representations for some classes of orthogonal polynomials (with respect to a positive measure $d\alpha$) have been discussed by Máté, Nevai, and Totik [106]. Such polynomials $\{y_n(x)\}$ satisfy a three-term recurrence equation of the form

$$xy_n(x) = a_{n+1}y_{n+1}(x) + b_n y_n(x) + a_n y_{n-1}(x), \qquad n \geq 0, \tag{9.121}$$

Here $\{a_n\}_{n=0}^{\infty}$ and $\{b_n\}_{n=0}^{\infty}$ are given sequences of real numbers such that $a_0 = 0$, $a_n > 0$ for $n \geq 1$,

$$\lim_{n\to\infty} a_n = \frac{1}{2}, \qquad \text{and} \quad \lim_{n\to\infty} b_n = 0, \tag{9.122}$$

and

$$\varepsilon_n = |a_{n+1} - a_n| + |b_{n+1} - b_n| \in l^1, \tag{9.123}$$

and x is a complex parameter. The results of particular interest here correspond to certain of their asymptotic formulae (including error estimates) for a particular solution $p_n(x)$ of (9.121) as $n \to \infty$. The approach taken here is first applying results from Chap. 3 to obtain asymptotic representations for a pair of linearly independent (more specifically, dominant and recessive) solutions $y_n^{(1)}(x)$ and $y_n^{(2)}(x)$ and next representing $p_n(x)$ as

$$p_n(x) = c_1(x)y_n^{(1)}(x) + c_2(x)y_n^{(2)}(x), \qquad p_{-1}(x) = 0, \quad p_0(x) = \gamma_0 > 0. \tag{9.124}$$

More detailed information concerning the connection coefficients $c_1(x), c_2(x)$ and their dependence upon the measure $d\alpha$ is beyond the scope of this book and we refer to [106] for an interesting analysis of that aspect.

Theorem 9.18 (cf. [106]; Theorems 1&2) *Consider* (9.121), *where the coefficients satisfy* (9.122) *and* (9.123). *Let* $x \in K_1$, *where* K_1 *is a compact set in* $\mathbb{C}$ *such that* $K_1 \cap [-1, 1] = \emptyset$. *Let* $\lambda_1(n, x) = \rho\left(\frac{x-b_n}{2a_{n+1}}\right)$, *where*

$$\rho(t) = t + \sqrt{t^2 - 1},$$

and $\lambda_2(n, x) = 1/\lambda_1(n, x)$. *Here, we make a fixed choice for a complex branch of the square root function so that* $\rho(t) > 1$ *if* $t > 1$. *Then the following statements hold for the solution* $p_n(x)$ *of* (9.121) *defined in* (9.124).

1. If $c_1(x) \neq 0$*, then* $\exists\, \delta > 0$ *such that*

$$\frac{p_n(x)}{\prod^{n}\lambda_1(k,x)} = c_1(x)\left[1 + \mathrm{O}\left(\frac{1}{(1+\delta)^n}\right) + \mathrm{O}\left(\sum_{k=[n/2]} \varepsilon_k\right)\right]$$

$$+c_2(x)\mathrm{O}\left(\frac{1}{(1+\delta)^{2n}}\right), \tag{9.125}$$

and

$$\frac{p_n(x) - \frac{1}{\lambda_1(n,x)}p_{n-1}(x)}{\prod^{n-1}\lambda_1(k,x)} = 2\sqrt{\left(\frac{x-b_n}{2a_{n+1}}\right)^2 - 1}\ \{c_1(x)[1+\mathrm{o}(1)] + c_2(x)\mathrm{o}(1)\}. \tag{9.126}$$

2. If $c_1(x) = 0$*, then*

$$\frac{p_n(x) - \lambda_1(n,x)p_{n-1}(x)}{\prod^{n-1}\lambda_2(k,x)} = -c_2(x)2\sqrt{\left(\frac{x-b_n}{2a_{n+1}}\right)^2 - 1}\left[1 + \mathrm{O}\left(\sum_{k=n}^{\infty}\varepsilon_k\right)\right], \tag{9.127}$$

where ε_k *is defined by* (9.123).

Proof We begin by establishing an asymptotic representation for $y_n^{(i)}(x)$ $(i = 1, 2)$. Let $\mathbf{y}(n,x) = [y_{n-1}(x), y_n(x)]^T$. Then (9.121) is equivalent to

$$\mathbf{y}(n+1,x) = \begin{pmatrix} 0 & 1 \\ -\frac{a_n}{a_{n+1}} & \frac{x-b_n}{a_{n+1}} \end{pmatrix}\mathbf{y}(n,x),$$

which we write as

$$\begin{aligned}\mathbf{y}(n+1,x) &= \left[\begin{pmatrix} 0 & 1 \\ -1 & \frac{x-b_n}{a_{n+1}} \end{pmatrix} + \begin{pmatrix} 0 & 0 \\ 1-\frac{a_n}{a_{n+1}} & 0 \end{pmatrix}\right]\mathbf{y}(n,x) \\ &=: [\quad A(n,x) \quad + \quad R(n,x) \quad]\,\mathbf{y}(n,x).\end{aligned} \tag{9.128}$$

Observe that from (9.123) and (9.122) it follows that

$$|R(n,x)| = \left|\frac{a_{n+1}-a_n}{a_{n+1}}\right| = \mathrm{O}(\varepsilon_n) \in l^1. \tag{9.129}$$

The eigenvalues of $A(n,x)$ are $\lambda_1(n,x) = \rho\left(\frac{x-b_n}{2a_{n+1}}\right)$ and $\lambda_2(n,x) = 1/\lambda_1(n,x)$ described above. Define K_1 to be a compact set in $\mathbb{C}$ such $K_1 \cap [-1,1] = \emptyset$. Then for all $t \in K_1$, there exist $M > 0$ and $\delta \in (0,1)$ such that $1 + 2\delta \leq |\rho(t)| \leq 2M$. Furthermore, since

$(x - b_n)/2a_{n+1} \to x$, it follows for for all n sufficiently large, say $n \geq n_1$, and $x \in K_1$ (by making K_1 slightly smaller) that

$$|\lambda_1(n,x)| \geq 1 + \delta \quad \text{and} \quad M \leq |\lambda_2(n,x)| \leq \frac{1}{1+\delta}. \tag{9.130}$$

From here on we will assume that $n \geq n_1$ and therefore (9.130) holds.

To diagonalize $A(n,x)$, we now apply the transformation

$$\mathbf{y}(n,x) = \begin{pmatrix} 1 & 1 \\ \lambda_1(n,x) & \lambda_2(n,x) \end{pmatrix} \hat{y}(n,x) =: T(n,x)\hat{y}(n,x).$$

Since

$$\lim_{n\to\infty} T(n,x) = \begin{pmatrix} 1 & 1 \\ \rho(x) & \frac{1}{\rho(x)} \end{pmatrix} = T(x),$$

$T(x)$ is invertible for $x \neq \pm 1$. Hence both $T(n,x)$ and $T^{-1}(n,x)$ are invertible for n sufficiently large, say $n \geq n_1$ and bounded for each fixed $x \neq \pm 1$, and uniformly bounded for $x \in K_1$. It follows for $n \geq n_1$ that

$$\hat{y}(n+1,x) = \left[\Lambda(n,x) + \hat{R}(n,x)\right]\hat{y}(n,x), \tag{9.131}$$

where

$$\begin{aligned} \Lambda(n,x) &= \begin{pmatrix} \lambda_1(n,x) & 0 \\ 0 & \lambda_2(n,x) \end{pmatrix}, \\ \hat{R}(n,x) &= T^{-1}(n+1,x)\Lambda(n,x)\left[T(n,x) - T(n+1,x)\right] \\ &\quad + T^{-1}(n+1,x)R(n)T(n,x). \end{aligned}$$

Clearly, $T^{-1}(n+1,x)R(n,x)T(n,x) = O(\varepsilon_n)$. Moreover, one can show as in the previous section that $\lambda_i(n+1,x) - \lambda_i(n,x) = O(\varepsilon_n)$. To indicate the argument, put $q(n) := (x - b_n)/2a_{n+1}$ and observe that

$$q(n+1) - q(n) = \frac{x(a_{n+1} - a_{n+2}) + a_{n+1}(b_n - b_{n+1}) + b_n(a_{n+2} - a_{n+1})}{2a_{n+1}a_{n+2}}.$$

Hence $q(n+1) - q(n) \in l^1$ by (9.123). It follows that

$$T(n,x) - T(n+1,x) = O(\varepsilon_n),$$

which together with $A(n,x)$ being bounded for $x \in K_1$ shows that

$$T^{-1}(n+1,x)A(n,x)[T(n,x)-T(n+1,x)] = \mathrm{O}(\varepsilon_n).$$

This together with (9.130) implies that

$$\frac{|\hat{R}(n,x)|}{|\lambda_i(n,x)|} = \mathrm{O}(\varepsilon_n) \in l^1, \qquad i = 1,2.$$

From (9.130), $\Lambda(n,x)$ satisfies an exponential dichotomy and therefore also an ordinary dichotomy for $x \in K_1$. Hence Theorem 3.4 can be applied and yields for n sufficiently large the existence of a fundamental matrix of (9.131) of the form

$$\hat{Y}(n,x) = \begin{pmatrix} 1+\eta_{11}(n,x) & \eta_{12}(n,x) \\ \eta_{21}(n,x) & 1+\eta_{22}(n,x) \end{pmatrix} \begin{pmatrix} \prod^{n-1}\lambda_1(k,x) & \\ & \prod^{n-1}\lambda_2(k,x) \end{pmatrix},$$

where $\eta_{ij}(n,x) \to 0$ as $n \to \infty$ $(i = 1,2)$, and therefore (9.128) has for n sufficiently large a fundamental solution matrix satisfying

$$Y(n,x) = \begin{pmatrix} y_{n-1}^{(1)}(x) & y_{n-1}^{(2)}(x) \\ & \\ y_n^{(1)}(x) & y_n^{(2)}(x) \end{pmatrix} = \begin{pmatrix} 1 & 1 \\ \lambda_1(n,x) & \lambda_2(n,x) \end{pmatrix} \hat{Y}(n,x). \tag{9.132}$$

In Lemma 9.19 below we will show that $\eta_{ij}(n)$ satisfy, as $n \to \infty$,

$$|\eta_{11}(n)+\eta_{21}(n)| = \mathrm{O}\left(\frac{1}{1+\delta}\right)^n + \mathrm{O}\left(\sum_{k=[n/2]}\varepsilon_k\right), \quad \eta_{22}(n) = \mathrm{O}\left(\sum_{k=n}^{\infty}\varepsilon_k\right). \tag{9.133}$$

Therefore

$$y_{n-1}^{(1)}(x) = [1+\eta_{11}(n,x)+\eta_{21}(n,x)]\prod^{n-1}\lambda_1(k,x)$$

and

$$y_{n-1}^{(2)}(x) = [1+\eta_{12}(n,x)+\eta_{22}(n,x)]\prod^{n-1}\lambda_2(k,x).$$

Focusing on the particular solution $p_n(x)$ defined in (9.124), it follows then that

$$p_n(x) = c_1(x)\,[1+\eta_{11}(n+1,x)+\eta_{21}(n+1,x)]\prod^{n}\lambda_1(k,x)$$
$$+c_2(x)\,[1+\eta_{12}(n+1,x)+\eta_{22}(n+1,x)]\prod^{n}\lambda_2(k,x)$$

or

$$\frac{p_n(x)}{\prod^n \lambda_1(k,x)} = c_1(x)\left[1+\eta_{11}(n+1,x)+\eta_{21}(n+1,x)\right]$$

$$+\; c_2(x) O\left(\prod^n \frac{\lambda_2(k,x)}{\lambda_1(k,x)}\right),$$

and using (9.133) one obtains (9.125).

On the other hand, (9.132) also implies that

$$\begin{pmatrix} p_{n-1}(x) \\ p_n(x) \end{pmatrix} = \begin{pmatrix} 1 & 1 \\ \lambda_1(n,x) & \lambda_2(n,x) \end{pmatrix} \hat{Y}(n,x) \begin{pmatrix} c_1(x) \\ c_2(x) \end{pmatrix},$$

from which it follows that

$$\frac{1}{\lambda_2(n,x)-\lambda_1(n,x)} \begin{pmatrix} \lambda_2(n,x)p_{n-1}(x) - p_n(x) \\ p_n(x) - \lambda_1(n,x)p_{n-1}(x) \end{pmatrix}$$

$$= \begin{pmatrix} 1+\eta_{11}(n,x) & \eta_{12}(n,x) \\ \eta_{21}(n,x) & 1+\eta_{22}(n,x) \end{pmatrix} \begin{pmatrix} \prod^{n-1} \lambda_1(k,x)c_1(x) \\ \prod^{n-1} \lambda_2(k,x)c_2(x) \end{pmatrix}. \tag{9.134}$$

Equating elements in the first position, we have

$$\frac{1}{2\sqrt{\left(\frac{x-b_n}{2a_{n+1}}\right)^2-1}} \left[p_n(x) - \frac{1}{\lambda_1(n,x)} p_{n-1}(x)\right]$$

$$= \prod^{n-1} \lambda_1(k,x) \left[c_1(x)[1+\eta_{11}(n,x)] + c_2(x)\eta_{12}(n,x) \prod^{n-1} \frac{\lambda_2(k,x)}{\lambda_1(k,x)}\right],$$

which implies (9.126), and hence

$$\lim_{n\to\infty} \frac{p_n(x) - \frac{1}{\lambda_1(n,x)} p_{n-1}(x)}{\prod^n \lambda_1(k,x)} = 2\sqrt{x^2-1}\, \frac{c_1(x)}{\rho(x)}.$$

If $c_1(x) = 0$, then $c_2(x) \neq 0$, and from the second element of (9.134) it follows that

$$\frac{-1}{2\sqrt{\left(\frac{x-b_n}{2a_{n+1}}\right)^2-1}} \left[p_n(x) - \lambda_1(n,x)p_{n-1}(x)\right] = c_2(x)[1+\eta_{22}(n,x)] \prod^{n-1} \lambda_2(k,x),$$

and therefore (9.133) yields (9.127). □

Lemma 9.19 *Let the assumptions of Theorem 9.18 be satisfied. Then* (9.133) *holds.*

Proof To derive the first estimate, we make in (9.131) the normalization

$$\hat{y}(n,x) = \left(\prod_{l=n_1}^{n-1} \lambda_1(l,x)\right) z_1(n,x).$$

Then

$$z_1(n+1,x) = \left[\begin{pmatrix} 1 & \\ & \frac{\lambda_2(n,x)}{\lambda_1(n,x)} \end{pmatrix} + R_1(n,x)\right] z_1(n,x),$$

where $R_1(n,x) = R(n,x)/\lambda_1(n,x) = \mathrm{O}(\varepsilon_n)$. Using the standard T-operator argument, $z_1(n,x)$ is for $n \geq n_1$ sufficiently large the solution of

$$\begin{aligned} z_1(n,x) &= e_1 + \sum_{k=n_1}^{n-1} \begin{pmatrix} 0 & \\ & \prod_{l=k+1}^{n-1} \frac{\lambda_2(l,x)}{\lambda_1(l,x)} \end{pmatrix} R_1(k,x) z_1(k,x) \\ &\quad - \sum_{k=n}^{\infty} \begin{pmatrix} 1 & \\ & 0 \end{pmatrix} R_1(k,x) z_1(k,x) \\ &= e_1 + \begin{pmatrix} \eta_{11}(n,x) \\ \eta_{21}(n,x) \end{pmatrix}. \end{aligned}$$

From (9.130) and (9.129) it follows that

$$\begin{aligned} |\eta_{11}(n,x) + \eta_{21}(n,x)| &\leq M_1 \left(\sum_{k=n_1}^{n-1} \left(\frac{1}{1+\delta}\right)^{2(n-k-1)} \varepsilon_k + \sum_{k=n}^{\infty} \varepsilon_k \right) \\ &\leq M_1 \left(\sum_{k=n_1}^{[n/2]} \left(\frac{1}{1+\delta}\right)^{2(n-k-1)} + \sum_{k=[n/2]}^{\infty} \varepsilon_k \right), \end{aligned}$$

which yields the first estimate in (9.133). Here M_1 is some positive constant. Similarly, setting

$$\hat{y}(n,x) = \left(\prod_{l=n_1}^{n-1} \lambda_2(l,x)\right) z_2(n,x),$$

one can show that there exists a solution satisfying

$$z_2(n,x) = e_2 + \begin{pmatrix} \eta_{12}(n,x) \\ \eta_{22}(n,x) \end{pmatrix} = e_2 + \sum_{k=n}^{\infty} \begin{pmatrix} \prod_{l=n}^{k} \frac{\lambda_2(l,x)}{\lambda_1(l,x)} & \\ & 1 \end{pmatrix} \frac{R(k,x)}{\lambda_2(k,x)} z_2(k,x),$$

from which the second estimate in (9.133) follows. □

9.5 Asymptotic Factorization of a Difference Operator

In [104], Máté and Nevai consider a kind of factorization of second-order linear difference operators

$$D = E^2 + \alpha(n)E + \beta(n),$$

where $Ey(n) = y(n+1)$ is the forward shift operator, $\alpha(n) = a + o(1)$, $\beta(n) = b + o(1)$, and

$$\varepsilon_n := |\alpha(n+1) - \alpha(n)| + |\beta(n+1) - \beta(n)| \in l^1. \tag{9.135}$$

Since the change of variables $y(n) = \left(\frac{1}{\sqrt{b}}\right)^n \tilde{y}(n)$ results in a normalization of the (nonzero) constant b, we can assume w.l.o.g. that $b = 1$.

Our purpose in this section is to show how asymptotic representations for solutions of the corresponding equation $Dy(n) = 0$ can be applied to yield a factorization of D in an asymptotic sense which will be described below.

In the limiting case (as $n \to \infty$) the operator $E^2 + aE + 1$ can easily be seen to factor as

$$E^2 + aE + 1 = (E - \lambda_2)(E - \lambda_1),$$

where $\lambda_{1,2}$ are the roots of the limiting equation

$$\lambda^2 + a\lambda + 1 = 0, \tag{9.136}$$

and factorization means that

$$(E^2 + aE + b)y(n) = (E - \lambda_2)(E - \lambda_1)y(n),$$

for all $y(n)$ and all $n \in \mathbb{N}$. Moreover, the factors $(E - \lambda_2)$ and $(E - \lambda_1)$ commute.

In the case of nonconstant coefficients $\alpha(n)$ and $\beta(n)$, it can be checked that in general

$$E^2 + \alpha(n)E + \beta(n) \neq (E - \lambda_2(n))(E - \lambda_1(n)),$$

where $\lambda_i(n)$ $(i = 1, 2)$ are the roots of

$$\lambda^2 + \alpha(n)\lambda + \beta(n) = 0. \tag{9.137}$$

We note, however, that

$$\begin{aligned}&(E-\zeta_2(n))(E-\zeta_1(n))y(n)\\&=E^2y(n)-\zeta_2(n)Ey(n)-E\zeta_1(n)y(n)+\zeta_2(n)\zeta_1(n)y(n)\\&=y(n+2)-[\zeta_2(n)+\zeta_1(n+1)]y(n+1)+\zeta_2(n)\zeta_1(n)y(n).\end{aligned}$$

Hence a factorization

$$D=E^2+\alpha(n)E+\beta(n)=(E-\zeta_2(n))(E-\zeta_1(n)) \tag{9.138}$$

requires that $\zeta_1(n)$, $\zeta_2(n)$ satisfy the system of non-linear equations

$$\begin{cases}\alpha(n)=-\zeta_1(n+1)-\zeta_2(n)\\ \beta(n)=\zeta_2(n)\zeta_1(n)\end{cases}. \tag{9.139}$$

This is equivalent to solving the Riccati difference equation

$$\beta(n)=-\zeta_1(n)\left[\zeta_1(n+1)+\alpha(n)\right], \tag{9.140}$$

for $\zeta_1(n)$ and then defining $\zeta_2(n)=-\zeta_1(n+1)-\alpha(n)$.

If a solution $y(n)$ of

$$y(n+2)+\alpha(n)y(n+1)+\beta(n)y(n)=0 \tag{9.141}$$

is known and if $y(n)\neq 0$ for all n, then defining $\zeta_1(n)$ by $\zeta_1(n)=y(n+1)/y(n)$ yields a solution of (9.140). On the other hand, a factorization $[E-\zeta_2(n)][E-\zeta_1(n)]y(n)=0$ gives a nontrivial solution $y(n)$ of (9.141), by first solving $[E-\zeta_2(n)]h(n)=0$ and afterwards $[E-\zeta_1(n)]y(n)=h(n)$. Hence factoring D *exactly* is equivalent to finding one nonzero solution of (9.141).

However, since we usually do not know exact nontrivial solutions of (9.141), a natural question is how to determine the factors $\zeta_1(n)$ and $\zeta_2(n)$ in (9.138) approximately, say up to l^1-perturbations. In [104, Theorem 2.1] Máté and Nevai show that if (9.135) holds and if the roots of the limiting equation (9.136) satisfy

$$\left|\frac{\lambda_1}{\lambda_2}\right|>1, \tag{9.142}$$

then there exist for n sufficiently large functions $\zeta_1(n)$ and $\zeta_2(n)$ satisfying (9.138), where $\lambda_1(n)-\zeta_1(n)\in l^1$. They do not give a corresponding statement regarding $\zeta_2(n)$.

In what follows, we give an independent proof of their result, which also allows us to show that $\zeta_2(n) - \lambda_2(n) \in l^1$. More specifically, we will show the following

Theorem 9.20 (cf. [104, Theorem 3]) *Suppose that* (9.135) *holds with* $b = 1$ *and that the roots of the limiting equation* (9.136) *satisfy* (9.142). *Then there exist sequences* $\{\zeta_1(n)\}$ *and* $\{\zeta_2(n)\}$ *for* n *sufficiently large satisfying* (9.138) *such that* $\zeta_i(n) - \lambda_i(n) \in l^1$ *for* $i = 1, 2$. *Here* $\lambda_i(n)$ *are the roots of* (9.137) *such that* $\lambda_i(n) \to \lambda_i$ *for* $i = 1, 2$.

Proof We write $[E^2 + \alpha(n)E + \beta(n)]y(n) = 0$ as the corresponding matrix system

$$\mathbf{y}(n+1) = \begin{pmatrix} 0 & 1 \\ -\beta(n) & -\alpha(n) \end{pmatrix} \mathbf{y}(n) = A(n)\mathbf{y}(n), \quad \mathbf{y}(n) = \begin{pmatrix} y(n) \\ y(n+1) \end{pmatrix}. \tag{9.143}$$

The eigenvalues of $A(n)$ in (9.143) coincide with the roots of (9.137) and are given by

$$\lambda_{1,2}(n) = \frac{-\alpha(n) \pm \sqrt{\alpha^2(n) - 4\beta(n)}}{2} \to \frac{-a \pm \sqrt{a^2 - 4}}{2} = \lambda_{1,2}, \tag{9.144}$$

as $n \to \infty$. By (9.142), there exists $\delta \in (0, 1/2)$ such that $|\lambda_1| \geq 1 + 2\delta$, $|\lambda_2| \leq 1 - 2\delta$, and

$$|\lambda_1(n)| \geq 1 + \delta \text{ and } |\lambda_2(n)| \leq 1 - \delta, \tag{9.145}$$

for all n sufficiently large, say $n \geq n_0$. We also note that $\{\lambda_i(n)\}$ are convergent sequences and hence bounded for $i = 1, 2$. In addition, $\lambda_i(n)$ stay bounded away from zero since $\lambda_1(n)\lambda_2(n) = 1$.

To diagonalize $A(n)$, we put

$$\mathbf{y}(n) = \begin{pmatrix} 1 & 1 \\ \lambda_1(n) & \lambda_2(n) \end{pmatrix} \hat{y}(n) =: T(n)\hat{y}(n). \tag{9.146}$$

Then $T(n)$ is invertible for $n \geq n_0$ and bounded. It follows that

$$\begin{aligned} \hat{y}(n+1) &= T^{-1}(n+1)A(n)T(n)\hat{y}(n) \\ &= [\Lambda(n) + V(n)]\,\hat{y}(n), \end{aligned} \tag{9.147}$$

where

$$\begin{aligned} V(n) &= \left[T^{-1}(n+1) - T^{-1}(n)\right]A(n)T(n) = T^{-1}(n+1)[T(n) - T(n+1)]\Lambda(n) \\ &= \mathrm{O}\left(|T(n+1) - T(n)|\right). \end{aligned}$$

We next claim that $\lambda_i(n+1) - \lambda_i(n) = \mathrm{O}(\varepsilon_n)$ $(i = 1, 2)$, which by the definition of $T(n)$ in (9.146) implies that $V(n) = \mathrm{O}(\varepsilon_n)$. To substantiate this claim, say for $i = 1$ (the case

$i = 2$ is done similarly), note that from (9.135) follows that

$$\begin{aligned}
&2[\lambda_1(n+1) - \lambda_1(n)] \\
&= -\alpha(n+1) + \alpha(n) + \frac{\alpha^2(n+1) - \alpha^2(n) + 4\beta(n) - 4\beta(n+1)}{\sqrt{\alpha^2(n+1) - 4\beta(n+1)} + \sqrt{\alpha^2(n) - 4\beta(n)}} \\
&= \mathrm{O}(\varepsilon_n) + \frac{[\alpha(n+1) - \alpha(n)][\alpha(n+1) + \alpha(n)]}{\sqrt{\alpha^2(n+1) - 4\beta(n+1)} + \sqrt{\alpha^2(n) - 4\beta(n)}} \\
&= \mathrm{O}(\varepsilon_n) + \mathrm{O}\left(|\alpha(n+1) - \alpha(n)|\right) = \mathrm{O}(\varepsilon_n).
\end{aligned} \tag{9.148}$$

Here we used that $\alpha^2(n) - 4\beta(n) \to a^2 - 4 \neq 0$, since otherwise (9.142) would not hold. Hence $V(n) \in l^1$ and also $V(n)/\lambda_i(n) \in l^1$ since $\lambda_i(n)$ are bounded away from zero for $i = 1, 2$. Due to (9.142), $\lambda_i(k)$ satisfy Levinson's discrete (even exponential) dichotomy conditions (3.8), (3.9). Then Theorem 3.4 implies that (9.147) has a fundamental matrix satisfying

$$\hat{Y}(n) = \begin{pmatrix} 1 + \varepsilon_{11}(n) & \varepsilon_{12}((n) \\ \varepsilon_{21}(n) & 1 + \varepsilon_{22}((n) \end{pmatrix} \begin{pmatrix} \prod^{n-1} \lambda_1(k) & \\ & \prod^{n-1} \lambda_2(k) \end{pmatrix}, \tag{9.149}$$

where $\varepsilon_{ij}(n) = \mathrm{o}(1)$. By (9.146), (9.143) has a fundamental system

$$Y(n) = \begin{pmatrix} y_1(n) & y_2(n) \\ y_1(n+1) & y_2(n+1) \end{pmatrix} = \begin{pmatrix} 1 & 1 \\ \lambda_1(n) & \lambda_2(n) \end{pmatrix} \hat{Y}(n). \tag{9.150}$$

In particular, $y_i(n) \neq 0$ for n sufficiently large.

We now define for such n

$$\zeta_1(n) = \frac{y_1(n+1)}{y_1(n)},$$

hence $\zeta_1(n)$ is a solution of (9.140). Then from (9.149) it follows that

$$\zeta_1(n) = \frac{[1 + \varepsilon_{11}(n) + \frac{\lambda_2(n)}{\lambda_1(n)}\varepsilon_{21}(n)] \prod^{n} \lambda_1(k)}{[1 + \varepsilon_{11}(n) + \varepsilon_{21}(n)] \prod^{n-1} \lambda_1(k)} =: \frac{1 + \sigma(n)}{1 + \tau(n)} \lambda_1(n),$$

where $\sigma(n)$ and $\tau(n)$ vanish at infinity. Hence

$$\zeta_1(n) - \lambda_1(n) = \frac{\lambda_1(n)}{1 + \tau(n)} [\sigma(n) - \tau(n)] = \frac{\lambda_1(n)}{1 + \tau(n)} \left[\frac{\lambda_2(n)}{\lambda_1(n)} - 1 \right] \varepsilon_{21}(n). \tag{9.151}$$

We will show in Lemma 9.21 that

$$|\varepsilon_{21}(n)| \leq M \sum_{k=n_1}^{n-1} \prod_{l=k+1}^{n-1} \left| \frac{\lambda_2(l)}{\lambda_1(l)} \right| |V(k)|. \tag{9.152}$$

From this it follows that

$$\begin{aligned}
\sum_{n=n_1}^{\infty} |\varepsilon_{21}(n)| &\leq M \sum_{n=n_1}^{\infty} \sum_{k=n_1}^{n-1} (1-\delta)^{n-k} |V(k)| \\
&= M \sum_{k=n_1}^{\infty} \sum_{n=k+1}^{\infty} (1-\delta)^{n-k} |V(k)| \\
&= M \sum_{k=n_1}^{\infty} \frac{|V(k)|}{(1-\delta)^k} \sum_{n=k+1}^{\infty} (1-\delta)^n \\
&= M \frac{1-\delta}{\delta} \sum_{k=n_1}^{\infty} |V(k)| < \infty.
\end{aligned} \tag{9.153}$$

where we used Fubini's theorem to change the order of summation. Thus $\varepsilon_{21}(n) \in l^1$, and (9.151) implies that

$$\zeta_1(n) - \lambda_1(n) \in l^1. \tag{9.154}$$

We next claim that the factor $\zeta_2(n)$ in the factorization (9.138) satisfies

$$\zeta_2(n) - \lambda_2(n) \in l^1. \tag{9.155}$$

To substantiate this, we note that (9.139) and the definition of $\lambda_i(n)$ in (9.144) imply that

$$\begin{aligned}
\zeta_2(n) &= -\zeta_1(n+1) - \alpha(n) = -\zeta_1(n+1) + \lambda_1(n) + \lambda_2(n) \\
&= [-\zeta_1(n+1) + \zeta_1(n)] + [\lambda_1(n) - \zeta_1(n)] + \lambda_2(n).
\end{aligned} \tag{9.156}$$

Furthermore,

$$\zeta_1(n+1) - \zeta_1(n) = [\zeta_1(n+1) - \lambda_1(n+1)] + [\lambda_1(n+1) - \lambda_1(n)] + [\lambda_1(n) - \zeta_1(n)],$$

which is in l^1 by (9.154) and (9.148), and now (9.156) implies (9.155).

Reversing the order of the two factors, we are next interested in a factorization of the form

$$D = E^2 + \alpha(n)E + \beta(n) = (E - \tilde{\zeta}_1(n))(E - \tilde{\zeta}_2(n)), \tag{9.157}$$

where we want to conclude now that

$$\tilde{\zeta}_1(n) = \lambda_1(n) + l^1 \text{ and } \tilde{\zeta}_2(n) = \lambda_2(n) + l^1.$$

This requires that $\tilde{\zeta}_1(n)$, $\tilde{\zeta}_2(n)$ satisfy the corresponding non-linear equations $\alpha(n) = -\tilde{\zeta}_2(n+1) - \tilde{\zeta}_1(n)$, $\beta(n) = \tilde{\zeta}_1(n)\tilde{\zeta}_2(n)$, or equivalently solving the Riccati difference equation

$$\beta(n) = -\tilde{\zeta}_2(n)\left[\tilde{\zeta}_2(n+1) + \alpha(n)\right], \tag{9.158}$$

for $\tilde{\zeta}_2(n)$ and then defining $\tilde{\zeta}_1(n) = -\tilde{\zeta}_2(n+1) - \alpha(n)$. Analogous to the calculations made above, it is straightforward to show that $\tilde{\zeta}_2(n)$ defined by $\tilde{\zeta}_2(n) = y_2(n+1)/y_2(n)$ is a solution of (9.158). From (9.150) follows that $y_2(n) \neq 0$ for n sufficiently large and then

$$\tilde{\zeta}_2(n) = \frac{y_2(n+1)}{y_2(n)} = \frac{[1 + \varepsilon_{22}(n) + \frac{\lambda_1(n)}{\lambda_2(n)}\varepsilon_{12}(n)]\prod^{n}\lambda_2(k)}{[1 + \varepsilon_{12}(n) + \varepsilon_{22}(n)]\prod^{n-1}\lambda_2(k)} =: \frac{1 + \tilde{\sigma}(n)}{1 + \tilde{\tau}(n)}\lambda_2(n).$$

Hence

$$\tilde{\zeta}_2(n) - \lambda_2(n) = \frac{\tilde{\sigma}(n) - \tilde{\tau}(n)}{1 + \tilde{\tau}(n)}\lambda_2(n) = \frac{\lambda_2(n)}{1 + \tilde{\tau}(n)}\left[\frac{\lambda_1(n)}{\lambda_2(n)} - 1\right]\varepsilon_{12}(n). \tag{9.159}$$

We will show in Lemma 9.21 that

$$|\varepsilon_{12}(n)| \leq M \sum_{k=n}^{\infty} \prod_{l=n}^{k} \left|\frac{\lambda_2(l)}{\lambda_1(l)}\right| |V(k)|. \tag{9.160}$$

Equation (9.160) and an application of Fubini's theorem shows that

$$\begin{aligned}
\sum_{n=n_1}^{\infty} |\varepsilon_{12}(n)| &\leq M \sum_{n=n_1}^{\infty} \sum_{k=n}^{\infty} (1-\delta)^{k-n} |V(k)| \\
&= M \sum_{k=n_1}^{\infty} \sum_{n=n_1}^{k} (1-\delta)^{k-n} |V(k)| \\
&\leq \frac{M}{\delta} \sum_{k=n_1}^{\infty} |V(k)| < \infty.
\end{aligned}$$

Thus $\varepsilon_{12}(n) \in l^1$, and (9.159) implies that

$$\tilde{\zeta}_2(n) - \lambda_2(n) \in l^1.$$

Similarly to the arguments used above, one can show that $\tilde{\zeta}_1(n) - \lambda_1(n) \in l^1$. □

Lemma 9.21 *In Eq.* (9.149), *the sequences* $\varepsilon_{21}(n)$ *and* $\varepsilon_{12}(n)$ *satisfy* (9.152) *and* (9.160), *respectively.*

Proof To show (9.152), we make in (9.147) the normalizing transformation $\hat{y}(n) = \prod^{n-1} \lambda_1(k) z_1(n)$. Then it follows that

$$z_1(n+1) = \left[\begin{pmatrix} 1 & \\ & \frac{\lambda_2(n)}{\lambda_1(n)} \end{pmatrix} + \frac{V(n)}{\lambda_1(n)} \right] z_1(n).$$

Theorem 3.4 implies the existence of a bounded solution $z_1(n)$ of the equation

$$\begin{aligned} z_1(n) &= e_1 + \sum_{k=n_1}^{n-1} \begin{pmatrix} 0 & \\ & \prod_{l=k+1}^{n-1} \frac{\lambda_2(l)}{\lambda_1(l)} \end{pmatrix} \frac{V(k)}{\lambda_1(k)} z_1(k) - \sum_{k=n}^{\infty} \begin{pmatrix} 1 & \\ & 0 \end{pmatrix} \frac{V(k)}{\lambda_1(k)} z_1(k) \\ &:= e_1 + \begin{pmatrix} \varepsilon_{11}(n) \\ \varepsilon_{21}(n) \end{pmatrix}, \end{aligned}$$

hence

$$\begin{aligned} |\varepsilon_{21}(n)| &= \left| \sum_{k=n_1}^{n-1} \prod_{l=k+1}^{n-1} \frac{\lambda_2(l)}{\lambda_1(l)} \frac{V(k)}{\lambda_1(k)} z_1(k) \right| \\ &\leq M \sum_{k=n_1}^{n-1} \prod_{l=k+1}^{n-1} \left| \frac{\lambda_2(l)}{\lambda_1(l)} \right| |V(k)|, \end{aligned}$$

for some positive constant M, which establishes (9.152). We also note that

$$|\varepsilon_{11}(n)| \leq M \sum_{k=n}^{\infty} |V(k)|. \tag{9.161}$$

Using similar arguments, to establish (9.160), we make in (9.147) the normalization $\hat{y}(n) = \prod^{n-1} \lambda_2(k) z_2(n)$. Then it follows that

$$z_2(n+1) = \left[\begin{pmatrix} \frac{\lambda_1(n)}{\lambda_2(n)} & \\ & 1 \end{pmatrix} + \frac{V(n)}{\lambda_2(n)} \right] z_2(n).$$

Theorem 3.4 implies the existence of a bounded solution $z_2(n)$ of the equation

$$z_2(n) = e_2 - \sum_{k=n}^{\infty} \left(\prod_{l=n}^{k} \frac{\lambda_2(l)}{\lambda_1(l)} \right)_1 \frac{V(k)}{\lambda_2(k)} z_2(k) =: e_2 + \begin{pmatrix} \varepsilon_{12}(n) \\ \varepsilon_{22}(n) \end{pmatrix},$$

hence

$$\begin{aligned} |\varepsilon_{12}(n)| &= \left| \sum_{k=n}^{\infty} \left(\prod_{l=n}^{k} \frac{\lambda_2(l)}{\lambda_1(l)} \right) \frac{V(k)}{\lambda_2(k)} z_2(k) \right| \\ &\leq M \sum_{k=n}^{\infty} \prod_{l=n}^{k} \left| \frac{\lambda_2(l)}{\lambda_1(l)} \right| |V(k)|, \end{aligned}$$

for some positive constant M, which establishes (9.160). □

9.6 Special Classes and Applications to Jacobi Operators

In studying properties of the spectrum of Jacobi matrices (semi-infinite, tri-diagonal matrices) and corresponding Jacobi operators, certain second-order linear difference equations of the form

$$a(n)y(n+1) + b(n)y(n) + c(n)y(n-1) = \lambda y(n) \tag{9.162}$$

arise. One method (attributed to Khan and Pearson [90]) for analyzing the spectrum relies on asymptotic representations for solutions. Here we focus solely on such asymptotic results and refer the reader to recent papers by Kiselev [92] and Janas et al. [83–85] for a discussion of how such results can be applied for a spectral analysis. As usual, in order to utilize results from Chaps. 3 and 5, we treat (9.162) as the equivalent system

$$\mathbf{y}(n+1) = \begin{pmatrix} 0 & 1 \\ -c(n)/a(n) & [\lambda - b(n)]/a(n) \end{pmatrix} \mathbf{y}(n), \tag{9.163}$$

where it is assumed that $a(n) \neq 0$ for $n \geq n_0$. This type of system is reminiscent of ones encountered above in Sect. 9.4.5 dealing with orthogonal polynomials in which the parameter λ in (9.163) corresponds to the variable x. As discussed there, depending upon various assumptions on the behavior of the coefficients in (9.163), there are many options for applying theorems from Chaps. 3 and 5 to analyze the asymptotic behavior of solutions. Some assumptions involving both the growth or decay as well as regularity conditions (involving various differences $\Delta^k a(k)$, etc.) give rise to what have been called "Stolz classes." We refer the reader to [83] for some results along these lines as well as to [84] for some other general types of asymptotic theorems similar to those in Eq. (9.163).

For the remainder of this section, we consider a special case of (9.163) treated by Kiselev [92] which is analogous to his results for one-dimensional Schrödinger operators found in Sect. 8.6.2. He considered the case of (9.162) corresponding to the system

$$\mathbf{y}(n+1) = \begin{pmatrix} 0 & 1 \\ -1 & \lambda - v(n) \end{pmatrix} \mathbf{y}(n), \tag{9.164}$$

where $v(n)$ is real-valued and vanishes at infinity, and $\lambda \in \mathbb{R}$. The limiting matrix $A = \begin{pmatrix} 0 & 1 \\ -1 & \lambda \end{pmatrix}$ has eigenvalues $\mu_{1,2} = \frac{\lambda}{2} \pm \frac{1}{2}\sqrt{\lambda^2 - 4}$, for a fixed branch of the square root function. In case $|\lambda| > 2$, A has an exponential dichotomy and Van Vleck's theorem [159] applies (see (3.72)). If, in addition, $v(n) \in l^2$, then Theorem 5.3 yields the existence of a fundamental solution matrix

$$Y(n) = [P + o(1)] \prod^{n-1} \operatorname{diag}\left\{\mu_1 + \frac{\mu_1 v(n)}{\mu_2 - \mu_1}, \mu_2 + \frac{\mu_2 v(n)}{\mu_2 - \mu_1}\right\},$$

where $P^{-1}AP = \operatorname{diag}\{\mu_1, \mu_2\}$. However, Kiselev was interested in the case when $|\lambda| < 2$ and $|\mu_i| = 1$, which corresponds to perturbations of a discrete harmonic linear oscillator. Observe that in this case the eigenvalues of A can be represented as $e^{\pm i\kappa}$, where $\kappa = \arccos(\lambda/2)$, and we chose the principal branch of arccos so that $\arccos(0) = \pi/2$. Then $P = \begin{pmatrix} 1 & 1 \\ e^{i\kappa} & e^{-i\kappa} \end{pmatrix}$ diagonalizes A and the transformation $\mathbf{y}(n) = P\hat{y}(n)$ leads to

$$\hat{y}(n+1) = \left[\begin{pmatrix} e^{i\kappa} & \\ & e^{-i\kappa} \end{pmatrix} + \frac{iv(n)}{2\sin\kappa}\begin{pmatrix} e^{i\kappa} & e^{-i\kappa} \\ -e^{i\kappa} & -e^{-i\kappa} \end{pmatrix}\right]\hat{y}(n), \tag{9.165}$$

which corresponds to (8.94). If $v(n)$ vanishes at infinity and is of bounded variation (i.e., $v(n+1) - v(n) \in l^1$) and if also $v(n)$ would be in l^2, then Theorem 5.1 could be applied to yield an asymptotic representation for solutions of (9.165) by first calculating the eigenvalues of the coefficient matrix and making an l^1-approximation to them. But when $v(n)$ is oscillating (and still vanishing at infinity), then as in the harmonic linear oscillator case, resonances may be encountered for particular values of κ, i.e., certain values of $\lambda \in (-2, 2)$. To identify possible exceptional values and to obtain asymptotic representations when these values are avoided, Kiselev proceeded analogously and first transformed (9.165) using

$$\hat{y}(n) = \operatorname{diag}\{e^{i\kappa n}, e^{-i\kappa n}\}\tilde{y}(n)$$

to obtain

$$\tilde{y}(n+1) = \left[I + \begin{pmatrix} 1 & e^{-2i\kappa(n+1)} \\ -e^{2i\kappa(n+1)} & -1 \end{pmatrix} \frac{iv(n)}{2\sin\kappa} \right] \tilde{y}(n). \tag{9.166}$$

Next, a conditioning transformation

$$\tilde{y}(n) = \begin{pmatrix} 1 & q(n,\kappa) \\ \bar{q}(n,\kappa) & 1 \end{pmatrix} w(n)$$

is sought in order to replace the off-diagonal terms in (9.166) by l^1-perturbations. For that purpose, a good candidate for $q(n,\kappa)$ is

$$q(n,\kappa) = -\sum_{l=n}^{\infty} \frac{iv(l)}{2\sin\kappa} e^{-2i\kappa(l+1)} \tag{9.167}$$

provided the series converges. For the class of $v(n)$ satisfying

$$|v(n)| \le Cn^{-3/4-\varepsilon} \qquad \text{for } n \ge n_0 \tag{9.168}$$

for some positive constants C, ε and n_0, the function $f(n) = v(n)n^{1/4}$ is in l^2. The discrete Fourier transform (defined by the l^2-limit)

$$\Phi(f)(n) = \lim_{N\to\infty} \sum_{-N}^{N} \exp(inl) f(l)$$

is used to construct a set S of κ-values for which $q(n,\kappa)$ as defined in (9.167) converges and, moreover,

$$\sum_{k=n}^{\infty} \exp(-2i\kappa l) v(l) = \mathrm{O}\left(n^{-1/4}\log n\right) \qquad \text{as } n\to\infty.$$

(We refer to [92, Lemmas 3.3, 3.4] for precise definitions and a proof.) This implies that the conditioning transformation exists and is invertible for all n sufficiently large and transforms (9.166) into

$$\begin{aligned} w(n+1) &= \left[I + \operatorname{diag}\left\{ \frac{iv(n)}{2\sin\kappa}, -\frac{iv(n)}{2\sin\kappa} \right\} + R(n) \right] w(n) \\ &= \left[\begin{pmatrix} \lambda_1(n) & \\ & \lambda_2(n) \end{pmatrix} + R(n) \right] w(n), \end{aligned} \tag{9.169}$$

where $R(n)/\lambda_i(n) \in l^1$ $(i = 1, 2)$. Since $v(n)$ is real-valued, $\Lambda(n)$ trivially satisfies Levinson's discrete dichotomy conditions (3.8), (3.9), hence a fundamental solution matrix exists for n sufficiently large satisfying, as $n \to \infty$,

$$W(n) = [I + o(1)] \prod^{n-1} \operatorname{diag}\left\{1 + \frac{iv(l)}{2\sin\kappa}, 1 - \frac{iv(l)}{2\sin\kappa}\right\}.$$

Kiselev also shows that there is an alternative representation by applying to (9.169) the diagonal transformation

$$w(n) = \operatorname{diag}\left\{\exp\left(\frac{i}{2\sin\kappa}\sum_{l=1}^{n-1} v(l)\right), \exp\left(-\frac{i}{2\sin\kappa}\sum_{l=1}^{n-1} v(l)\right)\right\} \tilde{w}(n),$$

which yields that

$$\tilde{w}(n+1) = \left[I + \tilde{R}(n,\kappa)\right] \tilde{w}(n), \tag{9.170}$$

where $\tilde{R} \in l^1$. Hence (9.170) has a fundamental solution

$$\tilde{W}(n) = I + o(1) \qquad \text{as } n \to \infty$$

and hence (9.164) has a fundamental solution matrix

$$Y(n) = P\begin{pmatrix} e^{i\kappa n} & 0 \\ 0 & e^{-i\kappa n}\end{pmatrix}\begin{pmatrix} 1 & q(n,\kappa) \\ \bar{q}(n,\kappa) & 1\end{pmatrix}\begin{pmatrix} e^{\frac{i}{2\sin\kappa}\sum_{l=1}^{n-1} v(l)} & 0 \\ 0 & e^{-\frac{i}{2\sin\kappa}\sum_{l=1}^{n-1} v(l)}\end{pmatrix}[I + o(1)].$$

Moreover, using estimates on $q(n,\kappa)$ and from the T-operators yields solutions $y_\lambda(n)$, $\overline{y_\lambda(n)}$ of

$$y(n+1) + y(n-1) + v(n)y(n) = \lambda y(n)$$

having the asymptotic representation (as $n \to \infty$)

$$y_\lambda(n) = \exp\left(i\kappa n + \frac{i}{2\sin\kappa}\sum_{l=1}^{n} v(l)\right)\left[1 + O\left(n^{-1/4}\log n\right)\right]$$

for all $\lambda \in \mathbb{R}$ except for a set of measure zero determined by the Fourier transform of $v(n)n^{1/4}$. For other results of this type, see [85], where Kiselev's procedure is applied to other Jacobi operators.

Chapter 10
Asymptotics for Dynamic Equations on Time Scales

10.1 Introduction

In the foregoing chapters, one sees that many asymptotic results for linear differential and difference equations as well as the methods used to derive them are closely related. So it is natural to ask for a framework which would encompass both sets of results as well as including some generalizations. One possibility for doing this has been discussed by Spigler and Vianello [143]. Their approach involves using Stieltjes integrals for the operators utilized in Chaps. 2 and 3 as a means to unify both the continuous and discrete cases.

Another approach for establishing such a framework is based on the concept of a "time scale." This relatively new theory was introduced by S. Hilger [75] and not only involves more general types of integral operators, but also their differential counterparts. This makes it possible to discuss more general types of equations referred to now as "dynamic equations on time scales" and construct a corresponding calculus that can be used for the analysis of their solutions. We refer the reader to the excellent book by M. Bohner and A. Peterson [25] for a thorough presentation of the basic theory as well as many interesting examples. We will mention in this chapter only the results which we will use for a discussion of the asymptotic behavior of solutions of such dynamic equations and refer to [25] for proofs of the basic results and a more complete and comprehensive treatment. Since we are interested in the asymptotic behavior of solutions as $t \to \infty$, we shall always make the assumption in this chapter that time scales are unbounded from above.

The fundamental idea behind this approach is to consider a non-empty closed subset $\mathbb{T}$ (called a *time scale*) of the real numbers and use it to create operators acting on functions defined on $\mathbb{T}$. These operators are defined by taking certain limits in $\mathbb{T}$ and for this purpose the closure of $\mathbb{T}$ is important. Two main examples of time scales are $\mathbb{T} = \mathbb{R}$, which gives rise to the ordinary differential operator, and $\mathbb{T} = \mathbb{N}$, which gives rise to the forward difference operator. To see how these operators arise as

S. Bodine, D.A. Lutz, *Asymptotic Integration of Differential and Difference Equations*, Lecture Notes in Mathematics 2129,
DOI 10.1007/978-3-319-18248-3_10

analogous "limits," the so-called forward jump operator $\sigma : \mathbb{T} \to \mathbb{T}$ ($\mathbb{T}$ unbounded from above) defined by

$$\sigma(t) = \inf\{s \in \mathbb{T} : s > t\}$$

is used.

Using σ, the important function $\mu : \mathbb{T} \to [0, \infty)$ is defined as $\mu(t) = \sigma(t) - t$ and is called the *graininess* of T. So for $\mathbb{R}$, $\mu(t) = 0$ for all t and for $\mathbb{N}$, $\mu(t) \equiv 1$. For more general time scales, μ measures in a way the distance to the "next" point. If $\mu(t) > 0$, there is a gap between t and the next point in which case t is called *right-scattered* , while if $\mu(t) = 0$ it is called *right dense*, indicating that there is a sequence of points in $\mathbb{T}$ converging to t from the right. Time scales where $\mu(t) > 0$ for all t are called *discrete*.

A function $f : \mathbb{T} \to \mathbb{R}$ is said to have a *delta derivative* at $t \in \mathbb{T}$ if

$$f^{\Delta}(t) := \lim_{s \to t} \frac{f(\sigma(t)) - f(s)}{\sigma(t) - s}, \qquad \text{where } s \in \mathbb{T} \backslash \{\sigma(t)\}$$

exists, where the limit is defined in the usual ε, δ sense. It follows immediately that for $\mathbb{T} = \mathbb{R}, f^{\Delta}(t) = f'(t)$ and for $T = N, f^{\Delta}(t) = f(t+1) - f(t)$, the forward jump operator. While we previously chose for convenience to express difference equations using the shift operator $Ef(t) = f(t+1)$, in comparing the results in this chapter with earlier ones for difference equations, this change should be noted.

Another important example of a discrete time scale is $q^{\mathbb{N}} = \{q^n : n \in \mathbb{N}\}$. For $q^{\mathbb{N}}$, the delta derivative is typically called the *q-difference operator* and is given by

$$f^{\Delta}(t) = \frac{f(qt) - f(t)}{(q-1)t}.$$

The function f is called *differentiable* on $\mathbb{T}$ if f^{Δ} exists for all $t \in \mathbb{T}$, and then the useful formula

$$f^{\sigma}(t) = f(t) + \mu(t) f^{\Delta}(t) \tag{10.1}$$

holds. The product and quotient rule for differentiable functions f and g are given by

1. $(fg)^{\Delta} = f^{\sigma} g^{\Delta} + f^{\Delta} g$;
2. $(f/g)^{\Delta} = (f^{\Delta} g - f g^{\Delta})/(g g^{\sigma})$, $\qquad g \neq 0$.

A function $f : \mathbb{T} \to \mathbb{C}$ is called *rd-continuous* provided it is continuous at right-dense points in $\mathbb{T}$ and its left-sided limits exist (finite) at left-dense points in $\mathbb{T}$. A function $F : \mathbb{T} \to \mathbb{C}$ with $F^{\Delta}(t) = f(t)$ for all $t \in \mathbb{T}$ is said to be an *antiderivative* of $f : \mathbb{T} \to \mathbb{C}$, and in this case we define $\int_a^b f(t)\,\Delta t = F(b) - F(a)$ for all $a, b \in \mathbb{T}$. However, theories in terms of Riemann and Lebesgue integrals on time scales have been established (see [26, Chap. 5]). If $\mathbb{T} = \mathbb{R}$, then $\int_a^b f(t)\,\Delta t = \int_a^b f(t)\,dt$, where the last integral is the ordinary integral. If $\mathbb{T} = \mathbb{Z}$, then $\int_m^n f(t)\Delta t = \sum_{k=m}^{n-1} f(k)$.

Proposition 10.1 *Let $f, g : \mathbb{T} \to \mathbb{C}$ be rd-continuous. Then the following hold:*

1. *f has an antiderivative;*
2. *$\int_t^{\sigma(t)} f(\tau)\,\Delta\tau = \mu(t)f(t)$ for all $t \in \mathbb{T}$;*
3. *(Hölder's Inequality) Let $p > 1$ and $1/p + 1/p' = 1$. For all $a, b \in \mathbb{T}$ with $a \leq b$ we have*

$$\int_a^b |f(\tau)g(\tau)|\,\Delta\tau \leq \left\{\int_a^b |f(\tau)|^p \Delta\tau\right\}^{1/p} \left\{\int_a^b |g(\tau)|^{p'}\right\}^{1/p'}.$$

A function $f : \mathbb{T} \to \mathbb{C}$ is called *regressive* if $1 + \mu(t)f(t) \neq 0$ for all $t \in \mathbb{T}$. The set of all regressive and rd-continuous functions $f : \mathbb{T} \to \mathbb{C}$ will be denoted by $\mathscr{R}$. The set $\mathscr{R}$ forms an Abelian group under the addition $\oplus$ defined by

$$(p \oplus q)(t) = p(t) + q(t) + \mu(t)p(t)q(t) \qquad \text{for all } t \in \mathbb{T}.$$

The additive inverse of p in this group is denoted by $\ominus p$, and is given by

$$(\ominus p)(t) = -\frac{p(t)}{1 + \mu(t)p(t)} \qquad \text{for all } t \in \mathbb{T}.$$

The "circle minus" subtraction $\ominus$ on $\mathscr{R}$ is defined by

$$(p \ominus q)(t) = (p \oplus (\ominus q))(t) = \frac{p(t) - q(t)}{1 + \mu(t)q(t)} \qquad \text{for all } t \in \mathbb{T}.$$

Given $t_0 \in \mathbb{T}$ and $p \in \mathscr{R}$, the function

$$e_p(t, t_0) = \exp\left[\int_{t_0}^t \xi_{\mu(\tau)}(p(\tau))\,\Delta\tau\right] \qquad \text{for all } t, t_0 \in \mathbb{T},$$

is the unique solution of the initial value problem (see, e.g., [25, Theorem 2.35])

$$y^\Delta = p(t)y, \qquad y(t_0) = 1.$$

Here

$$\xi_{\mu(\tau)}(p(\tau)) = \begin{cases} \frac{1}{\mu(\tau)} \operatorname{Log}[1 + p(\tau)\mu(\tau)] & \text{if } \mu(\tau) > 0, \\ p(\tau) & \text{if } \mu(\tau) = 0, \end{cases}$$

where Log is the principal logarithm function. Such a function $e_p(t, t_0)$ has been called an "exponential function on $\mathbb{T}$" associated with $p(t)$ and can be re-written in a more unified way as

$$e_p(t, t_0) = \exp\left[\int_{t_0}^{t} \lim_{u \to \mu(\tau)+} \frac{1}{u} \operatorname{Log}[1 + p(\tau)u]\,\Delta\tau\right], \quad \text{for all } t, t_0 \in \mathbb{T}, \tag{10.2}$$

and hence

$$\left|e_p(t, t_0)\right| = \exp\left[\int_{t_0}^{t} \lim_{u \to \mu(\tau)+} \frac{1}{u} \operatorname{Log}|1 + p(\tau)u|\,\Delta\tau\right], \quad \text{for all } t, t_0 \in \mathbb{T}. \tag{10.3}$$

The following identities (see [25, Theorem 2.36]) will be useful.

Proposition 10.2 *If $p, q \in \mathcal{R}$, then*

1. $e_p(t, t) \equiv 1$;
2. $e_p(\sigma(t), s) = [1 + \mu(t)p(t)]e_p(t, s)$;
3. $\dfrac{1}{e_p(t, s)} = e_{\ominus p}(t, s)$;
4. $e_p(t, s) = \dfrac{1}{e_p(s, t)} = e_{\ominus p}(s, t)$;
5. $e_p(t, s)e_p(s, r) = e_p(t, r)$;
6. $e_p(t, s)e_q(t, s) = e_{p \oplus q}(t, s)$;
7. $\dfrac{e_p(t, s)}{e_q(t, s)} = e_{p \ominus q}(t, s)$.

We call a $d \times d$ matrix-valued function A *rd-continuous* if each entry is rd-continuous, and A is called *regressive* if $I + \mu(t)A(t)$ is invertible for all $t \in T$. The following result concerning the existence of a fundamental matrix can be found in [25, Thm. 5.8].

Theorem 10.3 *Assume that $A : \mathbb{T} \to \mathbb{C}^{d \times d}$ is rd-continuous and regressive. Then $y^{\Delta} = A(t)y$ has a fundamental system of solutions.*

In the special case that $A(t) = \Lambda(t) = \operatorname{diag}\{\lambda_1(t), \ldots, \lambda_d(t)\}$ (rd-continuous and regressive), we will use the notation $e_{\Lambda}(t, t_0)$ as the unique solution of the initial-value problem

$$y^{\Delta} = \Lambda(t)y, \qquad Y(t_0) = I.$$

Standing hypothesis: To avoid repetitive notation, we assume throughout this chapter that the independent variable is in the time scale $\mathbb{T}$. For example, the notation

$$t \geq t_0 \quad \text{means} \quad t \in [t_0, \infty) \cap \mathbb{T} \text{ for some } t_0 \in \mathbb{T}.$$

10.2 Levinson's Fundamental Theorem

This section is first concerned with a generalization of Levinson's fundamental theorem to time scales, an extension of the treatment in Sects. 2.2 and 3.2. This important result is due to Bohner and Lutz [23] (see also [22, 24]), and we present it here with minor modifications to adjust it to the exposition chosen in Sects. 2.2 and 3.2. We then give a time scales generalization of Kooman's improvement of the discrete Levinson result (see Theorem 3.8) based on relaxing the growth condition on the perturbation.

Let $A(t)$ be regressive and rd-continuous. We will consider the unperturbed system

$$x^{\Delta} = A(t)x, \qquad t \geq t_0. \tag{10.4}$$

Definition 10.4 For a regressive and rd-continuous matrix $A(t)$, let $X(t)$ be a fundamental matrix of (10.4). Assume that there exists a projection P ($P^2 = P$) and a positive constant K such that

$$\left.\begin{array}{ll} |X(t)PX^{-1}(s)| \leq K & \forall \;\; t_0 \leq s \leq t, \\ |X(t)[I-P]X^{-1}(s)| \leq K & \forall \;\; t_0 \leq t \leq s. \end{array}\right\} \tag{10.5}$$

Then (10.4) is said to have an *ordinary dichotomy.*

Theorem 10.5 (Time Scales Version of Coppel's Theorem) *Consider* (10.4), *where $A(t)$ is a $d \times d$ matrix which is regressive and rd-continuous for all $t \geq t_0$, and a perturbed system*

$$y^{\Delta} = [A(t) + R(t)]\,y, \qquad t \geq t_0, \tag{10.6}$$

where $R(t)$ is rd-continuous $d \times d$ matrix satisfying

$$\int_{t_0}^{\infty} |R(t)|\Delta t < \infty.$$

Assume that (10.4) *satisfies an ordinary dichotomy* (10.5) *with projection P and constant K. Then there exists for t sufficiently large a one-to-one and bicontinuous correspondence between the bounded solutions of* (10.4) *and* (10.6). *Moreover, the difference between corresponding solutions of* (10.4) *and* (10.6) *tends to zero as $t \to \infty$ if $X(t)P \to 0$ as $t \to \infty$.*

Proof Fix $t_1 \geq t_0$ such that

$$K\int_{t_1}^{\infty} |R(s)|\,\Delta s =: \theta < 1.$$

Let $\mathscr{B}$ be the Banach space of bounded d-dimensional vector-valued sequences with the supremum norm $\|y\| = \sup_{t \geq t_1} |y(t)|$. For $t \geq t_1$, define an operator T on $\mathscr{B}$ by

$$(Ty)(t) = \int_{t_1}^{t} X(t)PX^{-1}(\sigma(s))R(s)y(s)\,\Delta s - \int_{t}^{\infty} X(t)[I-P]X^{-1}(\sigma(s))R(s)y(s)\,\Delta s. \tag{10.7}$$

Then, for y and $\tilde{y} \in \mathscr{B}$, it follows for $t \geq t_1$ that

$$|(Ty)(t)| \leq K\|y\| \int_{t_1}^{\infty} |R(s)|\,\Delta s \leq \theta\|y\|,$$

and

$$|(T[y-\tilde{y}])(t)| \leq K\|y-\tilde{y}\| \int_{t_1}^{\infty} |R(s)|\,\Delta s \leq \theta\|y-\tilde{y}\|,$$

i.e., T maps $\mathscr{B}$ into $\mathscr{B}$ and is a contraction. Hence, given a bounded solution x of (10.4), the equation

$$x = y - Ty \tag{10.8}$$

has a unique solution $y \in \mathscr{B}$, which using the product rule on time scales can be shown to be a solution of (10.6). Conversely, given a bounded solution y of (10.6), the function $x(t)$ defined by (10.8) is a bounded solution of (10.4). Equation (10.8) establishes therefore a one-to-one correspondence between the bounded solutions of (10.4) and (10.6) for $t \geq t_1$. The bicontinuity of this correspondence as well as the claim that the difference between these corresponding solutions, $x(t) - y(t)$, tends to zero if $X(t)P \to 0$ as $t \to \infty$ can be shown as in Theorem 2.2. □

Applying Theorem 10.5 will allow us to show the following times scales generalization of Theorem 2.7 (see [23, Theorem 4.1]):

Theorem 10.6 (Time Scales Version of Levinson's Fundamental Theorem) *Consider*

$$y^{\Delta} = [\Lambda(t) + R(t)]\,y, \qquad t \geq t_0, \tag{10.9}$$

where $\Lambda(t) = \operatorname{diag}\{\lambda_1(t), \ldots, \lambda_d(t)\}$ *is rd-continuous and regressive. Assume that* $R(t)$ *is a rd-continuous* $d \times d$ *matrix-valued function such that*

$$\int_{t_0}^{\infty} \left| \frac{R(t)}{1+\mu(t)\lambda_k(t)} \right| \Delta t < \infty, \qquad \forall\ 1 \leq k \leq d. \tag{10.10}$$

Suppose that there exist constants $K_l > 0$ $(l = 1, 2)$ such that for each pair (i, j) with $i \neq j$, either

$$\left.\begin{aligned} &\lim_{t\to\infty} e_{\lambda_j \ominus \lambda_i}(t, t_0) = 0 \\ &\textit{and} \;\; \left|e_{\lambda_j \ominus \lambda_i}(t, s)\right| \le K_1, \qquad \forall\ t \ge s \ge t_0, \end{aligned}\right\} \tag{10.11}$$

or

$$\left|e_{\lambda_j \ominus \lambda_i}(t, s)\right| \ge K_2, \quad \forall\ t \ge s \ge t_0. \tag{10.12}$$

Then (10.9) *has for t sufficiently large a fundamental solution matrix Y of the form*

$$Y(t) = [I + \mathrm{o}(1)]e_{\Lambda}(t, t_0) \qquad \textit{as } t \to \infty. \tag{10.13}$$

Proof Fix $i \in \{1, \ldots, d\}$ and put $y = e_{\lambda_i}(t, t_0)z$. Then (10.9) implies that

$$z^{\Delta} = \left[\frac{\Lambda(t) - \lambda_i(t)I}{1 + \mu(t)\lambda_i(t)} + \frac{R(t)}{1 + \mu(t)\lambda_i(t)}\right] z. \tag{10.14}$$

We also consider the unperturbed shifted system

$$w^{\Delta} = \left[\frac{\Lambda(t) - \lambda_i(t)I}{1 + \mu(t)\lambda_i(t)}\right] w. \tag{10.15}$$

Observe that $\Lambda(t)$ being regressive shows that $\frac{\Lambda(t)-\lambda_i(t)I}{1+\mu(t)\lambda_i(t)}$ is also regressive since

$$1 + \mu(t)\frac{\lambda_j(t) - \lambda_i(t)}{1 + \mu(t)\lambda_i(t)} = \frac{1 + \mu(t)\lambda_j(t)}{1 + \mu(t)\lambda_i(t)} \neq 0.$$

Hence (10.15) has a fundamental matrix of the form

$$W(t) = \operatorname{diag}\left\{e_{\lambda_1 \ominus \lambda_i}(t, t_0), e_{\lambda_2 \ominus \lambda_i}(t, t_0), \ \ldots \ , e_{\lambda_d \ominus \lambda_i}(t, t_0)\right\}.$$

Let $P = \operatorname{diag}\{p_1, \ \ldots \ , p_d\}$, where

$$p_i = \begin{cases} 1 & \text{if } (i, j) \text{ satisfies } \ (10.11) \\ 0 & \text{if } (i, j) \text{ satisfies } \ (10.12). \end{cases} \tag{10.16}$$

Then (10.15) satisfies the ordinary dichotomy condition (10.5) and $W(t)P \to 0$ as $t \to \infty$. Now, by Theorem 10.5 and since $w_i(t) = e_i$ is a bounded solution of (10.15), there exists for t sufficiently large a bounded solution $z_i(t)$ of (10.14) such that

$$z_i(t) = e_i + \mathrm{o}(1), \qquad \text{as } t \to \infty.$$

Thus (10.9) has a solution

$$y_i(t) = [e_i + o(1)]\, e_{\lambda_i}(t, t_0) \qquad \text{as } t \to \infty. \tag{10.17}$$

Repeating this process for all $1 \leq i \leq d$ and noting that this set of solutions is linearly independent for t sufficiently large by (10.17), gives the desired result (10.13). □

We note that if the conditions for Theorem 10.6 are satisfied for *one fixed* value of $i \in \{1, \ldots, d\}$, then (10.13) has to be replaced by (10.17) as the conclusion of the theorem.

We proceed with a generalization of Theorem 3.8, which is an improvement due to Kooman of Theorem 3.4, the discrete Levinson theorem for systems of difference equations. The following time scales generalization appears to be new.

Theorem 10.7 *Let $\Lambda(t)$ and $R(t)$ satisfy the assumptions of Theorem 10.6 with the exception of* (10.10)*. Instead of* (10.10)*, assume that the weaker hypothesis*

$$\int_{t_0}^{\infty} \left| [I + \mu(t)\Lambda(t)]^{-1} R(t) \right| \, \Delta t < \infty \tag{10.18}$$

holds. Then (10.9) *has for t sufficiently large a fundamental solution matrix Y of the form* (10.13)*.*

Proof The proof follows the proof of Theorem 10.6 up to the definition of the projection matrix P in (10.16). With $K = \max\{K_1, 1/K_2\}$, fix t_1 sufficiently large ($t_0 \leq t_1 \in \mathbb{T}$) such that

$$K \int_{t_1}^{\infty} \left| [I + \mu(t)\Lambda(t)]^{-1} R(t) \right| \, \Delta t \leq \theta$$

for some $\theta \in (0, 1)$. Let $\mathscr{B}$ be the Banach space of bounded d-dimensional vector-valued sequences with the supremum norm $\|z\| = \sup_{t \geq t_1} |z(t)|$, and where the vector norm $| \cdot |$ is the maximum norm. For $t \geq t_1$, define an operator T on $\mathscr{B}$ by

$$\begin{aligned} (Tz)(t) = &\int_{t_1}^{t} W(t)PW^{-1}(\sigma(s)) \frac{R(s)}{1 + \mu(s)\lambda_i(s)} z(s) \, \Delta s \\ &- \int_{t}^{\infty} W(t)[I - P]W^{-1}(\sigma(s)) \frac{R(s)}{1 + \mu(s)\lambda_i(s)} z(s) \, \Delta s. \end{aligned} \tag{10.19}$$

Observe that for $1 \le j,k \le d$ and $t_1 \le s \le t$,

$$\left| \left(W(t)PW^{-1}(\sigma(s)) \frac{R(s)}{1+\mu(s)\lambda_i(s)} \right)_{jk} \right|$$
$$\le p_j \left| e_{\lambda_j \ominus \lambda_i}(t,\sigma(s)) \frac{r_{jk}(s)}{1+\mu(s)\lambda_i(s)} \right|$$
$$= p_j \left| e_{\lambda_j \ominus \lambda_i}(t,s) \frac{r_{jk}(s)}{1+\mu(s)\lambda_j(s)} \right|$$
$$\le p_j K \left| \frac{r_{jk}(s)}{1+\mu(s)\lambda_j(s)} \right|$$
$$= p_j K \left| \left\{ [I+\mu(s)\Lambda(s)]^{-1} R(s) \right\}_{jk} \right|. \qquad (10.20)$$

Similarly, for $t_1 \le t \le s$

$$\left| \left(W(t)[I-P]W^{-1}(\sigma(s)) \frac{R(s)}{1+\mu(s)\lambda_i(s)} \right)_{jk} \right|$$
$$\le (1-p_j) \left| e_{\lambda_j \ominus \lambda_i}(t,s) \frac{r_{jk}(s)}{1+\mu(s)\lambda_j(s)} \right|$$
$$\le (1-p_j) K \left| \left\{ [I+\mu(s)\Lambda(s)]^{-1} R(s) \right\}_{jk} \right|. \qquad (10.21)$$

Therefore for $z \in \mathscr{B}$,

$$|(Tz)(t)| \le K\|z\| \int_{t_1}^{\infty} \left| [I+\mu(s)\Lambda(s)]^{-1} R(s) \right| \Delta s \le \|z\|\theta,$$

and, similarly,

$$|(Tz_1)(t) - (Tz_2)(t)| \le \theta \|z_1 - z_2\|.$$

Thus T maps $\mathscr{B}$ into $\mathscr{B}$ and is a contraction. Hence, given the bounded solution $w_i = e_i$ of (10.15) $w = z - Tz$ has a unique solution $z \in \mathscr{B}$, which using the product rule on time scales can be shown to be a solution of (10.14). The rest of the proof follows the proof of Theorem 10.6 . □

10.3 Weak Dichotomies

In this section, we will show that a Levinson-like result holds even when the leading diagonal matrix $\Lambda(t)$ does not necessarily satisfy the dichotomy conditions of Theorem 10.6, provided that the perturbation is sufficiently small. The next theorem, which appears to be new, can be interpreted as a time scales generalization of Theorems 2.12 and 3.11, formulated, however, in terms of a fundamental matrix instead of a single vector-valued solution. For reasons of brevity, we do not prove first a Coppel-type of result for weak dichotomies, but prove directly the main result on perturbations of diagonal systems.

Theorem 10.8 *For $t \geq t_0$, consider* (10.9), *where* $\Lambda(t) = \operatorname{diag}\{\lambda_1(t), \ldots, \lambda_d(t)\}$ *is rd-continuous and regressive. Suppose there exists a rd-continuous function $\beta(t) \geq 1$ for $t_0 \leq t \in \mathbb{T}$ such that for each pair (i, j) either*

$$\left.\begin{aligned} &\lim_{t\to\infty} e_{\lambda_j \ominus \lambda_i}(t, t_0) = 0 \\ &\text{and} \quad \left|e_{\lambda_j \ominus \lambda_i}(t, s)\right| \leq \beta(s) \quad \forall\ t \geq s \geq t_0, \end{aligned}\right\} \tag{10.22}$$

or

$$\left|e_{\lambda_j \ominus \lambda_i}(t, s)\right| \leq \beta(s), \quad \forall\ t_0 \leq t \leq s. \tag{10.23}$$

Assume that $R(t)$ is a rd-continuous $d \times d$ matrix-valued function such that

$$\int_{t_0}^{\infty} \beta(s) \left|[I + \mu(s)\Lambda(s)]^{-1} R(s)\right| \Delta s < \infty. \tag{10.24}$$

Then (10.9) *has for t sufficiently large a fundamental solution matrix Y of the form* (10.13).

Proof The proof is a variation of the proof of Theorem 10.7, and we will just indicate the necessary modifications here. For fixed $i \in \{1, \ldots, d\}$, a projection matrix $P = \operatorname{diag}\{p_1, \ldots, p_d\}$ is defined by $p_j = 1$ if (i, j) satisfies (10.22), and $p_j = 0$ if (i, j) satisfies (10.23).

Fix $t_1 \geq t_0$ such that $\int_{t_1}^{\infty} \beta(s) \left|\{I + \mu(s)\Lambda(s)\}^{-1} R(s)\right| \Delta s \leq \theta \in (0, 1)$. We note that for $1 \leq j, k \leq d$ and $t_1 \leq s \leq t$, it follows from (10.20) that

$$\begin{aligned} &\left|\left(W(t)PW^{-1}(\sigma(s))\frac{R(s)}{1 + \mu(s)\lambda_i(s)}\right)_{jk}\right| \\ &\leq p_j \left|e_{\lambda_j \ominus \lambda_i}(t, s)\frac{r_{jk}(s)}{1 + \mu(s)\lambda_j(s)}\right| \\ &\leq p_j \beta(s) \left|\left\{[I + \mu(s)\Lambda(s)]^{-1} R(s)\right\}_{jk}\right|. \end{aligned}$$

Similarly, for $t_1 \leq t \leq s$, it follows from (10.21) that

$$\left| \left(W(t)[I - P]W^{-1}(\sigma(s)) \frac{R(s)}{1 + \mu(s)\lambda_i(s)} \right)_{jk} \right|$$

$$\leq (1 - p_j) \left| e_{\lambda_j \ominus \lambda_i}(t, s) \frac{r_{jk}(s)}{1 + \mu(s)\lambda_j(s)} \right|$$

$$\leq (1 - p_j)\beta(s) \left| \left\{ [I + \mu(s)\Lambda(s)]^{-1} R(s) \right\}_{jk} \right|.$$

Therefore it follows for the T-operator defined in (10.19) that for $z \in \mathscr{B}$,

$$|(Tz)(t)| \leq \|z\| \int_{t_1}^{\infty} \beta(s) \left| [I + \mu(s)\Lambda(s)]^{-1} R(s) \right| \Delta s \leq \|z\|\theta,$$

and, similarly, that

$$|(Tz_1)(t) - (Tz_2)(t)| \leq \theta \|z_1 - z_2\|,$$

and the rest of the proof follows that of Theorem 10.7. □

Remark 10.9 Note that a function $\beta(s)$ satisfying (10.23) for all $1 \leq i, j \leq d$ always exists, e.g.,

$$\beta(s) = \max_{1 \leq i \leq d} \max_{1 \leq j \leq d} \sup_{t_0 \leq t \leq s} \left| e_{\lambda_j \ominus \lambda_i}(t, s) \right|.$$

However, this choice of $\beta(t)$ is usually too large to be useful in applications.

10.4 Reduction to Eigenvalues

The next result is due to Bohner and Lutz [23, Theorem 5.3]), who published the following time scales generalization of Theorems 4.1 and 5.1.

Theorem 10.10 *Consider*

$$y^{\Delta} = [\Lambda(t) + V(t) + R(t)]\, y, \qquad t \geq t_0, \tag{10.25}$$

where $\Lambda(t)$, $V(t)$, and $R(t)$ are rd-continuous $d \times d$ matrix-valued functions. Suppose that $\Lambda(t)$ is a diagonal and regressive matrix satisfying $\Lambda(t) \to \Lambda_0$ as $t \to \infty$, where Λ_0 is a diagonal matrix with distinct eigenvalues. Assume that $V(t) \to 0$ as $t \to \infty$, so that $\Lambda(t) + V(t)$ has distinct eigenvalues $\hat{\lambda}_1(t), \ldots, \hat{\lambda}_d(t)$ for sufficiently large t. Put $\hat{\Lambda}(t) = \operatorname{diag}\{\hat{\lambda}_1(t), \ldots, \hat{\lambda}_d(t)\}$, and suppose that $\hat{\Lambda}(t)$ satisfies the *dichotomy conditions* (10.11) *and* (10.12). *If* $V(t)$ *is differentiable and if*

$$\int_{t_0}^{\infty} \frac{\left|V^{\Delta}(t)\right| \left|I + \mu(t)\hat{\Lambda}(t)\right|}{\left|1 + \mu(t)\hat{\lambda}_i(t)\right|} \Delta t < \infty, \tag{10.26}$$

and

$$\int_{t_0}^{\infty} \left| \frac{R(t)}{1 + \mu(t)\hat{\lambda}_i(t)} \right| \Delta t < \infty \tag{10.27}$$

for all $1 \leq i \leq d$, *then there exists for* t *sufficiently large a fundamental matrix* $Y(t)$ *of* (10.25) *satisfying*

$$Y(t) = [I + o(1)]e_{\hat{\Lambda}}(t, t_0). \tag{10.28}$$

Proof Since the eigenvalues of $\Lambda(t) + V(t)$ are distinct for sufficiently large t, there exists a matrix-valued function $Q(t)$ for such t such that

$$\hat{\Lambda}(t) = [I + Q(t)]^{-1}[\Lambda(t) + V(t)][I + Q(t)].$$

Also, since $\Lambda(t) + V(t) \to \Lambda_0$ as $t \to \infty$,

$$I + Q(t) \to I \qquad \text{as } t \to \infty, \tag{10.29}$$

and both $I + Q(t)$ and its inverse are bounded for t sufficiently large, say $t \geq t_1$. Putting, for such $t \geq t_1$,

$$y = [I + Q(t)]z, \tag{10.30}$$

and observing

$$y^{\Delta} = (I + Q^{\sigma})z^{\Delta} + (I + Q)^{\Delta}z,$$

it follows that

$$\begin{aligned} z^{\Delta} &= (I + Q^{\sigma})^{-1}\left[(\Lambda + V)(I + Q) + R(I + Q) - Q^{\Delta}\right]z \\ &= (I + Q^{\sigma})^{-1}\left[(I + Q)\hat{\Lambda} + R(I + Q) - Q^{\Delta}\right]z \\ &= \left(\hat{\Lambda} + (I + Q^{\sigma})^{-1}\left[R(I + Q) - Q^{\Delta}\{I + \mu(t)\hat{\Lambda}(t)\}\right]\right)z, \end{aligned} \tag{10.31}$$

where (10.1) was used in the last identity.

It can also be shown using the same argument as in [69, Remark 2.1] that

$$Q^\Delta = \mathrm{O}(|V^\Delta|). \tag{10.32}$$

From (10.27) and (10.29) follows that

$$\int_{t_1}^{\infty} \left| \frac{(I+Q^\sigma)^{-1}R(I+Q)}{1+\mu\hat{\lambda}_i} \right| \Delta t < \infty, \qquad 1 \le i \le d.$$

Furthermore, (10.26) and (10.32) imply that

$$\int_{t_1}^{\infty} \left| \frac{(I+Q^\sigma)^{-1}Q^\Delta\{I+\mu(t)\hat{\Lambda}(t)\}}{1+\mu\hat{\lambda}_i} \right| \Delta t < \infty, \qquad 1 \le i \le d.$$

Since $\hat{\Lambda}$ satisfies the dichotomy conditions (10.11), (10.12), Theorem 10.6 implies that (10.31) has, for t sufficiently large, a fundamental matrix

$$Z(t) = [I + \mathrm{o}(1)]e_{\hat{\Lambda}}(t, t_0), \qquad t \to \infty,$$

and hence (10.30) and (10.29) imply (10.28). □

10.5 L^p-Perturbations with $p > 1$

This section is concerned with a generalization of the continuous and discrete Hartman–Wintner theorem (Theorems 4.11 and 5.3) to time scales. In both of these earlier results, the dichotomy conditions involve the standard (continuous or discrete) decreasing exponential function. In the context of time scales, a natural question addresses the type of exponential function to be used. Bodine and Lutz [16] decided to use exponential functions associated with $\mathbb{T}$, more specifically non-trivial solutions of $y^\Delta(t) = -\alpha y(t)$, where α is a positive constant. This choice was based on the fact that their proof techniques require the integration of the weight function.

In [12], Bodine and Lutz studied the asymptotic behavior as $t \to \infty$ of exponential functions $e_{-\alpha}(t, t_0)$ with $\alpha > 0$ for several classes of time scales and found that the graininess of the time scale had a major impact on the behavior of the exponential function. Vaguely speaking, exponential functions on time scales with small graininess behave similarly to $e^{-\alpha t}$,whereas such functions can behave quite differently when the graininess becomes large or even unbounded. For example, on the time scale $\mathbb{T} = q^{\mathbb{N}}$ ($q > 1$) one finds that

$$e_{-\alpha}(n, n_0) = \prod_{k=n_0}^{n-1} \left[1 - \alpha(q-1)q^k\right],$$

which does not approach zero as $n \to \infty$. In fact, it exhibits oscillating behavior for all n sufficiently large and its absolute value goes to infinity. To avoid such situations, they considered in [16] only time scales with bounded graininess [see (10.33)]. Here, we will just state the main results from [16] and refer to the original publication for proof and details.

Theorem 10.11 (Bodine/Lutz) *Assume that there exists a positive constant* α_0 *such that*

$$1 - \alpha_0 \mu(t) > 0 \qquad \forall \ \geq t_0. \tag{10.33}$$

Suppose that

$$\Lambda(t) = \operatorname{diag}\{\lambda_1(t), \ \ldots \ , \lambda_d(t)\},$$

is rd-continuous and regressive. Suppose that there exist positive constants α *and* K *with* $0 < \alpha \leq \alpha_0$ *such that for each pair* (i, j) *with* $i \neq j$ *either*

$$\left| e_{\lambda_j \ominus \lambda_i}(t, s) \right| \leq K e_{-\alpha}(t, s), \qquad \forall \ t \geq s \geq t_0, \tag{10.34}$$

or

$$\left| e_{\lambda_i \ominus \lambda_j}(t, s) \right| \leq K e_{-\alpha}(t, s), \qquad \forall \ t \geq s \geq t_0. \tag{10.35}$$

Assume that $V(t)$ *is an rd-continuous* $d \times d$ *matrix-valued function on* $\mathbb{T}$ *such that*

$$\int_{t_0}^{\infty} \left| \frac{V(t)}{1 + \mu(t)\lambda_k(t)} \right|^p \Delta t < \infty \quad \text{for some } 1 < p \leq 2, \quad \forall \ 1 \leq k \leq d.$$

Then

$$y^\Delta = [\Lambda(t) + V(t)]y, \qquad t \geq t_0, \tag{10.36}$$

has for t *sufficiently large a fundamental solution matrix of the form*

$$Y(t) = [I + o(1)]e_{\Lambda + \mathrm{dg}V}(t, t_1), \qquad \text{as } t \to \infty. \tag{10.37}$$

In applications, the dichotomy conditions (10.34), (10.35) might be difficult to verify. For that reason, they are replaced in the following corollary by pointwise conditions, which are more restrictive than their averaged counterparts, but possibly easier to verify.

Corollary 10.12 *Assume that there exists* $\alpha_0 > 0$ *and* t_0 *such that* (10.33) *holds for all* $t \geq t_0$. *Suppose that* $\Lambda(t)$ *and* $V(t)$ *satisfy the conditions of Theorem 10.11, except that* $\Lambda(t)$ *satisfies the following dichotomy conditions [instead of* (10.34) *and* (10.35)*]: There exists a positive constant* $\alpha \in \mathbb{R}^+$, $0 < \alpha \leq \alpha_0$ *such that for each pair* (i,j) *with* $i \neq j$

$$\text{either} \quad \left. \begin{array}{l} \operatorname{Re}\{\lambda_j(t) - \lambda_i(t)\} \leq -\alpha, \ \forall\, t \geq t_0 \text{ such that } \mu(t) = 0, \\ \left|\frac{1+\mu(t)\lambda_j(t)}{1+\mu(t)\lambda_i(t)}\right| \leq 1 - \alpha\mu(t), \ \forall\, t \geq t_0 \text{ such that } \mu(t) > 0, \end{array} \right\}$$

$$\text{or} \quad \left. \begin{array}{l} \operatorname{Re}\{\lambda_i(t) - \lambda_j(t)\} \leq -\alpha, \ \forall\, t \geq t_0 \text{ such that } \mu(t) = 0, \\ \left|\frac{1+\mu(t)\lambda_i(t)}{1+\mu(t)\lambda_j(t)}\right| \leq 1 - \alpha\mu(t), \ \forall\, t \geq t_0 \text{ such that } \mu(t) > 0. \end{array} \right\}$$

Then (10.36) *has for* t *sufficiently large a fundamental solution matrix of the form* (10.37).

The first result on L^p-perturbations for arbitrary time scales appeared in [23, Theorem 5.5]. Here the same growth condition (10.11) was assumed for the perturbation, but significantly stronger assumptions on $\Lambda(t)$.

Another point-wise Hartman–Wintner result for time scales with bounded graininess was published in [132, Theorem 3.1].

10.6 Conditionally Integrable Perturbations

As a basic result on conditionally convergent perturbations, Bohner and Lutz showed in [23, Theorem 5.2] the following time-scales generalization of Theorems 4.18 and 5.6.

Theorem 10.13 *Let* $\Lambda(t)$, $V(t)$, *and* $R(t)$ *be* $d \times d$ *rd-continuous matrices for* $t \geq t_0$ *such that* $\Lambda(t) = \operatorname{diag}\{\lambda_1(t), \ldots, \lambda_d(t)\}$ *is regressive and satisfies Levinson's dichotomy conditions* (10.11) *and* (10.12)*. Suppose that*

$$Q(t) = -\int_t^\infty V(\tau)\,\Delta\tau \tag{10.38}$$

exists for $t \geq t_0$ *and such that for all* $1 \leq i \leq d$

$$\int_{t_0}^\infty \left|\frac{G(t)}{1+\mu(t)\lambda_i(t)}\right| \Delta t < \infty, \quad \text{with } G(t) \in \{VQ,\ \Lambda Q - Q^\sigma \Lambda,\ R\}, \tag{10.39}$$

for all $1 \leq i \leq d$*. Then*

$$y' = [\Lambda(t) + V(t) + R(t)]\,y \tag{10.40}$$

has for t *sufficiently large a fundamental solution matrix satisfying*

$$Y(t) = [I + o(1)]\,e_\Lambda(t, t_0), \qquad \text{as} \quad t \to \infty. \tag{10.41}$$

Proof In (10.40), put

$$y = [I + Q(t)]z,$$

where Q is defined in (10.38). By (10.38), $Q(t) = o(1)$ as $t \to \infty$, hence $I + Q(t)$ is invertible for all t sufficiently large, and $I + Q$ and its inverse are bounded for all t sufficiently large. Then for such t follows that

$$\begin{aligned} z^{\Delta} &= (I + Q^{\sigma})^{-1} [\Lambda + \Lambda Q + V + VQ + R(I + Q) - V] z \\ &= \left\{\Lambda + (I + Q^{\sigma})^{-1} [\Lambda Q - Q^{\sigma}\Lambda + VQ + R(I + Q)]\right\} z \\ &= \left\{\Lambda(t) + \tilde{R}(t)\right\} z. \end{aligned} \qquad (10.42)$$

Since $\Lambda(t)$ satisfies the dichotomy conditions (10.11), (10.12), assumption (10.39) and Theorem 10.6 imply that (10.42) has for large t a fundamental matrix satisfying

$$Z(t) = [I + o(1)]\, e_{\Lambda}(t, t_0) \qquad \text{as } t \to \infty.$$

By (10.6), (10.40) has for t sufficiently large a fundamental matrix $Y(t)$ which satisfies (10.41). □

We note that the condition (10.39) in the special case $A(t) = \Lambda Q - Q^{\sigma}\Lambda$ is unlikely to be satisfied if $\Lambda(t)$ is not "small" for large t, e.g., $\Lambda(t) \equiv 0$ or at least vanishing at infinity.

In the context of perturbations of the trivial system $x^{\Delta} = 0$, Mert and Zafer [108] generalized results by Wintner and Trench (see Theorems 4.19, 4.23 and 5.7) to nonlinear dynamic systems on time scales of the form $y^{\Delta} = A(t)y + f(t, y(t))$. Following the approach taken in [152], they use a time scales version of the Schauder-Tychonov theorem to derive their results.

We will not give their result here, but offer the following Theorem 10.14, which is concerned with "mildly nonlinear systems." The reason for this choice is that in the important case of linear systems, it constitutes an improvement of the results in [108]. Note that it is a generalization of Theorems 4.24 and 5.9, and we will again work with the Banach fixed point theorem.

Theorem 10.14 is a combination of two statements. Given a constant vector c, in the first approach one only assumes the convergence of various improper integrals to show the existence of a solution $y(t) = c + o(1)$. The more quantitative version gives an estimate on how fast some convergent improper integrals vanish as $t \to \infty$ and uses this information to find an estimate of this error term $o(1)$. To combine these two statements, we follow an idea of Trench (see [152]) and use a nonnegative function $w(t)$ such that either $w(t) \equiv 1$ or approaches zero at infinity. To simplify the exposition, we assume without loss of generality that $0 \le w \le 1$.

Theorem 10.14 (Bodine/Lutz) *Consider*

$$y^{\Delta} = [A(t) + B(t)]y + f(t, y), \qquad t \geq t_0, \tag{10.43}$$

where $A(t)$, $B(t)$ and $f(t, y)$ are rd-continuous for $t \geq t_0$ and all $y \in \mathbb{C}^d$. Let $w : [t_0, \infty) \to [0, 1]$ be a rd-continuous function such that $w(t) \to 0$ as $t \to \infty$ or $w(t) \equiv 1$. For an integer $k \geq 0$, suppose that the integrals defined recursively by

$$A_j(t) = \int_t^{\infty} A_{j-1}^{\sigma}(s)A(s)\,\Delta s, \qquad (A_0(s) \equiv I) \tag{10.44}$$

are convergent for $1 \leq j \leq k+1$ and satisfy

$$\left|A_j(t)\right| = O\left(w(t)\right) \text{ as } t \to \infty; \quad 1 \leq j \leq k+1. \tag{10.45}$$

Define

$$\Gamma_k(t) = \sum_{j=0}^{k} A_j(t), \tag{10.46}$$

and suppose that

$$\left|\int_t^{\infty} \Gamma_k^{\sigma}(s)B(s)\,\Delta s\right| = O(w(t)), \qquad \text{as } t \to \infty, \tag{10.47}$$

and

$$\left|\int_t^{\infty} \Gamma_k^{\sigma}(s)f(s,0)\,\Delta s\right| = O(w(t)), \qquad \text{as } t \to \infty. \tag{10.48}$$

Assume that for all x and $\tilde{x}$

$$|f(t,x) - f(t,\tilde{x})| \leq \gamma(t)|x - \tilde{x}|, \tag{10.49}$$

where $\gamma(t)$ is rd-continuous for $t \geq t_0$, and satisfies

$$\int_t^{\infty} \gamma(s)\,\Delta s = O(w(t)). \tag{10.50}$$

Finally, suppose that

$$\int_t^{\infty} \left(|A_k^{\sigma}(s)A(s)| + |B(s)|\right) w(s)\,\Delta s = O(w(t)). \tag{10.51}$$

Then, given a constant vector c, there exists for t sufficiently large a solution y(t) of (10.43) *satisfying*

$$y(t) = \begin{cases} c + o(1) & \text{if } w(t) \equiv 1 \\ c + O(w(t)) & \text{if } w(t) \to 0 \text{ as } t \to \infty. \end{cases} \tag{10.52}$$

Proof With t_0 given in (10.43), we fix a constant $M \geq 1$ such that (10.45), (10.47), (10.48), (10.50), and (10.51) all hold with "$= O(w(t))$" replaced by "$\leq Mw(t)$ for all $t \geq t_0$." As an example, we thus rewrite (10.45) as the assumption that

$$\left|A_j(t)\right| \leq Mw(t) \qquad \text{for all } t \geq t_0, \quad 1 \leq j \leq k+1.$$

With $\Gamma_k(t)$ defined in (10.46), we make in (10.43) the change of variables

$$y_1(t) = \Gamma_k(t)y(t). \tag{10.53}$$

Note that

$$\Gamma_k^\Delta = \sum_{j=1}^k A_j^\Delta = -\sum_{j=1}^k A_{j-1}^\sigma A = -\Gamma_{k-1}^\sigma A.$$

Then it follows from (10.43) that

$$\begin{aligned} y_1^\Delta &= \Gamma_k^\sigma y^\Delta + \Gamma_k^\Delta y \\ &= \Gamma_k^\sigma \left[(A+B)\Gamma_k^{-1} y_1 + f(t, \Gamma_k^{-1} y_1)\right] - \Gamma_{k-1}^\sigma A \Gamma_k^{-1} y_1 \\ &= \left[\left(\Gamma_k^\sigma - \Gamma_{k-1}^\sigma\right) A + \Gamma_k^\sigma B\right] \Gamma_k^{-1} y_1 + \Gamma_k^\sigma f(t, \Gamma_k^{-1} y_1) \\ &= \left[A_k^\sigma A + \Gamma_k^\sigma B\right] \Gamma_k^{-1} y_1 + \Gamma_k^\sigma f(t, \Gamma_k^{-1} y_1). \end{aligned}$$

Since $\Gamma_k(t) = I + O(w(t))$ by (10.46) and (10.45), it can be shown that $(\Gamma_k(t))^{-1} = I + O(w(t))$. By making M larger, if necessary, we can assume that

$$|\Gamma_k(t)| \leq 1 + Mw(t), \text{ and } |\Gamma_k^{-1}(t)| \leq 1 + Mw(t), \qquad t \geq t_0.$$

It will also be convenient to introduce the notation $\Gamma^{-1}(t) = I + H(t)$, thus

$$|H(t)| \leq Mw(t), \qquad \text{for all } t \geq t_0. \tag{10.54}$$

Hence for $t \geq t_0$

$$y_1^\Delta = \left[A_k^\sigma A + \Gamma_k^\sigma B\right][I + H]y_1 + \Gamma_k^\sigma f(t, \Gamma_k^{-1} y_1), \ |H(t)| \leq Mw(t). \tag{10.55}$$

We define for $t \geq t_0$

$$Q(t) = -\int_t^\infty \left[A_k^\sigma(s)A(s) + \Gamma_k^\sigma(s)B(s)\right] \Delta s$$
$$= -A_{k+1}(t) - \int_t^\infty \Gamma_k^\sigma(s)B(s)\,\Delta s.$$

Then (10.45) and (10.47) imply that $Q(t)$ is well-defined and satisfies $Q(t) = \mathrm{O}(w(t))$. In particular, $I+Q(t)$ is invertible for large t. Making t_0 larger, if necessary, we can assume that $I + Q(t)$ is invertible for all $t \geq t_0$. And, making once more the value of the constant M larger, if necessary, we can also assume that

$$|Q(t)| \leq Mw(t) \quad \text{and} \quad \left|(I + Q(t))^{-1}\right| \leq 1 + Mw(t), \quad t \geq t_0. \tag{10.56}$$

We make a final transformation

$$y_1 = [I + Q(t)]y_2. \tag{10.57}$$

Then it follows from (10.55) that

$$y_1^\Delta = \left[A_k^\sigma A + \Gamma_k^\sigma B\right][I + H][I + Q(t)]y_2 + \Gamma_k^\sigma f(t, \Gamma_k^{-1}[I + Q(t)]y_2)$$
$$= [I + Q^\sigma]y_2^\Delta + Q^\Delta y_2.$$

Solving for y_2^Δ leads to

$$y_2^\Delta = (I + Q^\sigma)^{-1}\left\{\left[A_k^\sigma A + \Gamma_k^\sigma B\right][Q + H(I + Q)]y_2\right.$$
$$\left. + \Gamma_k^\sigma f(t, \Gamma_k^{-1}\,[I + Q(t)]y_2\right\}, \tag{10.58}$$

which we consider as a perturbation of the trivial system $z^\Delta = 0$.

With $t_1 \geq t_0$ to be determined below (see (10.62) in case that $w(t)$ vanishes at infinity or (10.63) if $w(t) \equiv 1$), let $\mathscr{B}$ be the Banach space of rd-continuous and bounded on $[t_1, \infty)$ equipped with the supremum norm

$$\|y\| = \sup_{t \geq t_1} |y(t)|.$$

For $y_2 \in \mathscr{B}$, we define for $t \in [t_1, \infty)$ an operator T defined by

$$(Ty_2)(t) = -\int_t^\infty [I + Q^\sigma(s)]^{-1}\left\{\left[A_k^\sigma(s)A(s) + \Gamma_k^\sigma(s)B(s)\right]\right.$$
$$\left.\times [Q(s) + H(s)\,(I + Q(s))]\,y_2\right\}\Delta s$$
$$-\int_t^\infty [I + Q^\sigma(s)]^{-1}\,\Gamma_k^\sigma(s)f(s, \Gamma_k^{-1}[I + Q(s)]y_2)\,\Delta s. \tag{10.59}$$

We will show in Lemma 10.15 that T maps $\mathscr{B}$ into $\mathscr{B}$ and is a contraction. Given a constant vector $c \in \mathbb{R}^d$, it then follows immediately that the equation

$$y_2 = c + (Ty_2)(t)$$

has a fixed point y_2 in $\mathscr{B}$ which is easily shown to be a solution of (10.58). It is clear that from its definition that $(Ty_2)(t) \to 0$ as $t \to \infty$. From Lemma 10.15 will also follow that $|(Ty_2)(t)| = \mathrm{O}(w(t))$, which of course is only meaningful if $w(t)$ vanishes at infinity. By (10.53) and (10.57), there exists a solution $y(t)$ of (10.43) given by

$$y(t) = \Gamma_k^{-1}(t)[I + Q(t)]y_2 = [I + \mathrm{O}(w(t))]y_2,$$

hence

$$|y(t) - c| \le |y_2(t) - c| + \mathrm{O}(w(t)),$$

which establishes (10.52) and completes the proof. □

In the next lemma, we establish the claimed properties of the mapping T.

Lemma 10.15 *Under the assumptions of Theorem 10.14, the mapping T defined in* (10.59) *maps $\mathscr{B}$ into $\mathscr{B}$, is a contraction, and satisfies* $|(Ty_2)(t)| = \mathrm{O}(w(t))$.

Proof To show that T maps $\mathscr{B}$ into $\mathscr{B}$, we use (10.54), (10.56), and the boundedness of $w(t)$ to first observe that for $s \ge t \ge t_1 \ge t_0$

$$|H(s)| \le Mw(s) \le M,$$

and similarly $|Q(s)| \le Mw(s) \le M$, and $\left|(I + Q^\sigma(s))^{-1}\right| \le 1 + M$. Let $y_2 \in \mathscr{B}$. Then it follows that

$$\begin{aligned}
&|(Ty_2)(t)| \\
&\le [1 + M]\,\|y_2\| \int_t^\infty \left[\left|A_k^\sigma(s)A(s)\right| + \left|\Gamma_k^\sigma(s)B(s)\right|\right] Mw(s)[2 + M]\,\Delta s \\
&\quad + \Bigg| \int_t^\infty (I + Q^\sigma(s))^{-1}\,\Gamma_k^\sigma(s) \\
&\qquad \{f(s, \Gamma_k^{-1}[I + Q(s)]y_2) - f(s,0) + f(s,0)\}\,\Delta s \Bigg| \\
&\le M\,[1 + M]\,[2 + M]\|y_2\| \int_t^\infty \left[\left|A_k^\sigma(s)A(s)\right| + (1 + M)\,|B(s)|\right] w(s)\,\Delta s
\end{aligned}$$

$$+ [1+M] \int_t^\infty |\Gamma_k^\sigma(s)| \gamma(s) \left| \Gamma_k^{-1}[I+Q(s)]y_2 \right| \Delta s$$

$$+ \left| \int_t^\infty (I+Q^\sigma(s))^{-1} \Gamma_k^\sigma(s) f(s,0) \Delta s \right|,$$

where (10.49) was used. Then (10.50) and (10.51) imply that

$$|(Ty_2)(t)| \leq M^2 [2+M]^3 \|y_2\| w(t) + [1+M]^4 \|y_2\| M w(t)$$
$$+ \left| \int_t^\infty (I+Q^\sigma(s))^{-1} \Gamma_k^\sigma(s) f(s,0) \Delta s \right|.$$

Integration by parts shows that

$$\int_t^\infty (I+Q^\sigma(s))^{-1} \Gamma_k^\sigma(s) f(s,0) \Delta s$$
$$= -(I+Q(t))^{-1} \int_t^\infty \Gamma_k^\sigma(s) f(s,0) \Delta s$$
$$+ \int_t^\infty \left[(I+Q(s))^{-1} \right]^\Delta \int_s^\infty \Gamma_k^\sigma(\tau) f(\tau,0) \Delta\tau \Delta s.$$

The product rule yields that

$$\left| \left[(I+Q(s))^{-1} \right]^\Delta \right| = \left| (I+Q^\sigma(s))^{-1} Q^\Delta (I+Q(s))^{-1} \right|$$
$$\leq [1+Mw(s)]^2 \left(|A_k^\sigma(s)A(s)| + |\Gamma_k^\sigma(s)B(s)| \right).$$

It now follows from (10.48) and (10.51) that

$$\left| \int_t^\infty (I+Q^\sigma(s))^{-1} \Gamma_k^\sigma(s) f(s,0) \Delta \right|$$
$$\leq [1+M]^2 M w(t)$$
$$+ [1+M]^2 \int_t^\infty \left(|A_k^\sigma A| + |\Gamma_k^\sigma B| \right) M w(s) \Delta s$$
$$\leq [1+M]^3 M w(t).$$

Hence there exist positive constants

$$c_1 = M^2[2+M]^4 + M[1+M]^4 \tag{10.60}$$

and $c_2 = [1+M]^3 M$, depending only on M, such that

$$|(Ty_2)(t)| \leq c_1 \|y_2\| w(t) + c_2 w(t), \tag{10.61}$$

which establishes that T maps $\mathscr{B}$ into itself and satisfies $(Ty)(t) = O(w(t))$.

To show that $T : \mathscr{B} \to \mathscr{B}$ is a contraction, note that similar estimates show for y_2 and $\tilde{y}_2$ in $\mathscr{B}$ that

$$\begin{aligned}
&|(Ty_2)(t) - (T\tilde{y}_2)(t)| \\
&\leq [2+M]^2 \|y_2 - \tilde{y}_2\| \int_t^\infty \left[|A_k^\sigma(s)A(s)| + |\Gamma_k^\sigma(s)B(s)| \right] M w(s) \Delta s \\
&\quad +(1+M) \int_t^\infty |\Gamma_k^\sigma(s)| \gamma(s) |\Gamma_k^{-1}(s)[I+Q(s)]| \, \Delta s \|y_2 - \tilde{y}_2\| \\
&\leq \left\{ [2+M]^3 M^2 w(t) + [1+M]^4 M w(t) \right\} \|y_2 - \tilde{y}_2\| \\
&\leq c_1 \|y_2 - \tilde{y}_2\| w(t).
\end{aligned}$$

In case that $w(t) \to 0$, with c_1 defined in (10.60), we fix $t_1 \geq t_0$ such that

$$c_1 w(t) \leq \theta < 1 \qquad \text{for all } t \geq t_1, \tag{10.62}$$

thus T is a contraction in this case. Similarly, if $w(t) \equiv 1$, then

$$\begin{aligned}
&|(Ty_2)(t) - (T\tilde{y}_2)(t)| \\
&\leq \|y_2 - \tilde{y}_2\| (2+M)^2 M \int_t^\infty \left(|A_k^\sigma(s)A(s)| + |\Gamma_k^\sigma(s)B(s)| \right) \Delta s \\
&\quad + \|y_2 - \tilde{y}_2\| (1+M)^4 \int_t^\infty \gamma(s) \, \Delta s
\end{aligned}$$

and in this case we fix $t_1 \geq t_0$ such that for $t \geq t_1$,

$$(2+M)^2 M \int_t^\infty \left(|A_k^\sigma(s)A(s)| + |\Gamma_k^\sigma(s)B(s)| \right) \Delta s + (1+M)^4 \int_t^\infty \gamma(s) \, \Delta s \leq \theta < 1. \tag{10.63}$$

Hence, for $t \geq t_1$,

$$|(Ty_2)(t) - (T\tilde{y}_2)(t)| \leq \theta \|y_2 - \tilde{y}_2\|,$$

which shows that T is a contraction mapping on $\mathscr{B}$, and completes the proof of the lemma. □

Applied to the linear system

$$y^{\Delta} = [A(t) + B(t)]y + g(t), \tag{10.64}$$

Theorem 10.14 has the following corollary, which is an improvement of [108, Cor. 4.2].

Corollary 10.16 *Suppose that $A(t)$, $B(t)$ and $g(t)$ are rd-continuous for $t \geq t_0$. Let $w : [t_0, \infty) \to [0, 1]$ be a rd-continuous function such that $w(t) \to 0$ as $t \to \infty$ or $w(t) \equiv 1$. For an integer $k \geq 0$, suppose that the integrals defined recursively by* (10.44) *satisfy* (10.45)*. With $\Gamma_k(t)$ defined by* (10.46)*, suppose that* (10.47) *holds and that*

$$\left| \int_t^{\infty} \Gamma_k^{\sigma}(s) g(s)\, \Delta s \right| = \mathrm{O}(w(t)).$$

Finally, suppose that (10.51) *is satisfied. Then* (10.64) *has for t sufficiently large a fundamental matrix satisfying*

$$Y(t) = \begin{cases} I + \mathrm{O}\,(w(t)) & \text{if } w \to 0 \\ I + \mathrm{o}(1) & \text{if } w \equiv 1. \end{cases}$$

References

1. R.P. Agarwal, *Difference Equations and Inequalities: Theory, Methods, and Applications* (Marcel Dekker, New York, 2000)
2. F.V. Atkinson, The asymptotic solution of second-order differential equations. Ann. Mat. Pura Appl. **37**, 347–378 (1954)
3. W. Balser, *Formal Power Series and Linear Systems of Meromorphic Ordinary Differential Equations*. Universitext (Springer, New York, 2000)
4. R.G. Bartle, *A Modern Theory of Integration*. Graduate Studies in Mathematics, vol. 32 (American Mathematical Society, Providence, 2001)
5. P.M. Batchelder, *An Introduction to Linear Difference Equations* (Dover Publications Inc., New York, 1967)
6. H. Behncke, Asymptotically constant linear systems. Proc. Am. Math. Soc. **138**, 1387–1393 (2010)
7. H. Behncke, C. Remling, Asymptotic integration of linear differential equations. J. Math. Anal. Appl. **210**, 585–597 (1997)
8. H. Behncke, C. Remling, Uniform asymptotic integration of a family of linear differential systems. Math. Nachr. **225**, 5–17 (2001)
9. Z. Benzaid, D.A. Lutz, Asymptotic representation of solutions of perturbed systems of linear difference equations. Stud. Appl. Math. **77**, 195–221 (1987)
10. O. Blumenthal, Über asymptotische Integration linearer Differentialgleichungen mit Anwendung auf eine asymptotische Theorie der Kugelfunktionen. Arch. Math. Phys. (Leipzig) **19**, 136–174 (1912)
11. S. Bodine, A dynamical systems result on asymptotic integration of linear differential systems. J. Differ. Equ. **187**, 1–22 (2003)
12. S. Bodine, D.A. Lutz, Exponential functions on time scales: their asymptotic behavior and calculation. Dyn. Syst. Appl. **12**, 23–43 (2003)
13. S. Bodine, D.A. Lutz, Asymptotic solutions and error estimates for linear systems of difference and differential equations. J. Math. Anal. Appl. **290**, 343–362 (2004)
14. S. Bodine, D.A. Lutz, Asymptotic analysis of solutions of a radial Schrödinger equation with oscillating potential. Math. Nachr. **279**, 1641–1663 (2006)
15. S. Bodine, D.A. Lutz, On asymptotic equivalence of perturbed linear systems of differential and difference equations. J. Math. Anal. Appl. **326**, 1174–1189 (2007)
16. S. Bodine, D.A. Lutz, A Hartman-Wintner theorem for dynamic equations on time scales. J. Differ. Equ. Appl. **14**, 1–16 (2008)

S. Bodine, D.A. Lutz, *Asymptotic Integration of Differential and Difference Equations*, Lecture Notes in Mathematics 2129,
DOI 10.1007/978-3-319-18248-3

17. S. Bodine, D.A. Lutz, Exponentially asymptotically constant systems of difference equations with an application to hyperbolic equilibria. J. Differ. Equ. Appl. **15**, 821–832 (2009)
18. S. Bodine, D.A. Lutz, Asymptotic integration under weak dichotomies. Rocky Mt. J. Math. **40**, 51–75 (2010)
19. S. Bodine, D.A. Lutz, Asymptotic integration of nonoscillatory differential equations: a unified approach. J. Dyn. Control Syst. **17**, 329–358 (2011)
20. S. Bodine, R.J. Sacker, A new approach to asymptotic diagonalization of linear differential systems. J. Dyn. Differ. Equ. **12**, 229–245 (2000)
21. S. Bodine, R.J. Sacker, Asymptotic diagonalization of linear difference equations. J. Differ. Equ. Appl. **7**, 637–650 (2001)
22. S. Bodine, M. Bohner, D.A. Lutz, Asymptotic behavior of solutions of dynamic equations. Sovrem. Mat. Fundam. Napravl. **1**, 30–39 (2003) (in Russian). Translation in J. Math. Sci. **124**, 5110–5118 (2004)
23. M. Bohner, D.A. Lutz, Asymptotic behavior of dynamic equations on time scales. J. Differ. Equ. Appl. **7**, 21–50 (2001)
24. M. Bohner, D.A. Lutz, Asymptotic expansions and analytic dynamic equations. Z. Angew. Math. Mech. **86**, 37–45 (2006)
25. M. Bohner, A. Peterson, *Dynamic Equations on Time Scales: An Introduction with Applications* (Birkhäuser, Boston, 2001)
26. M. Bohner, A. Peterson, *Advances in Dynamic Equations on Time Scales* (Birkhäuser, Boston, 2003)
27. M. Bohner, S. Stević, Trench's perturbation theorem for dynamic equations. Discrete Dyn. Nat. Soc., Art. ID 75672, 11 pp. (2007)
28. M. Bohner, S. Stević, Linear perturbations of a nonoscillatory second-order dynamic equation. J. Differ. Equ. Appl. **15**, 1211–1221 (2009)
29. V.S. Burd, *Method of Averaging for Differential Equations on an Infinite Interval: Theory and Applications.* Lecture Notes in Pure and Applied Mathematics, vol. 255 (Chapman & Hall/CRC, Boca Raton, 2007)
30. V.S. Burd, V.A. Karakulin, Asymptotic integration of systems of linear differential equations with oscillatory decreasing coefficients. Math. Notes **64**, 571–578 (1998)
31. V.S. Burd, P. Nesterov, Parametric resonance in adiabatic oscillators. Results Math. **58**, 1–15 (2010)
32. J.S. Cassell, The asymptotic integration of some oscillatory differential equations. Q. J. Math. Oxford Ser. (2) **33**, 281–296 (1982)
33. J.S. Cassell, The asymptotic integration of a class of linear differential systems. Q. J. Math. Oxford Ser. (2) **43**, 9–22 (1992)
34. S. Castillo, M. Pinto, Asymptotic integration of ordinary different systems. J. Math. Anal. Appl. **218**, 1–12 (1998)
35. S.Z. Chen, Asymptotic integrations of nonoscillatory second order differential equations. Trans. Am. Math. Soc. **327**, 853–865 (1991)
36. N. Chernyavskaya, L. Shuster, Necessary and sufficient conditions for the solvability of a problem of Hartman and Wintner. Proc. Am. Math. Soc. **125**, 3213–3228 (1997)
37. K. Chiba, Note on the asymptotic behavior of solutions of a system of linear ordinary differential equations. Comment. Math. Univ. St. Paul **13**, 51–62 (1965)
38. K. Chiba, T. Kimura, On the asymptotic behavior of solutions of a system of linear ordinary differential equations. Comment. Math. Univ. St. Paul **18**, 61–80 (1970)
39. S.K. Choi, N. Koo, K. Lee, Asymptotic equivalence for linear differential systems. Commun. Korean Math. Soc. **26**, 37–49 (2011)
40. M. Christ, A. Kiselev, Absolutely continuous spectrum for one-dimensional Schrödinger operators with slowly decaying potentials: some optimal results. J. Am. Math. Soc. **11**, 771–797 (1998)
41. M. Christ, A. Kiselev, WKB asymptotic behavior of almost all generalized eigenfunctions for one-dimensional Schrödinger operators with slowly decaying potentials. J. Funct. Anal. **179**, 426–447 (2001)

42. E.A. Coddington, N. Levinson, *Theory of Ordinary Differential Equations* (McGraw-Hill Book Company Inc., New York/Toronto/London, 1955)
43. W.A. Coppel, *Stability and Asymptotic Behavior of Differential Equations* (D. C. Heath and Co., Boston, 1965)
44. W.A. Coppel, *Dichotomies in Stability Theory*. Lecture Notes in Mathematics, vol. 629 (Springer, Berlin/New York, 1978)
45. N.G. de Bruijn, *Asymptotic Methods in Analysis* (Dover Publications Inc., New York, 1981)
46. S.A. Denisov, A. Kiselev, Spectral properties of Schrödinger operators with decaying potentials. Spectral theory and mathematical physics: a Festschrift in honor of Barry Simon's 60th birthday. Proc. Symp. Pure Math. **76**(Part 2), 565–589 (2007)
47. A. Devinatz, The asymptotic nature of the solutions of certain linear systems of differential equations. Pac. J. Math. **15**, 75–83 (1965)
48. A. Devinatz, An asymptotic theorem for systems of linear differential equations. Trans. Am. Math. Soc. **160**, 353–363 (1971)
49. A. Devinatz, J. Kaplan, Asymptotic estimates for solutions of linear systems of ordinary differential equations having multiple characteristic roots. Indiana Univ. Math. J. **22**, 355–366 (1972/1973)
50. O. Dunkel, Regular singular points of a system of homogeneous linear differential equations of the first order. Proc. Am. Acad. Arts Sci. **38**, 341–370 (1902/1903)
51. M.S.P. Eastham, The Liouville-Green asymptotic theory for second-order differential equations: a new approach and some extensions, in *Ordinary Differential Equations and Operators (Dundee, 1982)*, ed. by W.N. Everitt, R.T. Lewis. Lecture Notes in Mathematics, vol. 1032 (Springer, Berlin, 1983), pp. 110–122
52. M.S.P. Eastham, The asymptotic solution of linear differential systems. Mathematika **32**, 131–138 (1985)
53. M.S.P. Eastham, *The Asymptotic Solution of Linear Differential Systems. Applications of the Levinson Theorem* (Oxford University Press, Oxford, 1989)
54. S. Elaydi, An extension of Levinson's theorem to asymptotically Jordan difference equations. J. Differ. Equ. Appl. **1**, 369–390 (1995)
55. S. Elaydi, *An Introduction to Difference Equations* (Springer, New York, 1999)
56. S. Elaydi, Asymptotics for linear difference equations I: basic theory. J. Differ. Equ. Appl. **5**, 563–589 (1999)
57. S. Elaydi, G. Papaschinopoulos, J. Schinas, Asymptotic theory for noninvertible systems, in *Advances in Difference Equations (Veszprém, 1995)*, ed. by S. Elaydi et al. (Gordon and Breach, Amsterdam, 1997), pp. 155–164
58. U. Elias, H. Gingold, A framework for asymptotic integration of differential systems. Asymptot. Anal. **35**, 281–300 (2003)
59. G. Freud, *Orthogonal Polynomials* (Pergamon Press, New York, 1971)
60. A.O. Gel'fond, I.M. Kubenskaya, On Perron's theorem in the theory of difference equations. (Russian). Izv. Akad. Nauk SSSR. Ser. Mat. **17**, 83–86 (1953)
61. J. Geronimo, D.T. Smith, WKB (Liouville–Green) analysis of second order difference equations and applications. J. Approx. Theory **69**, 269–301 (1992)
62. J. Geronimo, D.T. Smith, Corrigendum to WKB (Liouville–Green) analysis of second order difference equations and applications. J. Approx. Theory **188**, 69–70 (2014)
63. J. Geronimo, W. Van Assche, Relative asymptotics for orthogonal polynomials with unbounded recurrence coefficients. J. Approx. Theory **62**, 47–69 (1990)
64. J. Geronimo, D.T. Smith, W. Van Assche, Strong asymptotics for orthogonal polynomials with regularly and slowly varying recurrence coefficients. J. Approx. Theory **72**, 141–158 (1993)
65. A. Ghizzetti, Un teorema sul comportamento asintotico degli integrali delle equazioni differenziali lineari omogenee (Italian). Univ. Roma. Ist. Nz. Alta Mat. Rend. Mat. Appl. **8**(5), 28–42 (1949)
66. H. Gingold, Almost diagonal systems in asymptotic integration. Proc. Edinb. Math. Soc. **28**, 143–158 (1985)

67. G. Green, On the motion of waves in a variable canal of small depth and width. Trans. Camb. Philos. Soc. **6**, 457–462 (1837)
68. M. Greguš, *Third Order Linear Differential Equations*. Translated from the Slovak by J. Dravecký. Mathematics and Its Applications (East European Series), vol. 22 (D. Reidel Publishing Co., Dordrecht, 1987)
69. W.A. Harris, D.A. Lutz, On the asymptotic integration of linear differential systems. J. Math. Anal. Appl. **48**, 1–16 (1974)
70. W.A. Harris, D.A. Lutz, Asymptotic integration of adiabatic oscillators. J. Math. Anal. Appl. **51**(1), 76–93 (1975)
71. W.A. Harris, D.A. Lutz, A unified theory of asymptotic integration. J. Math. Anal. Appl. **57**(3), 571–586 (1977)
72. P. Hartman, *Ordinary Differential Equations* (Birkhäuser, Boston, 1982)
73. P. Hartman, A. Wintner, Asymptotic integrations of linear differential equations. Am. J. Math. **77**, 45–86, 404, 932 (1955)
74. D. Henry, *Geometric Theory of Semilinear Parabolic Equations*. Lecture Notes in Mathematics, vol. 840 (Springer, Berlin, 1981)
75. S. Hilger, Analysis on measure chains—a unified approach to continuous and discrete calculus. Results Math. **18**, 18–56 (1990)
76. R.A. Horn, C.R. Johnson, *Topics in Matrix Analysis* (Cambridge University Press, Cambridge, 1991)
77. P.F. Hsieh, Y. Sibuya, *Basic Theory of Ordinary Differential Equations* (Springer, New York, 1999)
78. P.F. Hsieh, F. Xie, Asymptotic diagonalization of a linear ordinary differential system. Kumamoto J. Math. **7**, 27–50 (1994)
79. P.F. Hsieh, F. Xie, Asymptotic diagonalization of a system of linear ordinary differential equations. Dyn. Continuous Discrete Impuls. Syst. **2**, 51–74 (1996)
80. P.F. Hsieh, F. Xie, On asymptotic diagonalization of linear ordinary differential equations. Dyn. Contin. Discrete Impuls. Syst. **4**, 351–377 (1998)
81. M. Hukuhara, Théorèmes fondamentaux de la théorie des équations différentielles ordinaires, I. Mem. Fac. Sci. Kyūsyū Imp. Univ. A. **1**, 111–127 (1941)
82. G.K. Immink, *Asymptotics of Analytic Difference Equations*. Lecture Notes in Mathematics, vol. 1085 (Springer, Berlin, 1984)
83. J. Janas, M. Moszyński, Spectral properties of Jacobi matrices by asymptotic analysis. J. Approx. Theory **120**, 309–336 (2003)
84. J. Janas, M. Moszyński, New discrete Levinson type asymptotics of solutions of linear systems. J. Differ. Equ. Appl. **12**, 133–163 (2006)
85. J. Janas, S. Naboko, Jacobi matrices with absolutely continuous spectrum. Proc. Am. Math. Soc. **127**, 791–800 (1999)
86. C. Jordan, *Calculus of Finite Differences* (Chelsea Publishing Co., New York, 1965)
87. W.B. Jurkat, *Meromorphe Differentialgleichungen*. Lecture Notes in Mathematics, vol. 637 (Springer, Berlin, 1978)
88. W. Kelley, A.C. Peterson, *Difference Equations: An Introduction with Applications* (Harcourt/Academic, San Diego, 2001)
89. R.B. Kelman, N.K. Madsen, Stable motions of the linear adiabatic oscillator. J. Math. Anal. Appl. **21**, 458–465 (1968)
90. S. Khan, D.B. Pearson, Subordinacy and spectral theory for infinite matrices. Helv. Phys. Acta **65**, 505–527 (1992)
91. T. Kimura, Sur le comportement des solutions d'un système d'équations différentielles ordinaires linéaires. Comment. Math. Univ. St. Paul **11**, 81–90 (1963)
92. A. Kiselev, Absolutely continuous spectrum of one-dimensional Schrödinger operators and Jacobi matrices with slowly decreasing potentials. Commun. Math. Phys. **179**, 377–400 (1996)
93. A. Kiselev, Examples of potentials and l^p estimates. Private E-mail Communication (2009)

94. K. Knopp, *Theorie und Anwendung der unendlichen Reihen* (Springer, Berlin/New York, 1964)
95. M. Kohno, *Global Analysis in Linear Differential Equations*. Mathematics and Its Applications, vol. 471 (Kluwer Academic Publishers, Dordrecht, 1999)
96. R.J. Kooman, Decomposition of matrix sequences. Indag. Math. (N.S.) **5**, 61–79 (1994)
97. R.J. Kooman, An asymptotic formula for solutions of linear second-order difference equations with regularly behaving coefficients. J. Differ. Equ. Appl. **13**, 1037–1049 (2007)
98. R.J. Kooman, Asymptotic diagonalization of matrix systems. J. Approx. Theory **171**, 33–64 (2013)
99. V. Lakshmikantham, D. Trigiante, *Theory of Difference Equations: Numerical Methods and Applications* (Marcel Dekker, New York, 2002)
100. N. Levinson, The asymptotic nature of solutions of linear differential equations. Duke Math. J. **15**, 111–126 (1948)
101. J. Liouville, Second mémoire sur le développement des fonctions ou parties de fonctions en séries dont les divers termes sont assujétis à satisfaire à une même équation différentielle du second ordre, contenant un paramètre variable. J. Math. Pures Appl. **2**, 16–35 (1837)
102. M. Lizana, J. Muldowney, Riccati differential inequalities and dichotomies for linear systems. J. Math. Anal. Appl. **271**, 313–332 (2002)
103. P. Locke, On the asymptotic behavior of linear systems. Proc. Am. Math. Soc **25**, 93–95 (1970)
104. A. Máté, P. Nevai, Factorization of second-order difference equations and its application to orthogonal polynomials, in *Orthogonal Polynomials and Their Applications (Segovia, 1986)*, ed. by M. Alfaro et al. Lecture Notes in Mathematics, vol. 1329 (Springer, Berlin, 1988), pp. 158–177
105. A. Máté, P. Nevai, A generalization of Poincaré's theorem for recurrence equations. J. Approx. Theory **63**, 92–97 (1990)
106. A. Máté, P. Nevai, V. Totik, Asymptotics for orthogonal polynomials defined by a recurrence relation. Constr. Approx. **1**, 231–248 (1985)
107. R. Medina, M. Pinto, Linear differential systems with conditionally integrable coefficients. J. Math. Anal. Appl. **166**, 52–64 (1992)
108. R. Mert, A. Zafer, Time scale extensions of a theorem of Wintner on systems with asymptotic equilibrium. J. Differ. Equ. Appl. **17**, 841–857 (2011)
109. H. Meschkowski, *Differenzengleichungen*. Studia Mathematica, Bd. XIV (Vandenhoeck & Ruprecht, Göttingen, 1959)
110. L.M. Milne-Thomson, *The Calculus of Finite Differences* (Macmillan and Co. Ltd., London, 1951)
111. R. Naulin, M. Pinto, Dichotomies and asymptotic solutions of nonlinear differential systems. Nonlinear Anal. **23**, 871–882 (1994)
112. R. Naulin, M. Pinto, Asymptotic solutions of nondiagonal linear difference systems. Tamkang J. Math. **31**, 175–191 (2000)
113. P.G. Nevai, *Orthogonal Polynomials*. Memoirs of the American Mathematical Society, vol. 18 (American Mathematical Society, Providence, 1979)
114. N.E. Nörlund, *Leçons sur les séries d'interpolation* (Gautiers-Villars, Paris, 1926)
115. F.W.J. Olver, Error bounds for the Liouville–Green (or WKB) approximation. Proc. Camb. Philos. Soc. **57**, 790–810 (1961)
116. F.W.J. Olver, *Asymptotics and Special Functions*. Computer Science and Applied Mathematics (Academic, New York/London, 1974)
117. O. Perron, Über einen Satz des Herrn Poincaré. J. Reine Angew. Math. **136**, 17–37 (1909)
118. O. Perron, Über die Poincarésche lineare Differenzengleichung. J. Reine Angew. Math. **137**, 6–64 (1910)
119. O. Perron, Über lineare Differentialgleichungen, bei denen die unabhängige Variable reell ist (Erste Mitteilung). J. Reine Angew. Math. **142**, 254–270 (1913)
120. O. Perron, Über lineare Differentialgleichungen, bei denen die unabhängige Variable reell ist (Zweite Mitteilung). J. Reine Angew. Math. **143**, 29–50 (1913)

121. O. Perron, Über Systeme von linearen Differenzengleichungen erster Ordnung. J. Reine Angew. Math. **147**, 36–53 (1917)
122. O. Perron, Die Stabilitätsfrage bei Differentialgleichungen. Math. Z. **32**, 703–728 (1930)
123. G.W. Pfeiffer, Asymptotic solutions of $y''' + qy' + ry = 0$. J. Differ. Equ. **11**, 145–155 (1972)
124. M. Pinto, *Dichotomies and Asymptotic Behaviour*, Contributions USACH (Universidad de Santiago, Santiago 1992), pp. 13–22
125. M. Pinto, Discrete dichotomies. Comput. Math. Appl. **28**(1–3), 259–270 (1994)
126. M. Pinto, Dichotomies and asymptotic formulas for the solutions of differential equations. J. Math. Anal. Appl. **195**, 16–31 (1995)
127. M. Pinto, Null solutions of difference systems under vanishing perturbation. In honour of Professor Allan Peterson on the occasion of his 60th birthday, part II. J. Differ. Equ. Appl. **9**(1), 1–13 (2003)
128. M. Pinto, A generalization of Poincaré's theorem for recurrence systems. Appl. Anal. **85**, 373–381 (2006)
129. M. Pituk, Asymptotic behavior of a Poincaré recurrence system. J. Approx. Theory **91**, 226–243 (1997)
130. H. Poincaré, Sur les equations linéaires aux différentielles ordinaires et aux différences finies. Am. J. Math **7**(3), 203–258 (1885)
131. I.M. Rapoport, *On Some Asymptotic Methods in the Theory of Differential Equations (Russian)*. Izdat. Akad. Nauk Ukrain (SSR, Kiev, 1954)
132. G. Ren, Y. Shi, Asymptotic behaviour of dynamic systems on time scales. J. Differ. Equ. Appl. **12**, 1289–1302 (2006)
133. G. Ren, Y. Shi, Y. Wang, Asymptotic behavior of solutions of perturbed linear difference systems. Linear Algebra Appl. **395**, 283–302 (2005)
134. J. Rovder, Asymptotic behaviour of solutions of the differential equation of the fourth order. Math. Slovaca **30**, 379–392 (1980)
135. W. Rudin, *Real and Complex Analysis* (McGraw-Hill, New York, 1966)
136. Y. Sibuya, A block-diagonalization theorem for systems of linear ordinary differential equations and its applications. SIAM J. Appl. Math. **14**, 468–475 (1966)
137. J. Šimša, Asymptotic integration of a second order ordinary differential equation. Proc. Am. Math. Soc. **101**, 96–100 (1987)
138. D.R. Smith, Liouville–Green approximations via the Riccati transformation. J. Math. Anal. Appl. **116**, 147–165 (1986)
139. D.R. Smith, Decoupling of recessive and nonrecessive solutions for a second-order system. J. Differ. Equ. **68**, 383–399 (1987)
140. T. Solda, R. Spigler, M. Vianello, Asymptotic approximations for second-order linear difference equations in Banach spaces, II. J. Math. Anal. Appl. **340**, 433–450 (2008)
141. R. Spigler, M. Vianello, Liouville–Green approximations for a class of linear oscillatory difference equations of the second order. J. Comput. Appl. Math. **41**, 105–116 (1992)
142. R. Spigler, M. Vianello, WKBJ-type approximation for finite moments perturbations of the differential equation $y'' = 0$ and the analogous difference equation. J. Math. Anal. Appl. **169**, 437–452 (1992)
143. R. Spigler, M. Vianello, Discrete and continuous Liouville–Green–Olver approximations: a unified treatment via Volterra-Stieltjes integral equations. SIAM J. Math. Anal. **25**, 720–732 (1994)
144. R. Spigler, M. Vianello, WKB-type approximations for second-order differential equations in C* algebras. Proc. Am. Math. Soc. **124**, 1763–1771 (1996)
145. R. Spigler, M. Vianello, Liouville–Green asymptotics for almost-diagonal second-order matrix differential equations. Asymptot. Anal. **48**, 267–294 (2006)
146. R. Spigler, M. Vianello, Liouville–Green asymptotic approximations for a class of matrix differential equations and semi-discretized partial differential equations. J. Math. Anal. Appl. **325**, 69–89 (2007)

147. R. Spigler, M. Vianello, F. Locatelli, Liouville–Green–Olver approximations for complex difference equations. J. Approx. Theory **96**, 301–322 (1999)
148. S.A. Stepin, Asymptotic integration of second-order nonoscillatory differential equations (Russian). Dokl. Akad. Nauk **434**, 315–318 (2010); Translation in Dokl. Math. **82**(2), 751–754 (2010)
149. S.A. Stepin, V.A. Titov, Dichotomy of WKB-solutions of discrete Schrödinger equation. J. Dyn. Control Syst. **12**, 135–144 (2006)
150. J.G. Taylor, Improved error bounds for the Liouville-Green (or WKB) approximation. J. Math. Anal. Appl. **85**, 79–89 (1982)
151. W. Trench, On the asymptotic behavior of solutions of second order linear differential equations. Proc. Am. Math. Soc. **14**, 12–14 (1963)
152. W. Trench, Extensions of a theorem of Wintner on systems with asymptotically constant solutions. Trans. Am. Math. Soc. **293**, 477–483 (1986)
153. W. Trench, Linear perturbations of a nonoscillatory second order equation. Proc. Am. Math. Soc. **97**, 423–428 (1986)
154. W. Trench, Asymptotic behavior of solutions of Poincaré recurrence systems. Comput. Math. Appl. **28**, 317–324 (1994)
155. W. Trench, Asymptotic behavior of solutions of asymptotically constant coefficient systems of linear differential equations. Comput. Math. Appl. **30**, 111–117 (1995)
156. W. Trench, Linear asymptotic equilibrium and uniform, exponential, and strict stability of linear difference systems. Comput. Math. Appl. **36**, 261–267 (1998)
157. W. Trench, Linear asymptotic equilibrium of nilpotent systems of linear difference equations. J. Differ. Equ. Appl. **5**, 549–556 (1999)
158. W. Trench, Linear perturbations of a nonoscillatory second order differential equation. II. Proc. Am. Math. Soc. **131**, 1415–1422 (2003)
159. E.B. Van Vleck, On the extension of a theorem of Poincaré for difference-equations. Trans. Am. Math. Soc. **13**, 342–352 (1912)
160. W. Wasow, *Asymptotic Expansions for Ordinary Differential Equations* (Dover Publications, New York, 1987)
161. J. Wimp, D. Zeilberger, Resurrecting the asymptotics of linear recurrences. J. Math. Anal. Appl. **111**, 162–176 (1985)
162. A. Wintner, On a theorem of Bôcher in the theory of ordinary linear differential equations. Am. J. Math. **76**, 183–190 (1954)
163. V.A. Yakubovich, V.M. Starzhinskii, *Linear Differential Equations with Periodic Coefficients 1, 2* (Halsted Press/Wiley, New York/Toronto, 1975). Israel Program for Scientific Translations, Jerusalem/London

Index

Symbols

A

B

C

D

E

F

G

H

S. Bodine, D.A. Lutz, *Asymptotic Integration of Differential and Difference Equations*, Lecture Notes in Mathematics 2129,
DOI 10.1007/978-3-319-18248-3

Editorial Policy (for the publication of monographs)

1. Lecture Notes aim to report new developments in all areas of mathematics and their applications - quickly, informally and at a high level. Mathematical texts analysing new developments in modelling and numerical simulation are welcome.
   Monograph manuscripts should be reasonably self-contained and rounded off. Thus they may, and often will, present not only results of the author but also related work by other people. They may be based on specialised lecture courses. Furthermore, the manuscripts should provide sufficient motivation, examples and applications. This clearly distinguishes Lecture Notes from journal articles or technical reports which normally are very concise. Articles intended for a journal but too long to be accepted by most journals, usually do not have this "lecture notes" character. For similar reasons it is unusual for doctoral theses to be accepted for the Lecture Notes series, though habilitation theses may be appropriate.

2. Manuscripts should be submitted either online at www.editorialmanager.com/lnm to Springer's mathematics editorial in Heidelberg, or to one of the series editors. In general, manuscripts will be sent out to 2 external referees for evaluation. If a decision cannot yet be reached on the basis of the first 2 reports, further referees may be contacted: The author will be informed of this. A final decision to publish can be made only on the basis of the complete manuscript, however a refereeing process leading to a preliminary decision can be based on a pre-final or incomplete manuscript. The strict minimum amount of material that will be considered should include a detailed outline describing the planned contents of each chapter, a bibliography and several sample chapters.
   Authors should be aware that incomplete or insufficiently close to final manuscripts almost always result in longer refereeing times and nevertheless unclear referees' recommendations, making further refereeing of a final draft necessary.
   Authors should also be aware that parallel submission of their manuscript to another publisher while under consideration for LNM will in general lead to immediate rejection.

3. Manuscripts should in general be submitted in English. Final manuscripts should contain at least 100 pages of mathematical text and should always include

   - a table of contents;
   - an informative introduction, with adequate motivation and perhaps some historical remarks: it should be accessible to a reader not intimately familiar with the topic treated;
   - a subject index: as a rule this is genuinely helpful for the reader.

   For evaluation purposes, manuscripts may be submitted in print or electronic form (print form is still preferred by most referees), in the latter case preferably as pdf- or zipped ps-files. Lecture Notes volumes are, as a rule, printed digitally from the authors' files. To ensure best results, authors are asked to use the LaTeX2e style files available from Springer's web-server at:

   ftp://ftp.springer.de/pub/tex/latex/svmonot1/ (for monographs) and
   ftp://ftp.springer.de/pub/tex/latex/svmultt1/ (for summer schools/tutorials).

Additional technical instructions, if necessary, are available on request from lnm@springer.com.

4. Careful preparation of the manuscripts will help keep production time short besides ensuring satisfactory appearance of the finished book in print and online. After acceptance of the manuscript authors will be asked to prepare the final LaTeX source files and also the corresponding dvi-, pdf- or zipped ps-file. The LaTeX source files are essential for producing the full-text online version of the book (see http://www.springerlink.com/openurl.asp?genre=journal&issn=0075-8434 for the existing online volumes of LNM). The actual production of a Lecture Notes volume takes approximately 12 weeks.

5. Authors receive a total of 50 free copies of their volume, but no royalties. They are entitled to a discount of 33.3 % on the price of Springer books purchased for their personal use, if ordering directly from Springer.

6. Commitment to publish is made by letter of intent rather than by signing a formal contract. Springer-Verlag secures the copyright for each volume. Authors are free to reuse material contained in their LNM volumes in later publications: a brief written (or e-mail) request for formal permission is sufficient.

Addresses:

Professor J.-M. Morel, CMLA,
École Normale Supérieure de Cachan,
61 Avenue du Président Wilson, 94235 Cachan Cedex, France
E-mail: morel@cmla.ens-cachan.fr

Professor B. Teissier, Institut Mathématique de Jussieu,
UMR 7586 du CNRS, Équipe "Géométrie et Dynamique",
175 rue du Chevaleret
75013 Paris, France
E-mail: teissier@math.jussieu.fr

For the "Mathematical Biosciences Subseries" of LNM:

Professor P. K. Maini, Center for Mathematical Biology,
Mathematical Institute, 24-29 St Giles,
Oxford OX1 3LP, UK
E-mail: maini@maths.ox.ac.uk

Springer, Mathematics Editorial, Tiergartenstr. 17,
69121 Heidelberg, Germany,
Tel.: +49 (6221) 4876-8259

Fax: +49 (6221) 4876-8259
E-mail: lnm@springer.com

Printed in the United States
By Bookmasters